DEVELOPMENTS IN
EARTH & ENVIRONMENTAL SCIENCES 2

MEDICAL GEOLOGY

EFFECTS OF GEOLOGICAL ENVIRONMENTS ON HUMAN HEALTH

Translation of the book MEDICINSKA GEOLOGIJA
Published by Tellur - Beograd, 2001

DEVELOPMENTS IN **EARTH & ENVIRONMENTAL SCIENCES** 2

DEVELOPMENTS IN **EARTH & ENVIRONMENTAL SCIENCES 2**

MEDICAL GEOLOGY

EFFECTS OF GEOLOGICAL ENVIRONMENTS ON HUMAN HEALTH

MIOMIR M. KOMATINA
SERBIAN GEOLOGICAL SOCIETY, BEOGRAD, SERBIA AND MONTENEGRO

2004

ELSEVIER

Amsterdam – Boston – Heidelberg – London – New York – Oxford
Paris – San Diego – San Francisco – Singapore – Sydney – Tokyo

ELSEVIER B.V.	ELSEVIER Inc.	ELSEVIER Ltd	ELSEVIER Ltd
Sara Burgerhartstraat 25	525 B Street, Suite 1900	The Boulevard, Langford Lane	84 Theobalds Road
P.O. Box 211, 1000 AE	San Diego, CA 92101-4495	Kidlington, Oxford OX5 1GB	London WC1X 8RR
Amsterdam, The Netherlands	USA	UK	UK

Translation of the book MEDICINSKA GEOLOGIJA published by Tellur – Beograd, 2001

First edition 2004

Library of Congress Cataloging in Publication Data
A catalog record is available from the Library of Congress.

British Library Cataloguing in Publication Data
A catalogue record is available from the British Library.

ISBN: 0-444-51615-8

♾ The paper used in this publication meets the requirements of ANSI/NISO Z39.48-1992 (Permanence of Paper).
Printed in The Netherlands.

Excerpts from critiques...

Close connection between living and non - living nature is unambiguous. That is why one of the principal goals of medical geology, new geological, in other words - medical discipline, is to study general regularities of geological environment influence on human health, more precisely - to distinguish geological factors which can be fundamental carriers of certain diseases or which have positive effect on man. On that basis, the author, for the first time in the world, has completely developed subject, tasks and methodology of health care. The monograph certainly is significant contribution to Yugoslav and the world science.

Nikola Pantić

This is the first monograph in Serbian language presenting in original way scientific synthesis referring to mutual relation between geological phenomena, causes and consequences of geological events, and different influences on appropriate causes of medical occurences, not only in the field of human medicine, but also of veterinary medicine.

My opinion is that the monograph Medical Geology by Dr Miomir Komatina, does present a major contribution to our science, deserving to be published on the whole.

Branislav M. Ćirić

Preamble

We are living in an epoch when numerous secrets of nature are being discovered and natural resources are being reclaimed by humans. At the same time, however, this is an epoch of considerable human efforts to protect their health against an increasingly destructive anthropogenic activity. According to the accomplishments of various sciences and scientific disciplines, new ways in that direction have been found. Geology, one of the basic natural sciences, is becoming one of outstanding importance in solving problems referring to agriculture, exploitation of the Earth's mineral resources, protection of the nature and environment, preserving soil, water, energetic and other resources, protection against natural disasters (landslides, floods, volcanic eruptions and earthquakes), as well as problems of health protection.

It was a great pleasure to the author to study geosciences during a significant part of his life, to have opportunity to participate actively in solving different practical tasks in that domain, to explore numerous areas of the Planet and finally to examine various geological concepts. As a result of the mentioned activities, the author's need has arised to give a contribution at the interface of geology and medical sciences, which has not so far been the subject of a sufficient attention, and to try as well to establish basic principles of a new, indeed necessary discipline. It may turn out to be useful to mention a discussion that took place during an important international meeting (XXI International Geological Congress, Rio de Janeiro, August, 2000), where the author reported fundamentals of medical geology, when one colleague put a question: „How have you come to idea to think and write about that subject?" The answer was based on the assumption that the whole life is being performed on the Earth, in the part of geological environment which plays an essential role for the life activity, and it is therefore certainly understandable that there must exist strong influences of various geological factors onto the human health and that is indeed the medical geology which is to be extremely exploitated in investigating causes of the disease appearance. That is why another question is worth of thinking about - namely, why such a book has not been written so far.

There are many advantages of the moment for appearance of the monograph. During the last century, as a major number of sciences, biology, medical science and geology were developed extensively and in an interdisciplinar way, and occurrence of new disciplines is receiving a complex significance. A sudden technical progress, particularly in the fields of electronics and automatics, including computers development, has reflected in appearance of numerous research methods and new possibilities of quantitative data processing. This includes a rich geological, medical and medico - geographical literature, as well. All these facts enable establishment of a solid basis for medical geology and its intensive development.

The **Medical Geology** monograph consists of the following five important segments:

- *Introduction;*
- *Geological and other factors and their influence on the human health;*
- *Subject, tasks and methods of geomedical discipline;*
- *Regional medical geology;*
- *Applied medical geology.*

The main objective of the monograph is to present general rules on effects of geological environment onto the human health and to define a subject, tasks and methodology of medico - geological research, in order to perform an efficient prevention in the field of human health protection. If such approach makes a shift to improving health of population possible, it could be concluded that the mission of the book has been fulfilled.

The monograph is dedicated to wide circles of readers, geologists, doctors, biologists, ecologists, planners, and all others who deal with the task of life and health protection. It is understandable that the author was not able to give a more complete competent review of modern world experiences in domain of medical science. The intention was rather to indicate importance of geological and other factors, as well as fields related to the human health, where these factors can give a significant contribution.

The author is grateful to the high representatives of Elsevier BV, Ms. Patricia Massar, Administrative Editor, and Mr.Friso Veenstra, Senior publishing editor responsible for Elsevier's geochemistry publications, who made publishing the English version of this book by Elsevier possible. The author is thankful to the reviewers, Dr.Olle Selinus, Sweden Geological Survey, and Prof.Dr.Catherine Skinner, Yale University, for suggestions, comments and notes. The same is for Dr.Snezana Komatina, as the author of the monograph part on effects of geophysical fields onto the human health (Chapters 2 and 4), and as translator of Chapters 8, 10 and 11. Many thanks go to Mr.Raymond Dooley, who translated the book into English, as well as to Mr.Zoran Petrović for technical preparation of the monograph. The author also appreciates all comments and suggestions obtained by Dr.Dejan Komatina. Special thanks go to numerous authors and publishing houses all over the world, for permission to use their photos, figures and tables in this book.

October, 2003 *Miomir M. Komatina*

Contents

CHAPTER 1

General Part

INTRODUCTORY REMARKS

Public health depends in great measure on environmental conditions. The father of modern medicine, Hippocrates (460 - 377 B.C.), wrote among other things: "*Disease does not come to us from out of the clear blue sky, but rather develops from everyday small sins against nature.*" In his two - volume work "**Pathologie regionale de la France**", the well - known French pathologist R. Marot (1958) in connection with the given relationship even concluded in a pessimistic tone: "*Man will always remain liable to the action of laws of the environment surrounding him. He will never be able to free himself completely from that environment, no matter how much he would like to; and his fight against aggressive diseases, in spite of indisputable successes (improvement of living conditions, prolongation of life expectancy), will not finally dismantle the yoke in which he is held*" (p. 413). It is certain that man faces great difficulties in studying nature and seeking to establish the true causes of disease or the positive influence of natural factors on health. Medicine in its research seems to have lost sight of this fundamental task; hence, rapid development of various alternative treatment methods is somewhat understandable.

But the current level of development of biology, medicine, physics, chemistry, geography, geology, and many other sciences makes it possible to uncover not only basic regularities governing the distribution of disease on concrete territories, but also the factors that determine the level of public health. All that is needed is greater cooperation and coordination of multi - disciplinary research efforts, together with utilization of the possibilities afforded by more or less neglected sciences, above all geology, as one of the basic natural sciences. We note that Aristotle (Fourth century B.C.) and Pliny the Elder (First Century A.D.) already called attention to the geological (lithological or material) composition of a region as one of the primary factors in human life. As Aristotle taught, "*Water is of the same quality as the rocks across which it flows.*"

The efforts of doctors and geographers to study the distribution of human diseases throughout the world without deeper examination of the causes and regularities of their incidence led to the very early appearance of a special discipline of geography or medicine - **medical geography**. The very long period from Hippocrates to the end of the 19th Century can be characterized as the period of formation and development of descriptive medical geography. This discipline enjoyed especially rapid development during the 60's and 70's of the 20th Century in the United States, the former Soviet Union, France, Germany, Great Britain, and China. Elaborated at that time were many new conceptual and procedural problems based on a synthesis of the ideas and observations of an increasingly wider range of sciences. Today we have an exceptionally rich world literature in this area of science.

Based on a number of natural and social sciences, medical geography understandably has been able to synthesize primarily general information about the natural conditions of some territory or another. In order to fill the gap in the borderline area between geography and geology in treating the relationship between *living and non - living nature*, Soviet pedologist in the middle of the 20th Century formulated the principles of a new scientific discipline - **landscape geochemistry** (environmental geochemistry), which regards the Earth's surface in terms of a dynamic system with interdependence and intermingling of elements of the lithosphere, hydrosphere, and atmosphere. In the given system, closely interconnected and interdependent living organisms and inorganic materials interact with each other (Fig. 1.1). The position of the soil in the system is central, since it is predominantly in the soil that the interaction is realized. Rapid development of landscape geochemistry and its practical implementation have occurred since the outset of the 1960's. At the same time, **geochemistry** as a whole - like all the atomic sciences - has undergone intensive development, and its ideas and methods have been increasingly incorporated into other geological disciplines, as well as into medicine, biology, and astronomy. This has enriched those old and thoroughly elaborated areas of knowledge with new concepts and opened up to them new prospects for practical application that promote resolution of the increasingly complex problem of health protection. An immeasurable contribution here was made by Vernadskii, who already at the end of the 19th Century laid the foundations of the study of trace elements, constructed the basis of geochemistry and biochemistry, and created the general science of the biosphere. In this he emphasized the interconnection of all processes in the lithosphere, hydrosphere, and atmosphere with processes occurring in the living matter of human beings, animals, and plants. "*The Earth's crust*," wrote Vernadskii among other things, "*includes several geological shells - the biosphere, the stratosphere, and the metamorphic and granitic shells. All of them were biospheres when found on the Earth's surface in the course of geological time. They are all genetically interconnected and taken as a whole represent a single phenomenon.*" Proceeding from one of his capital laws - which holds that the elementary composition of organisms is inseparably linked with chemical composition of the Earth's crust - and regarding living material as a regular function of the biosphere, Vernadskii more than once stressed that organisms must be studied in close connection with chemical properties of the Earth's crust. "*Like geochemistry, biogeochemistry can be studied in three directions:* ***first****, from the biological aspect, in connection with its significance for understanding the phenomenon of life;* ***secondly****, from the* ***geological aspect*** *(boldface mine - M.K.), in connection with its significance for understanding the environment, i.e., primarily the biosphere; and* ***thirdly****, from the applied aspect, in connection with its practical significance, which can be scientifically reduced to the biogeochemical role of mankind*" (V. I. Vernadskii, 1934). Working on these principles, Vernadskii's distinguished student A. P. Vinogradov (1938) created the study of **biogeochemical provinces**, special natural zones characterized by elevated or reduced content of certain chemical elements in soil and water. Less than a full two decades later, V. V. Koval'skii intensively developed the study of trace elements and their biological significance. "*The diversity of geological and geochemical processes leads to geographic changes in the chemical composition of soils, natural waters, and plant and animal organisms, as well as to variability in exchange of substances, the emergence of morphological variability, periodic incidence of endemic diseases, sharpening of natural selection, and formation of new taxonomic units. In this way, the geochemical environment can gain significance as an important factor in organic evolution*" (Koval'skii, 1963).

In the past four decades, more than 50 substances of organic composition but inorganic (abiotic) origin have been discovered. These substances have been found in calcareous meteorites, warm volcanic ash, and storm clouds highly charged with electricity at the moment of discharge. A new term - *abiogenic organic compounds* is now used in chemistry. In contrast to *biogenic organic compounds*, their molecules are symmetrically constructed and therefore do not react to polarized light rays.

According to A. L. Yanshin (1988), this discovery makes it possible to reduce the problem of the origin of life on Earth to the time when molecules of abiogenic organic matter (which always existed) arose. In keeping with the Curie principle, their appearance must have had an asymmetrical cause.

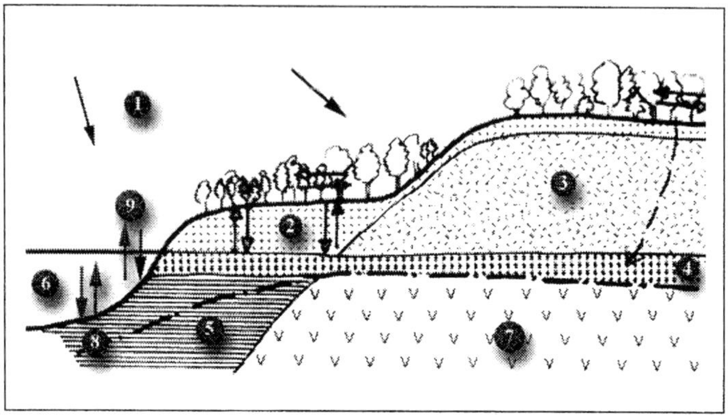

Fig. 1.1. Simplified plan of geochemical landscape. 1. lower atmosphere; 2. pedological soil; 3. weathering zone; 4. aquifer; 5. continental deposits; 6. water surface; 7. principal rocks; 8. landscape lower boundary; 9. connections within the system.

"*Comparative planetary study of the solar system supports the conclusion that asymmetry was caused by the Earth's strong magnetic field. It has been established that such a magnetic field is a specific characteristic of our planet. It is absent on the Moon, is 2,000 times weaker on Mars, and 20,000 times weaker on Venus than on Earth. But such a strong magnetic field has not always existed on Earth. It appeared in the course of a specific stage in the development of our planet, when differentiation of its original material resulted in the formation of a heavy core, and the upper shell of the Earth started to rotate around this core. According to existing data, the Earth's magnetic field was considerably stronger in archaic times than it is today. The reconstruction of molecules of abiogenic organic substances and their acquisition of asymmetric structure should probably be linked with the period when the Earth's heavy core was formed and a strong asymmetric magnetic field arose.*

Whatever the case, if we accept the influence of a strong magnetic field on reconstruction of the inorganic molecule, then the possibility exists that such phenomena can occur at the submolecular level and new ways or forms of life can arise at the present time also.

Could this be the cause of the appearance of new types of previously unknown viruses such as the AIDS virus?" (Yanshin, 1988).

In his inaugural address to the first students of the Faculty of Agriculture in Belgrade at the beginning of 1921, J. Zujovic (the father of

Serbian geology) had this to say: "*Allow me to cite in conclusion one more interesting example of dependence, an example from terrain composed of the oldest crystalline schists, of gneiss. Upon decomposing, this rock yields an inconsiderable quantity of infertile soil containing unusually small amounts of the phosphate and lime so useful for the development of organisms. Not only is farming poorly developed on such terrain, but animals are also sickly. It has been observed that snails are scarce there and their shells very thin; that hens lay very thin - shelled eggs; that the skeletons of livestock are quite weak; and finally that even the recruits are very small.*"

Given below are a few more examples illustrating the significance of certain geochemical factors for man and human health. This question will be discussed in greater detail in the corresponding chapter.

a) Already in the middle of the 19th Century, the incidence of goiter - a disease very widespread in various forms throughout the world - was linked with a deficit of iodine in the drinking water. It has also been known since ancient times that a deficiency of fluorine creates a predisposition for the development of caries. On the other hand, a fluorine surplus - especially pronounced in volcanic regions and in zones with water - bearing horizons rich in phosphorus deposits - causes endemic fluorosis of the teeth and skeleton. Natural zones of endemic fluorosis with severely diseased people and animals have been described in North and South America, India, Sri Lanka, Vietnam, and China.
b) It has been established that low rates of stomach cancer incidence and mortality are registered on the magnesium - rich chernozem of Armenia and sandy soil of Egypt (K. L. Bazikyan, 1957; V. A. Chervyakov, 1964; E. S. Feldman, 1977; N. V. Lazarev, 1966). On the other hand, Japanese authors link increased mortality from this disease with muddy peat - rich marsh soils characterized by a high level of groundwater, as well as with soil formed on alluvial material derived from broken - up volcanic rocks.
c) Acute anemia in domestic animals has been shown to be caused by deficiency of cobalt in the grass in some biogeochemical provinces and by copper deficiency in others. It is known that strontium rickets in animals and chronic molybdenum toxication are pronounced in cattle feeding on plants from molybdenum - rich soil. Endemic blindness of animals is caused by excessive soil nickel content on the arid steppes in some regions of the former Soviet Union.
d) The correlation between high nitrate content in drinking water and increased frequency of stomach cancer is especially clear in Columbia.

It is not difficult to ascertain that in practically all of the listed examples, the main cause of disease is some geological factor (rocks, soil, groundwater, mineral raw material) with its geochemical, geophysical, hydrogeological, and other characteristics (Table 1.1).

TABLE 1.1.
Influence of geological factors and socio - ecologic data on human mortality (after Mazac, ed., 1986). Mean correlation coefficients R* between geological (symbol TZ), geophysical (RN, RA), socio - ecologic (SE) factors and mortality (A to K) on various diseases or mortality index (L) calculated as the average of three independently found R - values. Coefficients R* > R* (0.05) = 0.264 (critical valid for n = 56 samples and significance level p = 0.05) proved statistically significant correlations between the analyzed parameters (where percentual impact is shown).

Symbols	Mortality on	Mean correlation coefficients R*			
		Radon RN	Tectonics TZ	Radioactivity RA	Index SE
A	Cancer generally	0.047	0.484 23.4%	0.143	0.375 14.1%
	Cancer of:				
B	bronchus and lungs	0.194	0.282 8%	0.186	0.488 23.8%
C	stomach	0.177	0.098	0.236	0.058
D	female breast	-0.170	0.057	0.054	0.090
E	rectum	0.182	0.136	0.227	0.208
F	large intestine	0.048	0.125	0.062	0.257
G	Blood circulation generally	0.010	0.159	-0.096	0.447 20.0%
H	Vessel, brain attack	-0.102	0.127	-0.065	0.078
I	Ischemic diseases	-0.008	-0.056	-0.146	0.386 14.9%
J	Heart attack	0.060	-0.017	0.072	0.235
K	Suicides	-0.140	0.158	0.046	0.063
L	Mortality index (MI)	0.064	0.297 8.8%	-0.028	0.469 22.0%

This confirms the earlier stated thesis as to the unambiguous close connection between living and nonliving nature. It can therefore be asked with justice: *Why has geology not been far better represented in study of environmental influence on man? More precisely, why has better use not been made of experienced geologists with an abundance of information and voluminous geological cartography at their disposal?* There can be no dispute as to the intensive development that geological science has undergone during the past century in the fields of regional geology, tectonics, geophysics, petrography, geochemistry, hydrogeology, geological engineering, research on nuclear and other mineral raw materials, etc. The very rich geological literature has been insignificantly utilized for the needs of medical geology and medicine. Despite the plethora of medical and medical - geographic atlases that have been compiled, there has been practically no comparison with diverse geological maps and atlases. We are inclined to the opinion that the advances in medical science, above all pathology, would have been even greater had there been more reliance on knowledge about numerous elements in the geological environment and direct interpretations given by geologists. In other words, *advances in the discovery and elimination of disease would have been far greater had medical geology developed parallel with medical geography.*

The first and one of the rare steps in involving geological science more significantly in multidisciplinary efforts of public health protection was taken by H. Zeiss, who introduced the term **geomedicine** in Germany already in 1931. Zeiss and his followers used this term to designate *a branch of science apart from medical geography that in studying the development of disease in space and time (geoecology, environmental ecology) focuses attention on geospheric factors causing one form or another of the studied phenomenon.* Both factors encompassed by the general geoecology of man (inhabitants of the Earth) and ones treated in the geoecology of human diseases are considered here. In terms of regional occurrence, geofactors affecting human health and disease are examined on the *micro-*, *meso-*, and *macroplans* (N. J. Jusatz, 1983). According to German authors, a leading role in realization of geomedical research must be played by multidisciplinary collaboration with specialists in the field, a view that is completely in order.

HISTORY OF MEDICAL GEOGRAPHY AND MEDICAL GEOLOGY

History of Medical Geography

The roots of medical geography reach back into the distant past. Already in Hippocrates, we find the idea that corporal and spiritual attributes of man and disease are linked with environmental factors. It is interesting that Hippocrates in considering their influence on man distinguished not only factors of physical geography (weather, climate, wind, water, soil, and relief), but also social and ecological factors, i.e., basic elements of the way of life, especially customs of the people, terrestrial laws, etc. In this sense, Hippocrates appears not only as the universally acknowledged father of the whole of modern medicine, but also as the first physician to foretell the future development of medical geography, geographical pathology, medical ecology, and even social hygiene.

Something more than 2000 years ago, the connection of certain endemic diseases with the natural environment was precisely indicated in the book "**The Spring and Autumn of Lu**". Later on, observations and discussions of this subject appeared in many ancient Chinese books on medicine and geography.

In comparison with this, the first stages in development of the medical geography derived from classical medicine (which lasted until the beginning of the 18th Century) were more comprehensive and based on increasingly modern research methods. In presenting a historical survey of the development of medical geography in individual countries, we have relied mainly on the "**Handbook of Medical Geography**" (edited by A. A. Keller, O. P. Shchepin, and A. V. Chaklin, 1993). Our aim is above all to get a picture that stresses the role of natural factors and enables us to discern the place of medical geology and possibilities for its development.

Russia. Medico - geographical studies in Russia were already being conducted in the 18th Century. We note several works from the first quarter of that century dealing with the properties of mineral waters, medicinal herbs, useful minerals, and poisonous animals. Also of interest is the program of education for army doctors, which included chemistry and balneology, anatomy and physiology, **mineralogy** (boldface mine - M.K.), and botany. During the relatively long period of its adoption (up until the 30's of the next century), army doctors made a significant contribution in the field of nosogeography, resolving questions of the distribution of diseases, primarily infective ones.

Medical geography in Russia achieved its greatest flowering in the 19th Century. A large number of studies in the domain of this discipline were published in the pages of the first Russian medical journals, and the Russian Geographic Society was formed in 1845. It can be stated that by the end of the first half of the century, medical geography had become an independent branch of medicine, elaborated both in the form of medico - geographical description of territories and in the guise of theoretical investigations. In his valuable work "**Attempt at Medical Geography of the Caucasus with Reference to Intermittent Fevers**," N.I. Toropov in 1864 wrote as follows: "*In order to be able to prevent any disease, it is necessary first of all to know from what and where it arises, i.e., to know the causes of its development in the organism and the area of its distribution on the Earth's surface. Study of the nature of the disease usually provides the answer to the first question, while the young science of medical geography gives the answer to the second.*"

The medico - geographical work of Russian physicians in the second half of the 19th Century was characterized by increased interest in study of the socio - economic conditions at concrete localities and their influence on health, sanitary conditions, and diseases. However, also of importance here were the fundamental works of Vernadskii, Dokuchaev, Sukachev, Vinogradov, and Pavlovskii, studies of decisive significance for the development of medico - geographical ideas, as has already been discussed in our introductory remarks. A great contribution to the development of this new discipline was made by A. A. Shoshin, at whose initiative was formed the Commission on Medical Geography (transformed into the Department of Medical Geography in 1963), which he headed over the course of the next two decades. Shoshin's scientific work represented a fundamental contribution to medical geography. He developed the theoretical principles of medical geography, explicated methods for medico - geographical investigation of territories and research on the geography of human disease (nosogeography), formulated the tasks and content of medical cartography, etc. The following approaches to the statement of theoretical questions and definition of the subject, tasks, and content of medical geography were fundamentally new:

a) Consideration of medical geography as an independent science on the interface between medical and geographical science;
b) Definition of the subject of medical geography as the regularities governing both negative and positive influence of natural and socio - economic factors at a given locality on human (public) health, incidence of human diseases, and their geographical distribution;
c) Postulation of a leading role for socio - economic factors in the complex of multiple influence erted by local conditions on public health; and
d) Development of several basic directions of medical geography:
   - medico - geographical estimation of natural and socio - economic factors;
   - medical management of the environment;
   - state protection of public health;
   - nosogeography; and
   - medical cartography.

Medico - geographical investigation of different regions of the Soviet Union and other countries, together with determination of the geographic distribution of diseases, represented an important aspect of work of the Commission.

The fact that the Commission dealt with an exceptionally vast territory made it necessary to organize conferences on different questions of medical geography, and the *first scientific meeting* was held in Leningrad already in November of 1962. This meeting treated pressing questions of medical geography and problems involved in using the results of medico - geographical research in the economy and public health. It also charted directions for the future development of medical geography and coordination of work in this scientific discipline. The next conferences were devoted to specific problems: *the second* conference (1965) treated *the geographical milieu and public health*; *the third* (1968) dealt with *the geography of human disease* and outlined a program for *compilation of a medico - geographical atlas*; *the fourth* conferencc (1973) was concerned with the establishment of a *medico - geographical cadastre*; *the fifth* (1979) was on *medico - geographical regionalization and forecasting*; and the sixth conference (1983) dealt with *procedural questions of medical geography*. Very visible results were achieved during those years in the field of medico - geographical regionalization: thus, for example, 78 medico - geographical atlases and maps were printed in 1977 alone. Significant results were also obtained in creation of a medico - geographical cadastre, which had to incorporate all reliable data about the influence of natural, economic, and sanitary conditions of a given territory on the level of public health in it and be compiled on the basis of multiple medico - geographical investigations of individual localities.

Processes of social renewal were considered at the *seventh conference of medical geographers* (1987), at which the following questions were examined:

- Theoretical and procedural questions of regional medical geography and medico - geographical management;
- Medico - geographical research in the framework of development and realization of complex regional programs;
- Regional medico - geographical atlases and maps; and
- Medico - geographical study of the World Ocean.

The next conference (held in October of 1991) was devoted to *realization and ways of increasing the effectiveness of medico - geographical research*. The main themes treated at the conference were as follows:

- The role of basic medico - geographical research in improving the ecological conditions of different regions;
- Utilization of the results of medico - geographical research in prevention of disease;
- Medico - geographical research in the domain of regional *public health* programs; and
- Ways of increasing the effectiveness of medico - geographical research and medico - ecological education of the population.

Considerable attention was devoted to *ecological aspects of medical geography* at this conference and in the years to follow.

United States. The first published works in the area of medical geography of the country appeared in the 18th Century.

We note a work published in 1776 on the relationship between *disease and weather conditions* in South Carolina (L. Chalmers) and a paper reporting the results of research on the interconnection of soil, climate, weather, and disease in that state (D. Ramsey, 1796). The year 1798 witnessed the publication of one of the first medico - geographical maps in the world, which showed the incidence of yellow fever in the vicinity of New York City. Communications in the area of medical geography continued to be published throughout the next century. As in other countries, this field at that time was most often referred to as medical topography. Up until the middle of the 20th Century, these studies were mostly of a descriptive nature and included such capital works as "**The Geography of Disease**" (E.B.A. McKinley, 1935) and the three - volume "**Global Epidemiology**" (J. S. Simmons et al., 1944 - 1954).

A turning point in the development of medical geography in the United States occurred when this science started to be viewed in terms of the *ecology of human disease*, a concept that was developed by the leading American medical geographer J. M. May (1950, 1954/55, 1958) proceeding from the premise that *groups of risk factors* - organic, inorganic, and socio - cultural - exist in the environment. Later on, the traditional direction involving study of disease ecology and interaction between the *environment* (as altered by man) and *public health* came to be increasingly supplemented by detailed consideration of what can be termed geographical health protection.

Socially oriented associative investigations represent a characteristic feature of medical geography in the United States, such investigations being useful in planning the network of medical institutions, their distribution, and financing. These investigations rely on the use of computers and methods of mathematical statistics (modeling).

Medico - geographical cartography has undergone intensive development, and its widespread automatization was initiated already in the 1960's. Automatized cartography is used to study the distribution of mortality, organize medical services, and process data on the environment (R. W. Armstrong, 1972), as well as to forecast disease incidence (G. F. Pyle, 1971) and assign doctors where needed (G. W. Shannon, G. F. A. Dever, 1974). Published in 1975 was the "***Atlas of Cancer Mortality in Districts of the USA: 1950 - 1969***" (T. J. Mason et al., 1975), which was compiled with the aid of computers.

Other characteristics of medical geography in the United States also need to be stressed. To be specific, appreciable growth occurred during the 1970's and 1980's in the fields of *epidemiological geography* (R. W. Armstrong, 1973; M. S. Meade, 1977), historical epidemiology (G. W. Shannon, R. G. Growley, 1980), and *geographical epidemiology* (J. M. Hunter, 1978; V. B. Robinson, 1978; M. L. Miller, 1979).

The first course in medical geography was introduced in 1969. During the period 1970 - 1984, 149 senior dissertations, 139 master's theses, and 51 doctoral dissertations (19 in the ecology of disease, 21 in the geography of health protection, and 11 in the geography of nutrition and hunger) were approved. American medical geographers, incidentally, have been constantly involved in work of the Commission on Medical Geography of the International Geographic Society and taken part in international symposia on problems of medical geography. The first symposium in this area was organized in North Carolina, and papers presented at it were published in an anthology entitled "***The Geography of Health and Disease.***" Published six years later were papers presented at another symposium, this one devoted to conceptual and procedural questions of medical geography.

Great Britain. The first medico - geographical investigations in the country appeared at the end of the 18th Century. Their development was primarily stimulated by the need to protect the health of British citizens scattered throughout the nation's many colonies. These studies were carried out exclusively by doctors. The first stage in the development of medical geography in Great Britain involved research on influence of the climate of tropical countries on human health (J. Lind, 1791). During the next stage, attention was focused on various diseases in England itself (J. Hennen, 1830; H. Marchall, 1832; J. Forbes, 1834). In the 19th Century, questions of the theory of medical geography began to be elaborated in the papers of English scientists, and the link between certain components of the geographic environment and the distribution of human disease was investigated. Works in this area of science were illustrated by maps of the distribution of individual diseases.

The first half of the 20th Century was characterized by further development of nosogeography and medico - geographical cartological studies, while the first decade since the end of the Second World War witnessed the appearance of works treating the connection of certain diseases with the environment, the role of specific environmental factors in mortality, and local medico - geographical phenomena. A series of papers dealt with geographical aspects of various problems such as the influence of background radiation on health (W. M. Court - Brown et al., 1960), significance of the type of water supply (E. Allen - Price, 1960), and the link between mortality and hardness of drinking water (M. D. Crawford et al., 1968). The year 1963 was marked by publication of "**The National Atlas of Mortality from Different Diseases in Great Britain**" (G. M. Howe), whose appearance for English investigators signified the end of the stage of accumulation of factual material and beginning of a new period in the development of medical geography of the country. The state of development of English medical geography in the given period is characterized fairly accurately by two books of the leading English geographer L. D. Stamp: "**Some Aspects of Medical Geography**" (1964a) and "**The Geography of Life and Death**" (1964b). Among other things, Stamp wrote as follows: "*Medical geography is still a no - man's land, a fertile soil waiting to be plowed. Whereas we have only scratched its surface in different places, future tilling will yield fruits of great significance to humanity.*"

The development of medico - social geography in Great Britain began in the middle of the 1970's (D. R. Phillips, 1971). There followed a gradual increase in the number of studies in this area employing various methods of modeling and characterized by more and more appreciable multidisciplinary collaboration among investigators. We note in conclusion the large number of works by A. Learmonth, including his capital monographs "**Disease and Hunger**" (1975), "**The Geography of Health**" (1981), and "**The Ecology of Disease**" (1988). It goes without saying that great contributions in the area of cartography were made by "**The Atlas of Cancer in Scotland**" (1975 - 1980); "**Incidence of Disease and the Epidemiological Perspective**" (J. Kemp et al., 1975); and the "**Atlas of Disease Distribution: An Analytical Approach to Epidemiological Data**" (A. D. Cliff, P. Hagett, 1988).

British geochemists of the 1980's made a significant contribution to knowledge about the geochemical characteristics of certain provinces. Their results were published in the anthology "**Environmental Geochemistry and Health**" (ed. by S. H. U. Bowie and I. Thornton, 1985).

France. Three periods can be isolated in the history of development of medical geography in France.

The first period, from the beginning of the 18th Century to the 80's of the next century, can be characterized as the time of formation of medical geography. During that period, more than 820 medico - topographical studies were carried out and a significant number of investigations were conducted that dealt with problems of pathology, acclimatization, and influence of natural conditions on human health. The term **medical geography** itself appeared in the French literature in 1817 in a paper of J. J. Virey, who gave the following definition: "*Medical geography - geografia medica - represents a description of the Earth's surface with reference to the influence of every region on the health, vital functions, and diseases of plants, animals, and (mainly) human beings inhabiting it.*" Among studies of this stage, the most attention was attracted by a capital two - volume work of J. Boudin entitled "**Manual of Medical Geography, Medical Statistics, and Endemic Diseases**" (1857).

The second period lasted from the 80's - 90's of the 19th Century to the 20's of the 20th Century. During this period, the discoveries of Pasteur and Laveran revealed the unknown causes of many mass diseases, and rapid development occurred in bacteriology, scientific epidemiology, and parasitology.

The third period is closely linked with the name of M. Sorre, one of the greatest geographers of France. Sorre marked the beginning of this period with a series of works dealing with influence of the geographic environment on the human organism. In 1928 he introduced the concept of the "*pathogenic complex*" to medical geography, treating it more fully in a special article entitled "**Pathogenic Comples and Medical Geography**" (1933). Published in 1943 was Sorre's major work "**Biological Principles of Human Geography. Section on Human Ecology**." With this work, Sorre was the first to take an ecological approach in medical geography. The indicated works represented the theoretical foundation of medical geography and cartography in France and were of great influence on its further development.

Unquestionably one of the most important medico - geographical works published in the country was the "**Regional Pathology of France**," the author of which was R. Marot (1958). On the basis of extensive data collected in collaboration with geographers, biologists, geologists, and ethnographers, the author was able to determine pathogenic comples characteristic of individual regions and create a medico - geographical regionalization of France. We note too the major work of H. Picheral "**Physical Space and Health: Medical Geography of the South of France**" (1976); and "**The Geography of Nutrition**" by R. Livet (1969).

Also of great significance for the development of medical geography in France was the symposium "**Geography and Health**," which was held in 1984 in Montpellier. Discussed at this meeting were a wide range of questions, including theoretical problems, nosogeography, regional medical geography, geography of health protection, and medical geography of cities.

Germany. The first study in medical geography was a work by F. Hoffmann on "diseases specific to certain nations and regions" (1705). Dividing diseases into endemic and epidemic, Hoffmann developed the concept of endemic diseases, basing it on the geographical conditions of localities, characteristics of nutrition, and the way of life led by the inhabitants of concrete regions. A fairly long time after this, the foundation of a new scientific discipline, **endemiology**, was laid in a fundamental work on endemic diseases by J. F. Cartheuser (1771).

Published between the years 1792 and 1795 in Leipzig was the major work of L. L. Finke "**Experiment in General Practical Medical Geography**." Medical geography in this work was treated as a new science taking as its subject study of influence erted by the geographical conditions of countries and way of life of their populations on public health, illness, prevention, and therapy. The fundamental work of A. Hirsch (1859 - 1864) "**Instructions for Historico - Geographical Pathology**" represented another major step in the history of medical geography. Practically all diseases known at the time were described in it by continents and countries.

Edited by H. Zeiss, an atlas of the distribution of endemic diseases in countries of Europe, the Mediterranean, and the Middle East was published in Germany during the period 1942 - 1945. The years from 1952 to 1961 saw publication of a three - volume atlas of the worldwide distribution of epidemic diseases. This atlas was edited by E. Rodenwaldt and H. J. Jusatz, and its publication not only represented a significant event in modern medical geography, but also gave a new impulse to medico - geographical cartography. The atlas did not simply indicate the ranges of diseases and their dynamics: it also considered various etiological, bio - geographical, physico - geographical, and ecological factors of influence on the distribution of diseases.

China. From the end of the 1960's, Chinese geographers took an active part in studying the distribution and natural settings of ailments (Keshan and Kashin - Beck diseases) caused by low environmental content of selenium (Tan Jianan, 1985). In general, the medical geographers of China have devoted considerable attention to diseases linked with *selenium and other trace elements* (in some regions of the Hubej Province, selenium content in the environment is hundreds and thousands of times higher than the usual concentration); *endemic goiter* (Yu Zhihen, 1987); and *fluoroses and caries* (Yue Songling, 1980). The major work "**Atlas of Endemic Environmentally - Linked Diseases in the People's Republic of China**" was published in 1989.

A second direction of nosogeographical research in China is concerned with *malignant neoplasms* (Wong Jaoping, 1983). Especially great attention has been focused on cancer of the esophagus and liver, as well as on the link between cancer and the environmental content of trace elements (silicon, cobalt, nickel, and selenium). The "**Atlas of Cancer Mortality in the People's Republic of China**" was published in 1981.

Investigation of infective and parasitic diseases represents a third direction of nosogeography in China, while study of cardiovascular diseases (Wu Yingkai, 1979) constitutes a separate direction.

Chinese medical geography enjoyed especially intensive development during the last decade of the millennium. A clear contribution to this was made by the Commission on Medical Geography, which was formed in 1990. The Commission organized the first symposium on problems of medical geography that same year and the second symposium the next year.

In conclusion, it must be stressed that medico - geographical research has been conducted and is being carried out not just in the countries mentioned above. Furthermore, an enormous role in development of medical geography in the world has been played by the

Commission on Medical Geography, with its international journal "**Geographia Medica**," which has been published since 1969.

History of Medical Geology

It is evident from the foregoing text that many factors of the geological environment were considered geographical and were studied predominantly by geographers and doctors throughout the three centuries of medical geography. With recognition of their indisputable contribution, several examples can be cited in support of this assertion.

1. During the period 1977 - 1983, Chinese medical geographers studied the distribution of endemic fluorosis on the territory of China and established a correlation between high fluoride content in groundwater and paleogeographic factors. In their opinion, local trace element - related medical problems are caused by arsenic, barium, and thallium. In study of the nosogeography of cancer, moreover, the correlation of this disease with trace element content in the external (geological) environment was carefully analyzed. Broadly developed in China, the given direction of research was termed *chemico - medical geography*[1].
2. Mongolian medico - geographers made a significant contribution to study of the geographical setting of endemic diseases by indicating unevenness of trace element content in soil of the country and characteristics of the distribution of rickets, hyperthyroidism, and diseases of the teeth in different regions.
3. A whole series of papers by Dutch authors have been devoted to investigation of the territorial correlation between the incidence and distribution of malignant diseases on the one hand and geochemical and geophysical factors (drinking water quality, occurrence of peaty - muddy soils) on the other.
4. A number of papers by Welsh medical geographers dealing with the influence of background radiation on health, the link between mortality and hardness of drinking water, etc., in fact had a geological slant.

That the exceptions prove the rule, i.e., that geologists were sporadically involved in multidisciplinary research, is indicated by the following examples drawn from the medico - geographical literature.

1. At the University of Trondheim in Norway, a multidisciplinary group for medical geography was formed in 1982 that included geographers, doctors, epidemiologists, geologists, and geochemists. With its Department of Environmental Chemistry, the Norwegian Board of Geology compiled a database on biologically active chemical elements in the environment of

[1] A more adequate term in our opinion would be *geochemico-medical geography*, since trace elements are for the most part constituents of rocks, soil, and groundwater.

concrete territories in the framework of the Geographical System of Information about Mortality in Norway.

2. Among the fairly few papers in the field of medical geography printed in Canada up to 1970, we note those of the geologist H. V. Warren on the geography of disease (1964, 1965) and an article by the geobotanist C. Laverdierr on the distribution of poison ivy (an agent of skin cancer) in Quebec.
3. The participation of geologists in creation of the "**Regional Pathology of France**" (Marot, 1958) has already been mentioned.
4. A wide range of specialists - geographers, doctors, geologists, biologists, cartographers, etc. - took part in the work of symposia on medical geography held in Prague in 1976, 1977, 1982, and 1985.

In addition, the term **geomedicine** has been used several times in the medico - geographical literature, but it can be seen to have a geographical connotation. For example, this was the case with the "**Geomedical Map of the World**," compiled in 1980 under the supervision of O. Kratochvil, and articles in which Y. Verhasselt (1976, 1977) employed the term **geocancerology**, meaning by it the use of territorial analysis of malignant diseases to establish their dependence on environmental factors. It is also interesting to note that at the scientific meeting known as XV Congres international d'Hydrologie, de Climatologie et de Geologie Medicales held in Belgrade in 1936, only one paper from the field of geology was presented (M.Lukovic, 1936).

From a more detailed analysis of the history of medical geography, it can be concluded that geographers and doctors in the course of long - term complex investigations have touched upon a whole range of geological factors and thereby created a good basis for accelerated development of medical geology. With the development of medical geology, what has nevertheless constituted an appreciable "omission" in world science as a whole will be corrected, since medicine will gain a much fuller and more adequate interpretation of the role played by inorganic (nonliving) nature in human disease.

An especially important proposal in this context was put forward already in 1931 by Zeiss, who considered it quite proper to develop **geomedicine** as a direction of science separate from medical geography. We have already spoken of this. Unfortunately, the given proposal did not meet with widespread acceptance.

In 1952 the Heidelberg Academy of Sciences formed the Center for Geomedical Research, staffed by 40 specialists. The Center belonged to the Department of Mathematical and Natural Sciences and for the most part carried out medico - cartographic studies. In addition to the "**Atlas of Worldwide Distribution of Epidemic Diseases**" edited by E. Rodenwaldt and H. J. Jusatz (1983), this institution published geomedical monographs devoted to Libya, Afghanistan, Ethiopia, Kuwait, Kenya, Korea, Sri Lanka, Thailand, Nigeria, Uganda, and Tanzania.

The Central Association of German Geographers in 1981 held a symposium in Mannheim that considered the geomedical consequences of environmental changes. Certain activities in the geologo - geographical field were being carried out already in Eastern Germany under the auspices of the Institute for Geography and Geoecology (I. Honsch), and in Austria at the Subdepartment of Geomedicine at the University of Graz, which was chaired by the well - known medical geographer W. Sixl.

The efforts of the German school of the 1930's had no repercussions in other countries and none at home after a certain time. Thus, the advent of a more integral elaboration of the principles of theory, subject matter, and procedure of medical geography had to wait until the last two decades of the 20th Century and beginning of the new millennium.

To be specific, two decades ago the Norwegian J. Lag promoted the term **geomedicine** in works published in 1980 and 1983. Broad international activity in this field was initiated some time later. Thus, a new international working group in the geosciences has been established to meet the needs of ecological planning (COGEOENVIRONMENT) and is now using a scientific approach to study the influence of external factors on the geographical incidence of health problems in man and animals. This approach consists of a multidisciplinary effort taking into account the results of different sciences, above all the geosciences and medicine.

Medical Geography and Medical Geology in Yugoslavia

The need to create and develop **medical geography** in Serbia was first discussed only in 1998 at a special meeting of the Serbian Geographic Society when a paper on the subject was presented by Z. Jovicic. However, this goal has not yet been fully realized. Worthy of mention here are the individual contributions of A. Radic ("**Principles of Medical Geography**," 1969), Z. Jovicic ("**Principles of Medical Geography in Serbia: a Theoretical - Procedural Concept**," 1998), and Z. Djordjevic ("**Medical Geography of Malignant Tumors of the Urinary Bladder in Serbia**," Master's Dissertation, 1980).

The justification for work in the area of **medical geology** was already noted by the great Serbian scientist J. Pancic, a popularly renowned physician, botanist with an international reputation, and author of the first book on geology in our country. Exceptionally knowledgeable in all three of the sciences mentioned, Pancic already in the middle of the 19th Century appealed in writing to local doctors calling on them to pay attention to "*mineral things*" *in the ground.* In the Introduction to his book "**Mineralogy and Geology According to Naumann and Beudant**" (published in 1867), Pancic wrote as follows: "*.....for what can be more impor-*

tant to man than to learn about everything concerning his existence and existence of the land on which he lives." Similar reflections of J. Zujovic have already been mentioned.

In spite of this, there has been no broader organizational work on formation and development of medical geology in Yugoslavia (or the rest of the world for that matter) up to the present day. The given gap in the system of science is partially filled by numerous contributions of geologists and doctors who have studied individual problems, some of which will be stressed here. Of undoubted particular significance have been symposia organized by the Serbian Academy of Sciences on *endemic nephropathy* (1970, 1977, 1995), *the role of selenium in health* (1993, 1995), and that of *magnesium* (1998, 1999). Pioneer steps in collaboration of geologists and doctors in our country were taken by Z. Maksimovic with papers on the close linkage of geochemical factors and endemic nephritis in Serbia. Maksimovic later led multidisciplinary meetings on selenium and magnesium, and published a number of well received works in the area of geochemistry and related problems of human health. Geochemical research on selenium content in geological formations and waters of our country was carried out for several years by A. Dangic (1993, 1995), and possible hydrogeological causes of endemic nephritis were investigated by J. Peric (1970, 1976/79, 1980, 1982). Important data on trace elements in our geological formations and soil are to be found in the papers of V. Jovic (1990, 1995, 1998, and 1999).

The origin, chemical composition, and balneological properties of mineral and thermomineral waters in Serbia have been investigated for more than a century (S. Daben, 1837; E. P. Lindenmajer, 1856; M. Leko et al., 1922; L. Nenadovic, 1936; V. Vujanovic & M. Teofilovic, 1983; D. Protic, 1995). It should be noted that geologists, chemists, and doctors have collaborated successfully in this. Investigating the content of trace elements in our mineral and thermal waters, Vujanovic and Teofilovic gave logical interpretations as to the genesis of these waters. Their results were published in a number of papers and synthesized in the book "**Spa and Mineral Waters of Serbia**" (1983). A complete review of geological, hydrogeological, geochemical, balneological, and other data on all our more important waters is given in the work of D. Protic "**Mineral and Thermal Waters of Serbia**" (1995). The distribution of the group of alkaline elements, calcium, magnesium, and fluorine in mineral waters of Serbia and their significance for human health were treated by M. Teofilovic (1999).

During the period after the Second World War, Serbia was one of the territories in the world where nuclear raw materials were intensively investigated, and the literature here is very rich. Significant knowledge has been gained in this way about the background state of radioactivity of our geological formations, and it is possible to speak fairly confidently of potential geological risks to public health.

The monograph of M. Babovic "**Geology and Environment Protection**" (1992) was the first book in our country to treat environment protection integrally from the standpoint of geology. Singling out useful components of the geological environment, the author elucidated the role of man in their genesis and exploitation, defined and classified geological wealth from the viewpoint of application, and isolated the most significant geological terms in an attempt to create a well - rounded eco - geological approach to environment protection. This work can accordingly be of greater or lesser usefulness in examination of separate questions in the domain of medical geology.

We note in conclusion that the author of the present book has spoken on *human health in relation to the geological environment* at many meetings of geologists, doctors, and agronomists held in Yugoslavia and abroad. He has in addition published a number of works on the subject (M. Komatina, 1993, 1997, 1998, 2000). Among others were the following: "**Geological Basis for Ensuring Healthy Production and Living Conditions**" (with B. Bozovic and M. Navala, 1997); "**The Geological Environment and Production of Healthy Food**" (1998); "**Principles of Medical Geology**" (1999); and "**Influence of Geological Factors on Human Health**" (2000).

METHODOLOGY OF MEDICO - GEOGRAPHICAL RESEARCH

Modern medical geography is the result of long - term development in the course of which theoretical principles, general questions of procedure, and concrete methods have been elaborated to resolve various practical problems. Having taken a complex path since its formation on the basis of several sciences, medical geography has today preserved certain features of a borderline discipline (science) whose *main purpose is to contribute to society by optimizing* (in the sense of sanitizing) *the human environment so as to achieve the highest level of public health.* Our survey of the methodology of medico - geographical research follows the treatment of Keller, Shchepin, and Chaklin in their book "**Handbook of Medical Geography**" (1993). There are two important reasons for concerning ourselves with the history and methodology of medico - geographical research: *first,* to be able to draw on previous experience for the needs of medical geology; and *second,* to single out interesting results obtained in studying the influence of natural (geological) factors of the environment on human health. In the further text, interaction of medical, biological, geographical, and sociological disciplines in resolving problems of medical geography will be discussed.

Biological and medical disciplines are treated as the subsystem of **geographic medicine** within the general system of contemporary medical geography. This subsystem consists of the following four disciplines:

1. **Human biology**, uniting disciplines whose subject of study is man as a biological species (normal anatomy, normal physiology, biochemistry, immunology, genetics, anthropology, developmental physiology, gerontology, etc.).
2. **Pathogens and sanogens**, with disciplines and specialties affording material for conclusions about biotic and abiotic elements of the environment. These elements play the part of disease agents (pathogens) or represent stimulants or stabilizers of health (sanogens). Included in this block are microbiology, virology, parasitology, vitaminology, biogeochemistry of macro- and microelements, radiology, bioclimatology, balneology, toxicology, etc.
3. **Adaptation and nonadaptation**, encompassing disciplines and specialties that provide medical geography with material for conclusions about the results of interaction of the human population with its environment. They include the science of adaptation, general pathology, pathological anatomy, epidemiology, infective pathology, and other clinical disciplines (ophthalmology, endocrinology, oncology, therapy, etc.).
4. **Disease prevention**, including disciplines concerned with prophylaxis, namely nutritional hygiene, dietology, industrial and communal hygiene, children's hygiene, etc. Of significance for medico - geographical research are data of these disciplines that have a bearing on the nature and extent of anthropogenic pollution of the environment, permissible limits of such pollution, measures for protection and optimization of the environment, etc.

Each of the listed disciplines and specialties has its own system of knowledge and scientific valorization dealing with questions of fundamental significance for it.

At the same time, some of the factual data obtained by each of these sciences are used by medical geography for spatial analysis to discover the nature of territorial distribution (dissemination) of disease agents, trace elements, adaptation indicators, etc. These data constitute the geographic aspect of each of the disciplines (geographic physiology, geographic pathology, etc.), which as a whole can be termed **geographic medicine**.

Use of concrete factual data from the arsenal of the given medico - biological disciplines demands preliminary detailed logical analysis, selection, and generalization of information based on the results of territorial study and cartography. In practical terms, this means that all of the selected facts must have a precise territorial link before they can be plotted on a *factographic map* depicting the whole range of available data on the studied phenomenon. The factographic map thus represents an important basic document.

While employing the factual data it needs from the domain of the medico - biological sciences, medical geography also utilizes conceptual positions of these disciplines. This primarily refers to the results of experimental, field, clinical, and other observations that can be used to develop criteria for anthropo - ecological evaluation of the environment and its influence on the human population. Such material includes, for example, data on the lifetimes and reproduction mechanisms of disease agents in various types of environments, information about the optimal and extreme concentrations of biologically active trace elements in food products and water, etc.

The concept of evaluation is also central to the second subsystem of scientific directions in the framework of the general system of medical geography. The structure of this subsystem includes *geographic, economic, and sociological disciplines and specialties*, and is likewise composed of four provisionally independent groups of disciplines:

1. **Environmental subdivisions**, as established by landscape science and physico - geographical regionalization. Both disciplines are complex and concerned with study of the genesis of geosystems, their contemporary structure and dynamics, and the existence of natural territorial comples of different ranks (natural zones, countries, provinces, landscapes, etc.). Data on differentiation of the Earth's surface into geosystems, their structure and types, and involvement of zonal and azonal factors in their formation are essential for medico - geographical investigations, including research of significance for anthropo - ecological evaluation of quality of the human environment.
2. **Environmental components**, as indicated by specialized disciplines providing information about inorganic and organic components of the Earth's shell (the lithological base, soil cover, floor rocks, climate, and waters), as well as data on migration of chemical elements, characteristics of vegetation, and the animal kingdom of different landscapes. The list of such disciplines includes geomorphology, pedology, climatology, hydrology and hydrogeology, landscape geochemistry, and biogeography. The data of these disciplines are needed for a thorough characterization of geosystems, discovery of correlations within them, and evaluation of positive or negative anthropo - ecological significance on the basis of criteria developed in the framework of the geographical aspect of medical and biological sciences (**geographic medicine**).
3. **Technogenesis and the environment**, as reflected by the data of disciplines treating anthropogenic transformation and economic exploitation of the natural environment. Economic geography and directions of physical geography dealing with construction projects are such disciplines.

Medical geography uses the data of these disciplines to establish conditions that promote the development of natural and technogenic diseases. This information is needed in order to devise measures for optimization of the human environment.

4. **Demography and social organization**, as deduced from the data of population geography, sociology, demography, health protection, and geography of services and culture.

Thus, the disciplines and scientific directions of the second subsystem—the subsystem of **medical geography**—serve medical geography as a source of knowledge about the natural environment, economy, and population of a studied territory. This knowledge is drawn from general scientific (fundamental) information on the basis of anthropo - ecological evaluation of natural and socio - economic phenomena and processes. On the whole, it forms the applied aspect of these disciplines, which cannot be called medical geography in the strict sense, since the examined second subsystem also includes demography, sociology, and health protection.

The methodology of research on the level of systems of medical geography as a whole is based on linked cartographic analysis of all data provided by the examined two subsystems of science in the form of cartographic depictions (factographic maps) of medico - biological content on the one hand and specialized maps of sociological - cartographic content on the other. It is noticeable that linked cartographic analysis and synthesis of data on negative influences of the environment on public health contributes to the resolution of questions of *nosogeography* or the *geography of disease*, which is one of the basic preoccupations of modern medical geography.

Together with data on negative influence, medical geography draws information about positive influence of the environment on public health from the arsenal of both scientific subsystems. Here we have in mind data on so - called *sanogens*, environmental factors that play the part of preconditions which strengthen or stabilize health (physiologically favorable climatic structure, the helio- and hydroresources of a territory, optimal content of trace elements in the environment, presence of vitamins and minerals in meat products, diet structure optimal from both the physiological and hygienic points of view, public water supply in keeping with hygienic norms and requirements, etc.). Sanogeographical research is an integral part of medical geography, and *sanogeography* or *geography of health*—together with *nosogeography* or *geography of diseases*—is considered one of the cornerstones of medical geography as a whole.

According to Keller, Shchepin, and Chaklin (1993), the strength and virtue of modern medical geography lies in its systematic approach to phenomena and processes. To quote from their writings: "*Modern medical geography does not lose itself in a sea of facts and petty details, but instead generalizes on broad geographical basis information about a territory obtained both from medical geographers themselves and from specialists in other areas of knowledge. This systemic complex approach was inherited by medical geography from the geographical sciences, where it was formed by the method of trial and error over the several centuries of existence of geography. Modern medical geography in this differs fundamentally from the narrow profiles of directions in a number of biological and medical sciences."*

Two opposed tendencies have caused growing interest in problems of the environment, the territorial - ecological aspect of public health, geographical distribution of human disease, and its pathological and prepathological state. On the one hand, disciplines concerned with the system composed of the environment, man, and health are undergoing a process of sharp integration. On the other hand, their differentiation is just as pronounced. Integration of sciences is dictated by the desire of scientists to obtain a complete picture of the most complex processes and phenomena, whereas differentiation is linked with complication of research methods, colossal increase in the volume of information, etc.

We have thus arrived at the starting point for examination of the relationship between geography and geology that is between medical geography and medical geology.

GEOGRAPHY AND MEDICAL GEOGRAPHY IN RELATION TO GEOLOGY

It is known that *geography* was for a long time the only science dealing with nature, the economy, and population of the Earth. In the course of its historical development, geographical science was transformed from a single unified science whose task was to describe physical space into a branching system of sciences with concrete technical and economic content. The differentiation first of all affected *physical geography* and was stimulated by research on various categories of natural resources (land, water, biological, and other), as well as by the development of landscape science. On the other hand, accelerated development of the economy was accompanied by formation of a number of disciplines in the area of *economic geography*. All of these separate geographical disciplines were closely intertwined with each other, as well as with many branches of allied sciences (geological, biological, economic, historical, etc.). In all of this, geography down to the present day has remained a link between the natural and social sciences, a science that studies various components of the very complex relationship between natural and social elements (I. P. Gerasimov, 1980).

The problem of the subject of geographical study is precisely that geography touches two spheres of existence: nature and society. Speaking of the geographical aspect of the relationship between the natural and the social, Z. Jovicic (1998) had this to say, among other things: "*Because geography treats a large number of different phenomena and forms of nature and society (relief, waters, climate, plant and animal life, the economy, population and settlements, etc.), it cannot be in equal measure competent for all those phenomena and forms, and criticisms as to both the subject and suitability of geographical studies are entirely in order. Some other sciences that also study natural or social phenomena (biology, meteorology, sociology, economics), with their more exact methods, resolve far more successfully than geography certain problems connected with these phenomena. Their basic nature cannot, therefore, be solely a subject of geographical research. This is the main reason why geologists in addition to geographers are involved in the study of relief, meteorologists in the study of climate, biologists in the study of plant and animal life, sociologists in the study of society, etc. Taking the landscape of a region or geographic environment as a subject of geographical research must be understood as an attempt to overcome indefiniteness, which necessarily gives rise to generalization, superficiality, and descriptivism in geographical science...*" Going further, R. P. Moss stresses that geography essentially is a science that for explanations uses laws of other sciences rather than discovering them.

The fact that the subject of geographical study has not been precisely enough defined is especially reflected in the *area of geological research.* Accelerated development of *fundamental geology* in the 20th Century gave rise to a number of precisely defined scientific disciplines: mineralogy, petrology, pedology, paleontology, geochemistry, geophysics, historical geology, tectonics, and their branches. The results of these disciplines in study of the structure, genesis, and historical development of the Earth's crust are certainly impressive. Here too are a whole series of somewhat younger disciplines of *applied geology,* such as the science of mineral deposits, the geology of coals, petroleum and gas geology, hydrogeology, geological engineering, and applied geophysics, with numerous research methods developed to such a point that they are able to resolve all scientific and practical problems connected with the geological environment. Geography would appear to have been arrested by the given developments in a state of greater or lesser inertia in regard to all factors of the natural environment, including geological ones. It goes without saying that **medical geography** has not been immune to such "behavior" and has largely eliminated geology from the chain of vital multi - disciplinary activity.

Overlooked was the fact that the geological environment is the carrier of all mineral aggregates, that all components needed for the development and survival of living nature emanate or are emitted from it. Perception of the geological structure of a certain terrain (or area) is of fundamental ecological significance, since it dictates the actual content, relations, and potentials of the geological environment as the basis for development of life, exploitation of resources, and use of space for production.

According to A. A. Shoshin (1961), *the task of medical geography is to study the natural and socio - economic characteristics of a territory in order to clarify their influence on the state of public health, the incidence of human diseases, and their global distribution.* Based on a number of sciences (including medicine, biology, geography, economics, and sociology), medical geography in treating the natural conditions of some territory has usually synthesized general information without delving far enough into factors of inorganic (nonliving) nature as possible causes of disease or agents of positive influence on human health. Their diversity accounts for the great difficulty encountered by geographers, doctors, biologists, and other specialists in attempting to emphasize some aspect of this large and important area of research. On the other hand, the practice of some epidemiologists to study only the distribution of human diseases without in - depth investigation of the causes and regularities of their incidence in a certain geological environment transforms medical geography into a purely descriptive science capable of recording the more important diseases (especially noninfectious ones), but not of analyzing, extirpating, and preventing them.

In his book "**Principles of Medical Geography in Serbia**" (1998), Z. Jovicic wrote as follows: "*The literature consulted indicates that two basic themes are recurrent in medical geography, namely the influence of the natural environment on health and the role of geographical variability in the distribution of diseases. Geographical analysis and cartographic presentation stand out as the main methods in medico - geographical research. It should be noted that two key expressions—health and disease—are unavoidable not only in connection with our discussion, but also much more widely, in medicine, sociology, biology, psychology, and even philosophy. In conclusion, it is perhaps possible to differentiate two variants of medical geography. Clearly, medically oriented medical geography cannot be denied the right to study the influence of the natural environment on health. On the other hand, geographically oriented medical geography cannot be declared incompetent to analyze the spatial distribution of various phenomena and manifestations, including diseases of course.*"

But the tendency of medicine to impose its approach to research has caused the importance of natural factors in the incidence of disease to be neglected to a greater or lesser extent. The impression is created that medicine in its rapid advancement has increasingly forgotten the inorganic (geological) environment, i.e., the main root from which life on the Planet developed and the ambience in which man lives and works. This should not greatly surprise us. At a time such as the present, when there are many branches of science and fields of scientific endeavors, geology - above all in the sphere of education - has been relegated to a side track, with the result that awareness of the geological environment and its significance is increasingly stunted among scientific workers. No wonder certain investigators of man and human health appear to feel it less and less important whether man as the subject of interest is an inhabitant of the Earth or a being from some other planet. The paths to diagnosis of disease have thus become increasingly more convoluted, while preventive action has been replaced by ever greater medicalization of society[2].

Knowledge of geology fosters a deeper understanding of all that surrounds man. It helps us to recognize the beauties of nature and enables us to overcome superstitious fear in the face of powerful, sometimes terrifying and destructive geological phenomena. It demonstrates the order of the Universe and the transient nature of seemingly eternal mountains, volcanoes, valleys, and lakes. It will help us to conquer many diseases and thereby create conditions for a safer life.

SUBJECT MATTER AND THEORETICAL PRINCIPLES OF MEDICAL GEOLOGY

A great physician of the past, K. Galen wrote that health is a state in which we do not suffer from pain and are not limited in our vital activities. Numerous later definitions of health supplement that thesis by linking the human state with the external environment. There also exist various attempts to classify the level of health and disease. Examining the level of health in the population of the Far North, N. S. Jagya (1980), for example, divided the whole of the examined population into the following five groups: 1) *healthy*; 2) *healthy, with functional and certain morphological changes*; 3) *ill, with long - term (chronic) diseases, but basic functions of the organism intact*; 4) *ill, with long - term (chronic) diseases; 5) bed - ridden seriously ill, invalids of groups I and II.* Such classifications are interesting for evaluation of health in different regions of the same territory and in different seasons of the year, and analyses of the ratio of groups with reliably different levels of health make possible the classification of territories in regard to that characteristic, which is of great practical as well as theoretical significance. A region with predominance of the first and sec-

---

[2] At the present time, there are more than 10,000 diseases in existence and about 400,000 different medicinal preparations for their suppression.

ond groups according to the indicated classification can be judged to be a *territory with a high level of public health.* On the other hand, a region where a great percentage of the population belongs to the third and fourth groups must be regarded as one marked by a *low level of public health.* It goes without saying that where medico - geographical maps are available, it is easier for medical and other services to take decisions and implement measures aimed at optimizing health of the entire population and its separate groups. Addition of geological elements to such maps makes it possible to discern correlations with geological composition of the terrain and isolate *negative or positive influence of inorganic factors on human health.* **Medical geology** can provide answers to questions in this domain.

Viewed more broadly, the level of public health of a certain territory is formed under the influence of **natural** and **anthropogenic factors**. **Natural factors** include the layer of the atmosphere close to the ground, solar radiation, rocks, soil, surface and groundwater, characteristics of relief, endogenous geological processes, radioactivity, geophysical fields, macro - and microelements in soil and water, vegetation, and animal life. **Anthropogenic factors** include soil pollutants, waste water and other water pollutants, air pollutants, radioactive pollution in the form of nuclear waste, pollutants arising from exploitation and processing of ore, complex harmful influences of urbanization and roads, etc. The influence of man on the environment grows stronger every day, and this leads to rapid alteration (and sometimes complete degradation) of its natural component, which in turn affects humanity itself, including the level of human health. In view of socio - economic conditions and their influence, it can be concluded that investigation of the causes of disease incidence is necessarily a multi - stage process, one full of discoveries, but also disappointments. The approach itself must take into account a number of factors, both ones of narrow interest and broader spatial considerations, which is to say that it should be complex and multi - disciplinary.

By the term natural settings of diseases and pathological states of man, we mean certain properties of the geosystem which acting together with the human organism can be manifested in the form of pathological states or diseases (E. I. Ignat'ev, 1964). A given setting can be used to discover the etiology (or ambience) of diseases that have still been inadequately studied. Here it is necessary to be on the alert for many still undiscovered relationships between diseases and some factor (or factors) of the natural environment. The field of geology is particularly challenging in this context, the more so inasmuch as it has been inadequately investigated, while the Earth's geological structure is variable in space and human life unfolds in the geological environment.

The diversity of natural and social factors affords the possibility of constructing different classifications of diseases. Based on differences in the main natural factors determining the distribution and characteristics of pathological processes, one such classification was given by A. P. Avtsin (1959). As insignificantly modified by A. G. Voronov (1965), it recognizes the following five groups of diseases:

1. *Diseases caused by geophysical (primarily climatic) characteristics of the environment* (sunstroke, heat stroke, snow blindness, altitude sickness, etc.);
2. *Diseases caused by geochemical characteristics of the environment* (endemic goiter, fluorosis, dental caries, etc.);
3. *Diseases associated with biochemical characteristics of food* (avitaminoses, hypovitaminoses);
4. *Diseases caused by poisonous plants and animals, and allergic states* associated with substances and smells of plant and animal origin (bronchial asthma, etc.); and
5. *Diseases* (infections and invasions) *caused by living agents.*

According to Avtsin, the question as to the link between diseases and geographical factors has inevitably confronted every investigator who ever delved more deeply into problems of geographical pathology and medical geography: "*Are we not acquainted with diseases of tropical territories and quite different diseases characteristic of territories with a cold climate? It is well - known that the pathology of high mountains has its own specificities... Have there not been described endemic diseases of human beings, animals, and plants where some disturbance of the balance of chemical elements in the geographic environment acts as the main cause of a certain form of damage? Is it possible to ignore the fact that catastrophic movements in the lithosphere and hydrosphere (earthquakes, floods, tsunamis, landslides, rock and mud slides) cause massive damage of a traumatic nature?*"

It is evident from the classification itself that **geological factors** can be of significance for practically all of the five isolated groups of diseases, being very significant for diseases of the first and second groups - diseases caused by geophysical and geochemical characteristics of the environment[3]. Let us now return to the earlier presented structure of medical geography in order to get a more complete picture of the areas in which geology can make an important contribution. The following three groups of disciplines are in question:

[3] It must be kept in mind that Avtsin relegated some purely geological factors to the geographical category. As indicated above, this was a generally accepted practice in the literature of medical geography and medical pathology.

- **Pathogens and sanogens**, where geology can provide important data on abiotic and (in some cases) biotic elements of the environment as pathogenic or sanogenic factors;
- **Environmental components**, where geology can provide important information about the lithological substrate, soils, groundwater (slightly mineralized, mineral, and thermal), exogenous and endogenous processes and phenomena, structural forms, deposits of mineral resources and their range of distribution, radioactivity in rocks and groundwater, trace elements (in rocks, soil, and water), geophysical fields, etc. Also about various geological risks such as seismic risks, volcanic activity, slips and slides, rock falls and mud slides, soil erosion, suffosion, and various geochemical risks; and
- **Technogenesis and the environment**, where information about the behavior of rocks, soils, and groundwater during anthropogenic action on the environment (various forms of pollution and destruction) can be of decisive significance in estimating the degree of environmental pollution and devising measures to reduce it, i.e., ways of protecting human health.

To conclude this examination, let us quote from Avtsin's book "**Introduction to Geographical Pathology**" (1972): "*We clearly see the interconnection of all living and nonliving nature, that enormous complex of the troposphere, hydrosphere, and lithosphere which the genius V. I. Vernadskii united in the concept of the biosphere.*" Geology will have performed its scientific and social mission if it carries out complex study of the geological component of the biosphere - the lithosphere - and thereby contributes fully to a better understanding of the hydrosphere and troposphere. The field of geological research is so wide, complex, and significant that there can be no doubt as to the justification of forming a scientific discipline linking *medicine and geology*. Filling the obvious gap in scientific multidisciplinary research along these lines will above all contribute much to medical science, particularly in the matter of disease prevention. It is one of the basic goals of science to bring about a state of optimization and sanitization of the human environment so as to achieve the highest possible level of public health. Medical geology will play its part only through far greater involvement of specialists, organization of international cooperation, and development of research theory and procedure. Two important initial steps can be taken in a relatively short time: 1) results of medical geography can be put to better use; and 2) valuable data of earlier geological research can be thoughtfully reinterpreted to create a modern basis for future research.

It is of indisputable importance to establish coordination of research efforts in geography and geology, i.e., to create agreement between medical geography and medical geology. That will probably be achieved when it is clarified in practice which discipline can better demonstrate both the positive and negative influence of environmental factors on public health, the incidence of human disease, and its distribution. *Time will show that there is no contradiction between medical geography and medical geology. On the contrary, there exists a single joint task - creation of a formula for maximally healthy conditions for human life - in whose accomplishment each science will participate in its own way.*

PROCEDURE OF MEDICO - GEOLOGICAL RESEARCH

As with most other sciences, medicine, biology, and geology experienced rapid and comprehensive development in the 20th Century, and the emergence of many new disciplines is significant in several ways. The precipitous post - war advancement of technology - particularly developments in the fields of electronics and automation, with the advent of powerful computers - has been reflected in the creation of numerous modern research methods on the one hand and the appearance of new possibilities in the matter of quantitative data processing on the other. All of this has resulted in the creation of a very firm foundation of medical geology and enabled it to enter a phase of accelerated development. Above all, medical geology has at its disposal a great many geological methods that have been developed to the point where they can provide answers to the majority of questions considered by practitioners of this new geological (or medical) discipline. Such methods include remote sensing, various field procedures, methods of fundamental geology (petrology, mineralogy, structural geology), geophysical methods, cartographic procedures, methods of geochemistry and other specialized geological disciplines (pedology, hydrogeology with hydrogeochemistry, geological engineering), nuclear methods in geology, etc. (M. Komatina, 1984). It goes without saying that many answers to the questions posed will probably also be obtained using methods of other scientific disciplines (biology, chemistry and bioinorganic chemistry, bioclimatology, pathology, physiology, epidemiology, toxicology, stomatology, veterinary medicine, sanitary hygiene, balneology, medical ecology, geomedical cartography and regionalization, statistics, information technology, mathematical modeling, and medico - geological diagnostics and forecasting).

Many spacious geological regions are known today in the world, and such units have been relegated to different so - called *biogeochemical provinces.* This term was introduced by A. P. Vinogradov in his work "**Biogeochemical Provinces and Endemics**" (1938).

The concept of such provinces makes it easier to understand both the distribution of regions with pronounced incidence of diseases resulting from an imbalance of elements and their significance for health. We note that the majority of geochemical diseases occur in developing countries in the tropical zone. By establishing more precisely the lithological composition and structural alignment of a given terrain, systematically analyzing its various components (rocks, soils, waters, and crops), compiling geochemical and geomedical maps, and ascertaining regional patterns in the incidence of diseases there, it is possible to create a good basis for finding their most important causes. This process includes the establishment of positive as well as negative influence of environmental factors, i.e., it involves their classification from the standpoint of health. Of special significance here is the method of correlational cartography of diseases, with which it is not only possible to establish the regions of their distribution, but also to gain an idea as to the link between pathological processes and certain environmental factors. One of the primary goals of research is to establish the link between natural characteristics of an area of interest and the occurrence of pathological states in it. It is therefore clear that results of medical and medico - geographical investigation of the area must exist parallel with geological data. For example, geology will treat abiotic components of the geological (natural) environment (rocks, soil, and geochemical composition of rocks, soil, groundwater, etc.) of a given territory, and study the migration of pollutants through them; while medical geography and medicine will supply statistical data gathered by health and veterinary services on the incidence of diseases in humans and domestic animals, results of field surveys by physicians, and information characterizing biotic components of the environment (data on plants, animals, agents of infectious diseases in them, etc.).

Inasmuch as multi - disciplinary investigations of complex objects are most often in question, development of a systematic approach is very important for the resolution of medico - geological problems. Such an approach enables multiple studies to be conducted on various processes and phenomena, and makes it possible for all specialists to work toward the achievement of a single common goal, namely joint interpretation (synthesis) of all important elements and forecasting of the behavior of a given system from the data of medical geology.

It is very important to limit the subject matter and task of medico - geological investigations of a given territory. In accomplishing the task posed, we have at our disposal many modern field methods, laboratory procedures, and methods of data processing and interpretation. Here too are the very useful experience and approaches of medical geography, developed in the course of centuries of research on environmental factors. The methods of medical geography and their usefulness in medico - geological research must be carefully analyzed. This applies to results of previous investigations in the areas of medical geography and cartography, which must be correlated with results of past geological research and cartography. Such analysis should enable us to judge how far geology, medicine, biology, and related sciences of interest can be used to estimate as authoritatively as possible positive and negative influences of the geological environment on public health and resolution of various problems in the domain of geo - ecology and medical geology.

Difficulties can arise in seeking the most rational (optimal) way of accomplishing the task at hand. Practice (experience) teaches us to select the investigative approach that is most appropriate in view of specific characteristics of the terrain of interest. Methods and approaches should be chosen that can give the most reliable results in studying the environment and accomplishing our purpose. *Optimization of the investigative process* is achieved through creative management of the project, with adherence to well - known basic principles:

- *Consistency;*
- *Exhaustiveness;*
- *Balance;*
- *Maximal information use; and*
- *Cost efficiency.*

Especially important is cooperation among specialists in different areas of science so as to ensure that they have a clear grasp of their part in the research and are able to accomplish their mission as fully as possible. The research must be planned in such a way that no participant in it is deprived of individuality (serving only as "bricks" in the overall construction), but rather that each is acknowledged to be of independent significance in contributing to the solution of a given problem.

The pattern of contemporary distribution of certain human diseases is so variegated that it is often impossible to establish with any confidence the factors that account for it. Where factors of the natural environment are in question, the only realistic course of analysis is to establish the boundaries of geographical or geological distribution of the suspected cause, characterize the medico - geological region isolated in this way, and finally compare the distribution of individual environmental factors with that of the disease. From the standpoint of geology, it is excusable to limit medico - geological research to relevant geological conditions of the terrain, i.e., to work within the confines of interesting *geological (hydrogeological) regions* or *geological formations*. Such an approach is all the more justified if it is also indicated by the geography of some endemic or other disease. For example, the incidence of endemic nephropathy in Yugoslavia is geographically linked solely with one type of geological formation, namely alluvial deposits, within which the cause of the disease should be sought. Or: the fact that human echinococcosis appears far more frequently than expected in the outer zone of the Dinaric karst is attributable to deficiency of surface water, as a result of which mass infection with *Echinococcus* occurs in the course of close contact with dogs under poor conditions of sanitation.

The investigative process unfolds in phases. The *first phase* involves acquisition and study of data obtained in previous investigations - geological, geographical, biological, medical, etc. A basis for planning and more precise guiding of research efforts is created in this way. Field and laboratory investigations are carried out in the *second phase*. The *third phase* involves data processing with textual and graphic interpretation. The goal of these activities is to discover basic correlations and regularities governing the link between *public health and the geological environment*, i.e., to determine which factors of the studied geological environment directly or indirectly affect the level of public health in a given region.

A distinction is drawn between fundamental and applied medical geology. It is logical first of all to resolve certain theoretical and procedural questions through fundamental investigations, create a medico - geological picture of an area or geological environment that reflects its regionalization, and establish geological locations (typical test grounds) on which it is relatively easy to study the pathogenic or sanogenic role of certain geological factors or geological formations. In other words, **fundamental medical geology** is concerned with development of theoretical principles and procedures of research, basic regional explorations of terrains, general analysis of the influence of individual geological factors (formations) on human health in a country or selected territory, etc. A foundation is thereby created for further (detailed) research directly linked with more restricted sites or certain practical problems.

Applied medical geology is used to study basic regularities governing influence of the geological environment or individual factors on public health in a concrete region, i.e., to establish factors which dictate the level of public health in that region. From the assembled data, measures can be devised for environment improvement and health protection. These ends are primarily realized as a result of *detailed medico - geological research*, whereas *specialized investigations* yield information about specific organic and inorganic components of the natural environment.

The very fact that medical geography has been so preoccupied with studying what are largely geological factors, while information about those factors has contributed to the discovery of pathogens and sanogens, justifies the existence of both medical geology as a whole and some of its disciplines. Inasmuch as the geochemical composition of rocks, soil, and groundwater has for years been at the center of interest of doctors, geographers, and to some extent geologists, it can now be considered that **medical geochemistry** has already existed for a long time. We note the significant work of C. B. Dissanayake and R. Chandrajith "**Medical**

Geochemistry of the Tropical Environment" (Sri Lanka, 1999), which deals with the influence of certain essential elements on the health of humans and animals. The authors assert that the etiology of diseases such as dental and skeletal fluorosis or diseases associated with iodine deficit lies in the geochemical environment. They further stress that significance of the chemical composition of soil, water, and rocks in the incidence of diseases in certain tropical regions presents new challenges in medicine and provides justification for seriously concentrating on geomedical research with participation of geochemists and epidemiologists.

The same justification can be cited for **medical hydrogeology**, owing to the plethora of data indicating the significance of slightly mineralized and mineral waters for human health.

Although the given two disciplines have thus found practical affirmation, adequate attention must be paid to their further development in keeping with the demands of medical geology and medical geography. In connection with this, an interesting proposal was put forward by J. S. Jean (Taiwan, 1999), namely that **biomedical hydrogeology** be formed as an independent discipline.

An epidemic of enterovirus broke out on the territory of Taiwan in 1998. About 300,000 persons contracted the infection in the course of that year and the next. This magnitude of the epidemic calls to mind several similar instances recorded previously in Hungary (1978), Bulgaria (1975), and Malaysia (1997). It is highly probable that the cause lay in pollution of the soil and groundwater occasioned by the burial of more than five million dead pigs infected with afthovirus throughout the entire country, mainly in coarse - grained sediments of plains along the banks of rivers. "*Research in locating the source of pathogens and devising measures to prevention and control the epidemic disease involve biology, medicine, and hydrogeology. Thus, I have coined the term* ***biomedical hydrogeology***" *(J. S. Jean, 1999).* In order to stress the significance of such a multi - disciplinary approach, the author cited the relationship between the hydrogeological environment and Blackfoot disease, an endemic disease known in southern Taiwan already before 1990, when arsenic was discovered in artesian well water used for drinking purposes. Tests confirmed the disease to be caused by elevated content of arsenic in groundwater and also indicated that its specific influence on health depends on the existence of other substances in the water, as well as on the hydrogeological conditions themselves. For example, negative effects were particularly pronounced in the case of artesian water occurring at a great depth, but were absent where water was drawn from shallow wells with a variable level.

GOAL, TASK, AND SIGNIFICANCE OF MEDICAL GEOLOGY

It is usually hard to answer the important question as to how favorable a certain area is for life. Investigation of the causes of disease represents a particularly complicated task full of uncertainty. The given problem can be successfully resolved only through a multi - disciplinary approach to it. But this does not mean that it is not necessary to extract the maximum from every contingent scientific discipline, especially in domains where that discipline can make a significant contribution. Finding the true (basic) cause of a disease or establishing the positive action of a given terrain on human health would certainly make it much easier to diagnose the disease itself and point to more effective ways of treating it, i.e., facilitate preventive work.

A favorable circumstance is that geology, as the science dealing with the inorganic world, offers diverse possibilities in this area. To be specific, the regularities of occurrence of certain geological environments and factors that can cause disease or strengthen the organism's defensive capabilities are most often well expressed. It is therefore possible within the framework of a geotectonic or structural unit (or geological formation) of interest to establish correlations between geological factors and the data of field and laboratory medical studies. In other words, the regionalization of an area of interest to us with respect to the degree of influence of geological factors on public health (geo - medical regionalization) can - in conjunction with maps, atlases, and regionalization schemes already created by medical geographers - be of great usefulness both for further medical research and practice, and in the areas of commerce, urbanization, and health protection.

The principal goal of medical geology is to study basic laws of influence of geological environment to human health, precisely to distinguish among geological factors those which can be the main causes of certain diseases or have a positive effect on human health. Medical geology contributes in this way to optimization of the environment and realization of the highest possible level of public health on the Planet. The existence of this new discipline and persons specializing in it will promote far better use of the information gained in voluminous fundamental and applied geological investigations, which represents another step forward in the efforts of world science in general. It goes without saying that the advances and experience of medical geography, medicine, biology, and other allied sciences have been utilized in laying the foundations of medical geology. Such a base allows for the fastest development of those branches of medical geology that treat factors of the geological environment, geochemical and other processes, and the procedure of research undertaken for special purposes.

It is the task of medical geology to study influence of the geological environment on the health of man as a social animal. To accomplish this task, it is necessary to develop methods and criteria for medico - geological evaluation of an area of interest to us, differentiate the geological environment from the standpoint of its influence on public health, and get a picture of the area's medico - geological regionalization. This is important for scientifically based optimal (coordinated) exploitation of geological resources and transformation of the geological environment in the interests of healthy living. The experience and results of medical geography can be of great assistance here, and analysis of the extensive data of past geological research from the geomedical angle is also useful. In addition to this, much can be gained through faster progress in geo - ecology, medical geochemistry, medical hydrogeology, and other disciplines. When developed to the level needed, medical geology will be able to produce results of great usefulness to the economy and health protection authorities. An especially important function is to indicate danger of a geological nature.

The significance of medical geology is manifold. This has already been discussed, and here we will stress only the importance taken on by medico - geological research in connection with the increased interest being shown by medicine in somatic (noninfectious) diseases, especially malignant neoplasms and diseases of the internal organs. A notable contribution in this field has already been made by geochemistry through the naming of certain geological elements (trace elements in rocks, soil, and water, for example) among etiological factors of primary significance. The very fact that the incidence of malignant tumors is unequal in different geographical regions of the world presents a challenging field for medico - geological research aimed at discovering correlations and regionalization of territory on the basis of established geological and medical parameters. Especially interesting here are microregions in which the carcinogenic situation is noticeably more expressed than in the surrounding territory. *Will medical geology be able to solve part of this global riddle?*

CHAPTER 2

Geological Factors

ENVIRONMENT OF LIFE, NATURAL ENVIRONMENT, AND GEOLOGICAL ENVIRONMENT

The level of public health is formed under the influence of an enormous number of factors, internal and external. External factors include natural conditions, the type of economic activity, cultural and sanitary - hygienic habits of the population, the level of medical science and nature of medical protection available, the existence of natural settings for disease and harmful substances of technogenic origin, etc. The whole aggregation of external factors in relation to organisms is defined as the **environment of life.**

There are two important components of this environment:

- Parts of the planet Earth on which real conditions exist for maintenance of life in any form;
- Natural conditions that can ert influence on the living world.

For this reason, the term **natural environment (geospace** - the space inhabited by man[4]) is used in the narrower sense.

---

[4] Russian naturalists use the term landscape, by which they mean the entire complex of natural conditions on a certain sector of the Earth's surface (the outward appearance of a locality, in addition to characteristics of climate, soil, vegetation, and groundwater) and other superficially unapparent characteristics of nature that play an important role in processes on the Earth (Fig. 1.1). This concept encompasses the great, complex, and unbalanced dynamic system of the Earth's surface, where elements of the lithosphere, hydrosphere, and atmosphere interact and intermingle with each other. Foundations of the science of landscape were laid at the end of the 19th Century by V. V. Dokuchaev in his well-known work "**The Study of Natural Zones**" (1948).

In simplified terms, the natural environment at the level of human life and productive activity is taken to mean the structurally complex system made up of the following five components: *atmosphere*, *hydrosphere*, *lithosphere*, *animal life*, and *plant life*. These components of the natural environment are closely interrelated and in constant interaction with each other to form a single whole material system. Any interference with this natural system created by human engineering activity always disturbs its dynamic equilibrium and gives rise to *technogenic processes* that negatively or positively affect changes in characteristics of the natural environment.

Three planetary functions of the natural environment (biosphere) play a special role in the life of man and his environment. They are: 1) *biological productivity*, which provides all forms of life with food products; 2) *maintenance of the optimal regime and balance of the Earth's aqueous envelope and gas composition of the atmosphere;* and 3) *natural biological self - purification.*

More precisely considered, the natural environment is made up of numerous components, including air, natural waters, soil, and rocks in the substrate, characteristics of relief, vegetation, and animal life. Taken altogether and in various combinations or individually, the given components ert positive or negative influence on human life. Except in the case of certain fundamental investigations, study of the influence of elements of the natural environment on public health is possible only inside certain territorial boundaries around areas where that influence is relatively equal. Moreover, classification of components (factors) of the natural environment creates conditions for their evaluation, as well as for their insertion into some comprehensive model of a medico - geographical (medico - geological) territorial system.

Components (factors) of the natural environment can be classified from various aspects. In the book "**Handbook of Medical Geography**" (1993), the following divisions of components are given:

- *On the basis of periodicity of influence on vital human activities*: 1) Constantly active (reduced air pressure in mountains, high solar radiation in the equatorial zone, etc.); 2) cyclic (seasonal circulation of the agents of certain infectious diseases, low winter air temperatures in nontropical latitudes, flooding of rivers, etc.); 3) acyclic (earthquakes, volcanic activity, rock and mud slides, landslips, etc.);
- *On the basis of preventability of negative influence on vital human activities*: 1) Preventable (for example, adverse effects of life in regions with a harsh cold climate); 2) harder to prevent (rock and mud slides, spring flooding); 3) unpreventable (earthquakes, volcanic eruptions, tsunamis);
- *On the basis of activity in relation to other components of the environment with resultant strengthening or weakening of influence on vital human*

activities: 1) Active (for example, strong wind at low air temperatures); 2) little active; 3) inert;

- *On the basis of mutual influence*: 1) Synergistic (for example, calm weather strengthens the influence of toxic substances on man); 2) antagonistic (for example, soil with a high capacity for self - purification disinfects impurities containing pathogenic microorganisms).

The term of special interest to us - the **geological environment** - is of relatively more recent origin. It refers to the upper layer of the lithosphere as an integral component of the environment of life. The part of the given environment accessible to human activity is made up of the following five components: solid rocks, loose soil, groundwater, natural gases, and microorganisms. All of these components interact with each other to create (under natural and disturbed conditions) a dynamic equilibrium. The surface (relief) of the Earth's crust can be taken as *the upper boundary of the geological environment.* Here it interacts with such components of the natural environment (or environment of life) as the atmosphere and the hydrosphere. This interaction is determined by a complex of constantly transpiring natural processes such as the following: penetration of solar heat into the upper layer of the Earth's crust; global circulation of moisture; hydraulic linkage of surface and groundwater; infiltration of atmospheric precipitation; reciprocal influence between rocks and moisture in the zone of aeration (the zone above groundwater) and in the zone of complete saturation; evacuation of groundwater on the surface; penetration of natural gases (oxygen, carbon dioxide, etc.) into the upper layer of the atmosphere; and geochemical processes in the system composed of *rocks, groundwater, natural gases, and microorganisms.* The indicated processes and other global processes of interaction between the outer spheres of the Earth's crust play a decisive role in formation of the structure of the geological environment, the characteristics of this environment, and its ecological quality.

The lower boundary of the geological environment apparently can be drawn to some extent arbitrarily, depending on the level of modern scientific - technological progress. That boundary today is for the most part determined by the depth of mined mineral deposits (to 4,000 m) or the depth of deep oil and gas boreholes (to 9,200 m)[5]. Whereas diverse processes resulting from interaction of the atmosphere, hydrosphere, and biosphere transpire actively at the upper boundary of the geological environment, global influence of endogenous processes caused by interior

[5] The deepest test borehole in the world (more than 11,000 m) was drilled on the Kola Peninsula (Russia).

temperature, pressure, and gases is erted on the geological environment itself at the lower boundary.

The part of the geological environment most interesting to study is naturally that part in which active forms of life occur and where human productive endeavors are enacted. That part of the geological environment plays a fundamental part in the vital activity of man. Concentrated in it is the main biomass of the Earth, the soil layer, all forms of mineral resources (including oil and gas reserves), various types of groundwater, rocks as building materials, etc. As a result of the engineering activity of man, anthropogenic (technogenic) processes are constantly transpiring in the given part of the geological environment, and their negative or positive influence causes changes in its state and properties.

BASIC FEATURES OF THE GEOLOGICAL ENVIRONMENT

In their major work "**Geology: The Science of the Changing Earth**" (1980), I. S. Allison and D. F. Palmer distinguish the following spheres of the Earth: 1) the *magnetosphere* - a geomagnetic field of great expanse; 2) the *atmosphere* - the gaseous envelope of the Earth; 3) the *hydrosphere* - the aqueous envelope of the solid Earth; 4) the *biosphere* - the envelope within which life unfolds; and 5) the *geosphere* - the solid sphere of the Earth, which is the main subject of the work of geologists. The geosphere is made up of three main parts that differ in their properties: a) the *lithosphere* or *Earth's crust* - the thin solid outer shell; b) the *asthenosphere* - the inner shell; and c) the *Earth's core.*

At depths of 35 - 70 km on the continents, the velocity of spreading of seismic waves increases abruptly from 6.5 - 7 to 8 km/s. Changes in both elementary and mineral composition and materials are presumed to occur at such depths. The given lower boundary of the lithosphere (Earth's crust) was discovered by A. Mohorovicic, after whom it was named. The lithosphere has its greatest thickness (up to 70 km) under mountain massifs and is thinnest (5 - 15 km) at the bottom of the oceans. On the territory of Yugoslavia, thickness of the Earth's solid crust varies between 22.5 km (the Pannonian Basin) and 50 km (the Dinarides of Montenegro).

Until relatively recently, it was considered that the lithosphere does not represent a homogeneous sphere, but is multi - layered instead. Thus, the following three main parts within the continental crust (the part of the lithosphere in the region of continents) are most often singled out:

1. The outer part or *stratosphere*, predominantly constructed of sedimentary rocks of Mesozoic, Tertiary, and Quaternary age, which occur on the surface of terrains and extend to a depth of the order of 12 km;

2. The middle part or *granitic layer* predominantly composed of acidic magmatic rocks (granites, granodiorites, etc.);
3. The lower part or *basaltic layer* composed predominantly of basalts.

The oceanic crust, on the other hand, is mainly constructed of rocks having a basaltic composition.

According to the theory of *new global tectonics*, the lithosphere is composed of six large and several small separated great tectonic plates[6]. These plates are subject to movement, approaching the others or moving away from them (Fig. 2.1). As we shall see later, processes transpiring in the lithosphere (including the greatest part of the asthenosphere) affect the disposal and relief of continents and ocean basins.

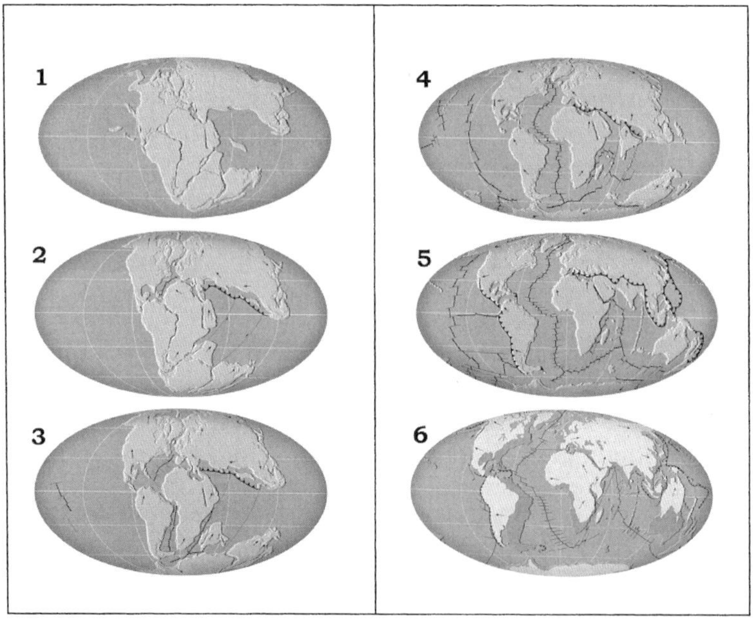

Fig. 2.1. Movement of continents from Pangaea period to nowadays.

In the course of the Earth's geological history, enormous expanses of the lithosphere were subjected to high pressures and temperatures arising as a result of reiterated processes of mountain massif formation. Under such harsh conditions, often accompanied by chemical reactions, magmatic and sedimentary rocks passed over into a new class - the class of

[6] Hence the expression "theory of plate tectonics".

metamorphic rocks. Thus, the lithosphere today is built of these three main classes of rocks, together with the loose cover formed by their decomposition.

The lithosphere, and especially the part of it called the geological environment, is under the influence of endlessly diverse processes. It is therefore characterized by properties variable in both space and time. Not only are we witness to the changes taking place before our eyes today, but we also have at our disposal the results of changes that occurred hundreds of millions of years ago. The more important among them are:

- *Crystallization of minerals;*
- *Formation, alteration, and decomposition of rocks; soil formation;*
- *Volcanic activity and other forms of magmatism;*
- *Processes of metamorphism;*
- *Deformation of rocks in the process of wrinkling and fault formation; recent tectonic movements;*
- *Movement of tectonic plates;*
- *Consequences of earthquakes;*
- *Formation of deposits of various mineral resources;*
- *Radioactive decay of elements;*
- *Geochemical and hydrogeochemical processes;*
- *Heat currents; magnetic field changes;*
- *Exogenous geological processes;*
- *Biogenic nonanthropogenic changes; and*
- *Anthropogenic geological processes.*

They are all the subject of geology, the science of the eternally changing Earth. These processes and their effects on the geological environment or environment of life will be treated in the corresponding chapters of the book.

Useful properties of the geological environment are of two main categories, namely ones inherent in the geological environment in the narrow sense or *geological ambient* and properties associated with *geological wealth*. The *geological ambient* is utilized by the whole living world, and its useful properties, according to M. Babovic (1992), are manifested in the following: a) the esthetic and sensual ambient; b) shelters and obstacles; and c) the fertile (pedological) ambient. *Geological wealth* refers to that part of the geological environment with its useful properties that can be separated, transported, and then utilized, whether in unmodified form or after certain transformations. The concept of geological wealth includes not only mineral raw materials, but also other mineral substances that are utilized without additional processing. Like the geological environment, geological wealth is characterized by variability of properties in space and time. According to M. Babovic (1992), geological wealth (solid mineral raw materials, energy raw materials, geological building materials) can be classified from the standpoint of application in the following way[7]:

1. ***geological wealth used for food, medicaments, agrotechnic purposes, and hygiene:***
   - *Substances that are consumed (in nutrition and treatment of man and animals);*
   - *Substances that are applied to the body in balneology (medicinal mud, etc.);*
   - *Substances used in agrotechnology;*
   - *Substances used for maintenance of hygiene.*
2. ***geological energy potentials:***
   - *Fossil fuels;*
   - *Nuclear mineral raw materials;*
   - *Pyrotechnical mineral raw materials;*
   - *Geothermal and geoelectric potentials.*
3. ***geological wealth in the guise of useful objects or raw material for their manufacture:***
   - *Rocks as primitive tools or weapons;*
   - *Rocks and minerals as cult objects and decorative elements;*
   - *Materials used in the construction industry;*
   - *Materials used in the manufacture and operation of machines, apparatuses, devices, and installations;*
   - *Mineral substances used in production of chemicals;*
   - *Mineral substances used as technological additives or processors;*
   - *Mineral raw materials used to obtain metals.*

This is a noticeably abridged form of the table given in the book "**Geology and Environment Protection**" (M. Babovic, 1992). Nevertheless, it is sufficient for us to conclude that the economy and indeed human existence itself is almost completely dependent on geological resources. Moreover, the exhaustibility of geological wealth is becoming more and more evident and already now represents one of the leading global problems facing man today.

The list of **harmful substances in the geological environment** for the most part consists of medium - heavy and heavy elements of Mendeleyev periodic system, some of which are strongly toxic or radioactive. The cycles of movement of heavy and harmful elements through the geological environment and through the hydrosphere are associated with inorganic substances, stabilization of such substances,

---

[7] The author did not include mineral and thermal waters and gases that are bottled (table mineral waters) or ones used for balneological purposes. Moreover, slightly mineralized groundwaters are utilized for drinking as a primary factor in maintenance of life on the Planet.

and their ability to become incorporated into crystal lattices or be liberated in the course of degradation of the inorganic matrix.

Pollutants originating from the geological environment (inorganic agents) are mainly indestructible, and special attention must therefore be paid to them. This is indicated by experience gained in human exploitation of certain inorganic mined forms of geological wealth (coal, oil, and gas, for example). As stressed by D. S. Veselinovic et al. (1995), it is unusually important to bear in mind the fact that all processes transpiring in the course of contamination by materials from the lithosphere are for the most part physico - chemical. The essence of pollution - whether it is from direct geological sources or from deposits of inorganic waste - is predominantly identical for both kinds of pollution. If the physico - chemical regularities governing processes of pollution with inorganic materials in the Earth are known, then those operative on the Earth are known as well. *In all of this, it is essential to have as complete as possible knowledge about the geological environment and its behavior, both in the process of pollution and in carrying out measures for its protection.*

For a long time, the relations of deeper parts of the lithosphere with the biosphere were interpreted as being one - way. It was held that magmatism and metamorphism - through the composition of their products (rocks), volcanic exhalations, rock folding, mountain formation, and other endogenous processes - ert considerable influence on weathering, rock formation, groundwater activity, and other exogenous processes and phenomena. The reverse influence - influence of external processes on endogenous ones - was not taken into account. However, the available data, especially material accumulated in the last quarter century, indicate unambiguously that the relations in question are two - way rather than one - way. For example, processes transpiring in the biosphere affect the composition of hydrothermal solutions, magmatism, etc. We note also the role played by vadose waters in feeding hydrothermal ones, gas breathing of the lithosphere, and sedimentary formation of sulfur. For all of these reasons, it is possible to speak of *unity of the lithosphere as a complex dynamic system with feedback relations that also include the biosphere* (A. I. Perelman, 1979). As Perelman wrote in the conclusion to his work "**Geochemistry**" (1979): "*Continuous arrival of solar energy and deep - seated energy of the Earth determines development of the tectonosphere and biosphere in a certain direction, increasing their complexity, diversity, and imbalance, while causing accumulation of free energy and reducing entropy.*" Vernadskii's remarks on unity of the lithosphere have already been cited.

The geological environment, like the environment as a whole, possesses *ecological properties*. Ecological properties are determined by a series of factors that can favorably or unfavorably affect development of the contemporaneous biosphere. Plant and animal life, microorganisms, and human life and activity develop differently in the upper layers of the geological environment in different climatic zones of the Earth. In connection with this, three types of regionally occurring natural geological environments can be singled out: arid, humid, and cryolithogenic (N. I. Plotnikov, 1989).

The geological environment possesses definite *physical, hydrogeological, engineering - related, geological, and biological properties.* The following can be considered fundamental general properties of the geological environment:

1. Variability of the environment in space and time;
2. Heterogeneity of the environment, which in the hydrogeological sense is manifested in the form of unequal filtration properties of different rocks;
3. Adaptive capacity, i.e., ability of the environment to change under new or disturbed natural conditions;
4. Discreteness of the geological environment, which is manifested in the guise of rock properties such as crack formation, karstification, etc.

The given properties of the geological environment can be altered in either a negative or a positive direction under the influence of anthropogenic processes.

By transforming nature, primarily through projects of waterpower engineering, man strives to improve ecological properties of the geological environment. Desert and steppe regions of arid zones are transformed into fertile oases under the influence of irrigation and drainage, plants are sprinkled in farming, new cities and settlements are built under favorable ecological conditions, etc.

CHARACTERISTICS OF GEOLOGICAL FACTORS AND THEIR INFLUENCE ON HUMAN HEALTH

In the multi - volume manual "**The Ecology of Human Disease**" published in 1958 by the American Geographic Society, J. May in the second chapter entitled "Stimuli" ("Etiological Factors") distinguishes *inorganic factors* (stimuli), *organic factors* (stimuli), and *social and cultural stimuli.* According to May, a disease can arise only where factors of two kinds coincide at a certain point in time and space: *first of all,* factors that take the form of an environmental stimulus (for example, poison in food); and *secondly,* factors that provoke a reaction of tissue in response.

In the category of *inorganic stimuli,* May includes heat, humidity, the regimes of wind and light, and trace elements in soil, food, and water. Other factors such as radiation, magnetic fields, cosmic rays, and static electricity must be considered important, but their action, according to May, has been inadequately studied.

Climatic and inorganic factors affect human health in two ways: directly and indirectly. Little is known about direct action because indirect action is incomparably more prevalent, a fact that befogs the picture of direct action.

Following Vernadskii (1965), Allison and Palmer (1980), and Perelman (1979) - whose outlooks on the Earth's spheres and their unity are presented above - we can relegate practically all natural factors to the category of geological in the broader sense. Similar thoughts were expressed by A. D. Howard and I. Remson (1978), who wrote as follows (among other things) in the Preface to their book "**Geology in Environmental Planning**": "*The geological environment refers to relief, soil and other loose materials, and basic rocks and soil substrates. That concept also includes natural processes that visibly alter the natural environment (landscape) and auxiliary factors which affect the development of these processes, for instance vegetation or the presence for many years of an ice cover.*"

However, we hold *factors of the geological environment* in the narrower sense to include only things that are predominantly the subject of geological research. Among them are: *minerals, rocks, geological formations, soil, geomorphologic factors, tectonic activity and structural forms, geophysical fields, thermal fields, geochemical fields, endogenous geological processes (earthquakes, volcanic activity), exogenous geological processes and phenomena, groundwater (unmineralized, mineral, and thermal), mineral raw materials and ranges of their distribution, radioactive elements and radioactivity, and macro - and microelements in rocks, soil, and water.* The influence of these factors on human health (geochemical risks, hydrogeochemical risks, and geomedical risks) is also considered. Owing to its geological predisposition and interconnection with geological factors, material that has traditionally been the subject matter of other sciences and scientific disciplines - meteorology, hydrology, and biology - will be treated in the next chapter. This refers to climatic, hydrological, and biological factors.

In broader terms, it can be stated that geological composition of the Earth has been studied on different levels: from the level of very small units (chemical elements and their components) to that of very large units (continental massifs and tectonic plates, all the way up to the planet as a whole). Between these extremes, geologists distinguish two important groups of formations - *minerals and rocks* - also counting here *geological formations* such as paragenetic associations of rocks. We shall devote longer attention to these components as basic structural elements of the geological environment.

Minerals

Minerals represent separate solid natural inorganic particles of crystalline structure. Almost all rocks are composed of them (Fig. 2.2). Several different minerals can be present in many types of rocks, but rocks can also be formed entirely of a single mineral, for example limestone or marble from the mineral calcite. The significance of minerals is twofold: first, they enable rocks to be classified; and *second*, they indicate the conditions of rock formation.

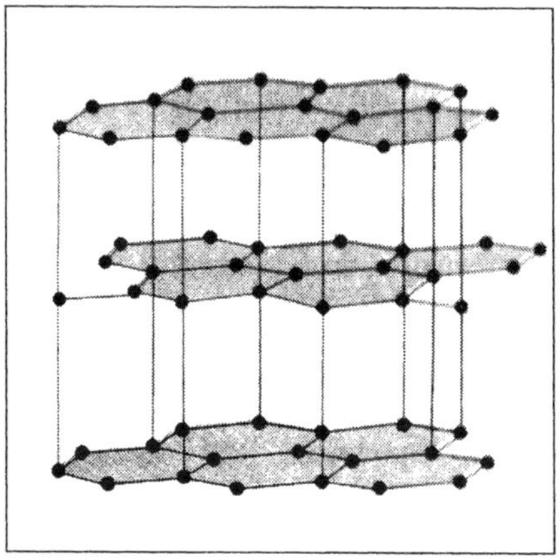

Fig. 2.2. Crystal grid of graphite.

Either a single element or a regular combination of elements can enter the composition of minerals. It has characteristic physical properties. Minerals of artificial or organic origin are excluded here, with the reservation that opal, a series of clay minerals, and some hydroxides that have a crystalline structure are exceptions.

Since more than 98% of mass of the Earth's crust is composed of only eight elements, the number of main rock - forming minerals is not great, regardless of the large number of possible combinations of chemical elements available for mineral formation. Some elements such as gold, silver, copper, sulfur, platinum, and carbon as graphite and diamonds can occur in pure form, but the majority are found in the guise of chemical compounds. Because the content of oxygen in the Earth's crust is about 47% by weight, chemical compounds of this element with other elements are common. Silicon and aluminum, which occupy second and third place, account for the formation of silicate minerals. Silicates are compounds of silicon and oxygen with other elements such as aluminum, sodium, potassium, iron, and manganese. They form the largest group of minerals, as can be seen from Table 2.1. Other mineral compounds include carbonates (salts of carbonic acid), sulfides (sulfur compounds), sulfates (salts of hydrochloric acid), chlorides, phosphates, hydroxides, nitrates, and borates.

The atomic structure of hundreds of minerals is known today. With rare exceptions, atoms or groups of atoms arrange themselves in orderly fashion to assume a regular disposition in conformity with the many mathematically possible modes of spatial grouping (Fig. 2.2.). The property known as the crystalline structure of a mineral is its most characteristic feature. The crystalline structure affects form of the crystal itself, how the mineral breaks up during destruction (weathering), hardness, specific conductivity, and many other physical properties. In general, the crystalline structure dictates the physical properties of a mineral.

Isomorphic mixtures represent an exception to the rule of structural stability established for the majority of minerals. Such mixtures are formed when some common components of minerals are partially replaced by certain chemical elements capable of taking their place. For example, iron can replace part of the zinc in the mineral sphalerite (zinc sulfide - a basic zinc ore).

Isomorphic compounds of silicate minerals - plagioclase, olivine, amphibole, and pyrone - are of the greatest significance. Plagioclases form a continuous series of minerals, from sodium albite ($NaAlSi_3O_8$) to anorthites with high calcium content ($CaAl_2Si_2O_8$), including albite, oligoclase, andesine, labradorite, bytownite, and anorthite.

The majority of minerals arise from several kinds of solutions. They are formed as a result of many processes: 1) cooling of magma (in the interior of the Earth) or lava (on the Earth's surface); 2) separation from deep - lying solutions containing hot water or hot gases (including water vapor), as is the case with many mineral veins and metal ores; 3) condensation of hot vapors during formation of minerals of the sublimation sulfur type near the openings of volcanoes; 4) chemical reactions with already existing minerals, for example during hydrothermal transformations of feldspar into mica or during oxidation of iron - containing minerals in the zone of chemical decomposition at the Earth's surface; 5) replacement of one primary mineral by another; 6) recrystallization of primary minerals with formation of new compounds under the influence of variable conditions of temperature and pressure; 7) evaporation of aqueous solutions.

A fact of special interest to us is that the number of minerals readily soluble in water (halite for instance) is small; most of the rest are either poorly soluble or completely insoluble. The mineral calcite dissolves with separation of carbon dioxide bubbles in dilute hydrochloric acid.

On the basis of stability in relation to chemical decomposition, it is possible to single out the following four groups of minerals:

1. *Very stable*: quartz, topaz, tourmaline, rutile; platinum, gold, zircon, corundum, etc.;

2. *Stable*: muscovite, orthoclase, microcline, acidic plagioclases; hematite, magnetite; disthene, barite, monazite, etc. (Fig 2.3.);
3. *Poorly stable*: amphibolites, pyrone; wolframite, celite, apatite, actinolite; epidote, etc.;
4. *Unstable*: basic plagioclases, feldspathoids, alkaline amphiboles, biotite, augite, olivine, glauconite, calcite, dolomite, gypsum; pyrrhotine, sphalerite, chalcopyrite, arsenopyrite, pyrite.

Chemical decomposition of rocks occurs in connection with many chemical reactions: hydrolysis, hydration, cation exchange, formation of comples, and simple chemical dissolution. Feldspars disintegrate mainly as a result of hydrolysis. Silicates with iron and most sulfide minerals oxidize fairly readily.

Fig 2.3. Antimonite from the Zajaca mine, Western Serbia.

The picture of mineral stability in relation to chemical decomposition is deceiving at first glance. To be specific, *poorly stable and unstable rocks are constituents of very widespread kinds of rocks and in large measure dictate the macro - and microcomponent composition of the soil. The effects of macro - and microcomponents on human health will be discussed at a later time.*

TABLE 2.1.
The most widespread silicate mineral of rocks.

Non - ferromagnesium (sialic) bright	Content
Quartz	SiO_2
Plagioclase	$NaAlSi_3O_8$ - $CaAl_2Si_2O_8$
Orthoclase	$KAlSi_3O_8$
Microcline	$KAlSi_3O_8$
Muscovite	$KAl_3Si_3O_{10}(OH)_2$
Ferromagnesium (maphite) dark or green	
Orthopyrone (hypersthene)	$(Mg, Fe) SiO_3$
Clinopyrone (augite)	$Ca(Mg, Fe)Si_2O_6$
Amphibole	$Ca_2(Mg, Fe)_5Si_8O_{22}(OH)_2$
Biotite	$K(Mg, Fe)_3AlSi_3O_{10}(OH)$
Olivine	$(Mg, Fe)_2SiO_4$

Rocks

Rocks are mineral aggregates that have definite composition, both mineral and chemical, and definite structural features. They are geological bodies, constituents of the Earth's crust. The term **rock** is used in petrography for all natural formations of a certain composition and structure, without reference to hardness and cohesiveness of materials. In the category of rocks, we include compact, hard formations (natural stone); plastic, sometimes semi - fluid aggregates; and unbound loose materials (sand, gravel, volcanic ash). In contrast to rocks, mineral deposits are restricted in extent and characterized by accumulation of certain elements of subordinate significance in structure of the Earth's crust.

The few minerals that determine their characteristics and represent crucial components of rocks are called *petrogenic minerals*. The most important petrogenic minerals belong to the silicate, carbonate, and oxide groups.

The oldest known rocks of the continental crust of the Earth are 3.9 billion years old (Antarctic), which is less than the age of some meteorites and certain moon rocks (4.7 billion years)[8]. It is important to stress that most rocks older than approximately 20 million years belong to the class of hard (rock) formations, whereas younger rocks are predominantly semi - bound and unbound (loose).

[8] The presence of detritic zircon in some archaic rocks indicates that there existed on the Earth an older continental crust, which probably has not been preserved.

All chemical elements known in nature are present in hard rocks, but the chemical compounds and minerals in which they occur are different, depending on the basic groups of rocks. Most of the minerals of magmatic rocks, for example, are noticeably altered during processes of chemical decomposition, so that sedimentary rocks formed from their fragments differ in composition from the primary magmatic rocks. The minerals of sedimentary rocks can in turn be transformed into new minerals if they reach depths where they are exposed to the influence of high temperatures, pressures, and mineralized solutions. In the process of these changes, certain chemical elements can be lost, while others increase their content. Table 2.2 gives the average chemical composition of several of the most widely disseminated sorts of magmatic and sedimentary rocks. Metam2orphic rocks are not included in this table, since they were formed by alteration of magmatic and sedimentary rocks.

TABLE 2.2.
Average chemical content of magmatic and sedimentary rocks (%).

Components	Magmatic rocks				Sedimentary rocks			
	Acid	Basic	Ultra basic	Average content	Sand-stones	Lime-stones	Clays	Average Content
SiO_2	68.9	48.2	43.8	59.14	81.23	5.19	53.10	57.95
TiO_2	0.5	1.9	1.7	1.05	0.25	0.06	0.65	0.57
Al_2O_3	14.5	17.6	6.1	15.34	2.77	0.81	20.40	13.39
Fe_2O_3	1.7	3.0	4.5	3.08	1.07	0.54	3.02	3.47
FeO	2.2	5.8	8.7	3.80	0.30	-	2.45	2.08
MgO	1.1	8.2	22.5	3.49	1.16	7.89	2.44	2.65
CaO	2.6	10.5	10.1	5.08	4.5	42.57	3.11	5.89
Na_2O	3.9	2.6	0.8	3.84	0.45	0.05	1.30	1.13
K_2O	3.8	0.9	0.7	3.13	1.31	0.33	3.24	2.86
H_2O	0.6	0.8	0.6	1.15	1.63	0.77	6.00	3.23
P_2O_5	0.16	0.3	0.3	0.3	0.08	0.04	0.17	0.13
CO_2				0.10	5.03	41.54	2.63	5.38
SO_3				-	0.07	0.05	0.64	0.54
BaO				0.06	0.05	-	0.05	-
O				-	-	-	0.80	-
MnO	0.07	0.17	0.18	0.14	-	-	-	-
Total				99.70	100.00	99.84	100.00	100.00

The chemical transmission of a health risk in fact begins from rocks - with enhanced or reduced content of elements - of greater significance for life. Further natural pathways are through water, soil, plants, and animals. The importance of knowing the chemical composition of rocks can be illustrated by the example of enhanced content of molybdenum, which is toxic in impermissible doses, in rocks of certain regions of the world. The increased concentration of that element in plants and wider incidence of toxicosis among domestic animals in those regions suggest that rocks with a high concentration of molybdenum are probably found under the loose soil. We have such a case in the vast expanse of the American West, where natural concentrations of molybdenum exceed average values for the Earth's crust. The majority of registered cases of toxicosis were encountered either in areas with high content of natural molybdenum or in regions downstream from them. Cases of toxicosis were also recorded on terrains with moist soil having high content of organic substances, where metals are retained (concentrated) by means of their compounding with organic particles.

Magmatic Rocks

The first stage of magmatic activity involves formation of magma, a fluid silicate solution of rock in the bowels of the Earth. Magmatic rocks are formed during cooling and crystallization (differentiation) of magma. Field investigations indicate that the bulk (9/10 and more) of magma does not flow out on the surface in the form of volcanic material, but is rather hardened under ground. Bodies formed in this way are called *intrusives*, and the largest of them are *batholiths* (Fig. 2.4.). Partially denuded by processes of erosion, systems of batholiths can extend without interruption over distances of the order of 8000 km, as is the case with the Andes mountain range in South America, where the length of individual batholiths exceeds 1300 km. The batholith of the Coast mountain range in British Columbia (North America) is about 2000 km long and from 130 to 200 m wide, with a surface of more than 300,000 km^2. *It can be imagined how greatly different are conditions for human life and health in such a geological environment from those in expansive Neogene basins, for example.*

On the basis of the manner of formation and position in the lithosphere, magmatic rocks are divided into the following three groups:

- *Deep - seated* (intrusive or plutonic);
- *Veined* - differentiated or lateral spurs of deep - seated magmatic rocks; and
- *Extrusive* (effusive or volcanic).

The chemical method of classification of magmatic rocks depends primarily on the content of SiO_2 (quartz) in rocks. On the basis of the content of this oxide, rocks are divided into the following classes:

- *Acidic,* with SiO_2 content above 66%;
- *Intermediate,* with SiO_2 content from 66 to 52%; and

- *Basic*, with SiO_2 content of less than 52% (some authors also single out a group of *ultrabasic rocks*, with less than 45% SiO_2).

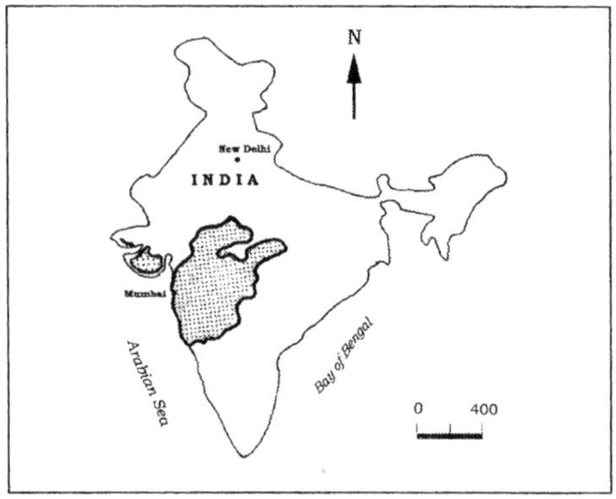

Fig. 2.4. Position of Deccan volcanic province, India.

On the basis of chemical nature and manner of association, magmatic rocks are divided into the following eight groups:

- *Granites and rhyolites* - acidic rocks characterized by the presence of quartz and alkali feldspars;
- *Granodiorites, quartz diorites, quartz latites, and dacites* - rocks of acidic to intermediate nature, with greater or lesser amounts of alkali feldspar and intermediate plagioclase;
- *Syenites and trahytes* - intermediate rocks with alkali feldspar and without quartz;
- *Alkali syenites and phonolites* - rocks rich in alkalis and poor in silica;
- *Monchonites and diorites, latites and andesites* - intermediate rocks with transitional plagioclase and without free quartz;
- *Gabbros and basalts* - basic rocks with basic plagioclase;
- *Alkaline gabbroides* - basic rocks rich in alkaline;
- *Peridotites* - ultrabasic magmatic rocks without feldspar.

We give below the main characteristics of the most widely disseminated representatives of *deep - seated rocks.*

Granites are among the most widely disseminated of deep - seated (plutonic) rocks (Fig. 2.5.). According to known statistics, granites make up as much as about 95% of upper levels of the continental crust. They are built of quartz and alkali feldspars, with the presence of mica or some other colored mineral that more closely determines the variety of rock (Fig. 2.6.). Weathering of granites begins with alteration of the colored minerals and oxidation of iron in them, after which feldspars are transformed into mineral clays. Beginning around the margin of the grain, these processes destroy the rock mass to such an extent that it passes over into a loose state, so - called granitic grus. In the sand - like loose mass, the only minerals not disintegrating are quartz (and muscovite if it was present in the rock).

Granodiorites are also very widely disseminated rocks. They are characterized by being silica - rich. In comparison with granites, however, they are poorer in alkalis and richer in calcium, magnesium, and iron. Together with quartz, intermediate plagioclase occurs as a characteristic mineral. During prolonged weathering in the surface zone, granodiorites also pass over into grus. They are very widespread in the world and in Yugoslavia (Boranja, Zeljin, Kopaonik, Besna Kobila near Surdulica, Jastrebac).

Gabbro consists of basic plagioclase and pyrone as essential components. If olivine is also present as an essential component, the rock is called olivine gabbro.

Fig. 2.5. Brajkovac granites, West. Serbia (Photo: D. Milovanovic).

Peridotites are very widespread in the world and in Yugoslavia (Zlatibor, Maljen, Kopaonik, Goles, Orahovac - Djakovica). They are poor in silica and alumina, but rich in magnesium and iron. Alkali and calcium levels are for the most part minimal. Olivine and pyrone are essential minerals, and mined minerals (chromite, magnetite, platinum, and diamonds) play a significant role in these rocks. As rocks, they are rarely fresh, and a more or less advanced process of serpentinization is regularly observed in them. Chrysotile asbestos, talc, magnesite, some chlorite, and opal are encountered as products of alteration, together with serpentine.

Among **extrusive (volcanic) rocks**, particularly widespread are basalts and basaltic rocks, which occupy as much as 98% of the surface of all extrusive rocks on the Earth (Fig. 2.4.). Andesites, quartz keratophyres, and quartz latites are significantly present in Yugoslavia. Of *veined rocks, granitic pegmatites are of interest, primarily because of elevated concentrations of radioactive elements (U, Th, K, and Rb).* Pegmatites occur very often in Yugoslavia, especially in the form of veins accompanying granodiorites.

It was mentioned earlier that each of the indicated types of magmatic rocks is defined by the kind and quantity of essential mineral components, the content of oxides present in them varying within definite limits (V. Knezevic - Djordjevic and P. Djordjevic, 1975), namely as follows:

- Silica (SiO_2) - between 35% (in ultrabasic rocks) and about 78% (in ultra acidic rocks);
- Alumina (Al_2O_3) - from 0.5% in peridotites to about 20% in basic and intermediate rocks (more than 20% in some alkali rocks);
- Oxides of iron (Fe_2O_3 and FeO) - present in amounts ranging from fractions of a percent (in granite) to maximally about 15% of total content (in gabbro and peridotite);
- Magnesium oxide (MgO) and magnesium - present in amounts ranging between fractions of a percent (in granite) and about 40% (in peridotite);
- Alkalis (Na_2O, K_2O) - variable both individually and in total content, their content declining from granite to peridotite;
- Other components - present in small quantities and not subject to any great variation.

For us it is very important that the magmatic masses of certain broader areas exhibit similarity of mineral composition and ratios of individual chemical components. Such areas are most often called **petrographic provinces**. For example, according to Niggli (1954), there are three basic provinces among younger magmatic masses, namely the *Pacific* (rocks of the calcium - alkali series), *Atlantic* (rocks of the sodium series), and *Mediterranean* (rocks of the potassium series) provinces. On the other hand, Russian scientists distinguish the *Atlantic, Pacific, Mediterranean, and Arctic provinces,* and newer research indicates the existence of several more petrographic provinces. For the broader territory of the Balkan Peninsula, L. Maric (1974) concludes that in addition to Pacific and Transitional petrographic provinces, a tendency is discernible toward development of rocks of the Atlantic and Mediterranean provinces (Eastern Serbia), i.e., a potassium tendency of development is evident.

Two separate **geochemical provinces** are clearly distinguishable in the Balkans: the Cu geochemical province of the Carpatho - Balkanides and the Pb geochemical province of the Serbian - Macedonian mass and Dinarides. In the region of Mt. Kopaonik - which is one of the largest and most interesting areas of magmatism and metallogeny in the Pb geochemical province - a regularity is discernible in the behavior of trace elements during intrusive magmatism (and the phase of volcanic effusion): 1) rocks are rich in lead and barium and poor in copper in relation to average rocks of the Earth's crust; 2) average content of Sr, Ba, Mn, and Ti is very high - more than 1,000 ppm; 3) trends of change in the content of trace elements in the direction of the most acidic rocks (quartz monzonites) indicate the following characteristics in the development of intrusive magmatism in this region: a) uniform decrease in content of Ni, Co, and Cu, together with less regular decrease in content of Ti, Cr, V, Mn, Sc, B, and Li; and b) uniform increase in content of Pb, Ba, Rb, Be, and La, together with less regular increase in content of Zr, Y, and Sr (A. Dangic, Z. Maksimovic, and I. Micic, 1995). However, geochemical study of the southeastern part of the Surdulica pluton indicated that granites are with lower average content of a number of trace elements (Ti, Ga, V, Cu, Y, Zr, Ni, Co, Cr, La, Ba, Sr, Yb, Li, and Rb), but higher manganese content in relation to acidic magmatic rocks (N. Vaskovic and V. Jovic, 1993).

It will be seen in the corresponding chapter how significant the study of trace elements in rocks and soil is for *medico - geological evaluation of the environment*. Here we give two examples of the negative and positive influence erted by magmatic rocks and their accompanying products.

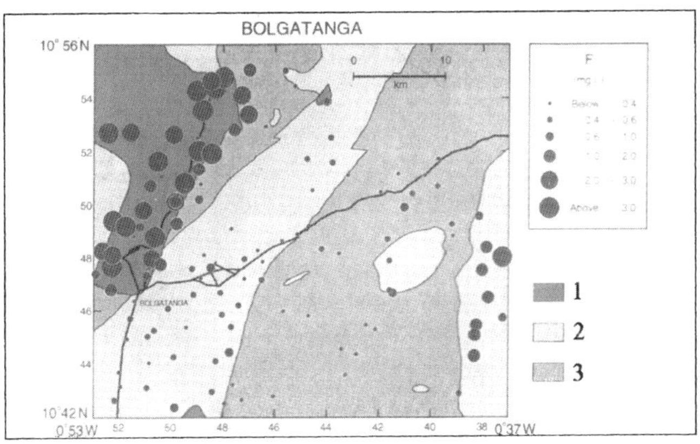

Fig 2.6. Distribution of fluoride in groundwater in the Bolgatanga area of northern Ghana. 1 - Bongo granite; 2 - Birmian magmatic and metamorphic rocks; 3 - Birmian green schists (BGS, ODA-funded program).

Negative consequences on the natural environment (environment of life) can be very pronounced in *regions of current volcanism and young volcanic rocks*. It is known that liberation of H_2O, HCl, H_2S, CO_2, SO_2, and other gases goes on for a long time after eruption of volcanoes in these regions. A portion of these gases enters the atmosphere directly, while another part dissolves in groundwater and causes formation of acid thermal springs. There are many such springs in Japan, the Kurile Islands, and other regions. For example, the water of some thermal springs in the Kurile Islands has pH values of from 0.2 to 1.0 and hydrochloric acid composition of as much as Cl 45 g/l. This water is very aggressive and profoundly alters the surrounding rocks: it extracts Fe, Al, Ti, Ca, Mg, Na, K, and Al from the rocks, which acquire the composition of almost pure SiO_2. Thermal waters contain up to 3 g/l Al and 1 - 0.5 g/l Fe. They are also enriched with titanium. Brooks, rivers, and lakes in such regions have very acidic water rich in aluminum and iron. Volcanic geological (natural) environments have not been studied enough geochemically. This applies particularly to biological circulation and migration of water in these regions.

Living conditions are completely different in regions of *peridotite massifs*. Let us take, for example, the Zlatibor peridotitic (ultramafic) massif, which - with an area of about 800 km^2 - is one of the largest in Europe[9]. Peridotites, the soil on them, and underground and surface waters in this region are rich in magnesium. Apart from widespread waters rich in magnesium bicarbonate, of interest are calcium hydroxide waters, as well as very rare alkaline waters (pH 11.4 - 11.9), which issue only from several peridotitic terrains. According to Z. Maksimovic et al. (1996), in the human population of the Zlatibor region, the content of magnesium in blood serum is in the highest part of the reference range (15.8 - 25.5 mg/l). Medical data indicate lower mortality from cardiovascular diseases in the area of Zlatibor than in other regions of Serbia (M. Djordjevic et al., 1992), which is in keeping with the magnesium status in the human population.

Sedimentary Rocks

Most sedimentary rocks represent the product of chemical and mechanical weathering of previously formed magmatic, sedimentary, and metamorphic rocks. Transported by river currents, glaciers, wind, and gravity, particles of primary rocks are eventually deposited (most often in an aquatic environment) in the form of layers on top of the rocks below them. Substances of organic origin can be deposited together with inorganic material. With the passage of time, the deposited unconsolidated sediment is subject to processes of diagenesis, consolidation (lithification), and transformation into a compact sedimentary rock. Three fourths of the land on our Planet is covered by sediments and sedimentary rocks, and only one fourth by magmatic and metamorphic products.

Sedimentary material can be divided into three large groups: 1) *clastic material*, which is transported in the form of hard particles; 2) *chemogenic*

---

[9] A comprehensive survey of ultrabasic rocks on territory of the former Yugoslavia was given by Z. Maksimovic (1975). Among other things, the author stressed the assumption that the interior (Vardar) and central (Bosno-Zlatibor) ultrabasic zones are with different chemism of peridotites, the central zone being particularly distinguished by elevated content of Na, Ti, and Li.

sediments, which are deposited directly from solution; and 3) *organogenic sediments*, which are created by organisms deposited together with inorganic (chemical and clastic) material.

1. *Clastic sediments.* Sorting by size of particles (grains) with the texture of gravel, sand, silt, and clay occurs in the course of their transport. If a river empties into a basin with calm water, gravel is deposited near the shore, sand settles somewhat farther away, and fine material can be carried a significant distance from the coastline. The natural binding matter or cement in semi bound or bound rocks of this group - which incidentally are the most widespread of sedimentary rocks - can be of a carbonate, siliceous, clay, marl, or iron nature.

Certain bound or semi bound rocks are formed, depending on texture of the unbound rock mass. The given regularity is illustrated by the following series of rocks, starting with those having the coarsest texture:

- *Breccia* is formed by binding or cementing of *detritus*, i.e., unbound material consisting of angular broken rock fragments measuring more than 2 mm in size. The detritus matter is deposited near the source of clastic material, so that breccia generally has a composition similar to that of the parent rock (limestone breccia, dolomitic breccia, marble breccia, serpentine breccia, etc.);
- *Conglomerates* are rocks composed of *gravel*, i.e., rounded rock particles with diameter greater than 2 mm, cemented by some natural binding matter;
- *Sandstone* is rock formed by natural binding of *sand*, fine - grained unbound material composed of rounded particles varying in diameter between 2 and 0.05 mm (from 2 to 0.5 mm in coarse - grained sand, 0.5 to 0.25 mm in medium - grained sand, and 0.25 to 0.05 mm in fine - grained sand). The most common component of sand (sandstone) is quartz, but muscovite is also fairly well represented. As in gravels, scattered veins of useful minerals (gold, monazite, ilmenite, garnets, diamonds, sapphires, etc.) can be formed in places in sands;
- *Aleurolite* is the bound rock that corresponds to *aleurite*; a term applied to unbound sediment with particles the size of dust (0.05 - 0.005 mm). Composition of the given rock formations includes particles of quartz, feldspar, mica, carbonate, and clay minerals, while the binding material is most often of carbonate (sometimes clay) nature;
- *Shale* is formed from mud, which represents the unbound pelitic sediment accumulating in water. Because of the high dissolving action of water, this finest of clastic material contains only remaining stable components, primarily clay minerals. Diagenesis of mud results in the formation of *clay*, or *shale* in the case of transformation into a hard stratified rock. Among clay minerals, the most important are kaolinite, hydro micas, montmorillonite, and other alumosilicates. Admixtures include hydroxides of iron, which give the rock a reddish, yellow, or brownish - red color; organic substances (especially bitumen), which give it a dark - gray or

even black color; and small amounts of manganese, which give it a greenish color.

Clay minerals make up the bulk of products formed in *rock - water* interaction under conditions of the Earth's surface. Owing to characteristics of their structure, they represent material that acts effectively during ion exchange. It turns out that the composition of many natural waters approximately corresponds to that resulting from equilibrium with secondary clay minerals, and with primary minerals of volcanic rocks in the exceptional case of potassium feldspar (J. I. Driver, 1982.).

2. *Chemogenic and organogenic sediments*. Rocks of a large group are created in various aquatic basins as a result of diverse chemical processes or accumulation of organic remains after the death of animals and plants (the shells of macro - and microorganisms or plant remains). *Chemogenic formations* include carbonate rocks (limestones, dolomites, and marls), silicate rocks, ferrous rocks, onyx marble, travertine, and sediments of salt deposits. The most important *organogenic sediments* are limestones and dolomites, including writing chalk, silicate sediments (cherts), diatomaceous earth, coal, petroleum (to some extent), etc. It is evident that the same rock, limestone for instance, can belong genetically to both the group of chemogenic formations and the group of organogenic formations.

Limestone consists mainly of calcium carbonate, manifested in the form of the mineral calcite. It usually contains chemical admixtures of iron, manganese, magnesium, and clay; organic matter; etc. It can be created inorganically by chemical accumulation of calcite, by accumulation of lime - rich shells, or by a combination of the two processes. If the rock is composed of 40 to 60% $CaCO_3$ and clay particles account for the remaining 60 to 40%, it is called *marl*. Clay minerals - kaolinite and montmorillonite - make up the noncarbonated part of marls. Chemically purer limestones are almost regularly affected by an intensive karst process.

Dolomite represents a carbonate rock composed of the mineral of the same name, the double salt of calcium and magnesium carbonic acid: $CaMg\ (CO_3)_2$. Pure dolomite is very rare. It usually contains many admixtures, and clay dolomites are distinguished, together with marl, iron, bituminous, and other dolomites.

Certain ***sediments with significant iron content*** accumulate chemically. Thus, for example, large amounts of iron are given up to the sea in the form of exhalations during intensive submarine effusion of basic volcanic rocks, and hematite deposits can be created in this way. Like all typical sedimentary rocks, iron deposits can be formed in the usual sedimentary manner.

Phosphate rocks represent sedimentary formations containing 12 - 40% P_2O_5. Very high content of phosphate minerals is characteristic of deposits of the *Phosphoria formation* in the states of Wyoming, Utah, and Idaho, while the phosphate admixtures present in limestones in the vicinity of Nashville (Tennessee) are wholly sufficient for industrial production of fertilizer. Phosphate minerals, by the way, can be present in the form of cement and other sediments.

Evaporites or saline deposits are formed by accumulation from solutions, usually in closed basins and lagoons. This group includes such single - mineral rocks as anhydrite ($CaSO_4$) and gypsum ($CaSO_4 \cdot 2H_2O$). Of special economic significance are rock salt (NaCl), salts of potassium, borates, and nitrates.

Caustobiolithes are combustible organogenic rocks. They include caustobiolithes of the coal - peat series, brown coal, hard coal or anthracite, and members of the bitumen - petroleum series.

An interesting fact is that only three types of rock account for nearly 100% of the total amount of sedimentary formations on the continents. Thus, according to Pettijohn (1957), claystones and related rocks constitute about 58% of all sedimentary rocks formed on land, sandstones about 22%, and limestones about 20%. On the other hand, such formations as coal, iron ore, gypsum, raw materials for fertilizer, etc., are relatively rare, but of great value.

According to Howard and Remson (1978), sedimentary rocks can be relegated to three *geochemical classes*. The *first class* consists of rocks such as sandstone, with grains of minerals first eroded from other rocks, then transported to regions of sedimentation without significant chemical changes. The *second class* includes rocks such as shale, which basically consists of clays and (or) aqueous compounds of quartz, aluminum, iron, or magnesium formed during chemical decomposition or interaction with water. Represented by rocks such as limestone, the *third class* encompasses rocks whose components fall into sediment from water as a result of a chemical reaction or biological activity. Table 2.2. shows average composition of the indicated three types of sedimentary rocks.

As for the chemical composition of sedimentary rocks, chemogenic and organogenic formations - being single - mineral aggregates - are characterized by very simple chemical composition. Limestones, dolomites, silicate rocks, and many other kinds of rocks are of this type, as are some sands and sandstones among clastic sediments. By way of contrast, most clastic deposits have a complex and varied chemical composition, depending on the kind of fragments and their heterogeneity, the kind of binding mass, and the quantitative ratio of fragments and binding mass. This type includes conglomerates, some sandstones, marls, shales, clays, and other similar sedimentary rocks. *Shales are especially rich in heavy metals and other environmentally important elements.*

Linkage between magmatic and sedimentary rocks is confirmed by data on their chemical composition indicating similar content of various components (Table 2.2.). At the same time, the following visible and basic differences are also present:

1. The ratio of Fe_2O_3 to FeO is opposite in sedimentary and magmatic rocks. Oxide iron is prevalent in sedimentary rocks. The reason for this lies in the fact that sedimentary rocks are formed in the biosphere in the presence of free oxygen, which oxidizes significant amounts of iron (and other polyvalent elements);
2. Levels of sodium expressed as Na_2O are noticeably (almost three times) lower in sedimentary rocks than in magmatic rocks, whereas the quantity of K_2O is approximately equal. This is attributable to the fact that sodium under conditions of the biosphere is readily soluble and is transported to the ocean, where it accumulates in great amounts and is also fid in pelagic colloidal sediments of the sea floor;
3. Sedimentary rocks are richer in H_2O and CO_2, components that in magmatic rocks occur in extremely small quantities; and
4. Sedimentary rocks contain organic carbon in varying amounts, whereas this component is completely lacking in magmatic rocks. Organic compounds in sedimentary rocks are products of photosynthesis, which has transpired in the Earth's biosphere since time immemorial.

A geological environment built of sedimentary rocks can ert positive and negative influence on forms of life. The following examples of this can be cited:

1. The concentration of selenium in black shales, hard coal, and petroleum is 10 to 20 times greater than the average content of this interesting element in the lithosphere. For example, selenium - bearing black shales (schists) represent the parent material of the widespread selenium - bearing soil on the Western Plains of the United States. Here certain fodder plants fed to cattle, horses, and sheep concentrate selenium in their tissues. The concentration can attain 15,000 mil^{-1}. Absorbed in such concentrations, selenium affects the motor system and senses of animals and can even cause their death. Luckily, cultivated plants and wild grasses do not accumulate amounts of selenium great enough to threaten the life of animals;
2. The Izhor Plateau near Saint Petersburg (Russia) is built of carbonate sediments rich in calcium carbonate. The richness of the rocks dictates richness of the soil itself in this calcium compound, as a result of which the soil is very fertile. Surface and groundwater are also calcium - enriched and have a neutral or weakly basic rather than acidic reaction. Humic materials and iron compounds do not dissolve in these waters, and

springs and rivers in such natural environments are especially clean and transparent. The vegetation has all the attributes of completely calcium - sufficient food. Farm animals rarely suffer from rickets, and the milk productiveness of cows is greater;

3. As an interesting example, J. Zujovic (1922) cites the famous cognac of France: "*The refined palates of alcohol lovers distinguish three kinds of French cognac, depending on the wine from which it is made. The first quality (Grande or Fine Champagne) is made from white wine from vineyards planted on layers of chalk, light and flaky. The second kind (Petite Champagne) is from vineyards on less chalky soil. The third and poorest cognac is made of wine from vineyards on hard limestone (like the one in the Topcider section of Belgrade) where forests once stood (this third class of cognac is called Pays des Bois). By looking at the geological map of the region in question, it is possible to determine the quality of cognac when the origin of the wine used to make it is known.*"

Metamorphic Rocks

Metamorphic rocks are formed from magmatic and sedimentary rocks by means of their profound alteration and transformation under the influence of high temperature, pressure, hot solutions, and gaseous components. This involves a complex process of recrystallization of minerals and rocks, changes of chemical composition, breakdown of old structures and formation of new ones, etc. Metamorphic rocks themselves can also be recrystallized anew if they meet with the corresponding thermodynamic conditions. One of the characteristic features of such rocks is their schistose structure, with pronounced parallel arrangement of components.

The following seven basic groups of rocks are singled out: 1) phyllites; 2) schists; 3) gneisses; 4) amphibolites and amphibolitic rocks; 5) marbles; 6) quartzes; and 7) other massive metamorphic rocks.

Phyllites are rocks of low crystallinity and well expressed schistose texture formed by metamorphosis of claystones. The leading mineral of these rocks is sericite, followed by quartz. The rocks are very subject to mechanical disintegration and pass over easily into a friable mass.

The **schist** group includes rocks of medium to high crystallinity with well expressed schistosity that is predominantly formed by metamorphosis of claystones. They are named for the mineral or minerals dominant in them: *mica schists* (if mica is dominant), *muscovite schists, and chlorite schists, talc schists*, etc. They readily undergo mechanical disintegration.

Gneisses are highly metamorphosed medium - to coarse - grained rocks composed mainly of quartz, alkali feldspar, and mica. *Mica, amphibolic, pyrone*, and other gneisses are distinguished, depending on the colored mineral present in them. Weathering of gneisses occurs in the same way as in granites, but due to facilitated circulation along the surface of their schistosity, the process is accelerated in relation to granitic rocks.

Amphibolites are rocks of high crystallinity built mainly of amphibole and plagioclase with or without quartz. Although they are massive in texture, there also exist partially schistose rocks of this group.

Marbles are rocks of massive structure formed by metamorphosis of limestones and dolomites. They are composed of calcite, more rarely of dolomite or calcite and dolomite together. Of accessory minerals, muscovite has somewhat higher content than others. Like limestone, marble is very subject to the karst process.

Quarzites are metamorphic rocks in which more than 80% of the rock mass consists of quartz. They are formed by metamorphosis of quartz sand and sandstones.

Serpentinites are singled out in addition to many kinds of regional metamorphic rocks. They represent an expanded mass formed by metamorphic transformation (serpentinization) of peridotites. The rocks are built of the mineral serpentine and as accessory ingredients can contain chlorite, talc, chromite, and magnetite. They are green to dark green in color.

A certain association of minerals or *metamorphic facies* is characteristic for every condition of pressure and temperature prevailing during the metamorphic process. For example, we note the zeolite facies, the blueschist facies, the greenschist facies, the amphibolite facies, and the granulite facies, which consistently reflect the degree of regional metamorphism under conditions of pressure and temperature increase.

Any rock, metamorphic rock as well, with concentration of an element or compound perceptibly greater than the average content acts as a potential source of pollution of soil, water, flora, or fauna. Whether it will act as a source of pollution depends on whether that element or mineral is present in exceptionally great quantities and in a *form that can be assimilated* (by plants for instance).

Mercury can be cited as an example of variability in the distribution of elements in the lithosphere. An interesting case is that of ultramafic rocks in California whose original composition included mercury - rich sedimentary rocks. The content of mercury in soil at a certain distance from these rocks is ten times greater than mercury content in other soil. If the soil is still closer, it can contain mercury in amounts obtainable on the premises of mines in the basic rocks.

Geological Formations

A geological formation represents an equal taxonomic unit in the series: *mineral - rock - formation - lithosphere - planet.* Every preceding unit in that series is elementary in relation to the next. Thus, a geological formation is a unified whole composed of a paragenesis of rocks. In the simplest case, a formation can be homogeneous and composed of a single kind of rock. Thus, it is possible to distinguish limestone, sandstone, evaporite, granite, or basalt formations. The Deccan basalt formation in southwest India, for example, which occupies an area about 700 km long and up to 500 km wide, is estimated to reach a depth of 1.6 km (Fig. 2.4.). It is very often the case that a formation consists of a number of regularly alternating rocks or is defined by extreme lithological heterogeneity that in itself is characteristic. It then gets its name from the main and/or characteristic petrographic members (for example, a flysch formation, a diabase - chert formation, etc.). To be specific, flysch represents a formation widely disseminated in many geosynclinal regions of the globe. It is a thick series of sedimentary structures with rhythmic alternation of interlayers of sandstones, shales, marls, and other rocks. Naturally, the geological environment in the given case cannot be formed solely in the sandstone or shale, but rather within the rock complex as a whole. It thus acquires the features of the complex itself.

The discipline devoted to study of geological formations is called **formation geology**. One of its final goals is to obtain a comprehensive knowledge (picture) of a formation. This requires information about lithological composition of the formation and the geological conditions under which it arose, its geochemical characteristics, the presence of mineral raw materials in it, and its hydrogeological and geological engineering features. Also significant is information about the type of soil arising on land built by a certain geological formation.

Every formation is distinguished by quite definite geochemical characteristics. For this and other reasons, *analysis of geological formations in areas of interest can be of significance for medico - geological evaluation of the environment* and eventual discovery of the causes of certain diseases. The following examples can be cited:

- Apambire et al. (1977) examined groundwater rich in fluorine and iron in the upper part of Ghana. They found that the content of fluoride (the most widespread inorganic compound of fluorine) ranges from 0.11 to 4.6 mg/l (even comprising up to 95 mg/l in one spring). The highest fluoride concentrations occurred in regions built of hornblende - granite and syenite, where dental fluorosis is outlined (Fig. 2.6.). It was concluded that the occurrence of F in granitoid groundwater is caused by dissolution of fluoride minerals and exchange of anions with minerals of micas and their clay products;
- The concentration of fluoride in Northern Tanzania is probably the highest in the world (with maximal levels of 12 - 26 mg/l in river waters and 63 mg/l in thermal waters). The elevated level of fluoride originates from marine sediments episodically enriched with ash and fluoride - rich subli-

mates deposited during volcanic activity from the Miocene to the present day;

- Very widespread on continental shields and platforms (as well as in Paleozoic orogenic belts), crystalline schist formations are generally poor in selenium and other important elements. This indirectly affects human health and growth in these regions. Due to inadequate intake of selenium, an increased risk of cancer, above all, is discernible in the human population;
- A formation of quartz sands created by weathering of moraine material during the Quaternary has a broad extent in Poland, the European part of Russia, and Western Siberia. As it is often composed almost exclusively of quartz, the content of biologically important elements (P, K, Ca, etc.) is negligible, which determines the geochemical and geomedical characteristics of the given terrains;
- Peridotite (serpentinite) formations occur in the guise of expansive massifs. The content of MgO in such rocks varies around 43% (in contrast to silicate rocks, with only 0.1%). Such terrains have soil, groundwater, and surface water rich in magnesium. As a result, in the human population the content of this element (which is biologically essential for plants, animals, and humans) in blood serum is in the highest part of the reference range. Under such natural conditions, cardiovascular diseases occur more rarely, as do hypertension and renal calculoses.

Rock Weathering

By *weathering*, we mean processes of mechanical disintegration and chemical decomposition of minerals and rocks caused by the action of temperature fluctuations, ice, water, oxygen, carbonic acid, and organisms. In the upper part of the Earth's crust, where rocks are found under conditions of close interaction with the atmosphere, hydrosphere, and biosphere, they undergo significant and diverse changes of their composition and state. To be specific, the overwhelming majority of rocks were created under specific thermodynamic conditions in the depths of the Earth, in zones of magma activity and metamorphic processes, or on the sea floor. On reaching the Earth's surface, these rocks are surrounded by new physicochemical conditions, become impermanent, and start to disintegrate under the influence of various factors. Their alteration itself occurs in different ways. In some cases, it amounts to breakdown of hard rock into fragments of different size or even into individual minerals. In others, radical changes of primary minerals and rocks occur, such changes resulting in the appearance of completely new minerals different from the primary ones. *The process of destruction of rocks by means of disintegration and formation of soil (the pedological cover) as a result of the process is of exceptionally great significance for mankind.*

Depending on the factors affecting rocks and the results of this influence, weathering processes can - with a certain degree of arbitrariness - be divided into the following two types:

- *Physical disintegration*; and
- *Chemical decomposition.*

The two types of weathering are closely linked with each other, acting together and at the same time. Only the intensity of manifestation of each is unequal. It depends on climate, relief, duration of the process, rock composition, and other factors.

Physical disintegration is caused by various factors. Still, the decisive role is played by factors that induce mechanical movement of rock particles, which leads to destruction of the mutual adhesion holding the components of rocks together. Physical disintegration occurs under the influence of: 1) *insolation*; 2) *frost* (ice); 3) *mechanical action of salt* crystallizing in depressions and cracks in rocks; 4) *capillary action of water*, and 5) *mechanical action of plant roots*. The most *widespread is disintegration under the influence of insolation* (temperature weathering), more precisely the weathering that takes place under the influence of daily temperature changes.

Chemical decomposition results from interaction between rocks of the outer part of the lithosphere and chemically active elements of the atmosphere, hydrosphere, and biosphere. The following materials are characterized by having the greatest chemical activity: oxygen, water, carbonic acid, and organic acids. Changes of volume occur during transformation of one mineral into another. This is of decisive significance for processes of rock destruction, since it frees space into which new amounts of water can enter. Processes transpiring during chemical decomposition are of several types.

- *Oxidation* of minerals and rocks. Oxidation occurs where free active oxygen exists, usually in the presence of water. Air and water in ionizing form destroy iron silicates such as olivine, pyrones, and amphiboles, and carry bivalent iron over into the trivalent form. Limonite - a compound most unstable under conditions of the Earth's surface - is formed from pyrite in this way;
- *Hydrolysis.* Chemical processes of the greatest complexity occur in the course of hydrolysis or decomposition primarily of silicates, which occupy about half of the volume of the outer part of the Earth's continental crust. Hydrolytic decomposition of feldspar results in the formation of clay, kaolin, micas, and other minerals, as well as in separation of free SiO_2;
- *Carbonization.* Crystals containing calcium, magnesium, sodium, and potassium ions enter into reaction with natural water saturated with carbonic acid, forming carbonates and bicarbonates of those elements. Such a process of ordinary chemical dissolution is also called carbonization. Dissolution of minerals in water is characteristic of limestone, marble,

dolomite, and (to a lesser extent) marl. The process results in formation of karst hollows and other forms of karstification. A process even simpler and faster than carbonate dissolution transpires in readily soluble mineral materials such as NaCl, KCl, anhydrite, and gypsum;

- *Hydration.* Hydration represents chemical annexation of water by minerals of rocks, with formation of new minerals, primarily hydrosilicates and hydroxides. Kaolinite is formed from feldspar during hydration, and various hydro silicates (chlorite, serpentine, talc, and zeolites) also arise in this way.

Organisms play a large part in the complex processes involved in chemical decomposition of minerals and rocks, and many weathering products are of biogenic origin. Weathering of rocks begins with action of microorganisms on their surface. These microorganisms create the necessary substrate for plant development, leading to further destruction of the rocks. Thus, *chemo - biological decomposition* is basically initiated by the work of bacteria, algae, and lichens, to be completed by the action of higher plants. Drawing inorganic substances from the ground, plants in this way accelerate and amplify hydrolytic decomposition. Certain organic acids (humic and others) and carbon dioxide are separated during decomposition of organic matter, and acids and gas increase the dissolving capability of water.

The nature and degree of weathering processes are for the most part determined by *climatic conditions.* The hot damp climate of the equatorial belt is especially favorable for complete and profound chemical decomposition. In temperate zones with a damp climate, chemical and chemo - biological decomposition usually occurs at a slower rate. In the former instance, thickness of the weathering crust is enormous and is measured in tens of meters (up to 100 - 120 m). The crust in this case exhibits readily discernible zonal structure. In regions of moderately damp climate in North America, Europe, and Asia, on the other hand, the crust in question is today not very thick (1.0 - 1.2 m at the most) and is practically identical with the pedological cover. According to N. M. Strahov (1963), slight thickness of the weathering crust is a consequence of short duration of the weathering process itself, since this zone in the north was only recently (12 - 15 thousands of years ago) freed of an ice cover.

There is a rich literature on weathering processes (physical disintegration and chemical decomposition) (Allison and Palmer, 1980; Gorshkova and Yakusheva, 1957; Knezevic and Djordjevic, 1975; Milovanovic, 1949; Perelman, 1975, 1979; Veselinovic et al., 1995).

We note that disappearance of a number of elements in the process of weathering and soil formation is usually incomplete. Thus, for example, soil formed from rock rich in copper - containing minerals will contain more copper than in other types of soils. Such soil is poisonous for plants[10]. By contrast, it happens not rarely that an original magmatic rock is almost completely transformed into clay minerals.

We cite in conclusion some interesting remarks of Perelman (1956) as to certain characteristics of life in damp tropical zones that depend on products of the weathering of various kinds of rocks.

It is known that loose clay material marked by a red color is the most widespread product of chemical decomposition in the tropics. This material contains much iron and aluminum because almost all other elements have been removed from rocks in the process of weathering. Since it is poor in calcium and other metals capable of neutralizing organic acids, the soil (and water) of the damp tropics has an acidic reaction. The given geochemical characteristic of landscapes in the damp tropics determines certain characteristic features of their plant and animal life. Perelman gives the following examples:

- One of the most familiar and obvious characteristics of such landscapes is their coloring. This is especially manifested in parrots, hummingbirds, and other tropical birds, which get appreciably more aluminum in their food than do their relatives in temperate geographic latitudes. The higher the content of aluminum in bird feathers, the more striking their color;
- Hollows in the stems of bamboo trees growing in tropical mud are sometimes found to contain fairly large excrescences of opal, an amorphous mineral composed of quartz and water. The opal was formed here in the course of weathering of granites and other rocks, and it has enriched soil water, as well as underground and river waters. Hence also the high silica content in many tropical plants, whose individual parts appear to be "petrified" as a result.

It may well be asked how such specific geological, geochemical, and climatic conditions affect man. How much influence do they ert on his anatomical structure and appearance, how great a trace do they leave on his organism, on its pathology, resistance, and adaptiveness? The complex and very numerous processes that transpire in rocks, the weathering crust, soil, and water can only be sketched in the present book. Greater attention must be devoted to these questions by investigators, including geologists, geochemists, geographers, biologists, and doctors. Without a good understanding of them, it will be difficult to attain the final goal: optimization of the

[10] Since many soils reflect the composition of rocks in the substrate, chemical analysis of soil (including useful elements in it) can be of great assistance in geological investigations of terrains. On the other hand, such analysis can be useful in resolving practical problems of agronomy, for example where we need to explain the causes of different natural soil fertility on the same space.

environment so as to achieve the highest level of public health in many territories of our planet.

Soil

Surface weathering of rocks and minerals is of exceptionally great significance for life, not only because it is the first (initial) stage in the process of soil formation, but also due to the fact that rocks represent the primary source of biologically essential macro - and microelements needed for life of plants and indirectly for that of animals and man.

Processes of soil formation (formation of the pedological cover) are closely associated with the process of rock weathering in the highest part of the Earth's crust. *By "soil" we mean the loose superficial horizon of the weathering crust capable of supporting plant life.* In this connection, soil formation is unquestionably the most important result of rock weathering from the human point of view. That active environment guarantees the entire biomass needed by humans, animals, and plants, predetermining (to a certain extent) the existence of all life on land, including human life[11]. The well - known Russian geologist and pedagogue V. V. Dokuchaev stressed that the soil constitutes a *fourth natural kingdom*: in addition to plants, animals, and nonliving nature, the soil generates (brings forth), alters, and under certain conditions disappears. Rarely more than several meters deep, this thin membrane envelopes a large part of the land of our planet, making it the fifth and youngest geosphere - the *pedosphere.*

Soil represents a natural mixture of inorganic mineral material, living and dead organic substances, air, and water. It has characteristic structure, texture, color, and other physical properties. From place to place, soil differs in thickness, composition, and ability to support plant life. It can be formed by surface weathering of rocks where they are found or from loose material transported from another region (alluvial soils for instance).

Basic Features of Soil

Factors that affect soil formation. The main factors affecting soil formation are the *bedrock* on which soil develops *climate, plant life, soil organisms, relief,* and *time.* A leading role in this complex process is played by the *biological factor* (mainly plants). Thus, soil formation (and rock weathering, as we have seen) occurs most intensively in the zone with

---

[11] In considering the *engineering properties of soil* as an environment for construction work, all loose materials above bedrock are interesting, regardless of whether or not they are capable of supporting plants. The same can be said of *secondary deposits of mineral raw materials or geological building materials.* The soil also plays a large part in formation of the hydrogeological conditions of a given region, especially in the feeding of groundwater.

activity of root systems and circulation of substances produced by organisms during their lifetime. An especially important role here is played by microorganisms, millions and billions of which are present in every gram of soil. The geochemical essence of soil formation consists of decomposition of organic substances by microorganisms.

The influence of different types of *rock as a source of material for the pedological cover* is evident in the early stages of soil formation, but bedrock has little effect on the end result of this process. Thus, almost identical soil can be formed from rocks of the greatest diversity under influence of the same set of conditions. Nevertheless, viewed overall, a link between soil types and different geological substrates is always present in greater or lesser measure. Somewhat greater genetic linkage is especially discernible in cases where some geological formation encompasses a large territory. The given regularity can be illustrated by several examples:

1. Analyzing the role of factors in creation of soil types, V. Nejgebauer (1952) stressed that the most widespread geological substrate of chernozem is *loess* (an Aeolian sediment of Quaternary age rich in fine quartz sand, clay, and calcium carbonate), since two thirds of Eurasian chernozem was formed on loess. In Vojvodina, a large part of Srem and almost half of Banat are built of loess covered by fertile chernozem, as is also the case with Macva, the vicinity of Belgrade, and lands along the Sava and Danube. Apart from it favorable physical characteristics as a parent rock, loess in our country has the additional advantage of being rich in calcium carbonate, which promotes better maintenance of this soil, its stability, and resistance to degradation;
2. *Neogene lacustrine sediments* of Serbia, Macedonia, and Bulgaria are covered by a dense sticky clayey soil having an almost completely black color and a resinous sheen, for which it is known as smonitza. Smonitza was formed after the withdrawal of Neogene lakes from residual lake and swamp clay greatly enriched with humus at the surface owing to rotted bog plants. This relict soil is widely disseminated in the regions indicated;
3. Soils of the *terra rossa* type are known to be characteristic of limestone terrains, where they were formed by dissolution of the limestones themselves and represent their insoluble residue composed chiefly of iron and aluminum. The Dinaric karst represents the typical region of their distribution. Similar soil in Eastern Serbia was formed by weathering of *red sandstones*;
4. In addition to chernozem and smonitza - whose black color arises from organic material (humus) - there is another kind of black soil whose color is dictated by color of the parent rock from which the soil was formed. Such black soil formed by weathering of *basic eruptive rocks* consists of undisintegrated or partially disintegrated ferromagnesian silicates - pyrone, amphibole, and chlorite. According to the pedological literature, it is found on several continents.

Climate determines not only the type of weathering, but also the kind of vegetation that takes part in soil formation. Infiltration of precipitation through the soil, dissolution of mineral salts, and their accumulation are linked with the climate. High temperature accelerates chemical reactions in the soil.

A terrain with broken *relief* usually erodes, and material is transported into morphological depressions. On plains, however, weathering products remain in place. Land relief also influences infiltration and the level of groundwater.

Time is also one of the key factors in the process of soil development, primarily because thousands and tens of thousands of years are needed for formation of a mature soil profile. In the case of long enough time and a favorable tropical or subtropical climate, ground is transformed into overly mature infertile laterite soil. Laterites are developed on large areas of India, Southeast Asia, countries of the Caribbean basin, Brazil, and Central Africa.

Soil components. *Mineral substances* of soil consist of secondary minerals formed by degradation of primary minerals of rocks and amorphous inorganic material. They are represented by sand, silt, clay, or a mixture of particles of these dimensions. Texture is determined on the basis of the content of these fractions, and clayey soils, sandy soils, dusty soils, and combinations of these basic soil types exist in nature.

Organic components of soil include living microbes (bacteria, fungi, etc.), soil nematodes, earthworms, insect larvae, tree roots, plant and animal remains, and various organic compounds formed in the course chemical decomposition of rocks. In spite of the fact that they represent from 2 to 5% of the soil at most, organic components are very significant for its quality. The colloidal matter of incompletely decomposed organic components is called *humus*. The role of humus is especially significant because it supplies microorganisms with nutrients, facilitates infiltration and retention of water and salts, increases the ion - exchange capacity of soil, affects soil structure, and improves plowed fields.

Water and *air* fill the space between soil particles. The amount of water retained in the soil for plant needs is determined by the relative content of sand, dust, and clay. Moreover, air and water in the soil are especially significant for lower plants, algae, and bacteria, since they affect whether oxidative or reductive degradation occurs in the soil.

The pedological profile. Plants are capable of concentrating various chemical elements. Acting like a pump, plant roots lift the elements P, S, Ca, and K - as well as many trace elements - from lower horizons of the soil to upper ones. Such so - called *biogenic accumulation* can bring about an improvement in the environment of plants. Together with biogenic accumulation from lower to upper horizons, descending migration of elements in aqueous solutions is also discernible in soil. Depending on the rates of these two processes, soil is divided into horizons with special physicochemical conditions. The *pedological profile* is formed in this way. From top to bottom, the following three horizons are more often than not present in the soil profile:

- *Horizon A* is rich in organic matter and organisms; it is characterized by accumulation of humus and many elements in its upper part (subhorizon A_1) and leaching of a portion of soluble salts in its lower part (subhorizon A_2);
- *Horizon B* is the middle horizon; by means of infiltration of water, clay and iron minerals accumulate in it, building a distinct crust in places; elevated content of Cu, Ni, Zn, Pb, and other minerals is discernible;
- *Horizon C* represents a loose and partially weathered rock mass, which in its lower layers gradually passes over into compact bedrock.

In the past decades, pedologists and geochemists have increasingly faced the problem of the role played by rare and dispersed chemical elements - trace elements - in the soil, as an invaluable and complex natural system that is open to degradation and pollution[12].

Here we shall analyze the natural soil content of trace elements as a function of bedrock and processes transpiring in soil (binding of elements as ion exchangers for clays, binding of Fe and Mn for oxides and hydroxides). Biogenic accumulation of Be, Co, Ni, Zn, Ge, AS, Cd, Sn, and other rare elements in the humus horizon of forest soils was discovered already at the outset of the 1930's by V. M. Goldsmith. Such phenomena were later confirmed in chernozem and other soils.

Table 2.3 presents the natural concentration of rare and dispersed elements in soil as compared with their content in magmatic (ultrabasic, basic, and acidic) and sedimentary rocks. It can be seen that the concentration of those elements in soil can be greater or lesser, depending on the bedrock in question. Ultrabasic and basic rocks are usually richer

---

[12] Of the total land surface of the Earth, arable land occupies only 10% (10% of the Earth's dry land is suitable for agriculture, but economically unprofitable for commercial production); 17% is occupied by pastures; 28% is under forests; and the remaining 45% is occupied by deserts, glaciers, inhabited settlements, and other types of covering in areas with little soil. This means that human beings feed themselves from a surface of about 1,450 million hectares. The given arable land provides 85% of the energy consumed by people in the form of food; 10% of this energy is from natural sources such as forests and pastures; and 2% is from the seas and oceans. It follows that arable land must be conscientiously conserved. The extent to which improper farming can contribute to erosion of arable land whose formation took hundreds or thousands of years is evident from an example in the United States that was analyzed by Howard and Remson (1978). The negative effects of unregulated surface courses and other causes are noticeable.

in Co, Cr, Cu, Ni, V, and Zn, but poorer in alkaline and earth - alkaline metals, whereas precisely the reverse situation is observed in granites and other acidic magmatic rocks. The difference in content of heavy metals can to a certain extent be attributed to greater content of oxides and hydroxides of Fe and Mn in ultrabasic and basic rocks and soil formed from them, as well as to their ability to bind these metals. According to R. L. Mitchell (1945), soils formed from basic rocks can contain cobalt in amounts of 20 to 100 ppm, whereas ones arising from acidic magmatic rocks contain less than 20 ppm. Some elements like B, Pb, and Mo are concentrated in soil, so that it is in most cases richer than the parent rock in these metals.

Let us examine the geochemistry of Co, Cr, Ni, and V in surface weathering of Cretaceous - Tertiary volcanites in Serbia according to V. Jovic (1990). The examined volcanites include basalts (basic rocks), andesites (neutral rocks), and dacites (acidic rocks). The given elements behaved differently in the course of surface weathering of these kinds of volcanites.

TABLE 2.3.
Content of trace elements in the Earth's crust, rocks and soil (ppm).

Element	Earth's crust	Basic rocks	Acid rocks	Sedimentary rocks	Soil
As	2	1.5	1.5	6	2
B	10	1 - 2	9	100	20 - 40
Co	22	40	20	20	3 - 15
Cr	150	200 - 2000	20	100 - 500	60 - 200
Cu	55	80 - 180	10	8 - 80	15 - 40
Mn	950	1500	700	600	300 - 500
Mo	1.5	1.2	2.5	1.2	1 - 2
Ni	80	50 - 800	5 - 25	52	20 - 90
Pb	65	3	24	17	18
Se	0.05	0.1	0.1	0.5	0.01
Ti	5600	9000	2300	3800	4000 - 5000
V	110	200 - 250	40 - 80	100	100
Zn	70	80	50	65	30 - 70

1. *Cobalt* is an element poorly represented in the Earth's crust (average content: 22 ppm; Krauskopf, 1979). Basalts contain 40 ppm of cobalt on average, while andesites show variations in the range of 10 - 30 ppm (Wedepohl, 1972). The content of this element is in large measure correlated with that of magnesium in rocks. Normal cobalt content in soils is 1

- 40 ppm, most often 3 - 15 ppm (Kabata - Pendias and Pendias, 1985)[13]. The content of cobalt varies from 2 to 24 ppm in volcanites of Serbia and is somewhat lower in soil. Although it is mobile under conditions of surface rock weathering, this element never migrates in the dissolved form, since it is adsorbed in significant amounts on Fe - Mn minerals and clay minerals (V. Jovic, 1989);

2. *Chromium.* The average content in continental basalts comprises approximately 200 ppm (Wedepohl, 1978). Andesites have lower chromium content (55 ppm on average), while dacites are poorer still (20 ppm). The average content of Cr in volcanic soils of the United States varies around a value of 85 ppm (Kabata - Pendias and Pendias, 1985). In Serbia, the basalt of Mt. Rudnik has the highest Cr content (250 ppm), while the andesite of Mt. Radan has the lowest (35 ppm). As for soils, Cr content remains practically unchanged in the case of soils formed from basalt and andesite, but is somewhat lower in ones formed from dacite (110 ppm in the rock, 50 - 55 ppm in soils formed from it);
3. *Nickel.* Average nickel content is 80 ppm in the Earth's crust (NAS, 1975) and 40 ppm in soil (Swaine, 1955), although there are very many localities with average content of as much as 90 ppm. Nickel is a typical element of basic and ultrabasic rocks: in basalts it occurs in the range of 25 - 530 ppm, in contrast to a range of only 3 - 28 ppm in andesites. In volcanites of Serbia, the content of nickel varies from 20 ppm (dacite) to 150 ppm (basalt). In soils formed from them, it ranges from 15 ppm (soils formed from dacite and andesite) to 240 ppm (soils formed from basalt). During weathering of basalt and formation of a soil cover, the content of Ni increases, whereas decrease in the content of this element is discernible in soils formed from andesite and dacite;
4. *Vanadium.* Average content of vanadium in the Earth's crust is 110 ppm (Krauskopf, 1979). It ranges from 200 to 250 ppm in basic rocks (basalt, gabbro), but is perceptibly lower (about 70 ppm) in acidic magmatic rocks (Kabata - Pendias and Pendias, 1985). Average content of vanadium in soil is 90 ppm, the range of its somewhat more frequent occurrence in soil being 3 - 500 ppm (Bowen, 1979). In Serbia, basalts and andesite - basalts have the highest content of V (220 - 225 ppm), while dacites have the lowest (44 ppm). The content of this element declines during weathering of basalts and is approximately the same in weathered andesites and dacites.

The author concludes from the given data that the parent rock erts the greatest influence on behavior of the analyzed elements.

Due to the possibility of increased concentration and great damage resulting from it, trace element content permissible for agricultural use of soil is limited by law (standards). As we have seen, cases of overstepping can be found in nature, and such a possibility has already been indicated

[13] The content of heavy metals tolerated by plants is given in Table 2.4.

by Table 2.3.[14]. However, testing of the content of Zn, Cu, Cd, Pb, and As in soils of Vojvodina (Ubavic et al., 1993) and analyses of heavy metals in land along the Morava River (M. Jakovljevic et al., 1997) and in the Sabac - Loznica region (P. Sekulic et al., 1997) have shown that (with rare exceptions in the cases of Cr and Ni) the content of heavy metals in our main agricultural regions is below the maximum permissible concentration (MPC), i.e., these soils satisfy criteria for production of safe food. It is interesting to note that all of the mentioned soils of Serbia arose over young sedimentary formations (loess, alluvial sediments, and comples of lacustrine sediments)[15].

The yield of any agricultural crop can be limited by the amount of nitrogen, phosphorus, or some other element present in the soil, potassium for instance. The given regularity was noted already in the middle of the 19th century by the well - known German chemist Ju. Libih, founder of agrochemistry and creator of the *theory of plant mineral nutrition.* Libih stated that the absence or deficiency of one of the essential elements in the presence of all others in the soil makes the soil infertile for all plants which require that element. Formulated later on this basis was *Libih's law of the minimum,* according to which the quantitative development of organisms (or "yield" in the broad sense of the word) is determined by those elements (or factors) present in relatively minimal amounts in the natural environment. Libih's law is most often referred to in the case of irreplaceable (incompensable) resources. Thus, for example, it is clear that if a plant lacks phosphorus, no increase in the content of nitrogen, potassium, or some other element can raise the yield of that plant beyond certain limits determined by the deficient amount of phosphorus.

Constant soil productivity depends on agricultural measures, including fertilizer use, calcification, melioration, and measures to prevent erosion. Nitrogen, phosphorus, and potassium - substances extremely necessary for plant growth - are introduced into the soil in the form of fertilizers. Moreover, acid soils react well to introduction of calcium in the form of slaked lime, high - quality limestone, or gypsum and the effects of calcification can be exceptional. Calcification sharply alters physicochemical and biological processes in the soil: not only does acidity decline, but so does the concentration of aluminum and manganese compounds harmful to plants, good conditions are created for life of

---

[14] In many regions of the world, anthropogenic pollution of soil with toxic elements today far exceeds the level of its natural pollution.

[15] The diversity of possible geological effects is illustrated by the following example. In the first hours of eruption of the volcano Hecla on 5 May 1970, ash covered 1/5 of the island's surface. Ash particles with fluorine content of up to 2,000 min^{-1} were scattered over a distance of more than 200 km and covered damp pasture regions. More than 100,000 sheep and cattle were poisoned by the ash-covered grass.

beneficial microorganisms, and there is an increase in content of mobile forms of N, P, K, Ca, Mg, Mo, and other elements important for plants. Repair of soils is increasingly being accomplished with the aid of zeolites, owing to their particular ability to retain soil moisture. Significant agricultural effects can also be obtained through drainage of lowland peat mud accompanied by application of potassium and copper fertilizers (plus boron and manganese in a number of cases) to replenish deficient trace elements[16]. The pronounced deficit of selenium in soil and cereals in Yugoslavia can be compensated by using selenium - enriched NPK fertilizer (Z. Maksimovic, 1990).

Soil and Risks

Many regions with agricultural soil are exposed to natural hazards such as erosion or deposition due to the action of wind or water, flooding, liquefaction, suffosion, land slippage, gas emission, sinking, cracking, seismic activity, and volcanism. They limit the exploitation of soil improved or threatened by human activities. Among human activities aimed at improving land quality, we note soil recultivation, other farming measures, and reforestation. Degradative human activities include exploitation of mineral resources, industry, transport, water management, urbanization, recreation, etc. We shall dwell in greater detail on problems of *soil in relation to life.*

One of the basic characteristics of life is the constant exchange of matter between organisms and the natural environment. In other words, plant and animal organisms constitute an indivisible whole with nature. The link between them is realized by various organic and inorganic substances that continuously enter all of the cells of organisms. Those substances needed for life, so - called nutrients, are introduced into the organism of animals and man by eating. An exception to this is oxygen, which is introduced by breathing. The introduced food serves as a source of energy for the organism on the one hand and as material for building of body tissues on the other. Energy is liberated and new living material is synthesized in the course of the very complex biochemical processes to which ingested food is subjected. Those processes are referred to as metabolism or turnover of substances (M. Jovanovic, 1989).

[16] Around the world and in Yugoslavia, intensive high-yield crop growing is causing removal of water-soluble forms of elements (including magnesium) from soil. Tests have shown that the amount of magnesium removed from soil varies from 1.05 to 6.3 kg/h or from 43 to 259 gmol/h. If the given values are compared with values of removed potassium (80-430 gmol/h) and phosphorus (73-547 gmol/h) - whose quantities are regularly and calculatedly replenished with each new planting - it is clear that the same measures must be undertaken in regard to magnesium.

Metabolism of mineral substances represents a significant segment in overall turnover of substances in the organism. Mineral salts are above all an important ingredient of all cells and bodily fluids. Every organ has its own characteristic mineral composition, which can in certain measure be influenced by diet. The physiological role of mineral substances is multifold. According to M. Jovanovic (1989), they are regulators of the following things: osmotic pressure in bodily fluids and cells; electrochemical reactions of the blood, lymph, tissue fluids, and tissues; cell membrane permeability and exchange of substances between cells and tissue fluids; and secretory and excretory processes and resorption. They are structural components of bones, teeth, cartilage, and other supportive elements of the body. In addition to this, they participate in the synthesis of substances of special significance for the organism (hemoglobin, coenzymes, hormones, and vitamins). The presence of certain ions is necessary for the functioning of muscles and nerves.

Let us now consider the question of trace elements in the soil and possible risks to human health. We have seen that trace elements can be found in different concentrations in the soil. Among them are the series of elements or *nutritive ingredients* needed for maintenance of life. However, every nutritive element can become toxic or even fatal in certain concentrations (Table 2.4.). A substance is considered poisonous if it prevents growth and metabolism of any organism when its concentration exceeds the norm. All elements are toxic in high concentrations, and some are toxic at low ones. For example, copper is very toxic at relatively low concentrations, and it is widely used in the form soluble compounds to eradicate algae. Especially toxic are so - called *cumulative poisons*, which are retained in organisms more easily than they are eliminated. Selenium and cadmium can serve as examples. This also applies to *combined action of two metals* when one metal enhances toxicity of the other. Especially important is the *form in which the poisonous substance is found.* For example, compounds of mercury and lead with hydrocarbons are far more toxic than inorganic compounds of mercury and lead. Conversely, organic compounds of copper are less toxic than inorganic ones[17].

Not all plants have the same threshold of sensitivity to certain elements or, more precisely, to accessible forms of elements. Increase in content of an accessible form of an element above the level of tolerance for a given plant acts toxically, blocking the metabolism of enzymes and vitamins.

In response to toxicity of metals, a certain degree of tolerance to them has developed in plants as a result of physiological processes. However,

---

[17] Soil is known to possess a certain capacity for self-purification in regard to biological and mineral pollution. This capacity can be estimated by recording the rate of increase of the biological component, the quantity of microorganisms, and the value of solar radiation.

the *mechanism of plant tolerance* has still not been adequately illuminated. There is a voluminous literature on biochemical changes, both as indicators of metal toxicity and as aids in estimating the tolerance of plants. Several mechanisms have been described whereby plants achieve tolerance. They include reduction of metal intake in the first place, processes of exclusion (a metal is either rejected or absorbed in limited amounts, its transport is limited, etc.), and processes of accumulation of a metal in certain parts of the plant (A.J.M. Baker and P. L. Walker, 1989, 1990). For example, tests on territory from Pancevo to Pozarevac - where soil was highly contaminated during the NATO aggression on Yugoslavia in 1999 - indicated that grain plants protected their fruit by accumulating heavy metals in their green parts.

The adaptive capacity of plants, animals, and man can be expressed to a greater or lesser extent, but it is generally limited in every case. Plants have adapted to soil conditions in the course of their evolution, and many varieties characteristic of definite biocenoses have been formed. In animals, adaptation or disappearance of extremities occurred (moles, etc.), together with enlargement or shrinkage of digits. Man became shorter or taller in stature, as well as more resistant to diseases. Meanwhile, the struggle for life on certain unfavorable soils often had visible negative consequences. This is examined below on several examples given by Perelman (1956) in his book "**The History of Atoms and Geography**".

TABLE 2.4.
Bearable content of heavy metals for plants
(Jakovljevic et al., 1990).

Metal	Total in mg/kg of soil		
	Often	Range	Bearable
Be	1 - 5		10
B	5 - 30		25
F	50 - 200		200
Cr	10 - 50	1 - 100	100
Ni	10 - 50	1 - 100	100
Co	1 - 10	1 - 50	50
Cu	5 - 20	2 - 100	100
Zn	10 - 50	10 - 300	300
Mo	1 - 5	0.2 - 10	5
Cd	0.1 - 1	0.01 - 1	3
Hg	0.1 - 1	0.01 - 1	2
Pb	0.1 - 5	0.1 - 10	100
As	2 - 20	1 - 50	20
V	10 - 100		50
Sn	1 - 20		50
Se	1 - 5	0.1 - 10	10

1. Plants on tundra soil contain small amounts of such elements as calcium, sodium, magnesium, and phosphorus. Lichens are especially poor in them. This solved the puzzle of the peculiar behavior of deer in such a landscape. Since precisely lichens constitute the basic food of deer in winter, they then suffer strongly expressed "mineral starvation": the organism of the animals is short of calcium, magnesium, phosphorus, and sodium. With the approach of spring, the deer for this reason start to exhibit "greedy" tendencies, eating bird's eggs, unfledged birds, etc. Such food at least partially compensates the deficiency of mineral elements, and it is not surprising that the animals eat bones, bird excrement, marine algae, and fish, and even drink seawater;
2. The acid swamp soil of very extensive forest belts in the European part of the former Soviet Union and Western Siberia is often composed of 98% quartz minerals. It is poor in mobile mineral compounds and so unfavorable for plant life that few species have adapted to such an environment (their number including pine, lichens, heather, and blueberry). Animals are especially acutely affected by the deficiency of calcium - an element that enters the composition of bones - and they for the most part suffer from rickits and are runty. The cause of an unknown disease that afflicted a considerable portion of livestock herds in these regions was for a long time impossible to ascertain. All attempts to isolate a microbic agent ended in failure, and various methods of treatment were of no help. It was finally established that the enigmatic disease is caused by reduced cobalt content in the food of livestock. The content of this rare metal was very low - below 0.0005% - both in the soil and in the food consumed by livestock. Since cobalt enters the composition of vitamin B12, a special vitamin vitally needed by animals, the deficiency of cobalt reduced the amount of it present in the organism, which led to sickness of domestic animals. Adding quite small amounts of a cobalt salt to food or using a special cobalt fertilizer completely eliminated acobaltosis (the name given to this disease, which was at one time also present in Australia, New Zealand, and Scotland) and raised the productivity of sheep and cows;
3. Podoconiosis (or non - filarial elephantiasis) - described by E. W. Price in 1988 - is frequent in Ethiopia, Kenya, Tanzania, Ruanda, Burundi, Cameroon, and the Cape Verde Islands. It was noticed that this disease occurs above all in regions with soil rich in red clay. Analysis of the lymph nodes of patients indicated the presence of micro - particles containing Al, Si, and Ti. It was concluded that the pathological agent is a mineral from the substrate of volcanic rocks (basalts), probably the amphibole eckermannite (Harvey et al., 1996);
4. In many developing countries (especially ones in tropical zones), negative influence of mineral deficits or imbalance in fodder is evident in addition to undernourishment of cattle. Areas with problematical deficits of trace elements or toxicity of trace elements in cattle are traditionally investigated by mapping of regional changes in the composition of elements in soil, fodder, and serum;

5. In tropical zones built of rocks containing clay minerals, the geochemistry of aluminum - because of the effects of this element on human and animal health - is particularly interesting for medical geochemists. In zones where weathering of such rocks occurs, the ability of mobile aluminum to be concentrated sometimes erts great influence on health, and its toxic effects have been the subject of many discussions (B. Smith et al., 1996; S. G. Epstein, 1988; C. N. Martyn et al., 1989; C. R. Harrington et al., 1994; etc.).

As can be seen, both high and low content of biologically essential elements in soil can have negative effects on living organisms, and a deficit or imbalance of trace elements - whatever its origin - represents a clear problem for the health of humans and animals. This problem is magnified by intensive agricultural activity, which leads to depletion of some important soil elements by plants grown on large areas. The influence of geochemical composition of the soil on human health can be best illustrated by the following examples.

1. Modern medicine has demonstrated the multiple action of selenium on various degenerative diseases of man. Among other things, it is known that cardiomyopathy occurs in extensive regions of China as a consequence of extreme deficit of selenium in the soil; that mortality from cardiovascular diseases in general increases with reduction of selenium intake by the organism; that a relationship exists between selenium content in agricultural products and human blood on the one hand and mortality from cancer on the other in different regions of the United States; etc.;
2. In the notable work "**Cancer and Soil**," R. M. Armstrong (1962) stressed the pronounced dependence of gastric cancer on the content of certain trace elements in soil. Writing of the indisputable link between gastric cancer and soil characteristics, S. W. Tromp assigned a certain role to copper, iodine, cobalt, boron, magnesium, and molybdenum. Soil - linked mortality of populations living on chernozem is significantly lower than in ones living on other types of soil;
3. Dental caries and endemic goiter are widespread in 70% of the inhabitants of West Africa living on the Bauta - Djalon Plateau, a situation linked with deficits of calcium and iodine in the soil. Insufficient iodine content in soil, drinking water, and food products is the cause of widespread endemic goiter in inhabitants of the tropics and subtropics. Goiter resulting from iodine deficiency is frequent among inhabitants of the Alps and Andes (J. R. Paul, 1972), as well as many other mountain regions. According to the results of investigations in Vietnam (L. N. Krepkogorskii, 1953) and Egypt, increased fluorine content in soil and water led to formation of fluorosis foci...

In forming medico - geological evaluations, it is also important to note the existence or nonexistence of microorganisms pathogenic and beneficial to man in the soil and establish whether the soil is habitat to parasites (fungi) and other possible agents of infectious diseases. An acidic reaction of the soil is favorable for development of the lowest fungi, and this can cause the various fungal diseases of the skin widely disseminated in the tropics and subtropics. Mountain soils ert various kinds of influence on human health, among other things creating conditions for anaerobic infection of wounds and incidence of tetanus or anthrax if microbic pollution of the soil is pronounced. According to Addington (1967), highly endemic zones in relation to histoplasmosis are situated in many river valleys of the world, and reddish - yellow podzolic soils with certain temperature, humidity, and wind conditions represent the environment needed for preservation and growth of this parasite.

Pedological Characteristics of the Territory of Yugoslavia

The diversity and wealth of soil types on the territory of Yugoslavia represents the main characteristic of its pedological cover. This applies above all to the territory of Serbia, where the diversity and physical differentiation of geological and other pedogenetic factors has contributed especially greatly to the complexity and diversity of that cover[18]. Thus, formation of a large number of classes and types of soil was possible on a relatively small area. A brief sketch of the classes and types represented is given according to J. Dinic (1997).

With respect to area, **automorphic soils** are dominant in Yugoslavia. Such soils are characterized by the fact that they obtain moisture exclusively from precipitation and precipitation is retained for a longer time in them. The *first class* of automorphic soils consists of undeveloped or poorly developed soils such as *lithosols*, *Aeolian sand*, and *colluvial soils*, which occupy an insignificant area of the country's territory.

The *second class* of automorphic soils is represented by the following types:

1. *Limestone - dolomitic dark soils* and *rendzinas*, which are developed in the carbonate terrains of Eastern and Western Serbia and Montenegro;
2. *Rankers*, soils on a silicate substrate, suitable for development of a forest cover;

[18] Most of the soil resources are found on the territory of Serbia, where the area of plowed land is 3,670,091 ha and there are 257,031 ha of orchards, 84,883 ha of vineyards, 668,633 ha of meadows, 1,011,884 ha of pasture land, 35,760 ha of land with fish farms and bogs, and 2,420,000 ha of forest land.

3. *Chernozem*, one of the most widespread and productive soils in Serbia (13.7% of its territory), formed predominantly on loess in Vojvodina, Macva, the vicinity of Belgrade, and Stig; and
4. *Smonitza*, developed on the Neogene lacustrine sediments of Central Serbia, as well as in basins of Eastern Serbia and in Kosovo and Metohija (8.6% of the territory of Serbia). Due to its richness in humus and other nutritive material and considerable distribution, smonitza is one of the most valuable of Serbia's soils.

The *third class* of automorphic soils is made up of so - called cambic soils:

1. *Brown forest soil*, formed on various rocks in Macva, Sumadija, land along the Sava River, land along the Morava River, the Timocka Krajina region, and Kosovo and Metohija (8.3% of the territory of Serbia). It is an excellent forest soil and very good agricultural soil;
2. *Acid soil* (distric cambisol), formed on acidic silicate rocks (granite, gneiss, phyllites) and sedimentary rocks (sandstone, shale), the most widely disseminated type of soil in Serbia (30.6% of its area) and the northeastern half of Montenegro. It is very suitable for development of a forest cover;
3. *Brown soil on limestones and dolomites* (calcocambisol), characteristic of karst (carbonate) terrains; and
4. *Terra rossa* spatially confined to karst of the Montenegrin Dinarides.

The *fourth class* of automorphic soils includes the following three types of eluvial - illuvial soils: *podzol*, formed on acidic rocks under conditions of a cold and damp climate; *leached soil* and *brown soil*, typical forest soils; and *rigosol* and *garden soil* (anthropogenic soils).

Apart from automorphic soils, **hydromorphic soils** also have a considerable distribution. They are soils whose evolution is influenced not only by atmospheric water, but by underground and flood water as well. The following are such soils:

1. *Pseudogleyic soils* are found in the valleys of the Kolubara, Jadar, and Western Morava Rivers. They are of low fertility;
2. *Alluvial soils* are characteristic of the alluvial plains of the larger river courses of Serbia (5.6% of its territory). They are often exceptionally fertile soils;
3. *Semigleyic soils* are found on alluvial plains and in marshes of Vojvodina;
4. *Marsh dark soils* and *swamp gleyic soils* (class of gleyic soils) are characterized by the constant presence of groundwater in the profile (Vojvodina, parts of Macva, the valley of the Sitnica River, the Pozega basin);
5. *Peat soils* (histosols) are of local distribution.

Halomorphic soils or **saline soils** are divided into two classes: *solonchaks* (acutely salinized soils) and *solonetzes*. Solonchaks contain more than 1% soluble salts, which is harmful for the development of higher plants. They are most widespread in Backa, somewhat less so in Banat and Srem. This type of soil is formed under the influence of groundwater, whose strong evaporation results in accumulation of salt in the surface layer. In contrast to solonchaks, solonetzes are characterized by low concentration of salt in the surface layer because it is leached and transferred to deeper parts of the profile. Solonetzes are present more in Banat than in Backa.

In considering the geographical distribution of soils, M. Ciric and Dj. Filipovski (1972, 1974) singled out four pedogeographic regions on the territory of Serbia. It can be seen that these regions to a considerable extent coincide with the main geotectonic units. Also characteristic is the great concentration of fertile soil (as much as 83.1%) in Vojvodina.

Geomorphologic Factors

Relief is a very important element of the natural environment. In the words of Jovan Cvijic (1922), "*Different factors are involved in determining the nature of a large natural region. Of primary importance are morphological characteristics, which constitute the basic features of such a region.*" The strong influence of relief on other elements of the natural environment and on human life results in noticeable coincidence between features of relief and geographic (economic) divisions, especially in morphologically more developed regions of the Planet. Thus, for example, the cartographic part of the "**Atlas of Prospective Development of Rural Health Protection in the Armenian SSR**" (1970) begins with an orographic map, since relief in Armenia very strongly affects many aspects of life of the population, including the organization of medical health protection.

By relief as an object of geomorphologic study, we mean the totality of all superficial lithosphere forms of different geological composition and origin. These forms can be at different stages of development, but they are connected in a complex way with each other and are in complex interaction with the natural environment. The type of relief represents a certain linkage of relief forms that have similar origin, geological composition, and history and which are repeated on large areas of the lithosphere's surface. Two groups of relief forms are distinguished: *positive forms* (mountains, mountain ranges, hills), which are convex in relation to the level horizon and surrounded by lower elements of relief; and *negative forms* (valleys, basins), which are concave and surrounded by elevated parts of relief.

One of the most significant factors in development of the morphology of terrains is *geological structure of the area on which the forms themselves are sculpted.* Regional geomorphologic characteristics are also determined by geological structure. To be specific, the tectonic pattern dictates the form of larger morphostructures, the position of such morphostructures, and their interrelations, while lithological composition determines the kinds and intensity of processes that is the features of relief forms. The karst process is dominant on terrains built of carbonate rocks, whereas fluvial, fluviodenudational, and slope forming processes are dominant on ones built of noncarbonated rocks. Thus, the relief of some area directly or indirectly tells us about the area's geological history in the last periods of its existence. We see in relief reflections of old denudational surfaces, forms of glacial accumulation, and traces of former rivers, rock slides, land slips, etc. By studying relief and its genesis, we get a rough idea of the geological composition of the upper parts or surface cover of the Earth's crust (usually the parts most important for building purposes and agriculture).

Apart from the influence of geological structure, development of exogenous processes and formation of individual types of relief are significantly affected by *climatic characteristics*, which determine the emergence of glacial, periglacial, and Aeolian relief. On the other hand, relief has an indirect effect on climate. Influence of elevation and exposure is evident in specific local climatic phenomena (air temperature, precipitation, cloudiness, winds, etc.), accompanied by corresponding characteristics of the water regime and particular aspects of the genesis of soil and animal life. To be specific, *elevation* is the most influential property of relief forms. It is usually expressed especially strongly in mountainous regions, where other characteristics - *slope inclination* and *exposure* - are also of great significance.

Sculpted in moraine and glaciofluvial material, *glacial relief* is characteristic of enormous areas in northern parts of Asia, North America, and Europe. Series of moraine bulwarks and hillocks stand out, together with gently rolling moraine plains and types of uniform relief. Forms of Pleistocene glaciation have been significantly altered under the influence of younger fluvio - denudational processes.

Several characteristics of relief need to be further considered as a direct factor in the disposition of *geochemical landscapes*. The given role of relief is essentially determined by the degree of its horizontal and vertical breakup. Relief above all dictates to a considerable extent the form of contours of autonomous and dependent landscapes, as well as the area between them. Thus, for example, regular alternation of hillocks, plateaus, lakes, and bogs is discernible in regions of hillocky moraine relief. Completely different is the picture of autonomous and dependent landscapes under conditions of typical erosional relief or under conditions of dunes, even if all these types of relief are situated in the same climatic zone. No less significant is the influence of *relief on the ratio between migration of matter in solid form and its migration in the dissolved state*: erosion of upper horizons of the soil and weathering crust, i.e., mechanical denudation, occurs under conditions of drained forms of relief, whereas chemical denudation is far more prevalent on undrained plains. Influence of relief is also manifested in other ways:

- *The rates of water circulation and redox processes in a landscape* depend on relief. If the relief is greatly broken up, then water circulates more intensively in the landscape, the oxidation ambient is better developed, and the reduction environment is less developed under otherwise equal conditions. Under conditions of less broken relief, the reverse is the case;
- On sectors with very broken relief, the Quaternary cover is usually thinner, and river valleys are completely or largely entrenched in different bedrocks. This results in *involvement of chemical elements of these bedrocks in recent migration*, and such elements can ert crucial influence on the landscape itself;
- Mountainous relief, with its sharply expressed vertical and horizontal breakup, can dictate landscape *heterogeneity*, whereas predominantly Quaternary formations are involved in recent migration under conditions of old continental shields and platforms, with relatively horizontal deposition of layers and level surfaces;
- *Diversity* of the soil - plant cover is characteristic of young, unbroken plains. Only future erosive breakup can bring about a more uniform landscape and its greater conformity with climatic conditions.

Relief and Living Conditions

Through modification of other elements of the natural environment, relief indirectly (and sometimes directly in several of its properties) influences the conditions of human life, exploitation of various resources, and other anthropogenic activities. The consequences of influence erted by relief can be negative for human life (worsening of agroclimatic conditions with increase of elevation, destruction of soil and vegetation by erosive processes, shortage of water in karst regions, impeded building and exploitation of roads and other construction projects, etc.), but also positive (the possibility of development of special forms of tourism in regions of high mountains, etc.).

High mountains (arbitrarily taken to mean elevations above 2,500 m) are characterized by a familiar affliction, namely *mountain sickness*. Already in 1590 the Spaniard Acosta, traveling in the Peruvian Andes, noted in himself and his companions strange symptoms of asthma, weakness, heart palpitations, headache, nausea, etc., which he lumped under the concept of *mountain sickness*. In his opinion, the given symptoms arose due to inhalation of rarified air. Only in 1887 did the well - known French physiologist Bert prove the link between *mountain sickness* and oxygen deficiency. Despite the long period of investigation of this question, serious elaboration of pathological processes transpiring during residence in high mountains is today still far from complete, although many physiologists have worked on the problem. Evidently, the whole gamete of pathological reactions encountered under conditions of high mountains cannot be covered by the single term *mountain sickness* or reduced to hypoxia (oxygen starvation of the organism) alone, even though the latter is of important pathogenic significance.

For most people, the climate of **high mountains** is deleterious, since it provokes overertion of the organism's system of accommodation. This does not apply to the extended zone of the Andean Plateau from Columbia to Chile, where more than 10 million people live at elevations above 2,500 m. That zone (up to the snowline above 5,300 m) is not merely favorable for life; it was even the center of the Inca state, one of the great civilizations of pre - Columbian America. In the present day, inhabitants capable of performing normal work live there, and their numbers are not declining. This applies especially to the Indians and "*gauchos*," who systematically endure *hypoxia* and the action of moderately low temperatures. As a result, development of the younger generation of "*gauchos*" has a hypoevolutionary character, with slowed growth and delayed appearance of teeth. The thoracic cavity in children of the local population has greater measurements in all dimensions than among children living at lower altitudes[19,20].

For climatic and other advantages, **low mountains** (with elevations of from 750 to 1,000 m) and **medium - high mountains** (with elevations of from 1,000 to 2,500 m above sea level) are of great value for rest, recreation, and therapy. Mountain spas are every year gaining in significance throughout the whole world. For example, it has long been considered that treatment of tuberculosis can be especially successful in mountains, and it is not without reason that such spa as Davos in Switzerland and many others on the shores of Swiss lakes, as well as ones in the mountains of Scotland and in the Caucasus, has enjoyed worldwide popularity. Moreover, it has turned out that not only lung ailments, but also cardiovascular diseases and atherosclerosis, are much more successfully treated in mountains than on plains. Mountains can also be used for recovery from more serious treatable diseases such as bronchial asthma in adults and children.

---

[19] Investigating Indians living in the high mountains of Peru, where oxygen pressure is reduced by 44%, scientists established that total ventilation of the lungs is only 22% greater than in the inhabitants of plains, and that the quantity of blood in aborigines there is 1.74 liters greater than in the sea-level population of Lima.

[20] Classification of mountain regions according to altitude above sea level can be arbitrary, since the same elevation in different mountain massifs of the world can act on man in different ways.

In study of natural biocenoses, special attention has been paid to the significance of mountain and inter - mountain territory, river valleys, and lowlands for diseases of man in areas with different morphological manifestations. They are the main points of orientation for workers engaged in drawing up an epidemiological regionalization or trying to classify the foci of diseases. For their part, those forms of relief dictate the nature of the soil and plant cover and characteristics of animal life. For certain regions in such areas, it is sometimes possible to strictly establish natural biocenoses with which natural foci of diseases are linked, i.e., to discover the pathways of circulation of disease agents. On the basis of her field studies of diseases with natural foci conducted over a number of years, P. A. Petrishcheva (1965) concluded *that river valleys in the medical sense represent the greatest danger, being distinguished by diversity of natural foci of diseases*, and gave 10 examples of such diseases.

Karst Areas and Living Conditions

Many extensive areas of the world (in countries bordering the Mediterranean, France, Great Britain, the former Soviet Union, the Southeastern United States, Canada, Mexico, Western India, the Malay Archipelago, Australia, and North Africa) are marked by karst phenomena sculpted in limestones, dolomites, gypsum, and other soluble rocks[21]. It is difficult to find in nature a similar geomorphologic formation or a type of relief so unusual. The main condition for development of the geologically predisposed erosion process is the presence of a specific lithological base, most often one composed of limestone. Dissolution of limestone, i.e., the process of karstification, gives rise to a rocky surface relief furrowed by sinkholes and crevices and marked by ponors, sinking streams, dry dolines, uvalas, enclosed karst poljes, and karst plateaus, with scarce tillable soil and degraded vegetation (Fig. 2.7.).

In addition to its specific surface morphology, a special underground morphology develops as the most conspicuous consequence of the karst process in the given geological environment, and this feature clearly distinguishes karst regions from non - karst terrains. The indicated underground morphology is expressed through an irregular network of karst galleries and caverns formed by means of dissolution and by the

---

[21] For example, in Asia karst terrains occupy an area of around 1.6 million km^2 or 3.6% of the continent's surface (the largest karst region in Southern China has an area of 500,000 km^2, of which 200,000 km^2 falls on the famous conical karst, with limestone elevations of conical and columnar form surrounded by valleys and rice paddies, while the karst of Asia Minor occupies an area of about 150,000 km^2). In the case of North America, karst regions occupy 15% of the territory of the United States, 60% of Jamaican territory, and 50% of the territory of Cuba. In Australia, the Narlabor Desert occupies an area of about 200,000 km^2.

mechanical work of water to great depths[22]. For this reason, karst is characterized by scarcity of water at the surface and abundance of groundwater. In contrast to non - karst terrains, most precipitation is filtered into a pervious environment in which circulation of water to the lowest drainage basins proceeds along privileged collectors and deformed cracks (Fig. 2.8). Runoff of water accumulated in karst poljes during rainy periods occurs through ponors and estavelles. For example, the Niksic polje in Montenegro is drained by more than 880 ponors and estavelles, and the swallowing capacity of the Slivlje ponor alone is as much as 120 m^3/s. Groundwater is evacuated through strong karst springs, mainly formed at points of contact between limestone and some impervious rock. The springs themselves are characterized by great variations between minimal and maximal output in the course of every hydrological year, depending on the level of precipitation. Thus, one of the strongest karst springs in the world, the Durmanli spring in Turkey, has a maximal output of more than 100 m^3/s, average output of 50 m^3/s, and minimal output of 20 m^3/s. The fountainhead of the Trebisnjica in Herzegovina has a maximal output of more than 300 m^3/s, average output of 80 m^3, and minimal output of only 2 m^2/s.

The described inorganic factors dictate the forms of life and human activity on karst. The thin layer of soil, hard rocky substrate, and special morphology of terrains has limited economic activity in the past, and indeed do so today. The specific natural conditions gave rise to a specific organization of settlements, road networks, and the like, which even today, with all the possibilities of modern technology, has remained virtually the same.

For a long time, it was believed that conditions for survival of living organisms are lacking in the dark karst underground. When the guide Luka Cec in 1831 discovered a cave beetle deep underground in the cave Postojnska Jama in Slovenia, it was the first discovery of such a species in Europe. Gradually other caves were found to be home to an increasing number of endemic and other species of animals and plants adapted to

---

[22] The deepest pit in the world to be investigated so far is the Reseau Jean Bernard in France with a depth of 1,535 m. In Yugoslavia, the pit Jama na Vjetrenim Brdima on Mt. Durmitor has the greatest depth (897 m) among known pits. The longest known cave systems are as follows: the flint Mammoth Cave system in the United States (530 km); the Optimisticheskaya system in Russia (153 km); and the Holloch system in Switzerland (133 km). One of the largest known cave systems on Earth is the system of Carlsbad Caverns in the national park named after it in the state of New Mexico, which is sculpted in limestones and gypsum of the Guadaloupe Mountains, with caves of giant dimensions (the Big Chamber has a length of 1,200 m, width of 190 m, and height of about 90 m). It is calculated that from 13 to 16 thousand pits and caves exist on the territory of former Yugoslavia, the longest of the caves being the cave Pecina nad Vrazjim Firovima, with the entrance in Montenegro and extending to the Pester plateau - 10.5 km.

life below the Earth's surface. The well - known tailed amphibian *Proteus anguineus* is certainly one of the most interesting inhabitants of groundwater of the Dinaric karst.

Fig. 2.7. Karstified limestone (Photo: D. Milovanovic).

Thirty thousand years have elapsed since the *Krapina prehistoric man* lived in the cave of Husnjakovo near Krapina (in the Zagorje region of Croatia). Tools of the Krapina man were fashioned of several kinds of silicate rocks or coarsely grained aggregates of quartz, opal, and chert. Prehistoric man of that era found shelter and sometimes took up permanent residence in the antechambers and darkness of many caves.

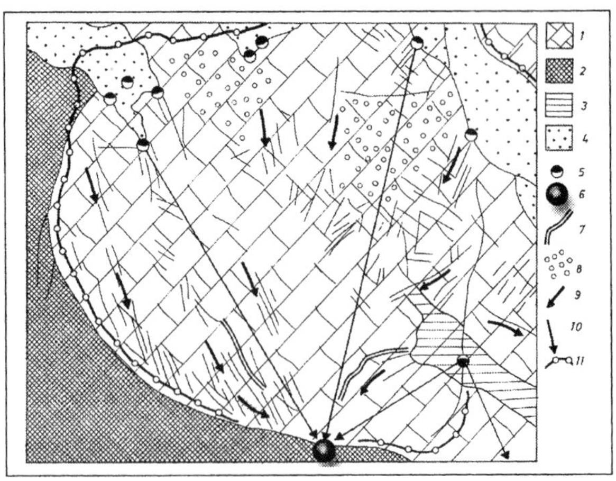

Fig. 2.8. Sketch map of a watershed in open karst. 1. limestone; 2,3,4. impermeable rocks; 5. ponor; 6. spring; 7. wet dolina; 8. pockmarked karst; 9, 10. flow directions; 11. water divide.

In the karst of Serbia proper, a number of Paleolithic cave finds have been made[23]. These finds are characterized by a small number of cultural artifacts and a large quantity of faunal remains (D. Mihailovic, 1998). In the late period of prehistory, caves were used as temporary residences by bands of humans. Discoveries in the cave Lazareva (Zlotska) Pecina near Bor in Eastern Serbia indicate that as early as 5,000 years ago, there existed true *metalworks* in this cave and that men of that time were already familiar with copper ore and the basic metallurgical processes needed to obtain *final products*. The drawings of a horse and rider discovered in a cave near the village of Gabrovnica (in the vicinity of Kalna) probably also date from that period.

Much later, during the Middle Ages, many caves served as shelters from various enemies. Due to exceptionally attractive calcite ornamentation, they have today become increasingly transformed into tourist attractions. We also note familiar cave drawings and engravings as important monuments of cultural heritage. Today millions of people annually visit the famous caves of the United States, Belgium, France, and countries of the Balkan Peninsula and other parts of the world.

Speleotherapy, i.e., curing in caves, is not a modern invention. Already in states of the ancient world and the Middle Ages, it was known that sojourns in caves - especially ones where the air is heated to 30 - 40°C - help to reduce pain in the bones and improve the feeling of general well - being. It was speculated by N. A. Danilova (1971) that the inaccessibility of caves to electromagnetic fluctuations and storms, together with increased temperature, air sterility, and microclimatic stability, can act as therapeutic factors for patients suffering from bronchial asthma, chronic bronchitis, whooping cough, rheumatism, and certain skin diseases. In the former Soviet Union, increasing use was made of *cave sanatoria* such as Kungurska in the Urals, Proval in the Northern Caucasus, Cufat, Kale, and Tepe Kamen in the Crimea.

One of the specific characteristics of karst lies in the wide range of *risks it holds for any form of human activity aimed at altering nature, particularly construction of hydrotechnological projects and structures for water capture and formation of artificial reservoirs.* We cite certain aspects of the given problem as discussed in the book "**Geological Engineering in Karst**" by P. Milanovic (1999).

1. *Breaches (collapses)* are spatially independent, unforeseen, and random events that represent grave danger for all structures built in karst. One of their most dangerous characteristics is the fact that they occur practically

[23] The Middle Paleolithic lasted from 130 to 40 thousand years before the period of recorded history, the Upper Paleolithic from 40 to 10 thousand years, and the younger period of prehistory from 10 to 0 thousand years.

instantaneously. They occur on terrains where karstified rocks are covered by loose sediments, most often after a long phase during which material from a ponor is carried away by underground streams (Fig 2.9). Since 1950 in the United States, for example, there have been thousands of such collapses, accompanied by the formation of pits measuring up to 100 m in diameter and 65 - 70 m in depth. In one such collapse in the vicinity of the Blyvooruitzig Mine in 1964, three residential buildings with five residents formally disappeared, and the Pennsylvania Bus Garage sank 60 meters in the course of several hours when the land under it collapsed into a cavern or karst sinkhole;

2. Despite very complex and often long - term studies employing the latest research methods, the *risks of building in karst* must be accepted as unavoidable. In dam construction, for example, the dominant risk is that of water loss, and cases are known where certain reservoirs (Lar in Iran, Montejaque in Spain, Vrutac in Yugoslavia, etc.) remained empty or were only partly filled with water. The majority of such reservoirs were sanated successfully or with partial success, but often with extensive sealing work. Especially great water losses were recorded in the cases of the Keban Reservoir in Turkey (26 m^3/s after the first filling, less than 10 m^3/s following sanation) and the Vrutac Reservoir in Yugoslavia (25 m^3/s after the first filling, sanation unsuccessful). Negative ecological consequences include changes in the regimes of downstream karst springs, sometimes with alteration of water quality;

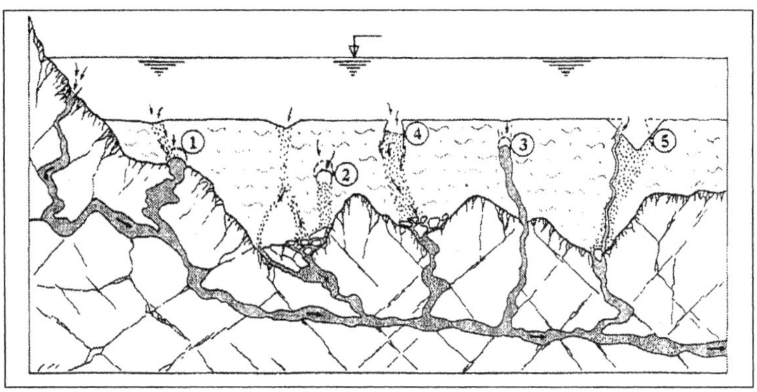

Fig 2.9. Scheme of important stages of collapses in alluvial cover (P. Milanovic, 1999).

3. *Gypsum* and *anhydrite* are especially sensitive geological environments for construction of dams and formation of reservoirs. After filling of the reservoir, the rate of dissolution abruptly increases and suffosional processes are manifested in such an environment. Several dams in the United States (the McMillan, Avalon, and Hondo Dams in Texas and New Mexico) were abandoned because of such problems. The enormous mass of dis-

solved gypsum on the shores of the Kamsk Reservoir near the city of Perm in Russia caused many collapses in the vicinity of the reservoir itself. Six months after filling of a reservoir on the Euphrates River slowed by the Mosul Dam in Irak, great amounts of gypsum and anhydrite (13,000 t) were dissolved and leached, which resulted in the formation of large cavernous spaces under the dam's foundation;

4. Undesirable consequences can also arise in the case of *underground accumulation*. The possible risks arising here can be illustrated by the two following examples: a) the underground stream Jijao in the province of Guangxi in China was blocked for irrigation purposes. Completely unexpectedly, fields upstream were flooded during the first rainy season as a consequence of this measure, and 240 persons were evacuated from the threatened region; b) on the outlet part of the channel of the Obod estavelle on the rim of Fatnicko Polje in Herzegovina, a concrete plug measuring 10 m in height and 3.5 m in average width of 3.5 m was constructed to prevent flooding of the polje. The first heavy rains after construction of the plug caused a sudden rise in the level of groundwater with appearance of several tens of springs on the slope 80 - 100 m above the level of the plug, these springs having a total output of around 11 m^3/s. This state in the discontinuous karst groundwater provoked strong local earthquakes in the vicinity, ground tremors, and slippage of rock masses. The experiment had to be stopped: the plate of reinforced concrete was mined, and water pressure dropped suddenly as a result[24].

Geomorphologic Characteristics of the Territory of Yugoslavia

Endogenous relief. Global geomorphologic features of the territory of Yugoslavia were determined by development and characteristics of the basic geotectonic units of the region. This is especially evident on the Geomorphologic (Morphostructural) Map of Serbia (M. Zeremski, 1990). Thus, the Inner Dinarides (Montenegro and Southwest Serbia) are characterized by karstic relief with dominant plicated morphostructures oriented in the direction of the given range. The same can be said about karst of the Carpatho - Balkanides, which is marked by two series of mountains running from north to south with a turn to the southeast. The Serbo - Macedonian mass in the central part of Serbia represents a zone of insular and peninsular mountains with a characteristic parquet structure. Negative relief forms are manifested in various depressed morphostructures - trenches, valleys, basins, and depressions. The Pannonian Basin in the north of Serbia stands out with respect to its dimensions.

---

[24] Under conditions of hard to foresee negative consequences of regulating the regime of groundwater, a possible solution to the problem in both cases could be to work in several stages, with partial closure of the channel initially, observation of effects during the first rainy season, and subsequent determination of the required scope of the next intervention.

Created over the course of their long geological history, positive and negative morphostructures resulted from the action of plicative and radial tectonics. The latter type was dominant during the Neogene and Quaternary, when the contemporaneous morphotectonic structure of Serbia was formed.

Exogenous relief. A number of morphological agents took part in the creation of exogenous relief. Some of them were active during earlier geological periods (abrasional, glacial, and Aeolian agents); creating various erosive and accumulative forms that today have a fossil character. Other natural agents (fluvial and karstic agents, forms of surface and linear aquatic erosion, landslips, and the Aeolian process) are active today, and the results of their work are recent. Creating various erosive and accumulative forms, all of the indicated agents have helped to make the relief of Yugoslavia become in every respect much more complex, varied, and discontinuous.

Fluvial relief is by all means the most strongly expressed and spatially widespread among all of the exogenous relief forms of Yugoslavia. A developed hydrographic network participated and is participating in its formation and development, effects of the erosive and accumulative work of this network having been determined by a series of geomorphologic factors (J. B. Dinic, 1997). As the largest forms of river erosion in Yugoslavia, river watersheds belong to three marine basins (Black Sea, Adriatic, and Aegean), among which the Black Sea basin embraces by far the greatest territory. Representing a second conspicuous form of fluvial erosion, river valleys here are characterized by considerable morphological diversity.

Created by the action of snow and frost, forms of *periglacial relief* are dominant above the tree line in higher mountain regions, while forms of *glacial relief* created by the powerful action of Pleistocene glaciers have been preserved in the highest parts of the Prokletije, Durmitor, and Sar Mountains. Enormous amounts of fluvioglacial material were accumulated in the Skadar Depression (Cemovsko Polje in Montenegro). Also preserved in the relief of Yugoslavia are forms of *abrasional relief*, which arose during the Neogene in the course of rhythmic sinking and withdrawal of the Pannonian Lake. Such regression resulted in the formation of stepped abrasional plateaus and terraces.

Genetically linked with the action of wind, *Aeolian relief* is spatially limited to the plain of Vojvodina and land along the Danube. Forms of this relief arose as a result of accumulation of loess and sand in the guise of expansive loess plateaus and two sandy regions.

Karst relief is linked with the occurrence of carbonate rocks and is most widespread in the Dinarides (Montenegro and Western Serbia). This is for the most part mountainous terrain marked by the presence of various karst forms. We have already discussed characteristics of karst geomorphology.

The main morphometric index - *hypsometric structure* - of the territory of Serbia is given according to J. B. Dinic (1997). The following hypsometric categories are represented here:

- Plains and lowlands up to 200 m above sea level, which occupy 32,540 km^2 or 36.83% of the territory of Serbia;
- Hilly land - 21,829 km^2 or 24.79%;
- Low mountains - 24,105 km^2 or 27.28%;
- Medium - high mountains - 9,681 km^2 or 10.96%; and
- High mountains - 206 km^2 or 0.23%.

The highest peak of Serbia, Mt. Djaravica in the Prokletije Mountains, is 2,656 m high, while the lowest point of Serbian territory, with an elevation of 30 m above sea level, is the place where the Timok River empties into the Danube. The difference in altitude between these two points comprises 2,623 m. The significant participation of low and hilly land in the hypsometric structure of relief (61.53%) can be considered a favorable circumstance for agricultural and other production.

Tectonic Movements and Structural Forms

Tectonic Movements

Throughout the entire geological history of the planet, and even in the historical period as well, the Earth's crust has been constantly subject to bending, leaning, rising, and falling. The relative position of rocks has been altered as a result of such movements. All such shifts are united under the term *tectonic movements*. Tectonic movements can have any direction: they can be upward, downward, at an oblique angle, or horizontal. They can be extremely slow and gradual or sudden and powerful. The effects of such activity are present everywhere. In the Himalayas, marine sediments are raised to more than 8,500 m above sea level. Deposits of similar origin with faunal remains are widespread in the interior of the United States, where the rim of the Grand Canyon of the Colorado River is more than 2000 m high. Over the last 10 million years on the territory of Yugoslavia, vertical movements have attained a maximum value of almost 8,000 m, with more than 2,000 m of uplifting (in the region of the Sar Mountains) and more than 5,000 m of downwelling (Segedin).

While on the subject of vertical movements, we note the fate of the Temple of Serapis, built around 2,000 years ago on the shores of the Bay of Naples in Italy. Due to long - term gradual sinking of the ground, this temple from the 13th to 16th Centuries was under water. The marble columns of the temple bear unmistakable marks of time spent in water, namely channels excavated by the marine mollusk *Lithdomi*, from which it can be seen that water reached a maximum level of 5.71 m above sea level. After that period, the temple "resurfaced" due to uplifting of the ground, only to suffer its old fate at the beginning of the 19th Century due to renewal of the sinking process. By 1954, water was already 2.5 m above the bases of the columns.

Another well - known example is sinking of the Adriatic shoreline. Sinking has resulted in shrinkage of the coastal zone, transformation of peninsulas into islands, and inundation of sand beaches and cultural - historical monuments, all of this taking place during the historical period, with obvious influence on the environment.

Two forms of tectonic movements are distinguished here: *epeirogenic* movements and *orogenic* movements. Epeirogenic movements (upliftings and downwellings) can affect rock masses on areas of continental size and are accompanied by only slight internal deformations. Some sectors of the Earth's surface rise above sea level, and processes of denudation are prevalent there. Others sectors sink, and sediment accumulation is prevalent in them. Generally regarded, the youngest period in the Earth's geological history has been marked by uplifting of continents and sinking of the ocean floor. Mountain formation on once geosynclinal areas is a result of orogenic movements. These movements create enormous mountain ranges, during whose formation both uplifting and compression of rock masses occur.

An *active tectonic regime* is peculiar to regions of Alpine and Mesozoic folding. Such regions extend in a belt along the Cordillera and Andes in America; along the Alpine - Mediterranean zone in Eurasia; and along a belt in East Asia embracing regions of Far Eastern seas and island chains (Fig. 2.10.). The relief of tectonically active regions is very broken. Activity of the tectonic regime is manifested as well in powerful seismicity and development of volcanism.

A *passive tectonic regime* is characteristic of a large part of Precambrian platforms. This applies to the entire territory of the North American and Russian platforms; the Siberian and Brazilian platforms; a large central zone of the African platform; the Indian platform; and others. The relief of such regions is level, low - lying in some and slightly elevated in others. Only on separate small sectors of the platform can the tectonic regime be somewhat activated and the sectors themselves marked by mountain ranges of limited dimensions (the region of the African rifts, for example). Shallow seas have developed on inundated sectors of some platforms. It goes without saying that all of the indicated territories are aseismic and without volcanism, except for a few active sectors (Fig. 2.10.).

During the 60's and 70's of the 20th Century, it came to be recognized that the outer shell of the Earth or lithosphere consists of seven large and several small plates. The large plates support both the American continents and the continents of Eurasia, Africa, Australia, and Antarctica, as well as a large part of the Pacific Ocean. The largest of them is the Pacific Plate, composed entirely of oceanic lithosphere. With a thickness ranging from 75 to 125 km, these plates migrate gradually in relation to each other.

Depending on the nature and relative movement of the plates, the boundaries between them can be: 1) *divergent* (boundaries between objects moving apart), when plates separate from each other, building a rift in the process; 2) *convergent* (boundaries of compression), when plates collide with each other or one descends beneath another in a zone of subduction; or 3) *transformational*, when plates slide against each other along a transform fault. The objects subject to the movements indicated are united under the term *tectonic plates*. Based on ideas of A. Wegener about continental displacement, this relatively new geological theory at the end of the 20's of the last century supplanted the *geosynclinal theory* of development of fold mountains, which up to that time had been the most generally accepted model of development of the Earth's crust.

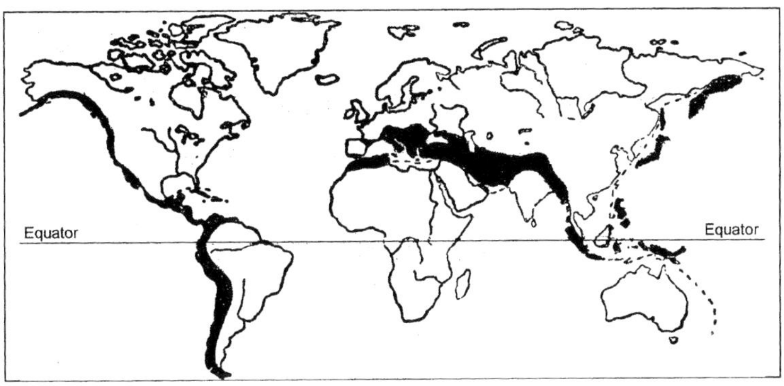

Fig. 2.10. Areas of geosynclines and continental masses during the end of Alpine orogeny (after Shatsky).

Recent Tectonic Movements

The history of the temple of Serapis serves as an excellent illustration of the nature and intensity of fluctuating recent vertical movements. A large number of other examples can be cited from different seacoast localities throughout the world. Precise geodetic measurements and other observations indicate that the rates of positive and negative vertical movements in different regions of the world are different. On the vast territory of the former Soviet Union, for example, the rate of lifting varies from 14.4 to 66 mm/year, whereas the maximum rate of sinking (52 - 65 mm/year) is recorded on the coast of the Black Sea, which for this reason is marked by active processes of abrasion and slipping. In the former East Germany, on the other hand, recent vertical movements of the Earth's crust are almost exclusively confined to sinking, which in certain regions exceptionally attains a rate of 4.0 - 4.5 mm/year, but throughout most of the country occurs no faster than 2.0 mm/year. Generally speaking, the average rate of recent tectonic movements on platforms is usually from 1 to 2 mm/year. The maximum intensities of both vertical and horizontal movements are controlled by faults, a fundamental type of environmental heterogeneity.

Influence of recent vertical tectonic movements is reflected above all in inundation of low - lying zones beside seas and oceans, where coasts have to be defended against encroachment of the sea by construction of dikes, as is done by the inhabitants of Holland. Land must also be defended against landslides and to some extent against flash floods. The indicated influence is realized indirectly, through such more important factors as relief, the level of groundwater, and the level of seas and oceans, which dictate changes of relief. In the case of landslides and material transported by flash floods, the influence in question is realized through the action of seismic factors (since it has been established that seismically active regions are characterized both by increased rates of recent movements and by significant differentiation).

Recent horizontal movements are characterized by faster rates in comparison with vertical movements. Thus, for example, throughout the system of the San Andreas Fault in California and that of the North Anatolian Fault in Turkey, the rate of such movements attains several centimeters a year. Precise geodetic observations have shown that these movements occur unevenly (precipitously) in time and space, and that recent activity in different parts of faults varies significantly.

A number of geodynamic testing grounds have been set up in regions of active rift structures. Repeated geodetic observations on Iceland established that an active rift process with intense horizontal movement (extension) of the order of 7.5 m and vertical movement of 3 m took place along the Krofla deformation zone during the period of 1975 - 1981. It has been established with a precision of up to 1 cm that the inner zone of the Asal Rift in East Africa is subject to the action of current forces of extension (movement) at a rate of 5 cm/year along the rift's axis. This is apparently caused by intrusion of deep - lying solutions into the axial zone of the rift.

Structural Forms

When rocks are subjected to pressure above their limits of elasticity, they start to bend plastically, forming series of folds in the process. Many fold belts tens and hundreds of kilometers wide have been formed in this way.

The main types of structures formed are: *anticlinal* and *synclinal flexures*; and *structural domes*, roughly symmetrical folds that are especially well expressed in regions with salt deposits (Fig. 2.11.).

Cracks and *faults* represent disturbances of the compactness of moveable parts of the Earth's crust. Sliding (movement) parallel to the fault surface occurs in faults, in contrast to cracks. Faults are encountered in all types of rocks, but share a common feature, namely that one wing of the fault has shifted in relation to the other. In some cases, faults are grouped into zones, when movement is distributed over several fault surfaces, and the total jump can comprise not more than 1 mm or attain several kilometers.

The rate at which the wings of faults move can vary greatly from place to place. Movement of the order of up to 12 m and more in only a few minutes has been observed along some faults. Earthquakes almost always occur during such sharp movements of the Earth's crust. On the other hand, movement along the fault surface is slow in many cases. Sometimes the occurrence of a very large total jump indicates that there were hundreds of smaller movements separated by long intervals of dormancy.

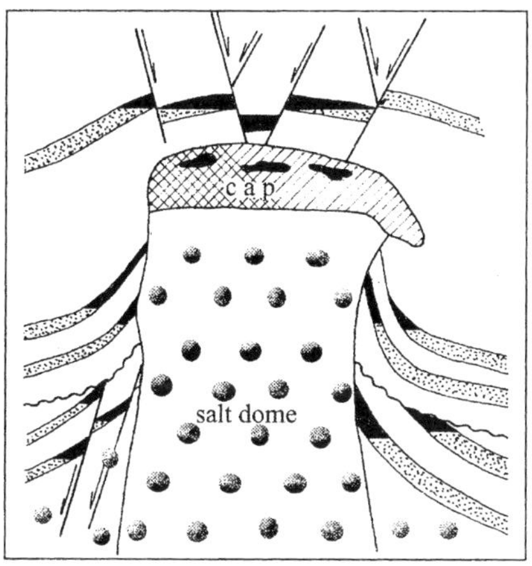

Fig. 2.11. Scheme of salt dome.

Some tectonic deformations are vertically oriented, but faults with an inclined fault surface are prevalent. Characterized by a steep drop, so - called *normal faults* are created in zones of shifting of the Earth's crust. On the other hand, zones of folding are characterized by reverse movement of one wing across the other at a slight angle, in which case *reverse faults* or *tectonic nappes* are formed. For example, the Alpine massif consists of a series of plates that have moved one over the other.

There are also horizontal faults, i.e., faults with horizontal movement of blocks parallel to their extension. One of the most famous examples of such a structural form is the San Andreas transform fault, which is characterized by generation of enormous seismic energy, i.e., it is known for causing the earthquakes of California. Total length of the fault is estimated at 800 km, while the total value of relative movement of one wing against the other from the Miocene to the present day comprises about 240 km, about 1 km in the Quaternary alone.

Sinking of a block between two bordering faults results in the formation of *tectonic trenches* (*grabens*), while uplifting produces *horsts*. Thus, due to sinking of the narrow trench in which they were formed, the Dead Sea and valley of the Jordan River are today below the level of the World Ocean: the surface of the shallow Dead Sea is as much as 430 m below the level of the ocean. Certain lakes were formed in the same way, for example Lake Tanganyika, which is 1,418 m deep and whose bottom is 620 m below sea level. The deepest graben in the world is Lake Baikal in Eastern Siberia, which has a depth of 1,713 m.

It was postulated in the 30's of the last century that there exists a network of faults deep in the Earth's crust, such faults being somewhat later called *deep faults* (A. V. Peive, 1945). Hundreds of kilometers long, these global tectonic breaks erted significant influence on the nature and localization of both tectonic and other deep - seated processes. They are usually manifested in a kilometers - wide zone of rock break - up. Among larger deep structures of this type, apart from the San Andreas Fault, is the Talaso - Fergana Fault in Central Asia, which is about 400 km long.

So - called *transform faults* represent a peculiar type of deep faults. They are structural forms for which an intensively crumbled oceanic crust is not unusual (Fig. 2.12.). Transform faults as a new type of tectonic plate boundary was proposed by J. T. Wilson in 1965. They are most numerous in mid - ocean ridges, 50 large structures existing in the region of the Mid - Atlantic Ridge. They generally extend parallel with each other and perpendicular to the ridge. Major transform faults attain a length of several thousands of kilometers. Regional transform faults are encountered much more often, the distances between them attaining hundreds of kilometers.

Deep faults between two moving plates are called *rift structures* or *rifts*. Basaltic magma pouring out along normal faults forms volcanic domes or volcanic plates (plateaus). Rifts are young complex tectonic structures characterized by great length measured in hundreds and thousands of kilometers and considerable vertical amplitude of movement (of the order of 2 to 5 km). The Afro - Arabian and Rhein Rifts are well - known, as is the Big Basin Rift in the Cordillera of North America. The width of rifts varies: from several tens of kilometers (the Rhein Rift, 40 km) to several hundreds of kilometers (the Big Basin Rift, 600 km). Among the rift belts of continents, the largest is the Afro - Arabian one, about 6,500 km long. Its composition includes deep faults, trenches, and rifts: the Syrio - Lebanon Fault, the Dead Sea, the Ethiopian - Kenyan (East African) Rift, and the Nyasa - Tanganyika Rift.

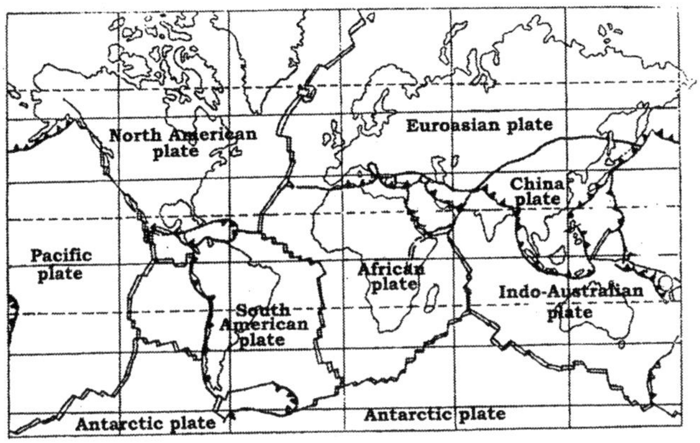

Fig. 2.12. Position of main tectonic plates.

Risks within Fault and Rift Zones

Many rifts, regardless of the depth at which they were formed, extend to the surface (Fig. 2.13.). Two main groups of risks can be present in regions with such structures: 1) *risks arising from the destructive power of faults*; and 2) *risks arising from so - called "gas breathing" of the Earth's crust through such structures*[25].

---

[25] Apart from risks, positive **indirect influence on human health** also exists without question. Above all, faults serve as conduits for circulation of mineral and slightly mineralized waters, as well as for gases of interest from the balneological standpoint (Fig. 2.13.), in spite of the fact that some faults perform a barrier function. Water-bearing faults are marked on the surface by lines of springs and concentrations of water-loving plants. High-quality waters of faults (privileged collectors) in karst regions are evacuated through numerous powerful fountainheads, and many faults in other types of water-bearing environments bring mineral and thermal waters and brines to the surface. For example, the mineral waters and brines of one of the meridional faults of the Lapland Tectonic Seam (within the framework of the Baltic Shield) have (among other interesting components) 288 mg/l of bromine and 73 mg/l of iodine. Fumaroles and hydrothermal phenomena in the Western United States follow well-known faults, and drilling along such structures resulted in the discovery of water vapor and hot water (the temperature of steam and thermal water in Yellowstone National Park at a depth of up to 350 m is more than 240°C, while highly thermal brines with a temperature of 360°C were discovered at a depth of 2,000 m on the southern extension of the San Andreas Fault).

Destructive effects are manifested in the case of active faults, i.e., faults with movement of blocks during the period of human history. Such faults can destroy structures in their path (buildings, roads, or dams), and the underground earthquakes caused by them are registrable at great distances away. As a rule, movement along faults is temporary, occurring after long pauses. For different faults, the periods of dormancy last from several years to several centuries and more rarely can exceed even a thousand years.

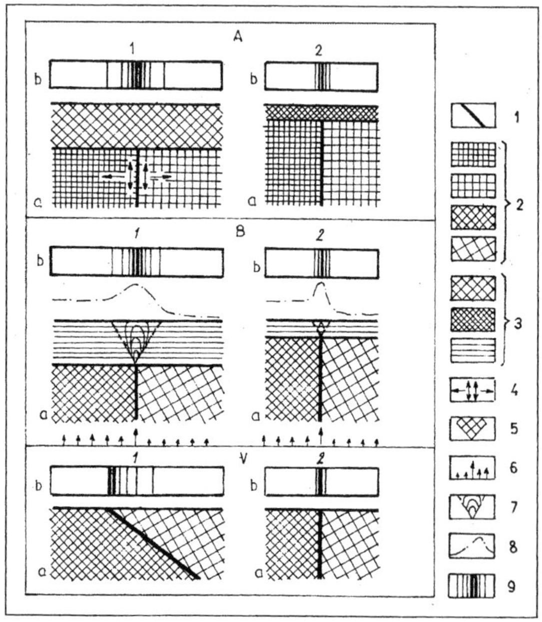

Fig. 2.13. General schematic map of development in the cross section of the Earth crust and their reflections at the surface. A,B - hidden faults (at deeper [1] or shallower levels [2]); V - open faults.

Small movements are incapable of causing earthquakes, but can be accompanied by great damage. Such movements have in recent time occurred along several large faults in California, including the Calaveras, Heyward, and San Andreas Faults (Howard and Remson, 1978).

Unfortunately, major construction is going on even today in zones of movement. For example, tectonic movements along the San Andreas Fault destroyed the concrete walls of a wine cellar, and measurements established an average movement rate of 1.25 cm/year. Discovery of movement along the Heyward Fault becomes especially alarming in the light of the destructive earthquakes that occurred in the zone of the indicated fault in 1836 and 1868 and in view of the fact that it passes through several highly urbanized regions. Although block movement itself can with difficulty cause physical damage to people, they can be hurt during loosening and destruction of buildings.

Accompanied by heavy human casualties and material damage, noticeably greater movements of rock masses along faults occur during earthquakes. The extent of such movements will be discussed in the appropriate chapter.

Zones of deep - seated and superficial faults, rift structures, and volcanoes are marked by *gas breathing* of the Earth's crust due to migration of gases along these structures from the depths of the Earth to the surface (Fig. 2.14.). Filtration of helium, argon, mercury vapor, CO_2, and other gases occurs in zones of deep faults[26] (Fig. 2.15.). After the Tashkent earthquake of 1966, increased content of radon, fluorine, uranium, nitrogen, and carbon dioxide (in addition to helium) was recorded in thermal waters of boreholes located in the zone of deep faults.

Content of mercury vapor can also be elevated on sectors with tectonic activity of faults, even in cases where structures are covered by young formations to depths of up to 1.5 km. Thus, for example, anomalous concentrations of mercury vapor were measured in the snow along a fault structure over the Srednje - Vetoubanski gas field in the former Yakut SSR. Regional investigations in Central Asia, Kazakhstan, and the European part of Russia have shown that mercury anomalies 20 to 100 km wide are formed in air of the soil cover over zones of deep faults (A. I. Perelman, 1979).

Broadly speaking, the underground atmosphere is rich in CO_2, H_2, CH_4 and heavy hydrocarbons, and N_2. The high capacity of gases for migration dictates their filtration and diffusion far beyond the boundaries of some gas field. The gas aureoles that are formed imitate the disposition of fault structures. For example, this is the case with aureoles of radon, a gas that is inert but very dangerous to man. Formed as a product of the radioactive decay of U_{238} in regions with uranium deposits, radon is especially present along fault zones and in adjacent air of the soil cover. It is also the case with aureoles of mercury in regions with mercury deposits or deposits and surrounding soil containing that very harmful element. Similar phenomena are encountered above sulfide deposits, as well as over coal and oil fields.

Radon content can be high in zones of faults and cracks. In certain waters above deep tectonic deformations in the Caucasus, Ural, and Tien - Shan Mountains, it attains 36,000 emans. Anomalous zones of nitrogen - containing waters in Trans - Caucasia, on the Kamchatka Peninsula, in

---

[26] We note the Jasenica Fault in the region of Smederevska Palanka, along which carbon dioxide gas migrates from a depth of the order 1,000 m and enriches shallower water-bearing horizons. The well-known "Karadjordje" mineral water here is drawn from the alluvial horizon.

the Altai Mountains, and in other regions of the world are likewise characteristic of tectonically deformed regions, where rare gases (Ar, Kr, , and He and Ne in somewhat smaller amounts) are also present.

In any event, tectonically disturbed zones can ert significant influence on mortality, especially on the incidence of cancer (of the bronchi and lungs). The SE index (Table 1.1.) directly affects mortality from disturbed circulation of blood and (especially) cardiac ischemia.

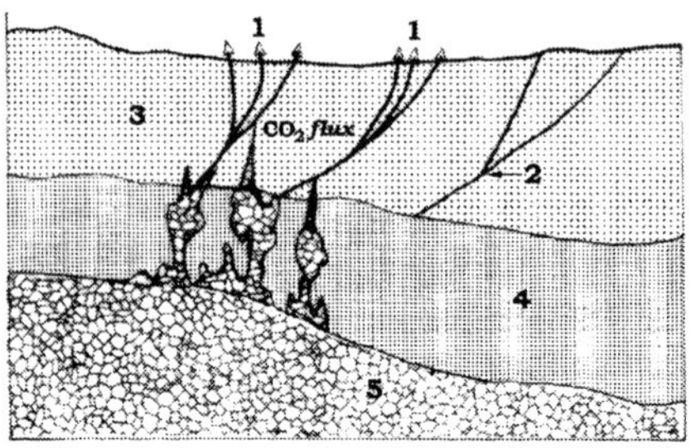

Fig. 2.14. Cross - section of a terrain rich in deep faults within the Earth's crust. 1. occurrence of CO_2 at the surface; 2. deep fault; 3. Betic detached crust; 4. Lower crust; 5. Upper mantle.

Formation of *thermal metal - bearing brines* is linked with rift structures. The most interesting ones are those from the Red Sea - East African rift zone and from the Imperial Valley around the Salton Sea in Southern California. Brines of the Atlantis II, Chain, and Discovery depressions within the Red Sea rift zone have been studied more than others. They are sodium - chloride brines with mineralization of about 255 g/l and chlorine content of about 155 g/l. The ore metals of brines are represented by sulfides, whose reserves in the upper layer measuring 10 m in thickness are estimated at 83 million tons. Also interesting are thermal springs on the shores of Lakes Magadi and Natron in the southern part of the East African Rift. On the shores of the former, more than 20 springs have been registered giving water with a temperature of 35 - 82°C and mineralization of 15 - 30 g/l, this water being rich in fluorine (100 - 162 mg/l), bromine (70 - 150 mg/l), and boron (up to 10 mg/l). A borehole 1,570 m deep located on the southern branch of the San Andreas Fault (in the vicinity of the Salton Sea) struck brines with a temperature of 270°C, mineralization of 332 g/l, and high content of a number of elements. Such high concentrations of potassium, lithium, silver, copper, and so many other elements had not been registered in natural waters before the middle of the 1970's. Later, on the southern part of the Turan Plate (east of the Caspian Sea), chloride brines were discovered along the Amu - Darya Fault and other structures that had mineralization of up to 350 - 370 g/kg and high content (mg/kg) of zinc (up to 80 - 100), lead (up to 30 - 80), iron (up to 2,000 - 2,500), and strontium (up to 1,000 - 4,000).

The negative and in some cases positive (from the balneological and industrial standpoints) effects of the given processes, gases, and waters on health of inhabitants of the regions where they occur unfortunately have not to date attracted the attention they deserve among scientists.

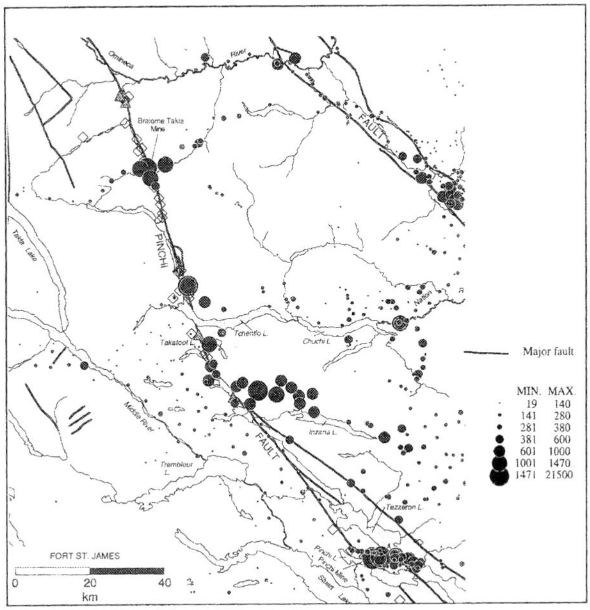

Fig. 2.15. Mercury concentrations in the clay - sided fraction of till of central British Columbia; modified from Plouffe (1995b). Sites with high mercury concentrations in trees from Warren and others (1983a).

Geophysical Factors

Recent years have witnessed a sudden growth of interest in research on geodynamics, i.e., study of deep - seated processes as causes of recent structural differentiation and contemporaneous heterogeneity of the lithosphere. The results of such research can be useful in resolving important practical problems, namely prediction of seismic and volcanic activity, investigation and utilization of mineral resources of the Earth's crust, and study of the effects of human activity on the environment.

Accelerated development of modern geodynamics has contributed especially in the following ways: *in the first place*, important scientific results have been obtained in measuring vertical movements of the Earth's crust on large territories; and *secondly*, recent movements of the Earth's crust have been found to be closely linked with structural differentiation of the lithosphere and the distribution of geophysical fields (magnetic, geomagnetic, thermal, etc.). Among other things, it has been demonstrated that the *problem of instability of geophysical fields manifested with the passage of time* takes on great significance in efforts to accomplish a wide range of scientific and practical tasks. Already, a great many reiterated high - precision gravimetric observations have been carried out in different regions where significant changes of gravitational force occurring with the passage of time have been shown to be linked with endogenous processes. For example, repeated measurements during the period 1974 - 1985 on the Izu Peninsula in Japan established changes of gravitational force of up to 40 mGal and 60 mGal in three years in the zone of an active fault. Magnetometric measurements repeated in the course of time in the Urals and other localities showed that zones of greatest field changes coincide with the disposition of active faults.

Long - term investigation of changes in geophysical fields has been carried out in seismically active regions to discover processes involved in preparation for a phase of earthquakes. Thus, in one region in California, a decrease of gravitational force by 50 mGal was established before the earthquake of 1978 (with a magnitude of 5.6 M). In Romania during the period 1977 - 1979, anomalous changes of gravitational force before several earthquakes ranging in magnitude from 2 to 4 M comprised 20 - 200 mGal at a distance of the order of 100 km from the epicenter. Similar results were obtained in measuring changes of the geomagnetic field.

Let us now direct our attention to electromagnetic, gravitational, and thermal fields (radioactive fields will be treated in a special chapter).

Magnetic, Electromagnetic, and Electric Fields

Earth, Jupiter, the Sun, i.e., the entire Solar System and Galaxy as a whole, all have magnetic fields. These fields play a significant role in many processes in that part of the Cosmos accessible to us for research, erting powerful influence on the movement of compacted particles. Earth is characterized by a powerful magnetic field that is believed to be generated by movement of material in the outer part of the core. In this way it acts a gigantic magnet. In rough approximation, the planet's magnetic field can be represented as a magnetic dipole with a momentum of 8.1×10^{25} $gs{\cdot}cm^3$ at a distance of about 340 km from the center of the Earth, the axis of the dipole cutting its surface in points known as geomagnetic dipoles.

Under the influence of terrestrial magnetism, the needle of a compass takes a position parallel with the geomagnetic field and points to magnetic poles of the Earth. The geomagnetic poles miss coinciding with the geographic poles by about 11.5° (Fig. 2.16.), so it is understandable that the needle of a compass does not indicate true north or south and that there exists a *line of zero deviation* or *magnetic declination*. Another characteristic of the geomagnetic field is that there exists an *angle of inclination*, which is assumed by a freely hung magnetic needle level with the horizon.

The old magnetization of rocks (i.e., the natural lag in expression of the magnetization of rocks) indicates position of the magnetic field at the time of formation of a given rock. It is therefore possible to establish the path of shifting of poles in the course of geological time by using many samples of rocks of different ages. The average rate of movement is 3 cm/year.

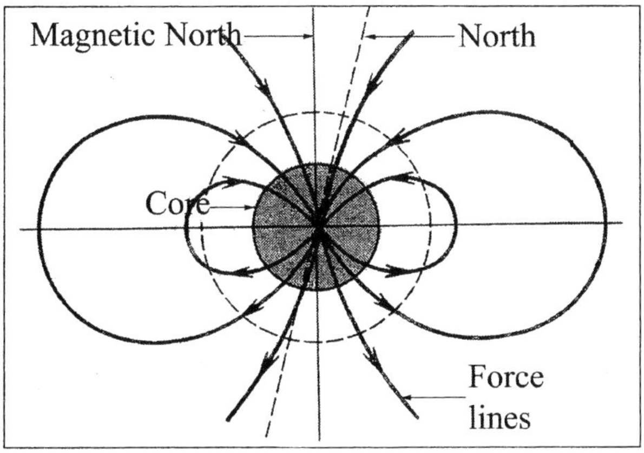

Fig. 2.16. Position of magnetic and geographic poles of the Earth.

During cooling in the Earth's magnetic field, magmatic rocks containing magnetite were magnetized in the direction of the general magnetic field prevailing at the time. In the course of erosion of such rocks, minute magnetized particles pass through the cycle of transport and sedimentation. During sedimentation, magnetic grains are oriented along the geomagnetic field. In this way, the formed sedimentary rock acquires natural old magnetization, from which it is possible to determine the time of that rock's formation. On the other hand, some magnetized rocks, tectonically brought out of their original position under influence of the geomagnetic field, take on new magnetization (*remnant magnetization*).

Since different rocks have different abilities to accept induced magnetism, stronger or weaker isolated magnetic fields are created in upper parts of the lithosphere. *Geomagnetic anomalies*, i.e., local deviations from regional average magnetic field values, are formed in this way. Such deviations can be considerable. For example, the well - known *Kursk Anomaly* in Russia has intensity of the order of 10,000 nT (Nikitin and Novikov, 1986). Of magnitude unique in the world is the *Zlot Magnetic Anomaly* near Bor (15,000 to 40,000 nT), whose main cause is still not known with certainty. Generally speaking, extensive masses of basalt and diabase, as well as granite massifs, can act as powerful magnets (*positive anomalies*), whereas fold regions with thick masses of sediments usually behave as weakly magnetized bodies (*negative anomalies*).

Differences in the magnetization of individual minerals or rocks are used to discover deposits of mineral raw materials, and geomagnetic (geophysical) methods are based on them. It goes without saying that geomagnetic methods give the best results in investigating magnetite deposits. On the other hand, investigation of the migrations of poles of the main magnetic field throughout the geological past (study of the magnetic field of past geological epochs or paleomagnetism) enables us to reconstruct the geometry of tectonic plate movement. Thus, for example, it is known today that during the Pleistocene, more precisely in a period of a million years, the North Atlantic between Europe and America widened by 23 - 25 km, Africa and South America moved 30 - 40 km away from each other, etc.

Natural electric fields are a consequence of geological action (telluric currents and self - polarization) or atmospheric electricity linked directly with ionization of the air.

Gravitational Field of the Earth

Gravitation represents the ability of bodies to attract each other. The force of gravity or gravitation of the Earth depends on its mass and decreases with removal from the Earth's surface to become nonexistent in the remote atmosphere. On the surface of the Planet, gravitation is stronger at the poles than in the region of the equator, due to the action of rotation. Moreover, due to the greater density of rock, it is stronger in oceanic zones than in continental regions.

Experimental study of gravitational force was initiated by Galileo, who already in 1590 conducted experiments with freefall of bodies. In 1673 Huygens established that gravitational force can be determined by measuring the period of a pendulum. The first indicators of change in gravitational force with latitude were discerned by J. Richer. The correct interpretation of this fact was given in 1687 by Isaak Newton, who formulated the law of universal gravitational attraction and determined volume of the Earth.

A deficit or surplus of mass in the point of measuring noticeably affects the value of gravitational force. A *negative gravity anomaly* is formed if rock masses have below normal density, whereas a *positive gravity anomaly* is formed if their density is above normal. Great *positive gravity anomalies* are recorded in South America, Europe, in the southeastern part of the Indian Ocean, and in the Western Pacific. On the other hand, great *negative gravity anomalies* occur in the northwestern and central parts of the Atlantic, in the eastern part of the Indian Ocean, and in the South and North Pacific. On the territory of Yugoslavia, the lowest values (110 mGal) are registered in the region of Podgorica in the framework of the Adriatic Coastal Anomaly, which runs in a SE - NW direction, whereas values generally increase moving toward the northeast of the country, the highest values (about 18 mGal) occurring in the region of Fruska Gora.

With the passage of time, changes can occur in the value of gravitational force in some examined point. The cause of this lies in geological and geophysical processes in the Earth's interior that lead to redistribution of mass. Also, changes in position of the Sun, Moon, and other celestial bodies affect the values of gravitational force. The given fluctuations can be divided into periodic fluctuations, which are linked with rotation of the Earth in relation to the Sun, Moon, and other celestial bodies; and non - periodic (time - linked) fluctuations, which are caused by geological and geophysical processes in the Earth's interior. As will be seen later, the dynamics of gravitational field changes can have significant effects on the environment.

Thermal Fields

People have observed for a long time that there exist on the surface of the Earth various phenomena possessing heat that originates from its interior. Around 355 B.C., the Greek philosopher Plato, in lecturing on the legendary Atlantis, mentioned that there were hot springs in the center of the city. It follows that more than 3,500 years ago, man already knew about the internal heat of the Planet.

Today it is estimated that the Earth gives off about 10^{21} J of heat annually. Relatively speaking, that is not a great amount, in view of the fact that our planet receives 6,000 times more energy from the Sun. On the other hand, it is still quite a lot, considering that annual production of electrical energy in the world is measured by a value of the order of 2×10^{19} J. Heat from the Earth's interior is today attracting increasing attention as a very promising energy resource that to date has been but little harnessed. Geothermal heat unquestionably represents the greatest source of energy at man's disposal at the present time.

A large part of the Earth's heat arises under the influence of *internal (planetary) sources*, above all at the expense of heat liberated during radioactive decay of uranium, thorium, and one of the isotopes of potassium. Another part originates from deep - seated zones where the temperature of rocks is in many places close to the melting point. Solar radiation is the most important *external (cosmic) heat source.*

In the Earth's crust, heat is for the most part transmitted in two ways: 1) by means of *conductive transmission*, i.e., heat transmission directly from particle to particle; and 2) by means of *convective transmission*, i.e., heat transmission by groundwater. The thermal conductivity of rocks represents the main mechanism of heat redistribution. This mechanism is expressed by the general rule that temperature increases with depth under a neutral layer. But owing to its practical significance, the phenomenon of heat conveyance from the Earth's interior by groundwater has attracted greater attention of investigators around the world. The given material is treated in more detail in the book "**Hydrogeological Explorations. Applied Hydrogeology, III**" (M. Komatina, 1995).

A natural thermal field is largely determined by the thermal properties of rock masses. For example, it is known that the thermal conductivity of sedimentary rocks is on average less than that of magmatic and metamorphic rocks, being greatest in peridotites and granites among magmatic rocks; and in compact carbonates, quartzes, rock salt, etc., among sedimentary rocks. Thermal fields can be disturbed by various local natural factors such as the exposure of slopes, form of relief and coloration of land surfaces, nature and thickness of the snow cover and vegetation, etc., or under the influence of human activity (building of reservoirs and irrigation systems, cutting down and planting of forests, etc.).

The average depth of a temperature increase of 1°C comprises 33 m. That relationship of temperature and depth is known as the *geothermal degree*. It can be greater or less than 33 m for 1°C, depending on the composition and structure of a region. For example, faster temperature increase is characteristic of volcanic regions, but appreciably slower increase is recorded in regions of continental platforms.

Of all characteristics of a thermal field, the *thermal flux* gives the most complete information about the energy regime of the Earth and separate sectors of the Earth's crust. This parameter is of decisive significance in estimating the geothermal energy potential of a studied region, and zones with a thermal flux greater than 100 mWm^{-2} are considered interesting for exploitation of ground heat.

It has been estimated that the amount of heat that rises from the interior to the surface of the Earth on average comprises 59 mWm^{-2} (about 77 mWm^{-2} according to A. M. Jessop et al., 1976). Regional fields with a low thermal flux (30 - 40 mWm^{-2}) are predominantly typical of old shields and platforms. On the other hand, zones surrounding younger fold mountains are characterized by high values of this parameter - from 100 to 600 mWm^{-2}. Anomalous values are associated with regions of current volcanism and to some extent with younger tectonic faults.

Geophysical Fields - Risks and Positive Effects

The past decade in the world of geophysics will be remembered for strengthening of a new branch of this applied science, namely

ecogeophysics (S. Komatina, 1996). Thus, instead of a depth range of several kilometers, geophysical methods of investigation are concerned with depths of only a few meters (or twoscore at most) below the surface of terrains. The questions resolved in this way can be represented in the form of two spheres of interest:

1. Investigation of natural and artificial physical fields arising in the Earth, as well as their action on the biosphere and living organisms (man, animals, microorganisms, and plants); and
2. Possibilities of quantitative determination and qualitative estimation of phenomena erting primary or secondary influence on the environment and its protection (D. Vogelsang, 1995).

In the region of the biosphere, interaction between physical (geophysical) fields and living organisms occurs due to:

- *Penetration of the organism by fields and their spreading throughout it;*
- *Primary interaction between fields and the organism; and*
- *Secondary response of the organism as a consequence of primary interaction.*

According to behavior of the organism, interaction can be of the following kinds:

1. **active** (when fields ert biological influence on the organism):
   - *Positive (stimulatory) influence,* where activity of the organism increases; and
   - *Negative (inhibitory) influence,* where activity of the organism decreases.

Inhibitory interaction can be reversible, in which case a return to the normal state is possible; and irreversible, when permanent damage is done to vital functions of the organism; and

2. **Passive** (when changes in the nature of a physical field are registered, but the organism itself is not affected by their influence).

I. **Influence of Natural Electromagnetic Fields.** Natural electromagnetic fields can be divided into the following categories:

1. *Fields of extraterrestrial origin.* In comparison with natural terrestrial or artificial fields, insignificant influence is erted on living organisms by radiation in the ionizing part of the electromagnetic spectrum (gamma-, X-, or certain parts of corpuscular radiation) and its no ionizing part (ultraviolet, visible, and infrared radiation, as well as radio waves);
2. *Fields of terrestrial origin.* A clear link is discernible between high - frequency electromagnetic radiation and weather fronts, mortality, births, traffic accidents, and industrial accidents.

Natural electric fields arise from geologically induced fields (telluric currents and self - polarization) or from atmospheric electricity directly linked with ionization of the air. A natural electric field arises due to the potential difference between positively charged ground and electricity produced by meteorological phenomena (130 V/m to maximum values of 10 kV/m due to lightning discharge). In the course of breathing, man daily inhales about 20,000 l of air and thereby introduces ions with electricity of the order of 10^{-14} A into the lungs. It has been established that dominance of negative ions affects organisms favorably, whereas positive ions cause fatigue, headaches, and indisposition.

Natural magnetic fields are formed wherever movement of charged particles occurs. The area of pathological reactions (and possibly diseases?) linked with fluctuations of terrestrial magnetism has been insufficiently studied. It is certain that geomagnetic storms can ert definite influence on living organisms. We note that the first symposium on *magnetobiology* (chaired by I. V. Toropcev) examined phenomena arising at the molecular level under the influence of magnetic fields. Magnetic fields can ert positive (favorable) and negative (unfavorable) influence on organisms, depending on their intensity, frequency, orientation, exposure time, and origin. Unfavorable influence is primarily manifested as disturbances in the central nervous system, glands, and sensory perception, but also in the respiratory, immune, and bone systems. Positive influence is manifested as magnetotherapeutic effects in treatment of certain diseases of the skin and nervous system. They are divided into the following classes:

1. *Heliogeophysical effects* (effects of magnetic storms and solar activity, the distribution of whose fields is variable in space and time within the range of 0.01 - 0.05 mT) have been insufficiently studied. However, it is known that the highest mortality occurs two days after maximum intensity of magnetic storms and solar activity. These fields also affect the number of traffic accidents, incidence of diphtheria, and worker absenteeism due to sickness (all of which increase under their influence); and
2. *Geomagnetic effects* are manifested in the following examples: it is known that the geomagnetic field is of assistance in the orientation of sharks and migratory birds (pigeons maintain their direction of flight with a deviation of up to 0.3 deg). Birds sense changes in magnetic field intensity of the order of 1 nT, while bees sense changes of the order of about 5 nT. In man, negative influence of geomagnetic storms is manifested in weakness of the heart and cerebral system (Miroshnichenko, 1987) (Fig. 2.17.). Also, geobiologists have established that man is sensitive to the Earth's magnetic field, and even to insignificant changes of it (like messenger pigeons). According to Rocquart, man became more insensitive and even adapted to greater field changes since he began to use iron. Nevertheless, it is known

that the body's ideal position is with the head pointing north and the feet pointing south.

The indicated factors together make up the constant magnetic component of the environment.

The Kursk Anomaly is the only geomagnetic phenomenon of geological origin proven to ert influence on human health. Nikitin and Novikov (1986) reported the registration of considerable influence of this anomaly on mortality of inhabitants of the Kursk Province. However, of much greater significance from the biological aspect are sudden changes of the geomagnetic field - pulsations and micropulsations with frequencies of 0.1 - 10 Hz. Negative influence of pulsations has been established in laboratory experiments, so their significance cannot be ignored. The situation is similar with periodic changes in the polarity of a natural magnetic field. The frequency of inversions in the last 10 - 20 million years is about every 700,000 years, which from the geological aspect is a relatively short period. Still, it is impossible to reject assertions that influence is erted on the genetic system in the sense of mutations and hereditary changes (Mares, 1997).

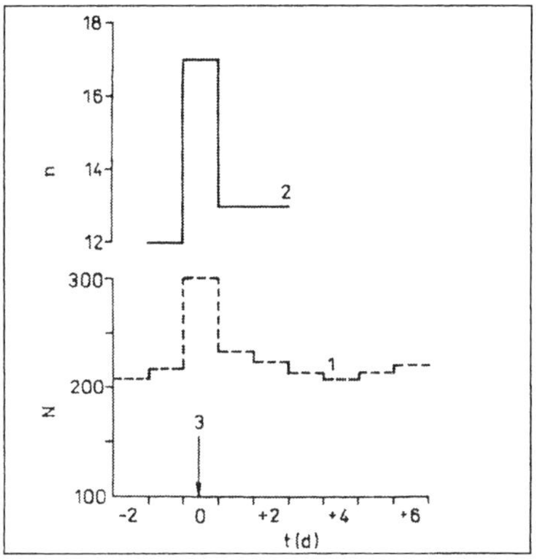

Fig. 2.17. Geomagnetic storms and heart - cerebral diseases. 1. mortality N on heart attack during 1960 to 1966 in Sverdlovsk (former Soviet Union); 2. daily averages n of morbidity on heart attack and on cerebral failures in Vilnius (former Soviet Union) during 1963 - 1966; 3. beginning of geomagnetic storms. t (d) - days before (-) and after (+) the storm started (after Miroshnichenko, 1987).

Above tectonically disturbed subvertical zones characterized by conductivity greater than that of the surrounding environment, a vertical air zone can appear with conductivity twice as great as the value for air in the immediate surroundings. It is therefore believed that both conductivity of the disturbed zone in the crust and that of air in the atmosphere simultaneously affect the people who live above such zones. Gruntorad and Mazac (1995) demonstrated that anomalous processes above such zones are a result of processes transpiring in the interior of the Earth. They also established that microclimatic conditions above these zones are different, not only in the geophysical respect, but in the chemical and physicochemical sense as well. Statistical processing of geologic - geophysical and socio - ecological factors and comparison with data on mortality in 56 municipalities in the Czech Republic (Table 1.1.) showed that:

- Tectonic and disturbed zones (TZ) ert significant influence on mortality, especially mortality due to increased frequency of cancer (primarily of the bronchi and lungs). The SE index also directly affects mortality from circulatory disturbances (cardiac ischemia in particular);
- No special correlation has been established between regional concentration of radon (RN) or radioactive field intensity (RA) and mortality. The same is true of the relationship between gravitational and magnetic field intensity and mortality; and
- Influence of TZ on mortality (in %) varies between 8.0 and 23.4%;

while that of SE is in the range of 14.1 - 23.8%, which corroborates the assertion of Gruntorad and Mazac (1995) that physiochemical elements from tectonic zones can be one of the reasons for increased mortality in some regions.

In the past ten years, a link has been detected between traffic accidents and the action of electromagnetic fields on man. For natural electromagnetic fields of extraterrestrial origin, a correlation has been established between magnetic field changes (the K index) and the frequency of traffic accidents. In regard to natural fields of terrestrial origin, it is postulated that the variable magnetic field with frequency of up to 10 Hz that is formed during driving at normal speed over a highway whose foundation contains a variety of rocks with different magnetic characteristics affects Alfa rhythms of the brain. A traffic accident is the consequence of this negative influence on the psycho - physiological state of the driver (Dohnal and Vacek, 1983). According to these authors, the occurrence of traffic accidents is affected not only by meteorological factors, but by electromagnetic ones as well. Thus, the number of accidents increases by 4 - 5% on days when an increase of the K index is noticed. Also, accidents are 4% more frequent during periods when pulsations of the geomagnetic field's horizontal component last more than eight hours for three days on end. In the event that pulsations lasted more than 16 hours over the next nine days, the frequency of accidents increased by 5%.

Natural ionizing radiation arises due to ionization of the environment during electromagnetic or corpuscular radiation (Dj. Sofrenovic, 1993). Ionizing rays have exceptionally great penetrability and chemical action, causing luminescence and ionization of gases and erting biological action on living organisms. Specific irritation of tissues occurs at insignificant intensities of radiation, while high intensities result in damage of the most severe kind to tissues. It has been established as a regularity that species of organisms at the lowest stage of development are most unresisting to radiation (a dose of 300 rads is enough to cause death of more than half of irradiated animals after a month's time, whereas for man that dose is 500 rads), and that local irradiation is easier to withstand than when the whole body is exposed. Moreover, not all tissues are equally sensitive: most sensitive are tissues where cells multiply rapidly, i.e., tissues with high metabolism. Inside the cell itself, the nucleus, cytoplasm, and cell membrane are more sensitive to radiation. In mammals the gonads, bone marrow, lymphoreticular tissue, and intestinal epithelium are most liable to damage, while the skin, kidneys, liver, and nervous and muscular tissue are exceptionally resistant. Lymphopenia (disturbed antibody formation) occurs due to radiation in lympathic tissue; agranulocytosis, anemia, and thrombocytopenia with a tendency to bleed, are recorded in bone marrow; desquamation of the epithelium with subsequent inflammation occurs in the stomach and intestines; etc.

Natural no ionizing radiation encompasses the segment of the spectrum of electromagnetic radiation that does not have photon energy sufficient to cause ionization in living tissue (12.4 eV) (B. Vulevic, 2000). According to the range of frequencies, i.e., wavelengths, no ionizing radiation in the integral electromagnetic spectrum is divided into the following categories: ultraviolet radiation (UV); visible light; infrared radiation (IR); radiofrequent radiation (RF); and electromagnetic fields of extremely low frequencies (ELF) (Table 2.5.). Laser radiation represents part of the spectrum of no ionizing radiation (UV, visible light, IR) that is manifested in the form of coherent bundles of radiation. This type of radiation also includes mechanical ultrasonic vibrations, i.e., ultrasound, since its biological effects and ways organisms protect themselves from it are similar to those observed in the case of electromagnetic radiation.

The *natural source of UV radiation* is the Sun, whose exceptionally broad continuous spectrum is in large measure absorbed in the atmosphere (especially in the ozone layer of the stratosphere), so that only wavelengths greater than 290 nm reach the Earth. People who spend a large part of their working day in the open - farmers, construction workers, sailors, fishermen, etc. - are especially exposed to UV radiation. Since the penetrating power of this radiation is limited, the most pronounced consequences of intense exposure to UV radiation are evident on the skin (mnemelanomic skin cancer, malignant melanoma of the skin, accelerated aging, enhanced pigmentation, freckles, etc.) and eyes (acute photokeretitis or snow blindness).

TABLE 2.5
Basic classification of electromagnetic spectrum (1 THz = 10^{12}Hz, 1peV = 10^{-12}eV) (B.Vulevic, 2000).

Type of radiation	Frequency (f)	Wavelength (λ)	Photon energy (E)
Ionizing radiation	> 3000 THz	< 100 nm	>12.40 eV
UV radiation	3000 - 750 THz	100 - 400 nm	12.40 - 3.10 eV
Light	750 - 385 THz	400 - 780 nm	3.10 - 1.59 eV
IR radiation	385 - 0.3 THz	0.78 - 1000 mm	1590 - 1.24 meV
RF radiation	300 GHz - 3 kHz	1 mm - 100 km	1.24 meV - 12.4 peV
ELF fields	< 0.3 kHz	> 1000 km	< 1.24 peV

The Sun is also the strongest *natural source of IR radiation.* Thanks to the presence of water vapor in the atmosphere, radiation with wavelengths above 1 μm does not reach the Earth. Workers who spend the greater part of their time outdoors are most exposed to this kind of radiation also. Due to its limited penetrating power, the consequences of exposure to IR radiation are similar to those of the previous kind. Like UV, IR radiation causes heat damage and can lead to sunstroke. Erythematose changes occur on the skin, together with appearance of unpigmented parts in the form of a network, atrophic changes, etc. Inflammation of the eyelids occurs in the eyes, in addition to conjunctivitis, cataracts, damage to the cornea, etc.

The *sources of RF radiation* are the Sun and other bodies in space.

Natural sources of ELF fields are natural electric (stationary and variable) fields and natural magnetic fields (internal and external).

II. Influence of Natural Radioactive Fields. Apart from electromagnetic fields, the fields that have been studied most thoroughly in regard to harmful action on man and other living organisms are radioactive fields.

Natural radioactive fields can be of extraterrestrial and terrestrial origin. Inasmuch as the intensity of cosmic radiation is insignificant and more or less constant, living organisms are used to its presence (even at an altitude of 10 km, at which passenger planes fly) (Table 2.6.). Under conditions of sunspot appearance, these values increase as much as 100 - fold. However, the most dangerous source of radiation for man and all other organisms is radon, specificically the products of its decay. Already in 1901, Elster and Geitel noticed that the elctroconductivity of air in caves and cellars is significantly higher than in the free atmosphere. They attributed this to the presence of natural radioactive gases - radon (Rn_{222}) and thoron (Rn_{220}) - which is products of the uranium or thorium series (I. Petrovic, 1998). The greatest concentrations of radon are registered in the part of the atmosphere next to the ground, and radon is liberated in high concentrations from uranium - containing soils and rocks: granites, clays, and phosphates.

TABLE 2.6.
Radon ratio received by humans (after Barnet et al., 1992).

Source of radiation	Quantity
Radon	55%
Natural radiation - cosmic and gamma radiation	26%
Medical sources	18%
Others	1%
Total radiation	100%

The main sources of radon are air originating from the ground, above all from U_{238} in rocks, where its content is highest in volcanic rocks, lower in metamorphic rocks, and lowest of all in sedimentary rocks (Fig. 2.18., Table 2.7.); groundwater, where it is present in insignificant amounts; and building materials, where its content is highest in concrete, lower in bricks, and almost nil in wood (Mares, 1997.).

Migration of radon, and consequently its influence on man, is dictated by permeability of soils and rocks, as well as by tectonic disturbances in rocks. Permeable rocks like sand and sandstone are much more suitable for migration than are less permeable ones (clays, shales, etc.). Also, tectonically disturbed rocks, as places suitable for occurrence of secondary minerals of uranium, are defined as zones favorable for migration of radon.

According to Matousek (1987), natural sources of radiation can from the biological aspect be considered passive in relation to active natural sources arising in every living organism (K_{44}, C_{14}, Ra and products of its decay, and Th). The muscles, skeleton, and skin of a man weighing 70 kg contain about 0.0166 g of K_{44}. The presence of the element Ca_{14} in the organism is a product of cosmic radiation, whereas Ra and Th enter the organism through breathing.

III. Influence of Natural Seismic Fields. Seismic fields are divided into natural (of terrestrial origin) and artificial categories. Natural fields are represented by the phenomenon of earthquakes. The probability and frequency of earthquakes can be estimated on the basis of magnitude M (Table 2.8.).

The number of earthquake casualties throughout the world during the period of 1970 - 1981 is known to have comprised 442,000, while material damage is estimated as having been at least 18.6 billion dollars. It is therefore not surprising that great efforts are being made to develop methods for predicting earthquakes, particularly in seismically active regions. Although certain animals (fish, snakes, cats) are capable of sensing an earthquake in advance, it still cannot be proved that some connection exists between changes in geophysical and physical fields and the occurrence of earthquakes. However, it must also be kept in mind that extraterrestrial natural fields affect the phenomenon of seismic activity (Fig. 2.19.).

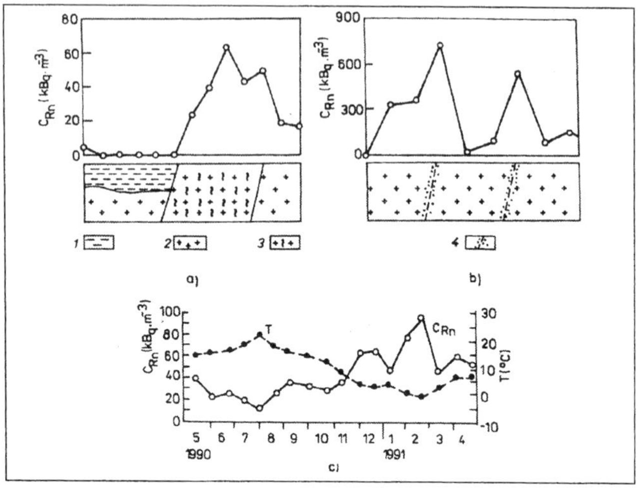

Fig. 2.18. Radon migration vs geology. a) Impermeable clayey cover decreasing the volume radon activity. b) increase of volume radon activity above uranium mineralization along tectonic disturbance; 1 - clay cover, 2 - granite, 3 - disturbed granite, 4 - tectonic disturbance with radon content. c) Seasonal variations of radon activity as a function of atmosphere temperature (after Barnet et al., 1992).

Endogenous Geological Processes

The regions in which we grew up and solid ground we live on give an almost permanent impression of exceptional stability and invariability. This stability creates the deceptive impression of constancy. In contrast to it, however, a number of occurrences in many regions of the world - often sudden and explosive, sometimes catastrophic for the population - indicate evident instability of the ground below us. Moreover, there are many areas where severe earthquakes devastate whole regions, where eruptions of huge volcanoes threaten many settlements, or where coastal regions sink so rapidly that the land must be defended by embankments from incursions of the sea.

All of these endogenous processes are linked with events in deeper parts of the endogeosphere. Very significant for life on our planet are sudden manifestations (earthquakes, volcanic eruptions), since defense against their unwanted action is a matter of great urgency. This does not lessen the importance of slow tectonic processes (neotectonic movements), which represent generators of the indicated sudden manifestations and are an important factor in the genesis of relief.

TABLE 2.7.
Radon sources in the geological environment (after Barnet et al., 1992).

Rocks	Radon content (kBq x m^{-3})		Ratio of fine-grained fraction F (0.063mm)
	Maximum	Average	
Durbachite	435	100	26
Granite	352	52	29
Diorite	126	40	32
Orthogneiss	155	30	35
Phonolite	59	25	38
Phonolite	126	24	41
Shale	61	23	44
Granulite	50	22	47
Terrace	183	21	50
Loess	75	18	53
Metasediments	106	17	56
Sandstone	73	16	59
Neovolcanite	162	15	62
Claystone	55	14	65

Earthquakes

In 1948 near Hollister in California, a vine grower built himself a new wine storehouse reinforced with concrete. After a few years, the walls on opposite sides of the storehouse started to shake. The cracks grew wider with the passage of time, and the concrete slabs on which the floor was constructed began to move. In fact, the building was so situated that one part of it lay on one side of the famous San Andreas Fault, the other part on the other side of it. Because no earthquake strong enough to destroy the building occurred in those years, the opposite sides of the fault slowly moved away from each other at an average rate of 1.2 cm per year. The vine grower's wonder was great because he didn't know that the Earth is never absolutely at rest. Far greater catastrophes than this have certainly been caused by powerful earthquakes, spreading indescribable terror among people with their appearance.

TABLE 2.8.
Frequency of earthquakes (after Mazac, ed., 1988).

Magnitude M	Number of earthquakes yearly	
3	> 100 000	Weak earthquakes
4	15 000	
5	3 000	
6	100	Intense earthquakes
7	20	
8	2	

In his book "**Earthquakes**," G. A. Eiby (1980) describes the atmosphere in London during an earthquake that occurred in February of 1750. "*Unlike the inhabitants of New Zealand or Japan, most Englishmen had up to then never experienced earthquakes, and the unrest was therefore considerable. This unrest increased following a second shock, which occurred four weeks later. After John Wesley declared that no divine punishment acts so strongly on sinners as an earthquake, it was no wonder the churches were overflowing the next Sunday and most preachers referred to the recent divine warning in their sermons.*"

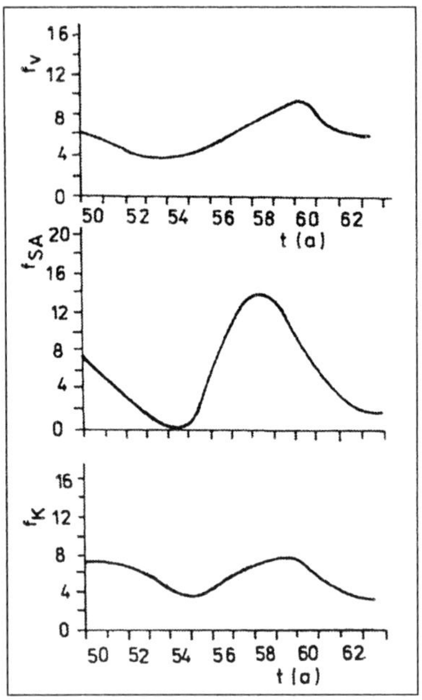

Fig. 2.19. Comparison of the frequency of earthquakes f_V, frequency of Sun activity f_{SA} and geomagnetic index f_K from January 1, 1950 to June 30, 1963 for mean monthly values (modified from Miroshnichenko, 1987).

People have known about earthquakes since ancient times and have feared them with reason. To be specific, earthquakes are among the limited number of natural phenomena that ert powerful destructive action which threatens the lives and property of human beings. Earthquakes are sometimes even more terrible than volcanic eruptions, possibly because they shake the Earth's surface, which most people are accustomed to consider stable. The feeling that the ground under foot is moving, that everything around is shaking and being destroyed, causes shock that can hardly be compared with any other. A map can best show how unquiet the Earth is and how its surface is only apparently stable (Fig. 2.20.). The series of black dots on the map would be transformed into completely black stains if it showed the epicenters of all shocks that have occurred from 1900 to 2000, since their number is truly enormous. Thus, during the period of earthquakes on the Hawaiian Islands in 1868, more than 2,000 clearly separate shocks were recorded only in the month of March; 300 to 320 strong shocks occurred from August of 1870 to August of 1871 in Fokidi (Greece); and more than 700 shocks were recorded on certain days in 1930 in northern Ido (Japan).

The initial cause of earthquakes is associated with liberation of heat from the Earth's core. However, the following are regarded as the most direct causes: 1) *formation of tectonic faults*; 2) *volcanism*; 3) *artificial arousal*; and 4) *summary action of different factors*.

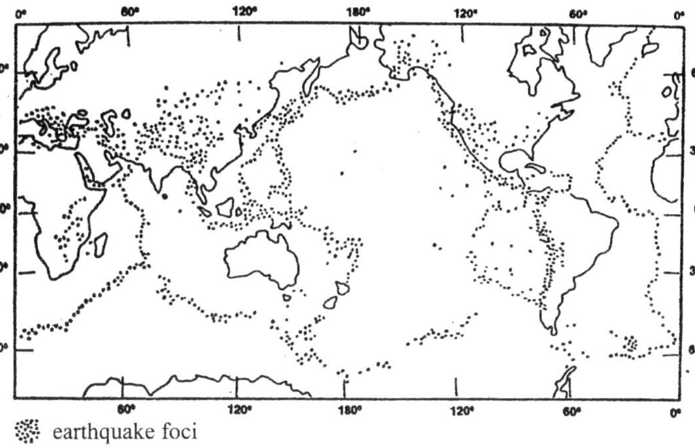

Fig. 2.20. Earthquakes distribution in period 1961 - 1967.

With the formation of *tectonic faults*, rock masses lying along such structures undergo crumbling and movement. The given processes cause earthquakes[27]. Rock crumbling can be linked with three main types of movement in the Earth's crust: *compaction*, *divergence*, and *horizontal movement*. The most numerous and powerful earthquakes occur in *compaction zones* along the shores of the Pacific Ocean, where the Pacific tectonic plate in subduction zones is gradually drawn beneath the continental boundary region and island arcs (Fig. 2.20.). In cold brittle submerging plates, faults are repeatedly formed from grooves in the process of downward movement, and this causes many shallow earthquakes (not deeper than 60 m).

[27] The potential energy of deformation increases to such a degree that the force of friction is overcome, causing movement along the fault's surface. The sudden break and release of energy cause seismic waves, which spread in all directions.

Only about 5% of earthquakes, usually shallow ones, are accompanied by a *regime of divergence.* The San Andreas structure in California (Fig. 2.21.) is most often cited as an excellent example of a seismogenerative transform fault. In just one earthquake (the California Earthquake of 1906), movement along the fault attained up to 7 m. We note that geologists know of the existence of thousands of faults, but only a smaller number of them are today active sources of earthquakes.

Volcanism is the cause of many local, predominantly weak oscillations of the ground. Both the famous eruption in 1883 of the volcano Krakatoa in Indonesia and the eruptions of many other volcanoes were accompanied by weak, but numerous earthquakes.

Artificial earthquakes occur due to human action on the existing state of equilibrium in upper parts of the lithosphere. For example, many weak earthquakes have in a number of cases been caused by overburdening of the ground with water during filling of reservoirs. We note also effects caused by the pressure of water polluted with radioactive waste that was pumped into deep boreholes drilled to receive it between 1962 and 1970 in the vicinity of Denver, Colorado: it caused more than 700 weak earthquakes around the boreholes.

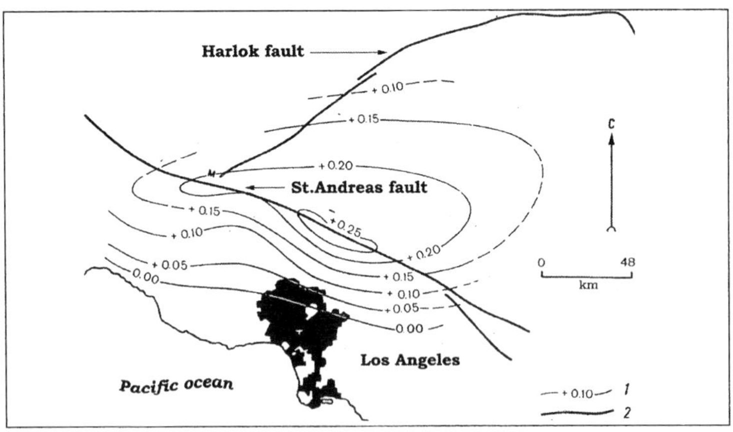

Fig 2.21. Regional deformation of the Earth's crust in the territory intersected by the St. Andreas fault, northern from Los Angeles. 1 - Shift during sinking or uplifting; 2 - fault (after USGS).

Most sources of seismotectonic waves are linked with two main belts:

1. The *Circum - Pacific belt,* which extends from Chile to Central America; passes through Mexico, California, and the Aleutian, Kurile, and Japanese

island arcs; and then runs through the Philippines and Indonesia to New Zealand; and

2. The *Alpine - Mediterranean - Trans - Asian belt*, which encompasses the regions of the Mediterranean countries, Iran, India, Burma, and China (Fig. 2.20.). Both belts are linked with the same basic cause: movement of plates of the Earth's crust. In contrast to regions located along the boundaries of plates, internal regions of geologically old continental platforms and shields (the eastern part of Canada, Brazil, Scandinavia, part of the territory of Siberia, a large part of Africa, Southern India, and West Australia) are relatively aseismic, with the sole exception of fault zones.

About 15 to 20 powerful and destructive earthquakes occur in an average year, but there are millions of relatively weak shocks annually. According to Howard and Remson (1978), a million people were killed by earthquakes in a hundred - year period (1875 - 1975), the victims being predominantly inhabitants of densely populated regions of the globe[28].

In certain localities, earthquakes repeat themselves with the passage of time, but obviously with no regularity[29].

Notes on Famous Earthquakes and Their Consequences.

1. The catastrophic earthquake that struck the *Shensi* Province of China in 1556 left behind it about 830,000 human casualties. The greatest number of deaths and enormous material damage were caused by strong quaking of the loose rock (filled with groundwater) of which the valley of the Yellow River is built, and by flooding that followed breaching of the river's embankments;
2. Although the earthquake in the Shensi Province was the most terrible seismic disaster recorded in the annals of history, primacy is usually granted to the *Lisbon Earthquake* of 1755. The magnitude of this earthquake possibly attained a value of 9, and the amount of energy released may have been equivalent to 17 million tons of TNT[30,31]. It is

[28] Seismic sea waves or *tsunamis* frequently accompany major earthquakes in marine or maritime regions. They often leave great devastation behind them. For example, a tsunami that arose in the Aleutian Islands on 1 April 1946 destroyed a lighthouse on Dutch Cape (Alaska), which was situated 15 m above the level of the sea. The wave reached the Hawaiian Islands, traveling a distance of 3,800 km at an average speed of 780 km/h and attaining a length of 150 km on the open sea. One hundred and fifty-nine people were killed in the narrow coastal zone, while the total material damage in Hawaii was estimated at 25 million dollars.

[29] The possibility of predicting earthquakes is being investigated today in California, Russia, Japan, and China, but so far with little success.

[30] The real energy of an earthquake released in its focus (hypocenter) is estimated according to the *standard scale of magnitude* proposed by C. F. Richter, with a range of from not much more than zero to 9. According to this scale, among the most powerful earthquakes (apart from the Lisbon Earthquake) were the Assam Earthquakes (India, 1950), with a

calculated that the epicenter of the Lisbon Earthquake was in the sea, 100 km off the coast of Portugal. The radius of territory on which the earthquake was registered perceptibly exceeded 1,500 km from the epicenter. The destruction of Lisbon was terrible. The city at that time had about 230,000 inhabitants, and according to the most modest counts about 30,000 were killed. The earthquake was followed by sea waves (tsunamis) around 7 m high, as well as by many fires. The catastrophe shook Europe, and moralists and sages of all stripes did not miss the opportunity to profit from that capital.

3. The history of *Chile* is marked by a large number of powerful earthquakes. The epicenter of the earthquakes that began on 22 May 1960 and had a maximum magnitude of 8.5 was characterized by a number of destructive aftershocks following the main shock. For this reason, it was difficult to establish which aftershocks were responsible for this or that damage done. The long series of several hundred earthquakes lasted several months and was accompanied by release of enormous energy. In addition to human casualties and enormous material damage, we note many geological consequences, namely the spreading of tsunamis over the entire Pacific Coast, destruction of river banks, landslides, torrential rock - and mud-slides, eruption of the volcano Puehu, local uplifting of terrains, general sinking of arable land and flooding of it with seawater, etc.
4. The 1964 earthquake in the vicinity of *Prince William Sound in Alaska* had a magnitude of 8.5 and 10 destructive aftershocks. Much of the city of Anchorage was destroyed, and the wider neighborhood (an area of 130,000 km^2) was severely damaged. Tsunamis 10 - 15 m high caused enormous damage, even in distant California, and spread to the Hawaiian Islands, Japan, and Antarctica. The elevation of terrains was visibly altered: one zone of land measuring 65 - 90 thousand km^2 in area was uplifted, while a neighboring region with an area of about 100,000 km^2 sank by an average of 1 m. Maximum uplifting of the sea floor comprised more than 16 m, which represents the greatest vertical movement in historical time.
5. A powerful earthquake in 1976 caused horizontal movement of up to 3.25 m along the fault that cuts across *Guatemala* between the North American and Caribbean Plates. The earthquake struck central regions of Southern Guatemala. According to official statistics, 22,778 people died and 76,504 suffered serious consequences. Material damage was estimated at 1.1 billion dollars. Movement of terrains along the fault caused thousands of landslides.

---

magnitude of 8.7; the Columbian Earthquake (1906), with a magnitude of 8.5; and the Great Alaska Earthquake (1964), with a magnitude of 8.5.

[31] According to Howard and Remson (1978), the energy released in the Great Alaska Earthquake (1964) was equivalent to 32 million tons of TNT; that released in the earthquakes in the Indian state of Assam (1950) was equivalent to 65 million tons; and the energy released in the earthquake in San Rikyu and the coastal region of Japan (1933) was equivalent to 125 million tons of TNT. The atomic bomb dropped on Hiroshima was equivalent to 20 thousand tons of TNT.

6. One of the greatest of modern earthquakes took place in March of 1933 not far from *San Rikyu* on the eastern coast of Japan 120 km north of Sendai on the island. Thanks to the fact that it occurred far from the shore, the earthquake did not possess great destructive power, but it caused sea waves with a height of 29 m above the water level in which about 3,000 people were drowned.

Seismic Characteristics of the Territory of Yugoslavia. The seismotectonic picture is marked by the region's very complex geological structure. It exhibits characteristics dictated by the meeting of a number of geotectonic units and existence of a network of many longitudinal and transverse neotectonic faults. Tectonic tensions caused by subduction of the African Plate beneath the European Plate are the main source of energy. The accumulated tensions are transmitted through a system of faults to the deeper interior in the northeastern part of the country. Depending on the distance from geotectonic contacts on the Adriatic Coast, the intensity of tectonic tensions, i.e., the value of magnitude, declines.

According to B. Sikosek and M. Petrovic (1992), the seismotectonic framework of Serbia is composed of four seismogenic (geotectonic) units: the Dinarides, the Pannonian Mass, the Serbo - Macedonian Mass, and the Carpatho - Balkanides, with 45 seismogenic blocks and 13 seismogenic faults. The territory of Montenegro belongs to the Dinarides, while the narrower region of the Negotin Depression is a component of the Misiska Plate.

The Dinarides Seismogenic Belt manifests the greatest seismic activity, with as many as 70% of all earthquakes. The central tectonic blocks are the site of 20% of them, while the Carpatho - Balkanides have the lowest earthquake frequency, with only 10%. As for depth of occurrence, 18% of all earthquakes on the territory of Yugoslavia had a hypocenter at up to 5 km of depth, 66% were in the depth interval of 6 to 10 km, 15.3% were in the range of 11 to 15 km, and only 0.7% was in the depth interval of 16 to 20 km.

A significant part of the territory of Yugoslavia is a region with a high degree of seismicity (VII to IX° MCS)[32]. According to historical data and research conducted to date, the following seismogenic regions can be singled out: 1) Montenegrin Coast and hinterland; 2) Kosmet; 3) Western Serbia; 4) Sandzak; 5) Central Serbia; 6) Pomoravlje; 7) Kopaonik; and 8) Fruska Gora. The strongest earthquakes attained an intensity of IX° MCS: Prizren (1456), Svilajnac (1893), Vranje - Ristovac (1904, epicenter around Berovo in Bulgaria), Pristina - Urosevac (1921), Lazarevac (1922), Rudnik

[32] The Mercalli-Cancan-Sieberg scale (MCS, with a range of I to XII°, is a measure of the damage caused by an earthquake on the surface.

(1927), Montenegrin Coast (1979), and Kopaonik (1980). The Berovo Earthquake is considered to have been the strongest of all earthquakes that occurred on the Balkan Peninsula in the past century[33].

In recent time, Yugoslav territory has witnessed the generation of several powerful earthquakes. The last two decades were marked by those on the Montenegrin Coast and in vicinity of Mt. Kopaonik. The former (15 April 1979, magnitude of 7.1) was registered at 227 seismological stations throughout the world and represents the strongest recent earthquake in this part of the Mediterranean. The earthquake on Mt. Kopaonik (15 May 1980, magnitude of 5.8) was reliably registered at 164 stations. The strongest earthquake on the Montenegrin Coast was accompanied by a series of powerful aftershocks, while a very strong earthquake (magnitude of 5.1) occurred in 1984 in the Kopaonik focal zone after the main earthquake in 1980.

The damage caused by earthquakes has been very great. Just during the period of 1963 - 1974, this damage on the territory of former Yugoslavia was estimated at about 2 billion dollars. Here it must be kept in mind that the famous catastrophic earthquakes on the Montenegrin Coast and in the Kopaonik seismic area occurred after the indicated period.

Volcanic Activity

Volcanic eruptions are one of the most impressive, destructive, and terrifying of natural phenomena. The very term *volcanism* arose from the word Volcano, the name of a small volcanic island (crater) of the Lipari Island group off the coast of Italy. It was believed that the smithy of Vulcan - the legendary ancient Roman god of fire and metal working - was located here. Better known than the island of Volcano is the neighboring active volcano Stromboli: this unique lighthouse of the Mediterranean Sea is a mountain 900 m high that ejects clouds of steam by day and whose surroundings are illuminated by the reflected light of hot lava from its crater at night (Allison and Palmer, 1980). The volcano Etna - a mountain nearly 3,300 m high on the nearby island of Sicily - began to manifest activity in the 12th Century. Lava from its crater (about 450 m deep) during a grandiose eruption of the volcano devastated the city of Catania. Other eruptions followed and have continued down to the present day.

Volcanoes can be divided into *effusive* volcanoes (with a quiet nature of emission), *explosive* or violent ones, and ones of the *transitional type.* Quiet emission, as in the Hawaiian Islands for example, results in the formation of gently sloping volcanic cones composed of basaltic lava. On

---

[33] The earthquake in the vicinity of Dubrovnik in 1667 was especially catastrophic; the city was for practical purposes completely destroyed, and Budva suffered the same fate on that occasion.

the other hand, fountaining lava produces cones of slag with a relatively great angle of slope inclination (from 30 to 35°), as in the case of Wizard Lake (Crater Lake in the state of Oregon, USA). Vesuvius and Etna represent volcanism of the transitional type, with medium angles of slope inclination (usually from 20 to 30°).

The majority of present - day volcanoes are conical hills with a funnel - shaped expanded *crater* on top, which continues downward in a *volcanic conduit.* The cones of some volcanoes have in the course of geological history been from time to time partially destroyed by powerful eruptions and violent explosions, and then built up again by newly ejected material. *Calderas* were formed in this way on many volcanoes. They are extremely expanded craters named after the enormous La Caldera Depression on the Canary Islands, with a diameter of more than 5 km and surrounded by rock walls nearly 1,000 m high. The term caldera is today used for larger depressions in volcanic regions that are sometimes of gigantic size (the caldera of Aniakchak Crater in Alaska has a diameter of 10 km, while that of Crater Lake in Oregon measures 8 km in diameter). However, the usual crater rarely exceeds 300 m in diameter.

Apart from flowing in the indicated way across the crater's central vent, lava can in certain cases also be emitted along some larger fissures from a series of smaller vents and cones that function simultaneously. For example, about 12 km^3 of lava and 2 km^3 of tuff flowed out in the course of only two months in 1783 from such openings along the famous 24 - km - long Laki Fissure on Iceland, and the given mass covered an entire region with an area of about 565 km^2. Basaltic ash carpeted a valley with its surroundings, destroyed thousands of head of livestock, and caused starvation and death of more than 10,000 people. The volcano Taravera on the North Island of New Zealand can be cited as an example of an explosive eruption along fissures. Here in 1886, volcanic products ejected along a fissure 30 km long covered an area of about 10 thousand km^2 with a layer up to 75 m thick (near the fissure).

Similar plates of consolidated lava formed by *fissure eruptions* exist in East Africa, Northern Syria, India, New Zealand, the Rocky Mountain region of the USA, etc.[34]. It is calculated that about $2.6 \cdot 10^6$ km^2 of dry land is covered by old, prehistoric products that arose due to emission of lava from fissures similar to the lava flows that can be seen today on Iceland.

Volcanoes eject material in three aggregate states: fluid, gaseous, and solid. To be more precise, there are three main kinds of products of volcanic eruptions: 1) lava; 2) gases and vapors; and 3) pyroclastic products. Volcanites - surface (effusive) magmatic rocks - are formed from lava, tuffs

---

[34] The lava plateau in the states of Washington, Oregon, and Idaho (USA) embraces an area of 520,000 km^2 and has a thickness of 1,200 to 3,000 m and more. The basaltic plateau in India is even larger.

and tuffites from pyroclastic material. Gases and vapors perform several more important geological functions, emerging in many places on the Earth's surface long after volcanic eruptions.

Lava is the main product of many volcanic eruptions and is sometimes emitted in truly enormous quantities. It is a superheated fluid mass with a temperature of 1,000 to 1,100°C. On the surface, lava cools relatively rapidly and at 700 - 600°C starts to solidify and pass over into rock. With respect to its chemical composition, lava forms a series of rocks, from basic to acidic. The very widely disseminated alkaline (basaltic) lavas, with low silica content, have relatively low density, are readily mobile, and can flow at a high speed (more than 30 km/h in the case of the lava of Hawaiian volcanoes). Length of the emitted mass can be more than 100 km, its width up to 4.8 km. Silica - rich acidic lavas are characterized by much greater plasticity, as a result of which they tend to accumulate on hard slopes, forming obstacles in the form of plugs. Blocking vents, they cause explosions. Masses of lava emissions can attain significant dimensions: 980 million m^3 in the case of Etna (1669) for instance. The total mass of consolidated volcanic lava of which the Hawaiian Islands are composed has a volume of more than 300,000 km^3. Put descriptively, the mass of Hawaiian basalt alone could blanket the whole of Europe with a cover about 32 - 33 m thick.

Great amounts of various gases and vapors enter the atmosphere as a result of volcanic eruptions. *Volcanic gases and vapors* consist mainly (75 - 90%) of water vapor. The remaining gases include nitrogen, oxygen, hydrogen, carbon dioxide, hydrogen sulfide, sulfur dioxide, chlorine, fluorine, sulfur, boric acid, ammonia, methane, argon, or products of their reactions. Some of the gaseous components are poisonous for animals and man. As for carbon dioxide, just its emission during volcanic eruptions has already for a long time maintained atmospheric equilibrium, and the whole diversity of plant life could not exist without it. If reserves of this gas were not replenished, it would in time be completely absorbed by plants, as well as by captive buried deposits of caustobioliths. We note that truly enormous amounts of gases and vapors can be released in a short time during volcanic eruptions (more than 1.5 km^3 by Vesuvius in 1929; 1.6 km^3 by Katmai in Alaska in 1912).

Loose *pyroclastic material* arises during explosive action of volcanic vapors and gases. It consists both of solidified lumps of lava and solid particles of different size arising from break - up of older rocks forming underground galleries and the volcano itself. *Volcanic ash* constitutes the finest fraction. This ash is spread by wind over great distances and falls still hot on the surface of the ground. For example, during eruption of the Katmai Volcano in Alaska in 1912, an area of 7.8 km^2 was covered by a layer of ash 30 to 90 cm thick, and thickness of the layer shrank by a few millimeters only at a distance of 6.6 km from the center of the eruption.

Activity of volcanoes in the late stages of cooling is for the most part reduced to exhalation of volcanic gases and vapors. On the basis of temperature, way of manifestation, and composition, all post - volcanic phenomena can be divided into four main groups: 1) solfataras; 2) fumaroles; 3) mofettes; and 4) geysers and hot springs.

Fumaroles are associated with vents from which water vapor and other superheated gases are released. Their name derives from the fact that they emit smoke ("fumare" in Italian means "to smoke"). They are encountered in the vicinity of both active volcanoes and volcanoes in the dormant state. Temperature of the vapor fluctuates between 100 and 650°C. Apart from water vapor, exhalations of fumaroles contain SO_2, HCl, and other compounds. They emit NaCl, Fe_2O_3, and other salts, and in this way form fumarolic deposits (Fig. 2.22.). Recent formation of mineral resource deposits in the vicinity of fumaroles is treated by E.A. Baskov and S.N. Surikov (1989) in the book "**Thermal Springs of the Earth**."

The term *solfatara*[35] is applied to phenomena characterized by exhalations of superheated water vapor, gaseous sulfur compounds, and the volcanic gases H_2S and CO_2 with temperatures ranging between 100 and 200°C. They get their name from the volcano Solfatara in the Phlegraean Fields (Italy), where water vapor with gases is emitted from fissures on the bottom of the crater. The category of solfataras also includes springs of water vapor and volcanic gases in Tuscany. Situated in groups on an area of about 30 km^2 or forming chains along faults (island of Ischia), these springs are well - known for yielding great amounts of boric acid, which has been obtained from them for more than 150 years. The indicated region in Tuscany is also famous for successful exploitation of volcanic (geothermal) energy, inasmuch as steam from the Larderello deposit has since 1904 been used to drive the first geothermal electric power plant on Earth. Also, the content of native sulfur in the vicinity of solfataras can attain industrial significance, and it is extracted in Italy, Mexico, and Japan.

Mofettes are gaseous exhalations of carbon dioxide in areas with almost completely cooled volcanoes. One of the best - known mofettes in Europe is the previously mentioned Dog Cave west of Naples.

---

[35] One solfatara exists on the territory of former Yugoslavia. This is Duvalo in the village of Koselje north of Ohrid. The local inhabitants call it that (Duvalo = Blower) because a jet of H_2S and CO_2 is constantly emitted from a miniature crater of elliptical shape. Inasmuch as the depression of the crater itself is filled with these gases, any animal that descends into it soon dies. The exhalations of H_2S are so intense that it can be smelled far around the village.

Geysers and *hot springs* are mainly characterized by an intermittent nature of emission of hot water. They are most numerous in New Zealand, Iceland, and North America. There are about 3,500 of them in Yellowstone National Park (USA) alone, including hundreds with periodic strong eruptions in the form of fountains several tens of meters high. The geyser Waimango in New Zealand in just one eruption ejects about 80 thousand liters of water in the form of a column as much as 400 m high.

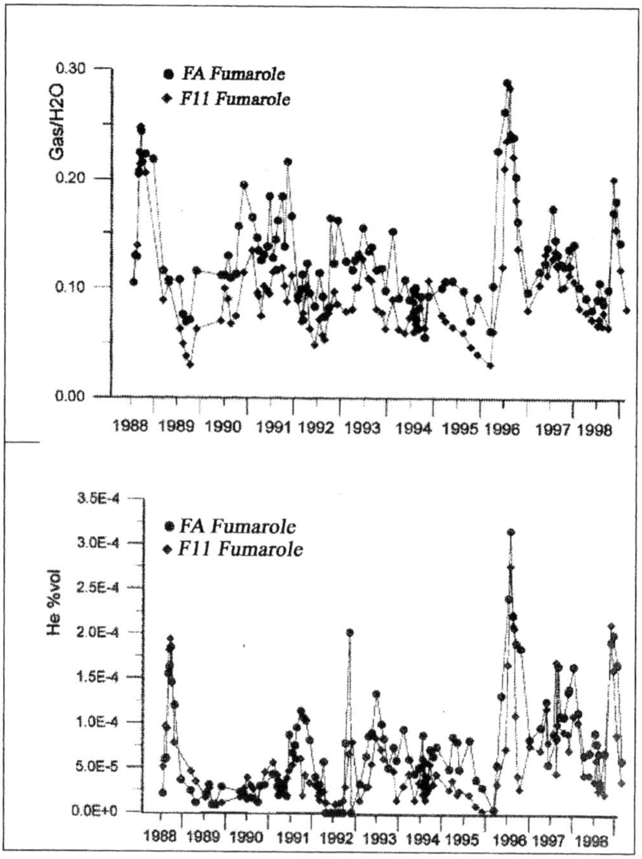

Fig. 2.22. Volcano island. Ratio gas/steam and time changes of helium content for fumarolas FA and F11 (Volpe et al., 1999).

Notes on Better - Known Volcanoes and Their Activity. The main active volcanoes of the Earth are situated on the boundaries of great tectonic plates. Like earthquakes, two thirds of the 600 known volcanoes active in historical time are situated in island chains around the Pacific

Ocean (Fig. 2.20.) or on the continental side of boundaries between plates. In other words, this largest mass of water on our planet is girdled by a ring of fire that has many times and in many places killed enormous numbers of people. A second belt with concentration of volcanoes (approximately a fourth of their total number) extends from Italy, Greece, and Turkey across South Asia to Indonesia. Apart from these dry - land geological phenomena, there are about 10,000 underwater volcanoes not less than 1,000 m high in the aquatorium of the Pacific Ocean.

Like earthquakes, volcanoes are a geological phenomenon that can cause many human casualties and great material damage. Twelve thousand people were suffocated during an eruption of the volcano Timbora on the island of Zumbawa in 1815. An enormous area was covered with a layer of ash 60 cm deep. All crops were destroyed and 44 thousand inhabitants died of starvation. An eruption of the volcano Mt. Pelee (Martinique) in 1902 destroyed the town of Ste. Pierre, and all 28 thousand of its inhabitants were killed.

Ejecting solid particles and gases into the atmosphere, volcanoes can alter the Earth's climate. Ash ejected by the eruption of the volcano Krakatoa (an island between Sumatra and Java in Indonesia) in 1883 circled the entire globe, causing evenings to be red - colored for several years, and mean - annual temperature on the planet declined by several degrees.

Not all manifestations of volcanic activity are harmful. Volcanoes act as safety valves (A. D. Howard and I. Remson, 1978). If pressure were not reduced by small eruptions from time to time in various places, the consequences would be far worse. Moreover, volcanic eruptions contribute necessary gases to the atmosphere and make the soil fertile; deposits of mineral raw materials are formed as a result of volcanic activity; and volcanoes serve as a source of geothermal energy, while lava acts as a collector of groundwater. Volcanoes are very picturesque: they beautify monotonous landscapes and offer considerable recreational possibilities.

Several of the world's better - known volcanoes are described below, together with the consequences of their activity for the environment and life.

Vesuvius (Italy). The most famous of the world's volcanoes, Mt. Vesuvius near Naples has been continuously active for more than 19 centuries (Fig. 2.23.). The first eruption in modern history (in 79 A.D.) was at the same time the strongest eruption of this volcano and was described in great detail by Pliny the Younger in his letters to the historian Tacitus. It is known that during this terrible eruption, volcanic lava and ash completely buried the up to then advanced cities of Pompeii, Herculaneum, and Stobia, with more than 20 thousand inhabitants. Something less than a tenth of the population met their deaths under a blanket of hot ash 6 m thick. From that time to the present day, about 12 additional powerful eruptions of Vesuvius have been recorded, and ash from one of them (in 1631) even fell in Istanbul, at a distance of about 1,300 km away.

Etna (Italy). This is the largest active volcano in Europe and one of the largest in the world. Rising 3,280 m above the sea, Mt. Etna is characterized by having on its sides more than 1,000 lateral or parasitic craters in addition to the main crater. The fissures along which the parasitic craters are situated can be more than 15 km long. One of them, with a length of 18 km, opened in 1869 on the southern side of the main crater. Its lava in the form of an enormous tongue covered an area of about 50 km^2, destroying the greater part of the city of Catania and 12 surrounding villages. Etna destroyed the city in several stages and completely buried it is 1669. This time, 49 other towns were destroyed in addition to Catania, and about 94 thousand people died. The last more powerful eruption of Etna devastated Messina in 1908, when about 100 thousand people lost their lives.

Volcanoes of Iceland. The island of Iceland is among the most volcanic areas on Earth. Two volcanic regions, eastern and southern, can be clearly singled out on it. Of nine dry - land and three underwater volcanoes in the eastern part of the island, two are especially significant: Sveinagja and Askja. The former volcano is famous for a powerful eruption in 1875, when 0.3 km^3 of lava was ejected along a single fissure 25 km long. With about 14 volcanoes, the southern region is considerably more active than the eastern and much better known for strong eruptions. Among volcanoes in the southern region is the famous Mt. Hecla, well - known for many powerful eruptions. Still, the greatest disaster Iceland suffered throughout its entire history was caused in 1783 by the volcano Laki. On that occasion, 12.3 km^3 of lava and 3 km^3 of loose material were ejected along one fissure with a length of 25 km, and the emitted lava covered an area of 565 km^2. Volcanic ash and sand blanketed meadows, pastures, and the small amount of arable land on Iceland, and a dense fog never seen before hung over the whole island for months. That was a great national disaster for Iceland: nearly a fifth of the island's population died of contagion and starvation, and the majority of domestic animals perished. The volcanoes of Iceland (like those of Kamchatka, the Aleutian Islands, and Alaska) are also interesting for a special phenomenon that accompanies eruptions in ice - covered regions. Thus, an enormous mass of ice melted during an eruption of the volcano Mt. Katmai in 1918, creating terrible steam - enshrouded torrents up to 70 m deep flowing toward the surrounding lowland. These torrents flooded extensive areas and covered them with mud, boulders, and large pieces of ice.

Tambora (Indonesia). One of the most terrible volcanic eruptions in the memory of man occurred at the beginning of the 19th Century on the island of Sumbawa east of Java. At that time, the volcano Mt. Tambora was lowered by approximately 1,500 m in an explosion that was clearly heard 3,300 km away, and a gigantic mass of pyroclastic material with a volume of 150 km^3 was ejected. More than 66 thousand people lost their lives, both during the eruption and as a result of the starvation and contagion that followed it. Wind carried ash over a considerable distance equal to that from Vesuvius to the Baltic Sea. On the island of Lomboko 120 km away, the ash destroyed all crops and caused terrible starvation of which 44 thousand people died, and severe epidemics appeared in places even farther away. The population of nearly the entire Sunda Archipelago long talked of that event as a kind of end of the world.

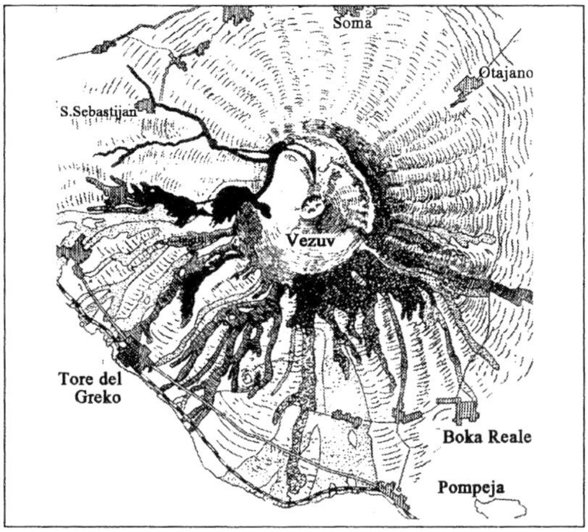

Fig. 2.23. Horizontal projection of Vesuvius with lava flows (dark).

Krakatoa (Indonesia). Among several active volcanoes between the islands of Sumatra and Java, a volcano on the deserted island of Krakatoa has for more than a century represented a symbol of the destructive power of volcanoes as a unique geological phenomenon. Following a 200 - year period of dormancy, Krakatoa showed signs of activity in May of 1883 and exploded powerfully on 27 August of that year, when the greater part of the volcanic island was blown to pieces and a depression in the sea about 360 m deep was created on an area of 72 km^2. Ash rose in the atmosphere to a height of about 30 km, and its depth on the remainder of the island attained 60 m. The total amount of ejected material is estimated at 20 km^3; due to evacuation of the underground reservoir, the surface sank suddenly to form a basin about 6 km wide and 230 m deep. The 40 m - high wave created in the sea pounded with enormous power against the shores of Sumatra and Java, washed over several parts of the islands, and drowned about 36 thousand people, then spread to all coasts of the Pacific and part of the Atlantic Ocean. Sea - going ships carried by the advance of this wave were later found in forests as much as 4 km away from the shore.

Extinct Volcanoes of Yugoslavia. In the not distant geological past, a large number of volcanoes were active in many parts of Yugoslavia. The products of those volcanoes are today found in several extensive volcanic regions such as the Timok Volcanic Region in Eastern Serbia, the Southern Morava and Rhodopo - Toplica Volcanic Regions in Southern Serbia, or the Kosovo - Kopaonik - Rudnik Volcanic Region in Kosmet and Western Serbia. The formation of many of the great ore deposits of Yugoslavia - the Majdanpek and Bor deposits in the domain of the Timok andesite massif, the Trepca deposit in the neighborhood of Kopaonik, and the Rudnik deposit in Sumadia - is linked with old volcanism, which was especially active in the Tertiary.

The country's old volcanoes have for the most part been considerably destroyed and modified by denudation and erosion. In the Timok andesite region, with length of about 50 km and greatest width of the order of 20 km, three main series of cones in the vicinity of the spa Brestovacka Banja and partly on the land between Mt. Rtanj and Tupuznica have been best preserved. Around Kosovo are found the more or less destroyed old craters Mrkonski Vis, Veletin, and Zvecan. Especially characteristic is the old crater Djurdjevi Stupovi near Novi Pazar. Modified by the work of exogenous forces, this crater is very deformed.

As the last gasps of volcanic activity, warm and hot springs occur in certain parts of the country, being in large part linked with great fault lines and fissures. Such phenomena include the spas of Vranjska banja (92°C), Sijarinska banja (71°C), Josanicka Banja (78°C), and many other springs in regions with younger volcanic formations. The numerous paleohydrogeothermal phenomena in Serbia have been treated by M. Milivojevic (1982).

Exogenous Geological Processes and Occurrences

On 28 June 1974 in Villavisensio on a sector of a road 150 km east of Bogotá (Columbia), an enormous mass of earth separated from a rocky slope and with a terrible crash started downhill. More than 750 m of roadbed was buried by pieces of rock, under which 200 workers died. The river below was clogged by the rock, forming a lake that today threatens to overflow in a catastrophic flood. In Northern Italy, a gigantic landslide in October of 1963 came rushing down into the deep but narrow reservoir formed by the Vajont Dam. Enormous waves breached the dam and caused great damage along the part of the valley downstream, on which occasion five towns were devastated and 2,100 people were killed. This was the greatest catastrophe in the world associated with economic activity. There have been many similar examples. Thus, people have died throughout the planet, great material damage has been done, and the fruits of human work have been destroyed as a result of shifting (movement) of rock masses caused by the action of gravitational force. According to predictions for the period of 1970 - 2000 in California (USA), damage from landslides and terrain subsidence could cost 10 billion dollars, which constitutes a significant part of losses from all forms of geological risks (55 billion dollars) in the state for that period.

Geological processes and occurrences (abrasion, river erosion and ravine formation, soil erosion, landslides, rockfalls, torrential rock - and mudslides, karstification) accomplish the basic work of relief modeling.

These exogenous (external) processes are caused by the action of gravitational force, atmospheric agents, or water. As we have seen, they often occur fast and energetically, and changes in the relief of slopes and shifting of enormous masses of rock and rock products from slopes act destructively on constructed objects, thereby altering the conditions of human economic activity.

Classification and Brief Survey of Exogenous Processes and Occurrences

According to the main factors causing them, exogenous geological (geodynamic) processes and occurrences can be relegated to the following six categories:

- *Processes under the influence of atmospheric agents and organisms* (rock weathering);
- *Processes caused by the action of surface water* (denudation, river erosion, abrasion, glacial erosion);
- *Processes caused by the dissolving action of surface and groundwater* (karstification);
- *Processes caused by wind* (eolian erosion);
- *Processes caused by groundwater* (suffosion, filtrational deformation of bedrock, etc.); and
- *Processes caused by the action of gravitational force and underground and surface water* (deformation of slopes and inclines).

In addition to processes of *sedimentation* (accumulation) of eroded material, processes occurring in water - saturated incoherent rocks in the presence of vibrations (*liquefaction processes*) are also worthy of attention. Moreover, effects of processes associated with *human activity* are becoming increasingly more pronounced with the passage of time, and their harmful consequences are being felt to a constantly growing extent.

Processes caused by the action of atmospheric agents and organisms (rock weathering) have already been discussed.

The geological action of surface water starts with atmospheric precipitation, which dissolves and erodes solid components from surface layers, performing *denudation* or *erosion* in the process. Atmospheric precipitation collects in the channels of surface streams, which then accomplish further downcutting of the stream bed and erosion of banks through the action of their own kinetic energy. This form of exogenous processes is called *river* or *fluvial erosion.* Also active in the world are processes associated with the denudational action of ice or *glacial erosion.*

Zones with a moderately damp climate outside annual isotherms of +10°C are characterized by weak and moderate mechanical erosion. The belt of damp tropical climate between isotherms of +10°C is characterized by great and very great *mechanical erosion* (especially in regions with temperatures above +20°C). Inside an area with a moderate tropical climate, the energy of mechanical erosion increases with increase in tectonic activity of the region and is noticeably greater in mountains than on plains. According to N. M. Strakhov (1963), minimal mechanical denudation (less than 10 tons per km^2) in the temperate zone is characteristic of the northern part of North America, the Baltic Shield, the Western Siberian Depression, and the Central Siberian Plateau between the Yenisei and Lena Rivers; mechanical erosion varies from 50 to 100 t/km^2 in lowland regions of South America, Africa, Hindustan, and humid sections of Australia; it is in the range of 100 - 240 t/km^2 in mountainous regions of the Caucasus, Iran, and Andes; and it comprises about 390 t/km^2 in Southeast Asia, attaining as much as 600 - 1,000 and 1,200 t/km^2 in certain regions of the basins of the Indus, Ganges, and Brahmaputra Rivers. With increase of mechanical erosion, there is also an increase of *chemical denudation*, whose action is on average half as great as that of mechanical erosion.

That mechanical erosion and the geological work of rivers can be of gigantic proportions is best seen from the amount of mud that certain great river courses introduce annually into seas and oceans:

- Danube - 82,000,000 t;
- Mississippi - 350,000,000 t;
- Indus - 446,000,000 t;
- Yangtse - 700,000,000 t; and
- Amazon - 1,000,000 t.

Coastal erosion (abrasion) is characterized by great intensity and continuous action. For example, sea waves during a storm shifted a quay wall weighing 8,000 KN in one Scottish harbor and a breakwater weighing more than 17,000 KN in the port of Bilbao in Spain. The rate of shoreline destruction due to the action of waves, marine currents, and other factors can be very different and depends on the conditions under which those factors act. For instance, on the coast of the Black Sea near Odessa, 0.93 m of shoreline is on average destroyed every year, while a coastal strip about 2 m wide disappears annually on the shores of the English Channel.

In limestones and other readily soluble rocks, geological processes are predominantly reduced to dissolution or *karst erosion*, which results in the formation of a specific geomorphologic and hydrogeological environment. Characteristics of karst regions, including risks to buildings, have already been discussed. We note also the great damage done by collapse of terrains, i.e., formation of recent sinkholes and ponors, in certain urban environments. Thus, more than 4,000 new collapses were recorded in the state of Alabama (USA) during the period of 1900 - 1976, and the given occurrences were reported on city streets, in the dooryards of buildings, under workshops, along highways, and in other places.

The velocity of wind and its influence on activity of the inhabitants of arid regions can be very great. During the drought of the mid - 1930's, the fertile upper layer of soil was carried away by wind, i.e., destroyed by *eolian erosion*, due to feebleness of the crops on a large area of the Western Plains of the United States. Something similar happened in certain regions of the Ukraine during a strong storm in 1928, when the top 12 cm of soil was carried away by wind. The potential destructive power of wind erosion can be illustrated by the example of large basins in Kazakhstan (up to 145 km long, from 2 to 10 km wide, and from 100 to 142 m deep) that were formed by the process of deflation[36]. There are hundreds of depressions formed in this way throughout the world. We mention ones in the Gobi desert of Central Asia (up to 50 km long and about 30 m deep) formed in weathered granites by winds dislodging tiny disintegrated particles of quartz and feldspar; and seven such depressions (from 200 to 300 m deep) in the Libyan Desert and Northwest Egypt. Rocky deserts from which all finer material has been blown away, so - called *hammadas*, represent the wildest and most desolate parts of deserts, "*regions of thirst and death*" in the words of a famous French geographer, vast areas on which caravan trails are lost as far as the eye can see[37].

Only 10 thousand years ago, in the Pleistocene Epoch, glaciers covered an enormous area of the planet. Although they withdrew soon thereafter, they had a significant effect on relief. In certain mountainous regions and in high latitudes of the Northern and Southern Hemispheres, glaciers continue to alter the appearance of nature. The process of *glacial erosion* is caused by expansion of water during freezing and by expansion and contraction during temperature changes. Glaciers act on land because they move under the influence of gravitational force, detaching fragments of rock and carrying enormous masses of morainal material. Pieces of morainal material can be very large: measuring 27 x 12 x 9 m, one of largest known was found in the United States near the city of Madison in the state of New Hampshire.

---

[36] Detachment and carrying away of particles from rocks by wind.

[37] Deposition of sand under conditions of a dry climate results in the formation of widely disseminated *loess* deposits. The enormously thick loess in northwestern regions of China was formed at the expense of materials from the deserts of Central Asia. Subsequent downcutting of riverbeds created an arena of catastrophic landslides caused by the earthquake of 1920, when about 100 thousand people met their deaths under the masses of material set in motion. Extensive regions of loess are found in Argentina, the Sudan, the northern part of Central and Eastern Europe, the basin of the Mississippi River, etc.

Shifting of rock masses on slopes under the influence of gravitational force represents a very widespread exogenous process, one that more or less deforms the slope and causes damage to man. *Landslides* are the best - known type of movement on slopes, but torrential *rock* - and *mudslides* are also frequently observed, as are *rockfalls* and *talus accumulations*.

The process of sliding represents movement of a portion of rock mass under the influence of gravitational force to a hypsometrically lower level with retention of contact between it and the immobile base (Fig. 2.24.). The process can occur at very different rates, from fractions of a millimeter per day to several tens and even hundreds of meters per hour. Formation and development of landslides can be affected by many factors, among which it is possible to distinguish geological factors (lithological composition, properties of rock masses, conditions under which subsidence of rocks occurs, and seismotectonic conditions) and hydrogeological conditions (properties of rocks as carriers of groundwater, manner and depth of subsidence of groundwater, characteristics of groundwater regimes, etc.). The mechanism of sliding processes can be different, and the following types are most often distinguished according to its nature: *sliding, out - pressing, flowing, subsidence,* and *softening* (M. Komatina, 1995).

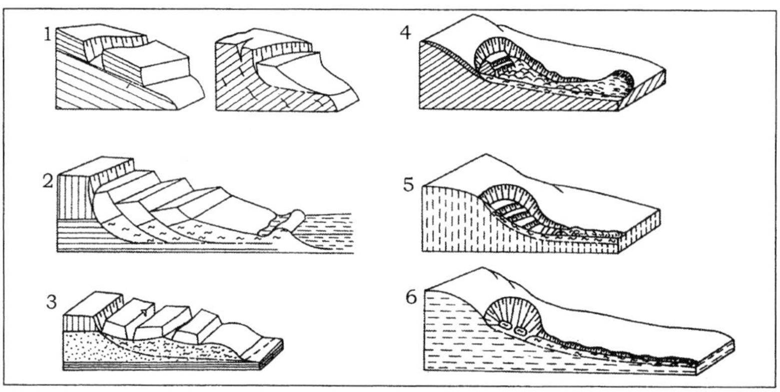

Fig. 2.24. Scheme of landslides types according to their mechanism.

In *rockfalls*, predominantly solid rock masses become detached and move down steep slopes, the process being accompanied by crumbling, rolling, and break - up into smaller pieces (Fig. 2.25.). Sometimes considerable amounts of snow and ice are set in motion together with rock fragments, and in this case the rockfall can be gradually transformed into an avalanche. In contrast to rockfalls, accumulation of *talus* is characterized by gravitational movement of rubble that has formed a slope in accordance with its own internal friction.

Torrential *rock* - and *mudslides* represent sudden currents copiously saturated with rocky material that arise during strong rains in highland regions or after intense melting of snow. They sometimes occur due to bursting of dams in valleys creating torrents with large amounts of coarse - grained material (M. Komatina, 1995). As in the case of landslides, years with extreme manifestation of the given geodynamic phenomena approximately coincide with solar activity and other cosmic factors.

Suffosion can be singled out among processes caused by the action of groundwater. It represents shifting and carrying - out of fine fractions of incoherent loose rock due to hydrodynamic pressure of filtered water, without destruction of the rock structure. The negative effects of suffosion and other kinds of filtrational deformations on rocks and building foundations can be obvious.

Water - saturated incoherent sediments (fine - grained sand, etc.) can pass over into a close to liquid state if they are subjected to dynamic forces caused by vibrations. This process is called *liquefaction*. The consequences of liquefaction can be manifested in different forms and with different intensities. For example, certain buildings sank as much as 1 m into the ground and some were destroyed as a result of liquefaction of sand in Nagata (Japan) in 1964. That same year, liquefaction caused the occurrence of landslides on a front 2,500 m wide on the shores of Cook Inlet in Anchorage (Alaska). Liquefaction caused by the Montenegro Earthquake of 1979 greatly damaged buildings (the Hotel Fjord in Kotor was destroyed), piers, and roads along the shoreline.

Fig. 2.25. Slope debris in limestone terrain (Photo: D. Milovanovic).

Caused by compaction of land under the influence of overburdening, *terrain subsidence* is pronounced in many large cities such as Tokyo, London, Venice, Houston, and Mexico City, and in numerous agricultural regions. It is most widespread in regions built of loose sediments - deltas, floodable valleys, etc. The Leaning Tower of Pisa (Italy) can serve as a typical example of the action of overburdening on a weak foundation. The tower started to lean in the early stages of construction; as a result of compaction, differential subsidence of the layer of clay under the foundation (8.5 m thick) continued long after completion of the tower, for all of 1,350 years. Attempts were later made to stabilize the tower, which was achieved only after very expensive work. It is generally the rule that subsidence occurs on a greater scale where deep water - bearing horizons are exploited, and many cases of such subsidence with catastrophic consequences are known in the world literature. Several examples of subsidence are as follows: 1) In 50 years of exploiting the groundwater reserves of Tokyo, the rate of ground subsidence attained 10 cm/year in some sectors of the city, and total subsidence amounted to as much as 3.34 m; 2) the rate of ground subsidence in Mexico City was perceptibly greater - 30 cm/year; and 3) in Long Beach it even attained 75 cm/year. Significant subsidence accompanied by great damage (total subsidence attaining 8.5 cm in places) is in many cases associated with pumping of gas and oil (at localities in Japan, in the region of Lake Maracaibo in Venezuela, and in California and Texas), as well as with evacuation of water from surface coal mines and underground exploitation of salt (salt deposits in England and in the vicinity of Tuzla in Bosnia - Herzegovina) or metal mineral resources. We note that during exploitation of gold from a deposit southwest of Johannesburg (Republic of South Africa), eight sinkholes measuring more than 46 m in diameter and 27 m in depth appeared on the surface in five years of drainage, causing much damage and loss of human life.

Risks from Exogenous Processes and Occurrences

Some of the possible risks from exogenous processes and occurrences have been indicated in the foregoing text. By his destructive action, man has every day raised those risks, i.e., constantly increased the danger of floods, landslides, mudslides, and avalanches. In general, movement of rock masses or various processes of erosion, together with possible disastrous consequences, can disturb the normal course of life and work, cause economic damage, hinder transport and communications, etc. If we were to compare the destructive action of landslides with that of other contemporaneous geological processes, we could conclude that they occupy a leading position, together with earthquakes and floods [38]. Catastrophic landslides, rock - and mudslides, and rockfalls take on especially grandiose dimensions at times of powerful earthquakes and volcanic eruptions. The greatest landslides on Earth occurred at such times. Among others, they include the following:

- Sliding of rock masses on the northern shores of the Peloponnesus, which occurred immediately after an earthquake in the 4th Century B.C.;
- Mass activation of landslide processes in the Konsu Province of China in 1920, which led to the destruction of tens of villages and the deaths of 200 thousand inhabitants;

---

[38] Floods are among the most serious of natural disasters and are well-known for causing enormous damage. Flooding of the Yangtze River in 1991 devastated more than 4 million dwellings, while the damage done by flooding of the Mississippi River two years later was estimated at more than 10 billion dollars.

- A landslide with a volume of 2.2 km^3 at the time of a strong earthquake in the Pamirs (Russia) in 1911, which completely dammed up the Murgab River; and
- Detachment of part of the mountain's slope with a volume of 2.8 km^3 due to eruption of the volcano Mt. St. Helena (USA) in May of 1980, which resulted in an avalanche that transported material 22 km to the west and covered an area of 60 km^2 in a river valley.

We mention three more characteristic examples of destructive movement of rock masses in the last century:

- Near the town of Frank in the Canadian province of Alberta in 1903, the whole frontal part of a mountain with a volume of about 30 million m^3 tumbled downhill to its foot, and was then raised up on the opposite slope to a height of 120 m. That landslide attained a length of 4 km. The whole sliding process lasted less than two minutes;
- An enormous landslide occurred in the valley of the Gros Ventre River south of Yellowstone Park in Wyoming in 1925. With a volume of around 38 million m^3, the landslide mass descended about 600 m, sliding along a sloping layer of water - saturated clays. The natural dam formed was about 75 m high and more than 750 m long; and
- The mudslide that covered the town of Hait in Tadzhikistan in 1949 moved at a speed greater than 360 km/h. A monument today marks the place where a town once stood at a depth of 30 m in relation to the present level of the terrain[39].

Disastrous landslides and rock - and mudslides have caused the deaths of hundreds and thousands of people throughout the world. According to F. O. Jones (1973), the direct material damage caused by landslides in the United States during the period of 1925 - 1971 was in the neighborhood of 75 billion dollars, which is three times higher than the damage caused by floods, hurricanes, tornadoes, and earthquakes in the same period of time. The annual damage from landslides and flash floods is about 1.5 billion dollars in Japan, while in India it is about a billion dollars in the case of the road network alone.

---

[39] Movement of material is also possible on the seafloor. The occurrence that took place in the region of the Grand Banks of Newfoundland is an excellent example of a so-called underwater landslide that was transformed into a mudslide. An earthquake with a magnitude of 7.2 and epicenter in the upper part of the continental slope served as the impulse for formation of several underwater landslides on an area of 150 x 300 km. One such great landslide instantly cut six trans-Atlantic cables and activated a mudslide that moved above the landslide and carried pieces of rock over distances greater than 350 km to cut the last cable as far as 730 km from the earthquake's epicenter.

Engineering - Geological Characteristics of the Territory of Yugoslavia

Engineering - geological characteristics of the territory of Yugoslavia were formed as a consequence of the geological structure of terrains as the main precondition, then as a result of the action of various physical (exogenous) factors and human activity affecting the geological environment. According to Lazic and Bozovic (1995), the following four main *engineering - geological categories of rocks* arose on this territory as a result of the given influences:

- *Unbound and weakly bound soft rocks* are widespread in the flat regions north of the Sava and Danube and along river courses. They are represented by fluvial and eolian formations that are highly to moderately deformable in regard to compactability. For such media, water represents the main unfavorable factor in evaluation of terrains from the engineering - geological standpoint;
- *Unbound and weakly bound soft rocks and poorly hardened rocks* of Neogene lacustrine basins represent a highly to moderately deformable medium, especially in the case of unstable slopes. They fill the Velika Morava Depression, the Kolubara Basin, the region of the Tamnava, the Valjevo Basin, the Cacak - Kraljevo Basin, the Zajecar - Knjazevac Basin, basins along the valley of the Southern Morava River, the Kosovo - Metohija Basin, etc. These are provisionally stable to markedly unstable terrains, with many landslides around the rims of the indicated Neogene basins;
- *Poorly hardened to hard rocks* are represented by clastic and sedimento - volcanogenic rocks, manifestations of a diabase - chert formation, and schistose metamorphic rocks of the Vardar Geotectonic Zone and southern part of the Serbo - Macedonian Mass. They constitute highly anisotropic media, ones that are very fractured and profoundly disintegrated, moderately to little deformable, and poorly permeable. Erosion processes and flood activity are very pronounced in the region of the Vranje Basin, the Grdelicka Gorge, and the watershed of the Vlasina River; and
- *Hard to very hard rocks* are predominantly disseminated in boundary - forming geotectonic units marked by limestone massifs such as the Inner Dinarides and Carpatho - Balkanides. They are characterized by great hardness and permanence. The limestone masses are karstified, with rockfalls on steep slopes.

The regionalization of terrains according to engineering - geological advantages for development of human activity, i.e., for human life, is based on a number of more important elements characterizing individual terrain categories. Such elements include:

- *Compactability*, in regard to which swampy and muddy terrains along larger river courses and in the vicinity of their deltas stand out for being highly to moderately compactable;
- *Instability of terrains*, marked by landslides and rockfalls, is characteristic of Neogene sediments, as well as of poorly hardened older rock masses that are greatly fractured and altered;
- *Fluvial erosion* and flooding along unregulated river courses;
- *Abrasion processes* along steep seacoasts;
- *Eolian processes* in unprotected parts of sandy regions; and
- *Karst processes* in the limestone terrains of Southwest and Eastern Serbia.

Terrains with relatively favorable engineering - geological characteristics are found outside regions with the indicated unfavorable elements, zones of increased seismic risk, and areas with more significant unfavorable technogenic influences. Such terrains include the extensive area of the Pannonian Basin and regions between the valley of the Southern Morava River and the Kopaonik Massif.

Groundwater

Water represents the most widely disseminated material in nature. It covers the surface of the Planet, permeates the Earth's crust, and takes part in natural processes. All forms of life require water, and it is found in the atmosphere, hydrosphere, cryosphere[40], biosphere, and lithosphere. Let us dwell on groundwater, i.e., water accumulated in permeable rocks of the lithosphere, as a very important reservoir of infiltrated atmospheric precipitation.

It is calculated that around 23.4 million km^3 of groundwater is found at a depth of up to 2 km, about half of it consisting of slightly mineralized (fresh) water[41]. Of the quantity of slightly mineralized water, something around 4 million km^3 lies in the zone of active exchange with surface water and precipitation. In any event, the amount of groundwater is enormous, 3,000 times greater than the quantity of water in all rivers of the Planet at any given point in time (B. J. Skinner, 1986). In terms of potential, groundwater is therefore unquestionably a primary resource for supplying the population, agriculture, and some industrial concerns with water of good quality.

The distribution of groundwater on our planet is very uneven, with the result that the amount of this essential raw material available to the population is very unequal. The geological structure of a territory is of great significance in the distribution of fresh (potable) groundwater. The

---

[40] That part of the Earth's shell with freezing temperature and ice.

[41] Not counting the groundwater of Antarctica, which is for orientational purposes estimated at 2 million km^3.

largest reserves of such water are characteristic of mountain and inter - mountain depressions built of coarse - grained loose materials, especially alluvial, alluvial - lacustrine, and fluvioglacial sediment, as well as coastal plains and fans built predominantly of sand and gravel. Very significant amounts of water are accumulated in regions with open karst and in artesian basins of the platform type.

It is also evident that the distribution of groundwater is subject to latitudinal zonality. Climate represents an especially significant factor here. The general rule is that regions with a humid climate have enough fresh water suitable for water supply or a surplus of it. On the other hand, there is a shortage of fresh groundwater in arid and semiarid regions, for example in dry regions of India, Afghanistan, China, and Mongolia, many African countries, America, and Europe.

Porosity and *permeability* are hydrogeological properties of rocks essential for accumulation of groundwater and its movement.

Total porosity of rocks refers to all spaces not occupied by solid mineral material, whereas the concept of *effective porosity* applies to those spaces in rock filled only by free (gravitational) water. Young loose material (gravel, sand, etc.) is characterized by intergranular porosity, hard carbonate sediments by cavernous (karstic - fracture) porosity, and most other kinds of hard rocks by fracture porosity (Fig. 2.26.).

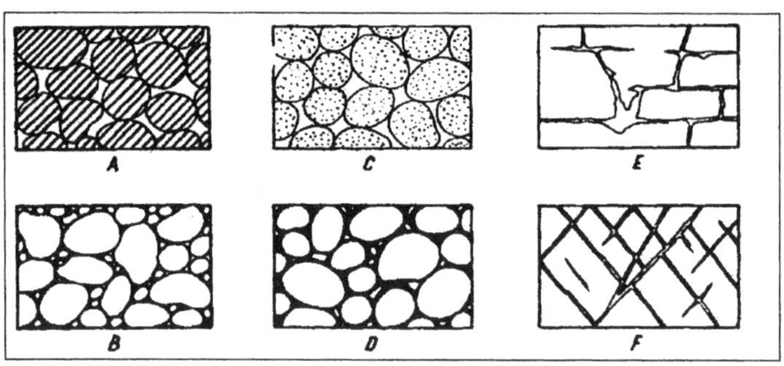

Fig. 2.26. Types of porosity after Meinzer. A, B, C, D - intergranural porosity (different grain - size); E - karst porosity; F - fracture porosity.

Permeability is the capability of rock to admit free (gravitational) water. According to the value of permeability, three *categories of rocks* are often singled out:

- *Highly permeable rocks* (gravel, sand, karstified limestones, dolomites, marbles);
- *Poorly permeable rocks* (deluvial and terrace sediments, marly limestones, serpentinites, peridotites, sandstones); and

- *Practically impermeable rocks* (most magmatic and metamorphic rocks, clays, flysch deposits, and diabase - chert formations).

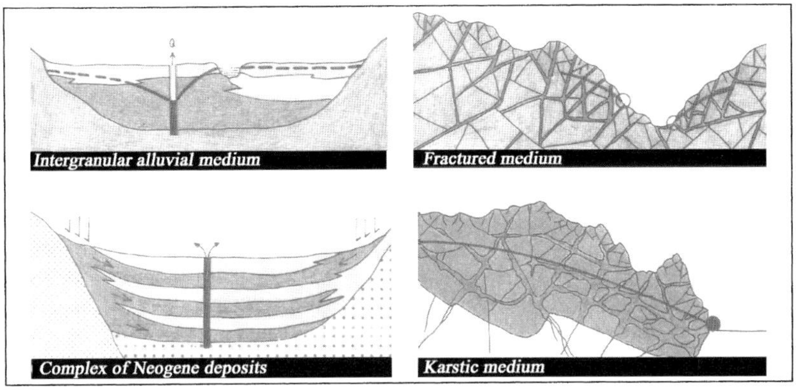

Fig. 2.27. Schemes of typical aquifers.

From the standpoint of water supply, the following four *types of water - bearing (permeable) media* are significant: 1) *alluvial media*; 2) *water - bearing horizons of artesian basins*; 3) *fractured media*; and 4) *karst media* (Fig. 2.27.). Most of the planet's sources of groundwater are linked with alluvial plains, and water of the very permeable gravel - sand layer is utilized (Fig. 2.27.). Karst water - bearing media are also very significant. Their water reserves are regularly evacuated through strong springs (Fig. 2.28.). Among other types of water - bearing media, sedimentary fans, fluvioglacial sediment, and permeable horizons of coastal plains stand out, in addition to artesian basins.

Fig. 2.28. Siphon karst spring Klokun (W. Herzegovina, Croatia).

The *composition of groundwater* is very diverse, since it depends on many factors: physico - geographic, geological, physico - chemical, physical, biological, and artificial. As a highly complex natural solution, groundwater contains practically all known chemical elements in the form of simple and complex ions and complex chemical compounds dissolved in the Earth's crust. More than 70 elements are found in groundwater. Exceptionally, the number of basic dissolved components in slightly mineralized (fresh) groundwater is relatively limited (Fig. 2.29.).

The complexity of composition of groundwater is dictated not only by the presence of a large number of chemical elements, but also by varying quantitative content of each of them - variable in different types of water - as well as by the diversity of soluble forms of each element. It is sufficient to say that groundwater regularly contains different ions of the bulk of chemical elements, dissolved molecules, gas - forming molecules, natural and artificial isotopes of many elements, involved organic compounds that form a completely unique class of dissolved substances, many living and dead microorganisms (bacteria), mechanical and colloidal materials of varying composition, complicated organominerals and other comples, etc. Different ratios of all these compounds, together with quantitative changes in the content of each element, determine the great diversity of groundwater, expressed in the existence of several thousand hydrogeochemical types.

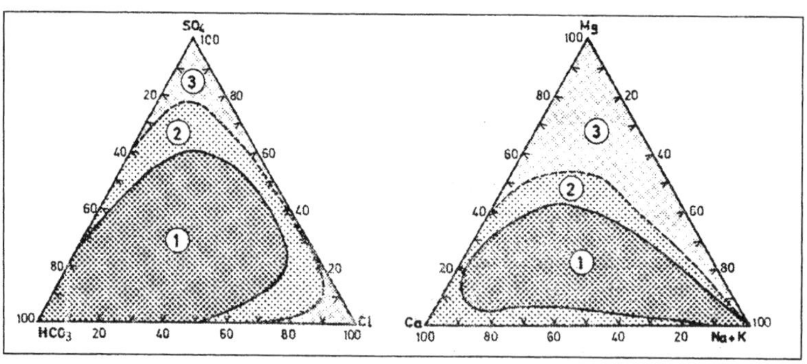

Fig. 2.29. Relative content of main dissolved components in drinking water (Davis, de Wiest, 1966).

Every genetic type of groundwater has its own characteristic properties. Thus, sedimentogenic waters differ from others in having relatively low content of SiO_2, high content of NH_4, and low K/Na, Li/Na, SO_4/Cl, F/Cl, and B/Cl ratios; on the other hand, volcanogenic and metamorphogenic waters are characterized by high content of SiO_2, low content of NH_4, and high Mg/Ca, K/Na, Li/Na, SO_4/Cl, F/Cl, and B/Cl ratios (D. E. White, 1960).

Because groundwater contains chemical elements in very different concentrations - from hundreds of grams per liter to billionths of grams and less - it is expedient to divide these elements into the following four groups depending on their content in water:

- *Macrocomponents*: Na, Ca, Mg, Cl, SO_4, and HCO_3; these components make up more than 90 - 95% of all dissolved salts and determine the chemical type of water;
- *Microcomponents*: Fe, Al, Br, I, F, B, Li, Rb, Ba, As, Sr, Mo, Cu, Co, Ni, etc; these components characterize the specific composition of water;
- *Ultramicrocomponents*: Au, Bi, Te, Cd, Se, etc.; these components are present in water in very small quantities; and
- *Radioactive elements*: U, Th, Ra, and Rn.

The mineralization of water and its pH value, redox potential (Eh), hardness, and aggressiveness represent the main indices of the chemical composition of water.

An especially significant index of the chemical composition of water is its *mineralization*, which represents the sum of all mineral substances in the water. This index has very different values, ranging from 0.1 to 650 g/l. Water with mineralization of less than 1 g/l is used for drinking purposes. Generally speaking, extremely low mineralization is characteristic of groundwater near larger glaciers and in tropical and subtropical regions (160 and 180 mg/l, respectively); groundwater in mountainous regions has low mineralization (about 200 mg/l); and groundwater in the region of old platforms is the most mineralized (having mineralization most often perceptibly above 500 mg/l). As a rule, mineralization increases more or less rapidly or slowly with depth. Thus, for example, in the central counties of Great Britain, it varies from 0.1 to 3.8 g/l in limestones nearer the surface; increases to 10.8 g/l at a depth of 3,200 m; and attains 100 g/l at depths of 5,000 - 5,100 m. The following types of water are distinguished according to mineralization:

- Slightly mineralized (fresh) water, up to 1 g/l;
- Brackish water, 1 - 10 g/l;
- Saline water, 10 - 100 g/l; and
- Brine, above 100 g/l.

Among *macrocomponents*, compounds of iron, manganese, nitrogen (ammonia, nitrates, and nitrites), phosphorus, and silicon (silicic acid) occur in addition to the three leading anions (chlorides, bicarbonates, and sulfates) and cations (calcium, magnesium, and sodium + potassium). As a rule, bicarbonate and calcium ions predominate in slightly mineralized water, whereas chlorine and sodium ions predominate in mineralized and highly mineralized water, with sulfate and magnesium ions occupying an intermediate position.

Although they do not determine the chemical type of groundwater, *microcomponents* can ert significant influence on formation of its specific characteristics and on biological processes in nature. The more important microcomponents belong to several groups:

- Microcomponents that form anions (I, Mo, As, and Se);
- Alkaline metals (Li, Rb, and Cs);
- Dispersed metals (Be);
- Chalcophilic elements (Zn, Cu, Pb, and Ag).

Table 2.9. gives hydrogeochemical characteristics of the more important microcomponents in groundwater.

The *pH value* represents a value that characterizes the activity or concentration of hydrogen ions in solutions. The concentration of hydrogen ions reflects the chemical (acidic or basic) reaction of water. The reaction of water is neutral at $pH = 7$, acidic at $pH < 7$, and basic at $pH > 7$. In groundwater, this value most often ranges from 5 to 8.

In addition to the pH value, the *redox potential (Eh)* erts significant influence on the form of elements in groundwater and conditions of their migration. Closely linked with the pH value, the redox potential fluctuates between 400 and 700 mV, Eh values of 100 to 300 mV indicating transitional redox conditions. Geochemical conditions of the environment can be estimated on the basis of this parameter.

The *hardness of water* is dictated by the content of certain salts such as Ca^{2+}, Mg^{2+}, Fe^{3+}, Mn^{2+}, Al^{3+}, Ba^{2+}, and Sr^{2+}, above all Ca^{2+} and Mg^{2+} ions. There are a number of classifications of groundwater on the basis of its hardness. Shown in Table 2.10., the classification of Klut (in °dH) is most often used in Yugoslavia. Water with hardness of up to 15°dH is best for drinking purposes.

The attention of investigators throughout the world is being increasingly drawn to the content of *natural organic components* in drinking water, as well as to the composition, distribution, and transformation of organic pollutants in different components of the hydrosphere. The influence of organic compounds in water on human health represents a special area of research.

The *microflora of groundwater* is a very important and active part of the hydrosphere. It is made up of tiny plant organisms, among which bacteria have been the main object of study. In the process of their vital activity, bacteria have performed and are performing enormous geological work on alteration of the chemical composition of groundwater.

TABLE 2.9.
Content of important trace elements in lithosphere and water ((μg/l).

Element	Content in lithosphere (without sedimentary cover) %	Content in water (μg/l)		
		Ocean	Low-mineralized drinking water (fon)	Maximal content in water
J	$5.0 \cdot 10^{-5}$	64	6	$120 \cdot 10^{3}$
Li	$2.0 \cdot 10^{-3}$	170	5	$320 \cdot 10^{3}$
Rb	$9.0 \cdot 10^{-3}$	120	3	$100 \cdot 10^{3}$
Cs	$2.0 \cdot 10^{-4}$	0.3	2	$20 \cdot 10^{3}$
Be	$1.5 \cdot 10^{-4}$	0.0006	0.2	$n \cdot 10^{3}$
Zn	$8.7 \cdot 10^{-3}$	5.0	74	$1 \cdot 10^{6}$
Cu	$6.5 \cdot 10^{-3}$	0.9	11	$1 \cdot 10^{6}$
Pb	$0.9 \cdot 10^{-3}$	0.03	6	$6 \cdot 10^{4}$
Ag	$9.0 \cdot 10^{-6}$	0.28	0.3	50
Mo	$1.3 \cdot 10^{-4}$	10	1.3	700
As	$1.9 \cdot 10^{-4}$	2.6	1	$n \cdot 10^{5}$
Se	$1.0 \cdot 10^{-5}$	0.009	1	300
U	$1.5 \cdot 10^{-4}$	3.3	3	$n \cdot 10^{4}$
Ra	$2.0 \cdot 10^{-10}$	$0.1 \cdot 10^{-6}$	$1.0 \cdot 10^{-6}$	$n \cdot 10^{-2}$

Slightly Mineralized (Fresh) Groundwater

Without doubt, slightly mineralized (fresh) groundwater represents the most important part of our water resources. In many regions of the world, this is the only source of water supply for the population and industry or for irrigation of agricultural land.

TABLE 2.10.
Classification of water according to hardness, after Klut.

Hardness ^{0}dH	Type of water
0 - 4	Very soft water
4 - 8	Soft
8 - 12	Moderate hard
12 - 18	Almost hard
18 - 30	Hard
> 30	Extremely hard

Slightly mineralized water can have different *chemical composition.* The overwhelming bulk of such water belongs to the HCO_3-Ca and HCO_3-

SO_4-Ca water types. It is characterized by pH values from 6.0 to 8.0 and redox potentials ranging from +500 to - 200 mV.

All slightly mineralized groundwater contains diverse *organic components*. The total content of organic substances expressed by the value of organic carbon on average comprises 25 mg/l. Among organic components that have been discovered, the greatest interest has been shown in various organic acids, phenols, and humic and bituminous components.

Slightly mineralized water contains a large number of *microorganisms* (bacteria, algae, fungi, protozoa, and viruses), but considerably less than in the soil. Water represents the normal place of residence for some of these tiny and generally very numerous organisms, while others (so - called contaminating microorganisms) enter water from the air or soil, from plants, and from excrement of human or animal origin. From the standpoint of human health, the most significant contaminants of drinking water are microorganisms entering it from human and animal excrement due to unhygienic disposal of sewage. Drinking water can contain pathogenic microorganisms that cause disease in plants and animals, and consumption of such water can be risky for humans.

With rare exceptions, all groundwater is to a certain extent radioactive. However, the *content of radionuclides* in it is considerably lower than in rocks. The greatest concentration of radionuclides is found in the groundwater of uranium deposits.

Hydrogeochemical zonality is expressed in hydrogeological structures with slightly mineralized water. It consists of regular changes in general chemical and gas composition, concentrations of microcomponents and organic substances, and pH and Eh of the water. *Horizontal (latitudinal, climatic)* zonality is characteristic of shallower water - bearing horizons. It is manifested through increase in the total mineralization of groundwater, as well as in the concentration of basic anions and cations, with transition from a humid zone to an arid one. The concentration of organic substances of the humic series decreases in this case, Eh and pH parameters rise, and increases are recorded in the concentrations of many anionic elements such as fluorine, molybdenum, arsenic, selenium, etc. Manifestation of *vertical hydrogeochemical zonality* depends upon the composition of hydrogeological structures. In artesian basins, such zonality is expressed in regular alternation of geochemical types of groundwater in the vertical profile of a given structure. In regions of hydrogeological massifs, regular transformations of geochemical types of groundwater are recorded with increase in the depth of their occurrence.

Under natural conditions, slightly mineralized groundwater is characterized by good *physicochemical properties*. Thus, the quality of water from fractured and karst media, for example, is usually excellent. An exception to this is observed in the case of salinization of water in coastal zones and arid regions, due to intrusion of seawater in the former instance and concentration of salt by means of evaporation in the latter. In water - bearing media with intergranular porosity such as alluvial media and media of artesian basins, the chemical composition of water is for the most part favorable; negative quality indices for such water are observed more rarely and involve content of iron, manganese, and sometimes humic substances in it and arsenic.

The content of *radionuclides and toxic chemical components* in slightly mineralized groundwater is predominantly low, below limits permitted by "Guidelines for Drinking Water." Risks to human health caused by higher (or lower) concentrations of these elements will be discussed in the corresponding chapter of the book.

The degree of *natural protection of groundwater* from pollutants depends mainly on geological (hydrogeological) factors. For example, water - bearing artesian horizons as a rule are protected from sources of pollution on the surface because they are covered by a thick overlying layer of clay (Fig. 2.27.). In the case of alluvial gravel - sand horizons, apart from possible lateral influence of a polluted surface stream, the danger of pollution from the surface also exists. The degree of protection then depends on thickness and permeability of the overlying layer of clay: if it is thinner than 3 - 4 m, there exists a greater or lesser possibility of influence from agricultural and other pollutants. Sporadic shallow groundwater in fractured rocks is very sensitive, a circumstance manifested in bacteriological contamination of the majority of springs through which such water is evacuated. Karst media are without doubt the most sensitive, since carbonate rocks in fold (geosynclinal) regions are for the most part exposed and very permeable, and the risk of pollutant infiltration is especially heightened by the direction of *permanent ponors* and *karst springs* (Fig. 2.8). The fact that karst water has so far predominantly retained its high quality is primarily due to the limited presence of industrial and other polluters in more inaccessible mountainous terrains.

In any event, to be used for drinking purposes, water must satisfy norms prescribed for physicochemical parameters, content of toxic chemical components, parameters of radioactivity, and microbiological and organoleptic properties. Special (preventive) *sanitary measures* are taken to protect sources, and appropriate purification systems remove undesirable ingredients from raw water. Laws require formation of the following *zones of sanitary protection* of water sources: *zone I*, where a strict regime of direct protection is established around the source from which water is obtained; *zone II*, where measures of "bacteriological" protection are implemented to prevent bacteria from entering the water;

and *zone III*, where chemical pollution is controlled over the entire area feeding the water source. In karst regions, the first or second zone must be expanded by the addition of a narrower zone encompassing privileged canals (faults) connected with the ponor zone.

Because of its quality, man has always been interested in groundwater for drinking purposes. It can therefore be freely stated that investigation of groundwater and its use as drinking water already began in the distant past. The first information about artificial structures for removal of water from wells date from the 3rd Millennium B.C. Large centralized systems of water supply already existed during the period of flowering of Ancient Greece and Rome. The first gravitational aqueducts of Paris and London appeared at the end of the 12th Century and in the 13th Century, respectively. In the year 1980, apart from 2,306 km^3 (or 73.1×10^3 m^3/s) of water used in agriculture, 1,194 km^3 (or 34.6×10^3 m^3/s) was used throughout the world to supply industrial concerns and the population, and groundwater accounted for a significant share of it. A high level of groundwater use is evident in most countries of Europe, Africa, North America, and Australia, and water supply in arid and semiarid zones of the globe is almost completely based on capture of groundwater. Thanks to the existence of deep water - bearing horizons, new oases have sprung up around boreholes that struck those horizons in many desert territories, the Sahara above all. The first borehole drilled in Gwardaja in Algeria struck a water - bearing horizon at a depth of 1,200 m, and water gushed in a jet 180 m high. Until the discovery of that rich horizon, oases in the Algerian Sahara were under threat of extinction due to lack of water and sandstorms. Limitation of resources of sprinkling water led to the appearance of salt on the surface of the ground, which resulted in the death of date palms. With the beginning of exploitation of the indicated horizon, salt was washed from the ground and dunes were consolidated by planting vegetation and in other ways. We note that the area of this horizon in North Africa is exceptionally great, extending to Libya in the east and for a distance of hundreds of kilometers from the Atlas Massif in the south.

Irrigation of desert regions has played a visible role in the history of mankind. Many states of the East, Egypt and Babylonia among others, arose and achieved a high degree of culture precisely on irrigated land. In most desert regions, however, reserves of groundwater are shrinking every day due to shortage of precipitation without renewal.

Until recently, man regarded water as a God - given "free" resource that can be utilized without limitation wherever it is found. But with the advent of the Industrial Revolution, living conditions changed suddenly: enormous urban populations arose, industrial and agricultural installations appeared, the population and standard of living grew at an accelerated rate, and the need for pure water increased accordingly. On the other hand, pollution of streams and reservoirs (and some groundwater too) has reached alarming levels, while exploitation of water resources has been irrational, sometimes literally greedy. We are thus confronted with one of the basic negative features of the modern world, namely the **global water problem** or the **problem of pure water**.

Due to the indicated problem and the anyway unequal distribution of water resources, the problem of supplying the population with water has become acute in large areas of our planet. Thus, in 1975 already, only 38% of the populations of developing countries had access to hygienically acceptable drinking water, with the result that nearly 800 million people suffered from gastroenteritis, malaria, schistosomiasis, and oncocercosis. The consequences of the shortage of pure water are even more serious today and are increasingly affecting countries of the developed world as well.

The global water problem must be resolved by a wide range of activities that cannot be postponed. Such activities should take two main directions: a) implementation of urgent measures for *protection of water resources from pollution*; and b) *protection of water from overexploitation*. In the distant future, man will be forced to employ more costly approaches to ensure an adequate amount of drinking water, namely desalinization of mineralized (sea and underground) water and transport of icebergs to regions of the Earth with a water deficit, and thereby shift the water balance in his favor. Until then, solutions to the problem of supplying the population with water of good quality will be found in more organized and intensive exploration, and in rational use and better protection of groundwater sources. Man is not even close to having exhausted the unlimited possibilities of regulating the regime of groundwater in especially widespread permeable rocks, and almost nothing has been done to reduce the loss of much good water into the sea (M. Komatina, 1990).

Mineral Waters

By mineral waters, the majority of investigators mean all waters with a dissolved ingredient content of more than 1,000 mg/l, as well as natural waters with low mineralization if they contain one or several specific components that ert therapeutic action on the human organism. In order for groundwater to be classified as mineral waters, they must contain specific components such as Li, Sr, Ba, Fe, Mn, Br, I, F, B, H_2S, $HSiO_3$, CO_2, and Ra. For therapeutically active components, the following values of content (mg/l) are prescribed: iron, 10; arsenic, 0.7; iodine, 1.0; titrated sulfur, 1.0; carbon dioxide, 250; and radon, above 185 Bq/l[42].

The very concept of mineral water differs to some extent in different countries. By mineral waters, French scientists most often mean waters with therapeutic properties and acknowledged as such by the Medical Academy of France. In Belgium, Spain, Italy, and Switzerland, it is also held that the term *mineral water* must be linked with its therapeutic properties. Water can be called mineral water if it erts pharmacological action on the human organism, which can be caused by the presence or absence of any specific elements (or element) regardless of their (its) quantitative content.

Mineral waters have been used to treat diseases since ancient times. Folk wisdom very early called attention to the medicinal properties of many mineral springs and elaborated practical means of using them. Moreover, mineral waters in the infancy of man, like other natural phenomena, were the object of blind admiration and the most fantastic speculations as to their origin. This can be in some measure ascribed to the fact that many sources of mineral waters were hot, gas - bearing, or fountaining springs. Especially grandiose occurrences of hot vapor and water are typical of regions with *"fire - bearing mountains"* or volcanoes, whose nature were unclear and gave rise to many legends. The views that arose as to the nature and origin of mineral waters were in Ancient Greece and Rome mechanically transferred to mineral springs situated far from the foci of volcanic activity.

In Hellenic science, a fairly clear idea prevailed as to the origin and properties of natural waters. Already in the First Century A.D., the well - known physician Archigenes divided mineral waters into alkaline, iron, saline, and sulfuric categories. The remains of marble basins and old caps of mineral water sources dating from the Roman epoch can be seen in many spas of Italy, Austria, Germany, France, Hungary, Romania, Yugoslavia, Croatia, Bulgaria, Libya, Algeria, and other countries.

In contrast to the period of the Roman Empire, massive utilization of mineral waters stopped in the Middle Ages, to be revived only in the 19th Century. The scientific study of mineral waters and their utilization was formed prior to the end of that century.

Distribution of mineral waters. The formation, renewal, and evacuation of mineral waters are processes closely linked with definite geological environments and certain structures in constitution of the Earth's crust, a knowledge of which is of great importance. A territory within whose boundaries certain mineral water groups are found is called

---

[42] In the former Czechoslovakia, appreciably greater minimal levels of specific components (mg/l) are prescribed: CO_2, up to 2,000; Li, 5; Br, 25; Iz, 10; H_2S, 5-10; and radioactivity, 36.2 Ki/l.

a *mineral water province.* To be more precise, a province represents space on which certain mineral water groups occurring there are linked by the entirety of structural, hydrogeological, hydrogeochemical, hydrothermal, and other characteristics. For example, three mineral water provinces are isolated on the vast territory of what was once the USSR (E. V. Posokhov and N. I. Tolstikhin, 1977). The *first province* is characterized by prevalence of the carbonated mineral waters and encompasses regions with current and recently extinct volcanic activity (the Caucasus, Pamirs, Kamchatka, etc.). The *second province* is marked by the presence of nitric thermal waters, includes regions with recent seismic occurrences in the East and South of the former USSR, and borders on the first province. The *third province* is linked with platform regions and boundary zones with geosynclines, where cold and thermal brines, saline waters, and N_2 - methane waters are formed.

Classification of mineral waters. A great number of mineral (medicinal) waters are present in nature. They can be classified on the basis of a number of criteria such as total mineralization, ion and gas composition, content of active therapeutic components, radioactivity, acidity or alkalinity, and temperature.

Among the many classifications of medicinal waters and mineral waters in general, the *genetic classification of V. V. Ivanov and G. A. Nevraeva* (1964, supplemented 1977) stands out by virtue of its completeness. The indicated classification makes it possible on the basis of the ionic composition of different mineral waters to gain an idea of their hydrogeological and balneological properties and ascertain from mineralization values their suitability for treatment by drinking or bathing. Medicinal waters are divided into four basic groups on the basis of mineralization:

1. *Waters with elevated mineralization* (1 to 5 g/l), which when consumed by drinking act on the human organism in a manner similar to the action of "ordinary" slightly mineralized water;
2. *Waters with medium mineralization* (5 to 15 g/l), whose osmotic concentration approaches that of blood plasma; these water are most suitable in balneology and are also imbibed as medicine;
3. *Waters with high mineralization* (15 to 35 g/l), mainly used for therapeutic bathing, but certain types of chloride - hydrocarbonate water of the sodium group are also imbibed;
4. *Brines* (water with mineralization of 35 to 150 g/l) in natural form are used exclusively for bathing purposes. In exceptional cases, brines with mineralization greater than indicated can also be used in balneology, but in this case they are preliminarily diluted with slightly mineralized water or water with elevated mineralization.

Certain gases, organic substances, trace elements, and radioactive elements can be balneologically active. Among gases, for example, CO_2, H_2S, and Rn have such properties. From the balneological standpoint, trace elements can be divided into the following four groups:

Group I - elements with expressed pharmacological action (Fe, Co, As, I, Br, and possibly B);
Group II - elements with precisely established influence on hormonal and enzymatic processes in the organism (I, Fe, Cu, Mo, Zn, Co, Mn, and possibly Ni and Ba);
Group III - elements toxic for the human organism (As, Pb, Se, He, V, and F); and
Group IV - elements that have been discovered in human tissues and fluids, but whose biological role has not yet been established (Ti, Zr, Ir, Cs, Ge, and many others).

In regard to content of specific components, Ivanov and Nevraev single out eight balneological groups of mineral waters, the more important characteristics of which are noted below.

1. *Medicinal mineral waters without specific components and properties.* Such mineral waters ert balneological action on the human organism as a result of ionic composition and mineralization. Chloride and sulfate ions are prevalent in their chemical composition. In regard to gas composition, these waters belong to the nitrogen and methane types. Their mineralization can attain 150 g/l;
2. *Carbonated mineral waters.* Such mineral waters are very widely disseminated in nature. Their medicinal effect is determined by high concentration of carbon dioxide, as well as by its ionic composition and mineralization value. They have varying chemical composition, HCO_3^- being dominant among anions. Mineralization of these waters fluctuates from fractions of a gram per liter to 90 g/l. Of all types of mineral water, they are characterized by the greatest saturation with gas. Apart from refreshment, carbonated waters are used to treat diseases of the stomach, intestines, bile and urinary ducts, etc. Today it is customary to isolate five types of carbon dioxide waters;
3. *Hydrogen sulfide (sulfide) mineral waters.* Their medicinal properties are determined by free hydrogen sulfide and hydrosulfide ions. Mineral waters of this group are characterized by great diversity of chemical composition, mineralization, and H_2S and HS^- concentration. They include hydrocarbonate, sulfate, and chloride waters with high mineralization, often above 500 g/l. In balneology, bathing in such waters is used to treat certain diseases (diseases of the skin, rheumatic diseases, nervous disorders, etc.). On the basis of concentration of hydrogen sulfide and hydrosulfide ions, these waters are divided into four subgroups. With

respect to the conditions of their formation, nitric, methane, and H_2S - CO_2 waters are distinguished;

4. *Radioactive mineral waters.* These waters have elevated content of radioactive elements. The following lower limits are prescribed: Ra, above $1 \cdot 10^{-11}$ g/l; U, above $3 \cdot 10^{-5}$ g/l; and Rn, above $1.85 \cdot 10^{2}$ Bq/l. Radium, radium - radon, radon, or uranium waters are distinguished, depending on which radioactive element is prevalent. Radon waters are the ones predominantly used in balneology. They are employed to treat diseases of the nervous system, cardiovascular diseases, skin ailments, and gynecological diseases;
5. *Iron waters, arsenic waters, and waters with elevated content of Mn, Al, Cu, and Zn.* The medicinal properties of this group of waters are determined by the presence of iron (lower limit of 20 mg/l), arsenic (lower limit of 0.7 mg/l), and other metals;
6. *Bromine and iodine waters.* Patients drink these waters and bathe in them. Such waters need to contain more than 25 mg/l of bromine or more than 5 mg/l of iodine, and their mineralization must be such that their drinking is permissible;
7. *Waters with high content of organic substances.* The basic balneological properties of such waters are linked with action of the complex of dissolved organic substances on the human organism. The total content of organic substances in these mineral waters varies within wide limits, from several mg/l to more than 400 mg/l and higher in exceptional cases. Mineral waters of peat bogs, mud flats, and petroleum deposits are especially rich in organic substances; and
8. *Silicic thermal springs.* They are represented by thermal and highly thermal waters with temperatures above 35°C containing not less than 50 mg/l of silicium in the form of silicic acid (H_2SiO_3). These waters often contain other medicinal components (Rn, CO_2, and trace elements) as well.

The frequently used *classification of mineral waters according to their gas composition* divides them into the following six groups: *carbonated waters; H_2S - CO_2 waters; hydrogen sulfide (sulfide) waters; nitric waters; N_2 - methane waters; and methane waters.*

There are a number of *classifications of mineral waters on the basis of temperature.* According to one of them, all mineral waters are from the balneological standpoint relegated to one of the following five groups: *cold*[43], with temperatures below 20°C; *warm*, 20 - 37°C; *thermal*, 37 - 42°C; *highly thermal*, 42 - 100°C; and *superheated*, above 100°C.

On the basis of *pH value*, one of the factors that in significant measure determines the physiological action of water on the human organism, the following groups of mineral waters are distinguished: *strongly acidic* (pH < 3.5); *acidic* (pH 3.5 - 5.5); *weakly acidic* (pH 5.5 - 6.8);

---

[43] More recently, cold water has been taken to mean groundwater with temperature below the average-annual temperature in a given locality.

neutral (pH 6.8 - 7.2); *weakly alkaline* (7.2 - 8.5); and *alkaline* (pH > 8.5). A strongly acidic reaction is characteristic of the sulfate waters of mineral deposits, highly thermal waters of active volcanic regions, and ultra - strong calcium chloride brines. All carbonated waters are characterized by a weakly acidic reaction. Neutral and weakly alkaline reactions are typical of sulfide waters and waters with elevated mineralization. Slightly mineralized waters of nitric thermal springs have an alkaline reaction.

Thermal Waters

Occurring primarily in volcanic regions, the numerous and diverse forms of thermal waters and vapors (hot springs, geysers, and fumaroles) have unquestionably attracted man's attention since prehistoric times for their medicinal properties. That thermal springs were used in the Stone and Bronze Ages for various purposes is indicated by an abundance of archaeological finds in the cultural layer of many sites on the territory of Iran, New Zealand, Kamchatka, and other regions of the world. In the Antic Period (Fifth to Fourth Century B.C.), the works of Democritus, Aristotle, and Pliny the Elder contain views on the genesis of mineral and thermal waters that in general features are largely consistent with modern views.

During the Middle Ages, the general shutdown of cultural and scientific development was reflected in the study and utilization of thermal waters, although a certain advance was perceived in the period of the Crusades (1096 - 1270) and during the Hundred Years' War (1337 - 1453). Interest in mineral and thermal waters grew during the 16th and 17th Centuries in Western Europe. In the 19th Century and first half of the 20th Century, numerous and diversified investigations of thermal waters were carried out, primarily for *healing purposes*. The thermal waters of popular spas were studied with particular care, and increase in the number of chemical analyses of water was especially rapid.

Starting from 1920 in the United States and predominantly in the 50's and 60's of the 20th Century in many other countries (the Soviet Union, New Zealand, Japan, Mexico, Kenya, Italy, France, Yugoslavia, etc.), exploratory drilling was initiated for the purpose of discovering geothermal resources in many localities marked by the occurrence of thermal springs. At the beginning of 1970, such drilling in the USA had been conducted in more than 35 regions and resulted in the discovery of six productive geothermal fields. Ten years later, 16 geothermal power plants were already in operation on this territory.

According to current ideas, the bulk of thermal waters and mixtures of steam and thermal water are formed at relatively great depths mainly from meteoric (infiltration) water, which is then (as a result of rising filtration along fault zones) evacuated on the Earth's surface or accumulated in permeable horizons of sedimentary basins. *Geosynclinal and platform regions* differ markedly with respect to conditions of distribution, formation, and evacuation of thermal waters.

In *geosynclinal regions*, thermal waters and steaming thermal springs are formed predominantly in areas of Cenozoic volcanism and more recent tectonic movements, where we find numerous sites of their evacuation linked with zones of large fault structures (M. Komatina, 1995). The waters usually have mineralization of the order of 1 g/l, more rarely 20 g/l, and in exceptional cases up to 100 g/l (New Zealand, Indonesia). The greatest possible hydrothermal activity occurs in regions with intensive manifestation of Cenozoic volcanism. Thermal waters and steaming thermal springs with temperatures of up to 100°C in the places where they are naturally evacuated and from 270 to 390°C in boreholes 2 - 3 km deep (California, Mexico, New Zealand) are found near active volcanoes, where they form deposits. The waters have diverse chemical composition and elevated concentrations of various microcomponents.

Platform regions are characterized by enormous reserves of thermal waters accumulated in deep permeable layers. The waters are very diverse in regard to temperature, composition, and mineralization. Young platforms are distinguished by high output of boreholes and high temperature of the water. On the other hand, brines with mineralization of up to 620 g/l (the Siberian Platform) and temperatures of 40 - 70°C are widespread in old platforms.

Depending on geological, geochemical, and geothermic conditions, thermal waters and steaming thermal springs can possess highly varied ion and gas composition and be characterized by a wide range of mineralization and very different temperatures. There are today a number of classifications of such waters. Thus, for example, V. V. Ivanov (1961) distinguishes the following five groups of thermal springs, primarily depending on gas composition:

- H_2S - CO_2 (fumarolic) thermal springs;
- Carbonated thermal springs;
- N_2 - CO_2 thermal springs (steaming thermal springs);
- Nitric alkaline thermal springs; and
- Methane thermal springs.

The Russian authors E. A. Baskov and S. N. Surikov (1975, 1989) systematized thermal waters according to acid - base properties (pH values), anion composition, and gas composition. With respect to acid - base properties, they distinguished two main groups, viz., *acidic* (pH < 4.5) and *alkaline* (and *weakly acidic*) (pH > 4.5).

On the basis of prevailing anion composition, **acidic thermal waters** are divided into chloride, sulfate, and silicic categories.

1. According to Baskov and Surikov, *acidic chloride thermal waters* are very widespread in areas with current volcanism (Eastern Kamchatka, the Kurile Islands, volcanic zones of Japan, the Philippines, Indonesia, the Cordillera de los Andes, etc.). The vents of thermal springs here appear on

fumarole fields of active volcanoes and emit natural condensates of fumarole gases. Temperatures of these thermal springs usually fluctuate from 70 to 100°C. Low pH values are characteristic: predominantly less than 2 - 2.5, often as low as 0.1 - 0.5. Mineralization is fairly variable, often from 3 - 5 to 10 - 50 g/l. Acidic chloride brines with mineralization of up to 300 - 400 g/l are formed on fumarole fields only in regions with developed halogenic formations (in the Danakil Depression in Africa, for instance). Acidic chloride fumarolic thermal springs are often characterized by high concentrations of the following components (mg/l): F, up to 400 - 800; H_2SiO_3, 200 - 400 (up to 800); H_3BO_3, 100 - 500 (up to 500 - 600); Fe, 100 - 600 (up to 5,000 - 10,000); Al, 100 - 500 (up to 2,000 - 8,000); and many other metals. With respect to gas composition, these waters are mainly of the H_2S - CO_2 and carbonated types.

2. Acidic chloride thermal waters represented by highly concentrated chloride, calcium - sodium, and calcium brines (with mineralization usually greater than 25 - 270 g/l, up to 400 - 420 g/l) are widespread in artesian basins rich in halogenic formations;
3. *Acidic sulfate thermal waters* usually have their vents on fumarole fields, in craters and calderas, and on the slopes of active volcanoes. The vents of these thermal springs for the most part occur in the form of "boiling" mud kettles and scattered ascending springs with water temperatures of 30 - 40, more often 70 - 90°C. Value pH usually varies from 1 - 2 to 3, mineralization from 1 to 3,5 g/l (92 - 106 g/l in exceptional cases on Java). The main anions are SO_4^{2-} and HSO_4^-. The cation composition is more complex, often with Fe^{3+}, Al^{3+}, NH_4^+, and H^+ as the main cation. These thermal waters are characterized by high concentrations of H_2SiO_3, usually 200 - 300 mg/l and higher. The following metals are found in them (mg/l): Cu, up to 0.1; Zn, up to 1 - 5; Mo, up to 0.3; and others. Gas composition is usually characterized by prevalence of CO_2 with a significant admixture of H_2S and (or) nitrogen; and
4. *Acidic silicic thermal waters* are known in regions of current volcanism on Iceland. The temperature of vapor condensates varies around 100°C, pH around a value of 4.4. The content of H_2SiO_3 comprises 234 mg/l at a mineralization of 307 mg/l. With respect to gas composition, these thermal waters are of the carbonated type.

The group of **alkaline (and weakly acidic) thermal springs** embraces thermal waters with pH values above 4.5. These waters are very diverse with respect to ion - salt and gas composition. They differ from the group of acidic thermal waters primarily in the presence of carbonate and hydrocarbonate compounds.

1. *Alkaline (and weakly acidic) chloride thermal waters* are very widespread, but have been most thoroughly studied in areas of current volcanism and artesian basins. In regions of current and early - Quaternary volcanism, CO_2 is prevalent in the gas composition of chloride thermal springs. Water temperature in springs situated at the foot of volcanoes generally

fluctuates from 40 - 50 to 80 - 90°C. Incidentally, chloride thermal springs have been discovered by many boreholes at depths of as much as 1,000 - 2,500 m, where temperatures attain 200 - 300°C. Mineralization of the water for the most part varies from 1 - 3 to 15 g/l. These waters are characterized by elevated content of the following components (mg/l): arsenic, up to 8 - 10, more rarely 40 - 60; boron (H_3BO_3), up to 50 - 100, more rarely 500 - 1,200; lithium, up to 2 - 10, more rarely 20 - 50; rubidium, up to 1 - 3, more rarely as much as 5 - 10; cesium, up to 2 - 5, more rarely as much as 10 - 20; H_2SiO_3, up to 100 - 200, more rarely as much as 300 - 500. Sodium is prevalent in their cation composition.
Nitric thermal saline waters and brines are widespread in many artesian basins. They are characterized by prevalence of chlorine and high content of calcium. Also represented here are chloride methane and N_2 - methane thermal springs, with mineralization usually from 15 - 20 to 200 - 300 g/l and relatively high content of iodine (up to 120 mg/l in artesian basins of Japan);

2. *Alkaline (and weakly acidic) sulfate thermal waters* occur relatively rarely. Springs of sulfate thermal waters are known in regions with active volcanoes on the Kamchatka Peninsula, in the Kurile Islands and islands of Japan, and in California. Water temperature attains 60 - 100°C, and mineralization varies predominantly within limits of from 0.8 to 1.5 g/l;
3. *Alkaline (and weakly acidic) hydrocarbonate (carbonate) thermal waters* are found on a large area in all regions with current and Quaternary volcanism. The conditions of their distribution in hydrogeological structures are characterized by great complexity and diversity, especially in regions with current volcanism. In New Zealand, the temperature of spring water attains 80 - 85°C, mineralization 1 - 2 g/l. Hydrocarbonate nitric thermal waters are widely disseminated in regions of fold mountains with more recent uplifts. As a rule, they are sodium waters with low mineralization (not more than 0.6 g/l) and elevated content of fluorine (up to 10 - 15 mg/l) and silicic acid (of the order of 50 - 100 mg/l). Among gases, nitrogen is by far the most prevalent; and
4. *Alkaline (and weakly acidic) silicic thermal waters* have greater distribution on Iceland and in regions with current volcanism. Water temperature at the Earth's surface is about 100°C. Silicic acid is markedly dominant in composition of the water, as is sodium among cations. The primary gases are hydrogen and carbon dioxide.

Bacteriological composition of thermal waters is diverse. Many investigators have shown that different groups of aerobic and anaerobic microorganisms adapted to extreme conditions exist in thermal waters even at fairly high temperatures (up to 100°C and higher). Moreover, activity is manifested by anaerobic bacteria (as shown by investigations of the microflora in thermal waters of deep horizons of artesian basins). The presence of microbes was established in acidic thermal springs with temperatures of 70 - 100°C in Iceland, Yellowstone Park, New Zealand,

and the Uzon Caldera in Kamchatka. Such microbes were called *caldera microbes* from the place of their finding (T. D. Brok, 1978). Certain data indicating even better the ability of microorganisms to adapt to high temperatures were obtained in investigating the bacteriological composition of high - temperature jets (with temperatures up to 380°C) in areas of mid - ocean ridges at depths of 2,000 - 2,500 m and more (J. A. Barros et al., 1982). A complex community of bacteria was established on that occasion.

Hydrogeothermal Energy

Heat accumulated in the Earth's crust unquestionably represents the greatest source of energy available to man today. The role of groundwater in redistribution and transfer of heat from the Earth's interior has taken on great significance in the past 30 years, and *hydrogeothermal energy* (the energy of thermal waters) is every day becoming a more important source of energy for mankind[44].

It is known that the heat regime of the Earth was formed and is being formed under the influence of many sources of heat energy, external and internal. Solar radiation is the most important external (cosmic) source of heat energy, while the heat released during radioactive decay is especially important among the diverse heat sources located in the Earth's crust. Particularly interesting to man at the present time are effects of current and young - Cenozoic volcanic activity, which erted crucial influence on the conditions of distribution and formation of thermal waters and vapors, as has already been discussed.

Fig. 2.30. shows zones with intensive hydrogeothermal manifestations in the Earth. This does not mean that other areas of the Planet (apart from the territory of old crystalline shields) are uninteresting from the standpoint of utilization of geothermal energy. For example, the expansive Pannonian Depression is characterized by anomalously high values of heat flow, from 100 to 120 mWm^{-2} (according to D. S. Chapman, H. V. Pollack, and V. Cermak, 1979, the world average of heat rising to the surface of the Earth from its interior comprises 59 mWm^{-2}).

In the majority of cases, thermal waters represent a complex mineral raw material: they often possess balneological properties; they act as a natural carrier of heat energy in the interior of the Earth; they serve as a source of valuable microcomponents (or their compounds), salts, and auxiliary gases; and they are used in technological processes. Because of the energy crisis in many countries, man in the last decades has

---

[44] We are today only in the phase of intensified work on development of technology for utilization of heat accumulated in rocks or *petrogeothermal energy*.

increasingly relied on so - called alternative energy sources, including hydrogeothermal energy.

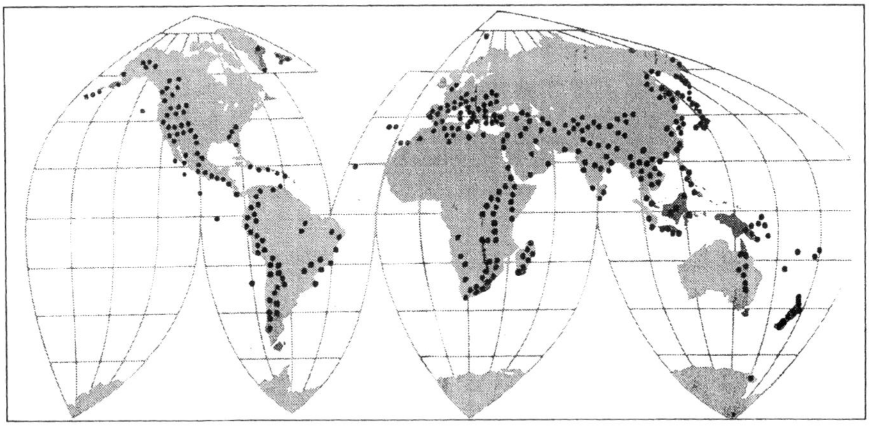

Fig. 2.30. World distribution of hydrogeothermal location.

In keeping with their potential for use as a source of energy, thermal waters can be relegated to one of the following classes: 1) *high - potential* waters (with temperatures above 150°C), which can be efficiently used for direct production of electrical energy; 2) *medium - potential* waters (with temperatures of 100 - 150°C), suitable for effective heating of industrial, agricultural, and communal - residential buildings, as well as for production of electrical energy in conjunction with isobutane, freon, and other liquids having a low boiling point; and 3) *low - potential* waters (with temperatures of 20 - 100°C), suitable for thermotechnological uses (warming of the aforementioned kinds of buildings), mainly with the aid of heat pumps.

Work on discovery and exploitation of high - potential reserves (steam - water mis or dry steam) has been carried out since the 50's of the last century in a number of countries, and 188 hydrogeothermal electrical power plants with total capacity of 4,764 MW were already operating throughout the world in 1986. With a capacity estimated at 1,000 MW, the largest plant of this type (The Geysers) was erected on a deposit of geothermal energy with an area of 52 km^2 in California.

Together with use of high - temperature geothermal resources to obtain electrical energy, hot waters with lower temperatures are being widely investigated and exploited. The captured thermal waters are used (sometimes with the aid of heat pumps) for balneology and recreation, to warm buildings, and (to a lesser extent) for heating of settlements.

Highly Mineralized Waters and Brines

As we have seen, highly mineralized waters and brines (with mineralization of 15 - 35 g/l and 35 - 650 g/l, respectively) serve as surface markers of many terrains throughout the world, especially areas of current and younger Cenozoic volcanism. They are present in enormous quantities in the deep horizons of great artesian basins.

Basins with halogenic (sulfate, halite, and potassium) formations contain groundwater with a wide range of mineralization: brines of sodium, calcium, and (more rarely) magnesium composition. In such cases, brines usually have sodium - chloride composition at mineralization below 280 - 350 g/l, and calcium - chloride or magnesium - chloride composition at higher mineralization. Both contain little sulfate, but are rich in various chemical elements.

Brines of rock salt deposits are characterized by low concentrations of calcium, magnesium, strontium, potassium, lithium, bromine, etc. The content of calcium and sometimes magnesium and strontium increases appreciably in carbonate and gypsum - bearing rocks. When potassium salts dissolve, the content of potassium and bromine increases sharply.

Calcium - chloride and magnesium - chloride brines differ in composition from the water - bearing rock itself. Due to the high degree of mineralization (from 290 - 350 to 600 g/l and higher), which exceeds the solubility of NaCl in water, these brines are called *concentrated brines* (E. V. Pinneker, 1968). They contain the following components in elevated concentrations: potassium, up to 40 g/l; bromine, up to 10 g/l; strontium, up to 8 g/l; hydrogen sulfide, up to 2 g/l; and lithium, barium, boron, and other elements.

Characteristics of highly mineralized thermal waters and brines formed in zones of current and younger - Cenozoic volcanism and in artesian basins have already been discussed. The *metal content of thermal waters* in different regions of the Earth is examined more closely below. To be more precise, high and extreme levels of certain metals are considered, using the book of Baskov and Surikov "**Thermal Springs of the Earth**" (1989). This question is interesting because it is important to establish the possibility of using such waters for balneological and industrial purposes, as well as their possible toxicity.

Copper represents a very widely disseminated metal in thermal springs of different regions. The content of this metal in different types of thermal waters usually fluctuates within limits of from 0.05 to 0.25 mg/l. The highest concentration (18.5 mg/l) has been established in acidic chloride thermal waters of the Dallol Dome (Ethiopia), and available information indicates anomalously high (up to 51 mg/l) content of copper in acidic fumarolic thermal springs of the volcano Ebeko (Kurile Islands).

Mercury is present in the greatest amounts in thermal waters and gases of regions with current volcanism. Its content in most localities does not exceed 0.002 mg/l. Elevated concentrations (up to 0.13 mg/l) are characteristic of the Geyser Springs in California. Conspicuously high content of mercury ().1 - 0.5 mg/l) is present in saline chloride methane waters of the Cimrik oil field in Southwest California. It is known that in regions with volcanic (magmatic) activity, ore occurrences and deposits of mercury are used as an excellent indicator of concealed accumulations of hydrogeothermal fluids and young paleohydrogeothermal systems.

The distribution of *cadmium* in thermal waters has been very little studied. According to existing knowledge, the maximal concentration of this metal in chloride - methane brines is up to 2.3 - 4.5 mg/l.

Zinc is present in different types of thermal waters in both fold and platform regions, but its distribution is very uneven. Anomalously high levels of zinc (even up to 300 - 500 mg/l) were established in brines discovered by boreholes in the vicinity of the Salton Sea in California, very high levels (up to 100 - 160 mg/l) were recorded in brines of the Siberian Platform, and high levels (up to 33 - 367 mg/l) occur in brines of the Central Mississippi Basin.

Potassium attains its highest concentrations in waters of springs in regions with current and Cenozoic volcanism, where concentrations from 0.02 to 0.36 mg/l have been established. The maximum content of this metal (0.65 mg/l) was established is carbonated chloride thermal brines in the vicinity of the Salton Sea.

In contrast to potassium, *lead* has significant distribution in thermal waters of different types. Very high concentrations of this metal have been established in chloride (1.6 mg/l) and sulfate (2.0 mg/l) springs in regions of volcanic thermal springs in the Kurile Islands, as well as in acidic chloride thermal springs of Japan (1.6 - 2.6 mg/l). Lead attains far greater content in highly concentrated chloride thermal brines of artesian basins rich with halogenic formations. For example, lead content comprises up to 115 mg/l in podsolic brines of Eastern Siberia and 125 mg/l in the Amu - Darya Basin.

Arsenic is widely disseminated in different types of thermal waters in many regions of the Earth, especially in ones with current and Quaternary volcanism. Anomalous concentrations of arsenic (up to 59 mg/l) are found in waters of the Latera geothermal system in Italy, while somewhat lower values occur in those of the El - Tataio geothermal system in Chile (up to 47 mg/l) and the Kizildere geothermal system in Turkey (up to 39 mg/l).

The distribution of *chromium* in thermal waters has been little studied. The content of this metal ranged from 0.0043 to 0.01038 mg/l (with a maximum value of 2.5 mg/l) in 38 springs of Japan and from 0.006 to 0.225 mg/l in 20 samples of acidic thermal waters in the Kurile Islands.

Vanadium has been unevenly studied and is unevenly distributed in thermal waters of different types. It achieves its greatest distribution in H_2S - CO_2 and carbonated thermal springs in regions of current volcanism. High concentrations of vanadium were established in natural condensates of chloride composition from fumarole fields of the volcano Ebeko (up to 7.5 mg/l) and in several springs in volcanic regions of Japan (6.1 to 7.0 mg/l).

Manganese is among the most widespread of metals in different types of thermal waters. Its content usually varies from 0.1 to 1 - 2 mg/l and attains 10 - 15 mg/l in chloride emissions of acidic fumarolic springs in regions of current volcanism. Anomalously high concentrations of manganese were established in acidic sulfate thermal springs in the neighborhood of the volcano Issaika in Japan (350 mg/l) and in highly concentrated chloride brines discovered by boreholes near the Salton Sea. Very high levels (up to 500 - 700 mg/l) of this element were also recorded in highly concentrated acidic brines of artesian basins constructed of salt deposits (E. A. Baskov, 1977).

Iron is a component of all thermal waters. It attains its greatest content (up to 1000 - 3000 mg/l) in chloride and sulfate thermal emissions of volcanic regions that for practical purposes represent natural condensates of fumarole gases formed in calderas and craters of many active volcanoes. The maximum content of iron (10,600 mg/l) in thermal springs of the given type was recorded in condensate of the fumarole Big Donald in New Zealand (S. H. Wilson, 1953). Very high concentrations (up to 1,000 - 9,000 mg/l) of this metal are achieved in highly concentrated brines of platform regions.

Cobalt is most often found in acidic thermal waters. In acidic thermal springs of the Kamchatka - Kurile region, for example, maximal cobalt content attains 0.48 mg/l in chloride waters and 0.12 mg/l in sulfate waters. Anomalously high content of cobalt (2.19 mg/l) was recorded in acidic sulfate waters of the Tentoku Spring on the Japanese island of Kyushu (Uzumasa, 1965). However, in the majority of tested samples from different localities throughout the world, this metal was not recorded.

Like cobalt, *nickel* has no great distribution in thermal waters. As in the case of cobalt, the highest concentration of nickel (9.38 mg/l) was discovered in acidic sulfate waters of the Tentoku Spring.

Molybdenum is widely, but unevenly, disseminated in different types of thermal waters. High levels of it most often occur in hydrocarbonate and sulfate nitric thermal springs in different fold regions of the world, where it can attain a concentration of 0.5 mg/l. Otherwise, the content of this metal in thermal waters from different localities throughout the world predominantly varies from 0.001 to 0.03 mg/l .

Tungsten is one of the least studied elements of thermal waters, primarily due to the very low sensitivity of existing analytical methods of detecting it. The most reliable results of testing the content of this element are linked with hydrocabonate and sulfate - nitric thermal waters in several fold regions. In regions of current volcanism, tungsten content in acidic and weakly acidic thermal springs is lower and usually does not exceed 0.001 - 0.01 mg/l. Exceptionally, it often attains a relatively high concentration (up to 0.3 - 0.4 mg/l) in thermal springs of Yellowstone National Park in Wyoming (USA). Anomalously high tungsten content (9.4 mg/l) was established in the Vibur Spring (D. E. White, J. D. Hem, G. A. Waring, 1963). It has also been established that tungsten can sometimes be concentrated in brines of salt strata in the northeastern part of the Mohave Desert in California (USA), where it even attains a content of 60 mg/l.

Lithium is one of the most thoroughly studied and widely disseminated of metals in thermal waters of different hydrogeological structures of the Earth. Its concentration in different types of water fluctuates within a very wide range: from thousandths and hundredths to hundreds of milligrams per liter. Very high concentrations of lithium are linked with weakly acidic and weakly alkaline thermal springs in regions of current volcanism (up to 48.9 mg/l in Chile, up to 25 mg/l in New Guinea, up to 10 - 24 mg/l in New Zealand, and up to 11 mg/l in Italy and Japan). Even greater values were established for thermal springs emitting carbonated waters in regions of current volcanism (the Arima Spring in Japan - 55.8 mg/l; and the Hervidea spring in Peru); and in regions of Cenozoic volcanism rich in carbonated waters (Caucasus, Turkey, Afghanistan, China, Yugoslavia, France, and other countries). Lithium is widely disseminated in methane and nitric thermal waters of artesian basins, but in these conditions its concentration varies greatly. Anomalously high concentrations (up to 100 - 300 mg and higher) have been recorded in brines of artesian basins of the platform type associated with salt formations.

Rubidium is fairly widespread in different types of thermal waters. Its content usually does not exceed 0.2 - 0.5 mg/l and more rarely attains 7.7 mg/l in acidic thermal springs in regions of current volcanism; more rarely, levels of the order of 3.5 - 10.0 mg/l are found in alkaline and weakly alkaline thermal springs. The greatest concentrations have been recorded in highly concentrated methane thermal brines of artesian basins in whose geological profile halogenic deposits are represented (up to 60 - 78 mg/l in the Western European Artesian Basin).

Cesium has been most thoroughly studied in thermal springs of regions with current and extinct volcanism. Its concentrations are somewhat greater in weakly alkaline carbonated thermal waters and in H_2S - CO_2 chloride thermal waters (maximum content of 20 mg/l was recorded in waters of the El - Tatio geothermal system in Chile). In most of the springs tested in Japan, the content of this element varied from 0.01 to 1.0 mg/l. In saline and briny methane thermal waters of artesian basins in platform regions, cesium content varies within wide limits (from hundredths of a milligram to several milligrams per liter) and generally is perceptibly lower than in volcanic regions.

Strontium belongs to the group of most widely disseminated elements in thermal waters, but its concentration varies within wide limits. Waters of hydrocarbonate nitric thermal springs are characterized by the lowest levels of strontium (below 0.1 mg/l). Strontium content often exceeds 4 mg/l (up to 24 mg/l) in carbonated waters of some fold regions, and relatively high concentrations also occur in sulfate nitric thermal springs of Jasper National Park in Canada (up to 13 - 15 mg/l). Strontium is usually found in higher concentrations (10 - 200 mg/l) in saline thermal waters and brines of artesian structures (anomalous content of the order of 5,000 - 7,000 mg/l have been recorded in greatly concentrated calcium chloride brines of artesian basins rich in salt deposits). It is interesting to note that relatively high concentrations are also characteristic of crystalline rocks (up to 1,600 mg/l within the framework of the Canadian Shield).

Barium is a very widespread element in thermal waters, but its concentration is highly variable. In fold regions its content can range from 2 to 6 mg/l. Appreciably greater concentrations of barium (62.4 mg/l) have been established in carbonated brines of the Arima Spring in Japan. Variability of barium concentrations is especially pronounced in thermal waters of artesian basins, where they range from tenths of a milligram to several tens of milligrams per liter and more rarely comprise as much as a couple of hundred milligrams per liter. Anomalously high barium content of 235 mg/l was established in very hot carbonated brines from boreholes in the neighborhood of the Salton Sea in California.

Aluminum is very widespread in thermal waters of regions with current volcanism, especially in acidic media (where its content depends mainly on pH and mineralization), sometimes attaining levels of several grams per milliliter (4.913 and 2.018 g/l in the fumarolic thermal springs Idjen on the island of Java and Big Donald on the Isle of Wight in New Zealand, respectively). Certain acidic springs discharge significant amounts of aluminum, which with neutralization of the water is precipitated in the form of the hydroxide of this metal. It is calculated that the thermal springs of some volcanoes every day discharge up to 200 kg and more of aluminum. Still, this metal is present in insignificant quantities in alkaline and weakly acidic thermal waters of different composition, more rarely attaining a few tens of milligrams per liter.

Groundwaters whose solutions contain elements and their compounds in concentrations making it profitable to extract them are called **industrial waters**. They are for the most part mineral waters of high mineralization (brines), which were discussed in the foregoing text. More precise names are given to these waters according to the element of interest, for example bromine water in the case of water containing bromine in industrial concentrations. Table 2.11. gives the classification of industrial mineral waters after Bondarenko and Kulikov (1984).

Among about 30 exploitable components of high - mineral waters (brines), iodine, bromine, and compounds of boron are most often singled out, as well as a number of others in recent years (lithium, rubidium, germanium, uranium, tungsten, copper, calcium and potassium salts, calcium fluoride, ammonium chloride, etc.).

Gases in Mineral Waters

All mineral waters - like all natural waters (surface and underground) in general - contain gases in various amounts. The main gases of these waters are O_2, N_2, CO_2, H_2S, CH_4, H_2, NH_3, He, Rn, Ar, , and Kr. The gas composition of water best indicates the circumstances of formation of its chemical composition, as well as the origin and medicinal value of the water. It is known that gases in groundwater can be of atmospheric, biochemical, radioactive, volcanic, and magmatic - metamorphogenic origin.

Mixtures of different gases are very often present in water. However, it has been demonstrated that the diversity of natural water composition can be reduced to the following three basic groups:

- *Acidic gases* (CO_2, H_2S, etc.);
- *Nitric gases* (N_2); and
- *Methane* or *hydrocarbonaceous gases* (Ch_4, etc.).

Radon and inert gases (Ar, He, Ne, Kr, etc.) always play the part of accessory gases.

TABLE 2.11.

Classification of mineral industrial waters (Bondarenko, Kulikov, 1984).

Type of water	Components	Minimal concentration (mg/l)
Iodine	J	18
Bromium	Br	200
Iodine - bromium	J+Br	J - 10 Br - 200
Boron	B	250
Iodine - boron	J+B	J - 65 B - 162.5
Strontium	Sr	500
Lithium	Li	10
Rubidium	Rb	5
Cesium	Cs	1
Radium	Ra	10 - 5

Carbon dioxide (CO_2), hydrogen sulfide (H_2S), nitrogen (N_2), methane (CH_4), and radon (Rn) are the gases of greatest interest in balneological evaluation of the gas composition of mineral waters.

Carbon dioxide (CO_2). This is a very widely disseminated gas in groundwater. Its greatest concentrations occur in mineral waters. Large reserves of this gas are formed in the Earth's crust as a result of dynamo - and thermometamorphism of carbonate rocks, as well as due to biochemical reactions in sedimentary rocks. It can also enter groundwater from the atmosphere. Carbon dioxide formed at great depths by the process of metamorphism migrates along fault structures to the surface and mis with groundwater of different origin. Depending on partial pressure and temperature, it attains a concentration of 1 to 3 g/l in such water and as much as 20 g/l in deep water - bearing horizons.

The presence of carbon dioxide in definite amounts gives water specific characteristics. From the balneological standpoint, such waters are judged to be medicinal. According to norms adhered to in our country, the content of CO_2 should be greater than 250 mg/l. If carbonated mineral water is used in baths, the content of CO_2 must be greater than 1,000 mg/l, since its solubility in water at a temperature of 33 - 37°C and normal atmospheric pressure ranges from 1,100 to 1,200 mg/l.

Hydrogen sulfide (H_2S). It is formed in zones of hypergenesis under the influence of combined action of water and organic material with the participation of sulfate - reducing bacteria. In deeper parts of the Earth's crust, the indicated gas is associated with thermocatalytic decomposition of sulfur compounds and reduction of sulfates under conditions of high temperatures and pressures. Very large amounts of H_2S are also emitted during volcanic eruptions. The content of hydrogen sulfide in groundwater rarely exceeds 50 mg/l, and only in exceptional cases can it attain 2,000 mg/l in petroleum deposits.

Like carbon dioxide, hydrogen sulfide is of very great balneological significance. This gas lowers sugar levels in the blood of diabetics, contributes to lowering of blood pressure in humans, etc.

Nitrogen (N_2). Nitrogen is the most widespread gas in groundwater. It is most often of atmospheric origin, especially in shallower water - bearing horizons, where its quantity is proportional to partial pressure of the gas. Nitrogen of biogenic origin is also widespread, but this gas is rarely formed during volcanic processes. Its concentration is of the order of tens to hundreds of milligrams per liter, but can attain 1,200 mg/l.

Slightly mineralized acrothermal - alkaline waters of atmospheric origin usually belong to the category of nitric thermal waters, such waters being very often enriched with silicic acid (the waters of silicic thermal springs).

Methane (CH_4). Dispersed organic substances in sedimentary rocks and coal beds serve as the main source of formation of this gas, which is very widely disseminated in groundwater. In addition to methane, heavy hydrocarbons occur in a high percentage (up to 40%) in gases of petroleum deposits. Hydrocarbonaceous gases in general achieve their greatest content (13,000 mg/l) in waters of petroleum deposits. Also characteristic is the presence of helium (from 0.001 to 0.005, more rarely 2%), as well as elevated nitrogen content.

Methane (hydrocarbonaceous) mineral waters are especially valued for medicinal action if they are formed in deep closed hydrogeological structures under reducing conditions with intensive development of biochemical processes. Waters of petroleum deposits and coal beds are of this kind. By the way, it should be stressed that hydrocarbonaceous gases are of great industrial significance.

Hydrogen (H_2). In groundwater, hydrogen is formed by the process of dissociation of water during decomposition of organic substances or by the process of hydrolysis of the sulfates of heavy metals (Fe, Cu, Al, etc.) in zones of oxidation of sulfide minerals. It is often present in deeper parts of the Earth's crust, in reducing environments, and in regions with current volcanic activity. Hydrogen in mineral waters is considered to be of volcanic origin.

Oxygen (O_2). This gas is primarily of atmospheric origin. In groundwater occurring at shallower depths, content of oxygen can attain 15 - 20 g/l. Its concentration as a rule declines with depth, so that oxygen in deeper water - bearing horizons can be found only in areas of faults or - in karst regions - in zones of privileged collectors connected with ponors. The presence of this gas in groundwater is of enormous biological and geochemical significance.

Inert gases. Of inert gases, helium (He) and argon (Ar) occur most frequently in groundwater. They are both predominantly of radiogenic origin and will be treated (together with radon) in the chapter dealing with radioactive elements.

Geological and Hydrogeological Characteristics of the Territory of Yugoslavia

The territory of Yugoslavia is characterized by varied lithographic composition and a complex structural framework. Several geotectonic units can be singled out within this territory. They are characterized both by specific geological composition and by special geomorphologic and hydrogeological relations. The following geotectonic, that is hydrogeological units are distinguished here (Fig. 2.31.):

- The *Pannonian Basin* (the regions of Backa and Banat);
- The *Sava Trench* (Srem, Macva, and land along the Sava and Tamnava Rivers);
- The *Inner Dinarides* (Montenegro and Southwest Serbia);
- The *Vardar Zone* (the bulk of Western Serbia);
- The *Serbo - Macedonian Mass* (Central Serbia); and
- The *Carpatho - Balkanides* with the *Dacian Basin* (Eastern Serbia).

The given primary hydrogeological units are marked by a certain type (or types) of basic water - bearing media. Thus, masses of karstified limestones are the main water - bearing media in morphologically broken - up regions (*Dinarides and Carpatho - Balkanides*), water - bearing media of the *Vardar Zone* consist almost exclusively of rocks with fracture porosity, and water - bearing media of young depressions *(Pannonian and Dacian Basins*, as well as depressions within the *Serbo - Macedonian Mass*) are represented by alluvial formations and Neogene lacustrine sediments, i.e., water - bearing media with intergranular porosity.

The *Pannonian Basin* is filled with Tertiary and Quaternary deposits up to 4,500 m thick. A clearly expressed geological, geomorphologic, and hydrogeological whole is thereby formed in the north of Serbia, one that is set off from the other units approximately by the valleys of the Sava and Danube Rivers. Two parts can be singled out in the Tertiary parcel: 1. the *lower part*, represented by rocks of Miocene and older Pliocene age, within which Lajtovac and Sarmatian limestones with highly mineralized waters constitute a rich water - bearing medium; and 2. the *upper part*, with significant content of permeable Pliocene and Pleistocene sands and gravels, in which several artesian horizons are formed. The main water - bearing medium of Pleistocene age is the so - called basic water - bearing complex in the upper part of the indicated parcel, whose slightly mineralized waters represent the main drinking water of the Backa and Banat regions. Alluvial formations of the Danube and Sava Rivers are also significant from the standpoint of water supply.

The *Sava Trench* represents a region with significant accumulations of groundwater formed in alluvial sediment of the Drina and Sava Rivers. Thickness of the sand - gravel sediment is greatest in the valley of the Drina, where it attains 85 m, so that the given zone is regarded as a potential source of water for Belgrade and part of Serbia.

On the territory of the *Inner Dinarides*, the most important formations are highly karstified limestones of the Middle and Upper Triassic (Southwest Serbia) and Jurassic and Cretaceous (Montenegrin Dinarides). The given karst environment is characterized by high infiltration of atmospheric sediments. Movement of groundwater is realized through systems of karst channels and fissures, while evacuation is through powerful karst springs. With minimal output of 400 - 1,000 l/s, 12 such phenomena are registered in the Serbian part of the Dinarides (the springs Vrelo Raske, Vrelo Bistrice, Podpec, Sopotnica, Vapa, Seljasnica, Grabovica, Susica, Bucje, Vrelo Belog Drima, Vrela Istok, and Vrelo). The springs Vidrovanska Vrela, Bijeli Nerini, Mareza, Oko Karuc, Dubravska Vrela, Alipasini Izvori, and Bistrica are located in the Montenegrin Dinarides. Most karst springs are characterized by great variation of output every year, from 1:4 to 1:100.

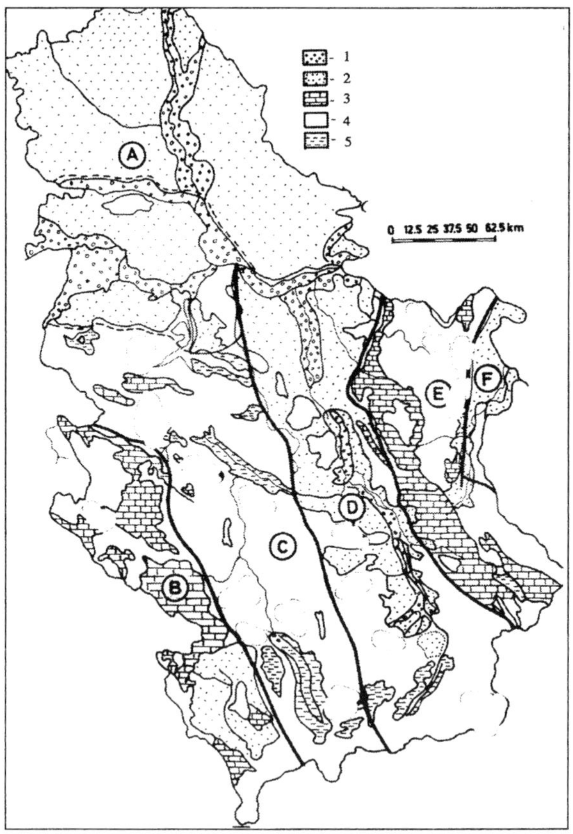

Fig. 2.31. Geotectonic units at the territory of Serbia: A - Pannonian Basin; B - Inner Dinarides; C - Vardar zone; D - Serbo-Macedonian mass; E - Carpatho - Balkanides; F - Dacian basin; 1 - Alluvial sediments; 2 - Neogene lacustrine sediments; 3 - Limestones; 4 - Fractured medium; 5 - Neogene clayey sediments.

The *Vardar Zone* is characterized by scarcity of slightly mineralized waters. Although very diverse, rocks of this exceptionally complex geotectonic zone represent poorly permeable to practically impermeable media marked by springs with low output.

The region of the *Serbo - Macedonian Mass* hydrogeologically stands distinctly apart from the other units. It has a relatively uniform hydrogeological framework. In the northern half, crystalline schists constitute a floor of permeable horizons, and water - bearing rocks are represented by sandy Mio - Pliocene layers and coarsely grained sediment of the Morava River. The bulk of terrains in the southern half are built of crystalline schists very poor in potable groundwater. Quaternary sand - gravel deposits in the alluvial plain of the Morava River represent the main water - bearing medium, while accumulations of slightly mineralized waters of greater significance from the practical point of view are formed in water - bearing sands of Pliocene age in the Leskovac and Jagodina - Paracin Basins. Fractured media in Southern Serbia are evacuated by numerous weak springs with output below 0.5 l/s, more rarely up to 3 l/s.

Masses of Upper Jurassic and Lower Cretaceous limestones give the *Carpatho - Balkanides* their main hydrogeological characteristics. The limestones are very cracked and karstified, and so represent an exceptionally permeable medium. The discontinuous groundwater formed in them is evacuated through karst springs that arise where the limestones come into contact with impermeable rocks. More than 70 springs with minimal output of more than 10 l/s are registered, 16 of them having an output greater than 100 l/s. The most powerful are the Ljuberadja Springs near Babusnica. Also with a very high output are the springs Vrelo Mlave, Krupajsko Vrelo, Radovansko Vrelo, Beljevinska Vrela, Modro Oko, Belopalanacko Vrelo, Krupac in the vicinity of Pirot, etc. From the standpoint of utilization of groundwater, an unfavorable circumstance is the great variation in output of most of the springs, which depends on the amount of precipitation.

Slightly mineralized waters on the territory of Yugoslavia represent the main resource for supplying the population and industry with water. It is estimated that the many sources of these waters satisfy close to 90% of requirements for high - quality water.

The extent to which slightly mineralized waters are protected depends mainly on geological factors. Artesian horizons as a rule are well protected, while karst water - bearing media are most sensitive in this regard.

Favorable geological and hydrogeological factors have dictated the formation of many springs of ***mineral and thermal waters***, which differ greatly among themselves with respect to mineral - chemical composition, temperature, presence of gases, etc. If we proceed from the density of occurrence of these waters and their diversity, we are able to conclude that Serbia in proportion to its area is the richest country in the Balkans and in Europe. Utilization of this natural wealth dates from the oldest times, and special attention was paid to them during the period of the Roman Empire, as is indicated by the baths that have been preserved in a number of spas.

The following three *hydrogeochemical provinces* can be singled out on the territory of Yugoslavia:

1. *A province with carbonated and carbono - nitric waters* mainly of atmospheric origin, a region with young magmatic and thermometamorphic processes;
2. *A province with nitric and nitro - oxygenated waters of atmospheric origin.* This province is predominantly represented by thermal waters linked with regions of younger tectonic movements; and
3. *A province with artesian nitric, nitro - methane, and methane waters* of atmospheric and marine or mid origin.

The first province encompasses central parts of the country's territory (the Vardar and Serbo - Macedonian geotectonic units); *the second* includes zones in the southwest and eastern parts of it (Dinarides and Carpatho - Balkanides); and *the third province* coincides with the Pannonian Basin and connected lacustrine sediments of Neogene age south of the Sava and Danube Rivers.

The significant explored (available) reserves of mineral and thermal waters have not been even close to fully utilized, either for bottling and balneological uses or for energy purposes. To be more precise, only a small percentage of the total reserves are utilized. We note the existence of 49 active spas, among which only a few are relatively organized (in the modern sense); and more than 230 tested springs, the waters of only 20 of which are bottled. The heat energy of thermal waters is mainly used in the traditional way, mostly for balneological and sport and recreational purposes. The use of this natural wealth unquestionably needs to be far more diversified.

Mineral Resources of the Earth

Since the Pleistocene, i.e., for the past million and a half years, man has used solid rocks and hard minerals for tools. Rocks and minerals for a long time remained the main material used by man in his work. In the second half of the Third and Second Millennia B.C. (at the transition from the Neolithic to the Metal Age), he began to use metals (copper, gold, and silver) as well. These metals are found in nature both in the native state and in minerals from which they have to be extracted and purified, and the development of metallurgy began in that period. Thus, copper and bronze started to be used (together with rocks and minerals) as constructive materials almost five thousand years ago, while gold and silver (in addition to copper and bronze) were used to fashion jewelry.

As we see, man has passed through several periods (the Stone, Copper, Bronze, and Iron Ages) prior to the advent of the atomic era. In each stage of his development, materials obtained from the interior of the Earth took on increasing significance, and it can be freely asserted that mineral resources were the starting base on which human civilization evolved in all phases of man's development.

With respect to their significance in improving the population's living conditions, mineral resources at the present time occupy second place behind agriculture. Thousands of ways of utilizing energy and mineral raw materials have been developed to obtain fuels, metals, building material, abrasives, chemical raw materials, etc. It is therefore not surprising that the demands for these raw materials increase every year, at an exponential rate for many of them (B. J. Skinner, 1986). For example, gold production increases by 4% annually and the period of doubling of production is equal to 10 years. Since this is a limited unrenewable resource very unevenly distributed in the Earth's crust, it is clearly high time for man to proceed far more cautiously with its consumption, in keeping with a strategy of sustainable (adjusted, long - term) development.

Energy Resources

Viewed historically, human utilization of energy sources for thousands of years was a slow and long - term process. Solar energy and wood as a source of energy have been used since the prehistoric period. Coal became the main energy source before about three centuries, as did oil and gas nearly 150 years ago. Atomic energy was discovered after the Second World War.

Energy is unquestionably an essential condition for the existence of modern society, and energy resources are for this reason the most highly prized of mineral products. Today we use only three forms of caustobioliths (fossil fuels): coal, oil, and natural gas. This "big three" supplies more than 95% of the world's energy[45]. If we analyze factual data on the rate of their exploitation, we are able to draw two conclusions: *first of all*, the time of their utilization has been exceptionally short[46]; and *secondly*, the rate of increase in production over the past several decades cannot be sustained too long, especially since caustobioliths make a visible negative contribution to deterioration of the environment. Use of uranium and production of nuclear energy can be counted on for a considerably longer time, but the risks from breakdown of nuclear power

[45] In terms of world reserves, coal is in the first place, with more than 73% of them, followed by oil (17%) and natural gas (10%). However, with respect to consumption, oil comes first (53%), followed by coal (26%) and natural gas (21%).

[46] One half of the coal exploited to date was burned in the last quarter of a century, while half of total world oil production falls on the 12-year period starting from 1956.

plants and the dangers of radioactive radiation are well - known. For this reason, constantly increasing attention is being devoted to noncombustible forms of energy - hydroenergy, solar and geothermal energy, and new energy sources - and new technologies are being developed for their exploitation.

Oil and natural gas. The wide dissemination and specific appearance and properties of oil and other (solid and gaseous) hydrocarbons have always made it possible for their extrusions to be easily recognized on the Earth's surface. For this reason, oil, asphalt, pitch, bituminous formations, and natural gases are mentioned in the oldest of written documents, and since ancient times they have played an important role in the religious beliefs, medicine, and even economics of many regions. However, only in the second half of the 19th Century - after the discovery of naphthids in significant amounts - did their potential economic significance become apparent.

The initial phase of oil utilization unfolded at a very slow rate. That time, more precisely the period from 1858 to 1900, is known as the "*kerosene age*." It was only approximately at the start of the 20th Century, with introduction of the internal combustion engine into practice, which we have, intensive development of the oil industry, and the rate of its growth has not abated down to the present day. It can be said that we live in the "gasoline age," since gasoline is the main product of natural petroleum. When it is also taken into account that we obtain thousands of other products from oil and gas, it is clear why it can be asserted that naphthids represent one of the main mineral resources (in the broadest sense) of modern society (Fig. 2.32.), and understandable that from 31 to 50 billion dollars were in the past decade invested annually throughout the world in exploration for oil and gas and their exploitation.

Oil and gas finds are very widespread, but unevenly distributed throughout the globe. For example, over vast areas of Asia, Australia, and Africa, no industrially interesting deposits of these raw materials have been discovered, or else the reserves of discovered deposits are insignificant. At the same time, certain territories are very rich in oil and gas. This is the case with the so - called "*oil axis*," on which the region of the Middle East (Iran, Iraq, Kuwait, Saudi Arabia, the Apsheron Peninsula in the former USSR) sits on the one hand, and the region of the Gulf of Mexico (the Caribbean Sea, including parts of the USA and Mexico, Columbia, Venezuela, and Trinidad) on the other side of the globe. Two thirds of the world's oil reserves are concentrated in those two regions. Smaller with respect to reserves, but also industrially significant regions are found in North America) parts of the USA, Mexico, and Canada).

Surface occurrences of naphthids are encountered in different forms. *Active naphthids* can migrate along faults from deposits all the way to the Earth's surface, where they are often associated with groundwater reserves. Such occurrences are sometimes large, as for example in California, Venezuela, or Trinidad, where they can cover more than 5 km^2 and be economically interesting. Solid naphthids are encountered on the surface either in the form of scattered impregnations in rocks or in the guise of accumulations of bitumen along fissures and faults.

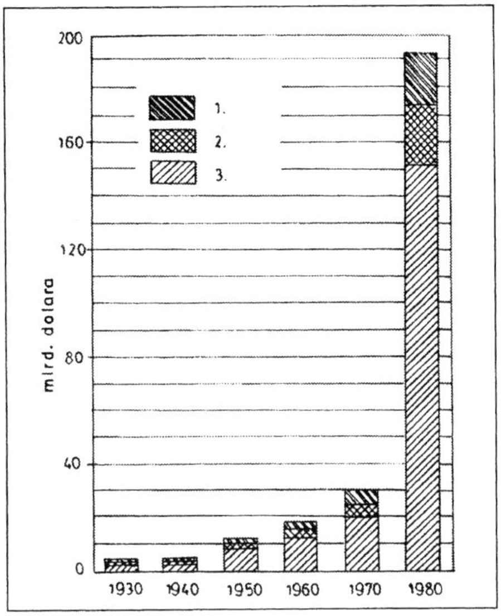

Fig. 2.32. Annual production of mineral raw materials in US. 1 - metals; 2 - nonmetals; 3 - oil and gas.

Oil is usually accumulated within porous sandstones, limestones and dolomites (Fig. 2.33.). Particularly rich oil deposits are formed in sandstones (the largest deposit in US East Texas, deposits of Venezuela, Kuwait, etc.)

A porous medium of permeable rocks in the vicinity of any deposit fills with water of increased salinity, oil, and gas. In natural reservoirs - so - called *stratigraphic traps* - gas, oil, and water are arranged in layers: as the lightest of the three, gas accumulates in the upper part, oil in the middle, and water (the heaviest) at the bottom. A close connection of oil and gas deposits with anticlinal structures is discernible.

There are three spheres of utilization of petroleum: as fuel, as a lubricant, and as a chemical raw material. Oil accounts for about half of all combustible materials used for energy production.

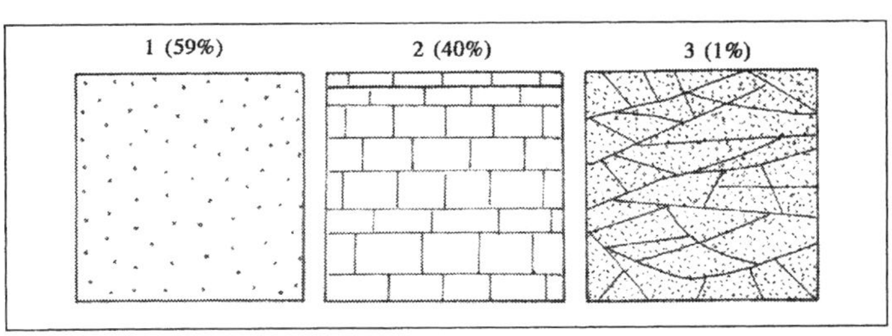

Fig. 2.33. Parts (percentages) of world oil production obtained from the main types of collectors. 1 - sandstones; 2 - carbonate rocks; 3 - fractured formations.

The total reserves of oil in the world amount to about 640 billion barrels, which at the current level of consumption is enough for approximately 30 years (according to Allison and Palmer, 1980). Known gas reserves are estimated at 65.1 trillion m^3 (potential reserves are estimated at not more than 138.8 trillion m^3). Since gas will probably replace oil when the latter is exhausted, it is calculated that all oil and gas resources will be used up by the year 2050.

Coal. Coal is formed from the remains of freshwater vegetation. Especially favorable conditions for coal formation prevailed 250 - 350 years ago. During that period of a hundred million years, particularly large deposits of coal were formed in North America (USA and Canada), Europe (Great Britain and Germany), and Asia (China), since those territories were for longer periods of time at equatorial and temperate latitudes, where a warm climate and copious precipitation favored the development of enormous swamps. Approximately 43% of the world's coal is found on territory of the former USSR, 29% in North America, 14.5% in Asian countries (without the USSR), and 5.5% in Europe.

Most large coal basins have already been discovered. The world's reserves of all forms of coal are estimated at 8,620 billion tons, with potential reserves of 15,270 billion tons, which is equivalent to heat energy of the order of $4.2 \cdot 10^{23}$ J. About 50% of the given amount can be exploited.

Wide use of coal as fuel began in the 12th Century A.D., when the inhabitants of the northwest coast of England found that combustible black rocks uncovered by wave action can nicely replace wood. Since that time, the new form of fuel was used to an increasing extent and on such a scale that dissatisfaction of the citizens of London was expressed already in 1273 because of air pollution caused by coal burning. Coal today represents the leading fuel in many countries of the world, and it is predicted that coal will again become the dominant fuel on the Planet around the year 2020 due to possible difficulties in oil and gas supply.

Coal is primarily used as a fuel to satisfy personal and industrial needs. About three fourths of all coal is consumed in production of electrical energy, a sixth of it is used to produce coke, and a tenth is burned in different branches of industry. Coke is especially valued as a high - quality fuel and is used for smelting of iron ore in blast furnaces.

Oil shales and bituminous sandstones. Oil shales contain solid components from which hydrocarbons can be separated by distillation. Bituminous sandstones, on the other hand, contain viscous asphalt, which is extracted by heating and used to obtain gasoline and other products.

World reserves of oil shales are predominantly concentrated in the USA, Brazil, Zaire, and the former Soviet Union. These resources are very large, but current conditions of production do not permit their more intensive utilization. Just on the territory of Wyoming, Utah, and Colorado in the USA, usable reserves are estimated at more than 1.6 billion barrels.

Resources of bituminous sandstones on the Earth have been inadequately studied, but the usable reserves in Canada (300 billion barrels), the USA (29 billion barrels), and Venezuela (79 billion barrels) indicate that this source of oil can take on considerable significance in the future.

Nuclear fuel. Uranium represents one of the main forms of fuel in the category of combustible raw materials. Used in practice for release of nuclear energy, initially for military ends and later for peaceful purposes as well, uranium (and to some extent thorium) is of first - class significance as a fossil material. During the period of 1950 - 1960, prospecting for uranium was carried out on a previously unseen scale similar to that observed during periods of *gold fever.* A new boom in prospecting for uranium and thorium deposits set in at the outset of the 1970's.

Uranium is a very mobile element and as a result forms its deposits in diverse environments by means of endogenous and exogenous processes. Nevertheless, deposits formed in sandstones account for nearly 30% of current world production. Uranium - bearing conglomerates occupy second place in terms of productive reserves and overall potential. It is also possible to single out uranium - bearing phosphates, schists, granites, and granodiorites.

This element is very widely distributed on all continents. The main areas of its occurrence (outside countries of the former Communist Block, for which only sparse data are available) are in the United States: Wyoming and the Colorado Plateau, including parts of Colorado, Utah, and New Mexico. The main deposits in sandstones were estimated in 1969 at more than 200,000 t of U_3O_8. In Canada, on the other hand, 85% of uranium is produced from deposits in conglomerates, the bulk of uranium being extracted from the Dennison Mine, which is one of the largest mines of this raw material in the world. Significant deposits have been discovered in the northernmost part of Australia, south and southeast of Darwin in the Northern Territory. Already in 1915, radioactive materials were discovered in gold - bearing conglomerates in the vicinity of Johannesburg in the Republic of South Africa, and the mineral uraninite was described in 1923. It is believed that conglomerates in South Africa contain nearly 20% of the world's uranium reserves. In Europe, the largest amounts of exploitable uranium have been found in France.

Uranium reserves in the world (without the former communist countries) exploitable at a cost of 80 dollars per kg at the beginning of 1977 comprised a total of 1,647,300 t, of which 31.7% fell on the USA, 18.6% on South Africa, 17.5% on Australia, 10.1% on Canada, 9.7% on Niger, 2.2% on France, 1.7% on India, 1.7% on Algeria, 1.2% on Gabon, 1.2% on Brazil, 1.1% on Argentina, and 3.1% on other countries. If calculations are based on a cost of more than 80 dollars per kg, then the exploitable reserves of 300,000 t possessed by Sweden would lift this country to first place in the world[47].

In the year 1976, a total of 22,325 t of uranium was produced in the world, 44% of that amount falling on the USA, 21.7% on Canada, 11.8% on the Union of South Africa, 9.2% on France, 6.5% on Niger, 3.6% on Gabon, 1.6% on Australia, and 1.6% on other countries.

If safe reactors were to be developed with the advance of technology, then nuclear energy would probably have a great future. In the meantime, the chief problem remains to be resolved, namely the deposition of wastes of nuclear reactions.

Geothermal energy. A steam - water mixture or dry steam is used to produce electrical energy. Techno - economic analyses indicate that geothermal energy

---

[47] Estimates of thorium reserves are unreliable, but it is known that this element is more widespread than uranium in the continental crust.

is the cheapest raw material for power plants, appreciably cheaper than energy of thermoelectric or hydroelectric power plants.

Larderello (in the Italian province of Tuscany) can serve as a classical example of a natural vapor deposit. The first turbine in Larderello was established already in 1904. Exactly a decade later, the first geothermal electric power plant in the world was put into service. Together with the Geysers in California and the Matsukawa geothermal deposit in Japan, Larderello represents a unique phenomenon in the world in that it yields dry steam, which is very suitable for production of electrical energy. The leading producers of this kind of energy today are the United States (2,012 MW), followed by the Philippines (894 MW), Mexico (645 MW), and Italy (519 MW).

In comparison with other sources of electrical energy, use of heat of the Earth's interior has both advantages (cheapness, renewability of reserves, minimal pollution of the environment, etc.) and disadvantages (technical problems, formation of incrustations in pipes, etc.). As the technology involved in hydrogeothermal systems improves and petrogeothermal reserves begin to be exploited, the importance of such systems will grow and their disadvantages shrink and become less of an obstacle.

Metallic Mineral Raw Materials

Deposits of mineral raw materials are exceptionally diverse in regard to ore composition, geological structure, and conditions of formation. If we proceed from their origin, two groups of deposits can be distinguished: 1) *endogenous*, formed under conditions of high temperatures and pressures; and 2) *exogenous*, formed at temperatures and pressures close to conditions of the Earth's surface. The origin of endogenous deposits is linked with geochemical processes in the deep part of the Earth's crust. In the series of endogenous deposits, the following can be singled out: magmatic, pegmatic, pneumatolithic, and hydrothermal. Among exogenous deposits, weathering crust deposits (bauxites, iron and manganese deposits) and disseminated deposits (gold) are distinguished.

Metals in nature occur in the following paragenetic groups:

1. The *iron group* (Fe, Mn, Cr, V, Ni, Co), genetically linked primarily with basic and ultrabasic magmatic rocks;
2. The group of *molybdenum* and *tungsten* (Mo, W, Ma, Re), mainly linked with acidic magmatic rocks;
3. The *platinum group* (Ru, Rh, Pd, Os, Ir, Pt), linked with ultrabasic and (sometimes) basic rocks;
4. The *group of elements of polymetallic deposits* (Fe, Cu, Zn, Pb, Au, Ag, Sn, Hg), formed by hydrothermal processes; and
5. The *group of As, Sb, Bi, Se, and Te,* isolated during hydrothermal processes in the form of sulfides, arsenides, antimonides, and sulfosalts of heavy metals.

According to industrial classification, metals can be relegated to one of the following six groups:

1. *Ferrous metals* (Fe, Mn, Cr);
2. *Rare metals* (V, Ni, Co, W, Mo, Sn, Bi, Sb, As, Hg);
3. *Non - ferrous metals* (Cu, Pb, Zn, Al, Mg);
4. *Precious metals* (Au, Ag, Pt, Ir, Os, etc.);
5. *Radioactive metals* (Ra, Th, U); or
6. *Metals of the rare earth group* (Nb, Hf, Zr, Ta, La, Ce, etc.).

Of all metals, ferrous metals are of the greatest industrial significance.

Local concentrations of metallic mineral raw materials formed under especially favorable geological conditions can be truly imposing. For example, the Chucvicamata deposit in Chile has a surface area of just 2 - 3 km^2, but contains reserves of as much as a billion tons of first - class ore or 21.5 million tons of copper. Butte Mountain (USA) - considered to be the richest mountain in the world - in the course of 82 years of exploitation yielded about 6 million tons of copper, 1.5 million tons of zinc, 0.6 million tons of lead, 25 thousand tons of silver, and 500 tons of gold.

Deposits of metallic raw materials are very unevenly distributed throughout the Planet. Iron ore is most plentiful in Africa and Asia, although Europe and North America are the largest producers of iron. Russia, China, and the countries of Central Africa are outstandingly rich in manganese, chromium, molybdenum, vanadium, nickel, and tungsten, which are important raw materials for production of special steel and for the arms industry in general. Always sought non - ferrous metals (copper, lead, aluminum, zinc, etc.) are primarily found in countries with vast territories - Russia, China, the USA, and India. In production of gold, the undisputed record holder is the Republic of South Africa, with more than 600 tons annually.

Widespread Metals.

1. *Iron* , the second most widespread metal in the Earth's crust, is one of the main pillars of civilization. Its share in consumption of all metals is 95%. On the other hand, significant amounts of such metals as nickel, chromium, molybdenum, tungsten, vanadium, cobalt, and manganese are added to iron to obtain high - quality steel.

Three types of iron deposits can be singled out: 1. deposits linked with magmatic rocks; 2. weathering crust deposits; and 3. sedimentary deposits. *Deposits of the first type* can arise when magma comes into contact with surrounding rocks, whose modification results in the formation of so - called aureoles of contact metamorphism with accumulations of iron ore; or when hydrothermal fluids moving along fissures gradually cool and dissolved materials settle with formation of hydrothermal deposits. *Weathering crust deposits* are formed in zones with intensive weathering of rocks when bivalent iron from the rocks is oxidized to insoluble trivalent iron, which is marked by a change of rock color to brown, yellow, or red. For example, laterites in the tropics contain iron in amounts of up to 30% and more, but such ores are not yet utilized in metallurgy. *Sedimentary deposits of iron ore* formed from chemical sediments in marine basins contain by far the largest reserves and represent the basis of current iron production. Thus, the United States obtains a large part of its iron from hematite sedimentary ores in the region of the Great Lakes. Enormous reserves of iron are found in the Transvaal (Republic of South Africa), Ukraine, Australia (Homersly), Canada, and Brazil. If it is assumed that the world's annual needs are less than 10^9 t, we can safely say that reserves of iron are so great that there will be no serious supply problems for many centuries to come.

2. *Aluminum* is more widespread than iron in the Earth's crust and owing to its lightness has a better ratio of mass to hardness, which constitutes its advantage over iron. This element was first separated in pure form in 1827. The separation process, by the way, is very difficult. Only at the end of the 19th and beginning of the 20th Century were methods developed for industrial production of pure aluminum, after which the area of its application expanded considerably. World production of aluminum was already above 15 million tons in 1981.

Products of rock weathering under tropical conditions, bauxites are the main mineral raw material of aluminum. Bauxite - bearing laterite soils are very widely disseminated in tropical regions, but large deposits with high bauxite content occur somewhat more rarely there. The Caribbean island of Jamaica is the leading supplier of this raw material. Large deposits are found in Guyana, Surinam, and Brazil (South America); Guinea and Cameroon (Africa); countries of the former USSR; Southern France; the Middle East; and Taiwan, Indonesia, and Australia (Northern Queensland). In 1982, world reserves of bauxite were estimated at $25 \cdot 10^9$ tons. According to B. J. Skinner (1986), these reserves (like oil reserves) are sufficient for the near future. Mineral raw materials potentially capable of later replacing bauxite include kaolinite, nepheline, alunite, anorthosite, andalusite, etc. Today, aluminum is already obtained from nepheline (Russia) and alunite (Japan, Russia, Mexico, and the USA).

3. *Manganese* is an extremely important additive to steel, one that mitigates the negative action of small amounts of sulfur and oxygen. The largest deposits of manganese - in which pyrolucite (MnO_2) and other manganese minerals enter the composition of *chemogenic sedimentary rocks* are the Chiatureko (Georgia) and Nikopol'sko (Ukraine) deposits. The reserves of those two deposits amount to hundreds of millions of tons, which constitutes 75% of the world's reserves of manganese ore. Mainly represented by pyrolucite and psilomelane, significant *weathering crust* deposits in the tropics are found in the Republic of South Africa, India, Australia, Brazil, Gabon, and China. Available reserves of manganese comprise $25 \cdot 10^9$ t, which is not so much in comparison with the level of its consumption ($25 \cdot 10^6$ t in the year 1981). As in the case of aluminum, it will probably be necessary to exploit non - traditional sources in the near future. Ferro - manganese concretions of the ocean floor are one of the greatest potential sources of this metal. Useful admixtures (copper, nickel, cobalt, and other rare metals) can also be extracted from such concretions.

4. Like aluminum, *titanium* has low specific gravity linked with great hardness and resistance to corrosion. It is difficult to separate this metal from ore, and its industrial production began only after the Second World War. Titanium reserves have not been precisely determined, but they are certainly considerable. Because the titanium minerals rutile and ilmenite are among the most resistant minerals of weathering crusts and at the same time are heavy minerals, they accumulate in debris. Large *disseminated deposits* were formed in this way in Australia, Sri Lanka, India, Sierra Leone, the Republic of South Africa, and the former Soviet Union. The reserves of marine debris on the Atlantic Coast of the United States attain 10^9 t, which in the opinion of experts is enough for the foreseeable future.

5. *Magnesium* is the lightest and most stable of widespread metals. Since it is hard, magnesium is needed in production of light alloys with high corrosion resistance, especially in alloys with aluminum. World production of metallic magnesium is relatively low, about 300,000 t (1981). However, magnesium is not generally used in the form of a metal, but rather in compounds, especially in oxide form (MgO), which is characterized by excellent thermo - and electro - insulatory properties. The main source of magnesium is the sea. It also enters the composition of very widespread minerals of the sedimentary rocks dolomite and magnesite. The reserves of this element are therefore practically unlimited, but the technological difficulty of extracting it is a limitation.

Rare Metals. These are metals whose distribution in the Earth's crust is under 0.1%. For this reason, together with gold, silver, and platinum, the category also includes copper, zinc, lead, and nickel, whose production is constantly increasing. All of them are vitally important elements, ones that accelerated the development of such technological wonders as production and transmission of electrical energy, the telegraph, radio, television, aviation, and rocket and nuclear energetics.

A large portion of rare metals in the Earth's crust (about 99.99%) forms part of the composition of rocks and minerals in the guise of admixtures, while only a tenth of a percent falls on ore minerals that can be enriched. For this reason, a portion of rare metals is obtained as a side product from ore minerals of copper, lead, and zinc. This is the case, for example, with silver.

Deposits of geochemically rare metals are formed under the influence of a number of factors. The following two classes of deposits containing concentrations of rare metals are especially important: 1. *deposits in which magma acts as the concentrating agent*; and 2. *hydrothermal deposits*, formed by the action of heated water rich in dissolved salts on sulfide and oxide ore minerals. Hydrothermal deposits are considered to be the most important class of deposits with concentration of rare metals.

Rare metals are mainly formed in sulfide deposits. In exceptional cases, chromium, antimony, tungsten, niobium, tantalum, and beryllium form oxide deposits, while rare native metals (platinoids and gold) can form disseminated deposits.

Copper has been used since ancient times and today represents the foundation of the electrical industry as the main material for energy transmission. Deposits of this element are very widespread, but (as hydrothermal deposits) are most often with relatively low reserves. Since the beginning of the 20th Century, underground exploitation of deposits with a high percentage of copper (above 10%) has been abandoned in favor of surface processing of porphyry copper deposits (likewise hydrothermal) with low ore quality (less than 2% copper). Hundreds of *porphyry copper deposits* have been discovered to date, and they are especially numerous in both North and South America, the former USSR, Iran, Yugoslavia, the Philippines, New Guinea, and islands of the Pacific. Deposits of this type in North and South America constitute a distinct metallogenic province parallel with the western boundary of the continents. So - called *stratiform deposits* represent another important type of copper deposits. They were formed during the Permian in the former shallow Zechstein Sea in Central Europe (Germany, Poland), as well as in the Copper Belt of Zambia and Zaire. More than 50% of world production is linked with exploitation of *porphyry copper deposits*, and 20% with *stratiform deposits* (mainly the Zambian deposit). Available reserves of copper comprised 511 million tons in 1982, when production was 7.78 million tons.

Ore minerals of *lead* and *zinc* make up a close association in nature. For practical purposes, only two mineral forms are represented: galenite (PbS) and sphalerite (ZnS), on which the world's production of lead and zinc is based. As in the case of copper, hydrothermal and stratiform deposits are industrially important. *Hydrothermal deposits* constitute an expansive metallogenic province from the states of Ohio and Missouri to Wisconsin (USA), and they also occur in Northwest Canada, a number of countries of Europe and North Africa, Northern Australia, and the former USSR. Production of these two metals from *stratiform deposits* is gradually increasing, and among the largest deposits of this type are the Sullivan and Kimberley deposits in British Columbia and three deposits in Australia. Total world reserves of lead comprised 146 million tons in 1981 (when production was 3.46 million tons). For zinc, these figures were 243 and 6.16 million tons, respectively.

Nickel is obtained from two classes of deposits: *magmatic deposits* (Canada, the former USSR, Australia); and *deposits of weathering crusts of certain volcanic rocks* (New Caledonia, the state of Oregon in the USA, Cuba, the Philippines, Greece). Available reserves of this metal comprised 56.1 million tons in 1981, when production was 784 thousand tons. It is considered that human needs for this element will be satisfied well into the distant future.

Molybdenum is present in the mineral molybdenite (MoS_2), which is very unevenly distributed. Molybdenum deposits are concentrated in a metallogenic province from Mexico to Alaska extending parallel to the porphyry copper province mentioned earlier. World production of this metal was 90 thousand tons of this metal in 1982 (when reserves comprised 6 million tons).

Silver usually occurs in *hydrothermal vein deposits* with copper, lead, and zinc. Large deposits are found in the Cordilleras from Alaska to Tierra del Fuego. These deposits account for more than 55% of the world's production of silver. The remainder falls on Australia and territories of the former USSR.

Platinoids are associated with rocks formed from magma of the Earth's inner shell. Their content is extremely low. Platinoid reserves are in large part concentrated in the Bushveld Complex (Republic of South Africa) and in the former USSR.

In contrast to platinoids, *gold* is associated with magmatic rocks formed in the Earth's crust. Among the best - known deposits of gold are those in the region of Witwatersrand (Republic of South Africa), where gold ore is concentrated in layers of conglomerates. Although many disseminated deposits of gold occur throughout the globe, the indicated deposit (also a disseminated deposit) has since 1984 represented the world's leading gold producer (with 52% of world production in 1982).

The only industrially important mineral of *chromium* is chromite, which has limited reserves. This to some extent does not apply to the stratified intrusive massifs characteristic of the African continent (the Bushveld Complex in the Republic of South Africa and the Big Dike of Zimbabwe). If chromium ores with low content of this metal were to be exploited, there would be no problem about securing the amounts of chromium needed in the future.

Antimony and *tungsten* often occur together in minerals. Their content in the Earth's crust is low and the number of deposits small. Two metallogenic provinces embrace a significant part of the world's antimony reserves. One of the provinces extends from Java through Malaysia, Eastern China, and Korea to Eastern Siberia. The other is located in the high Andes of Bolivia and Peru. Approximately half of the world's production of tungsten is realized in the East Asian metallogenic belt.

Non - Metallic Mineral Raw Materials

In prehistoric times, man incorporated selected stones into his dwellings and fashioned monuments of culture and works of art from attractively shaped stones and costly materials. Approximately 30 thousand years ago, the Krapina Prehistoric Man made tools out of several kinds of silicate rocks and tuffs, and he subsequently used coarse - grained aggregates of quartz, opal, and dense chert for this purpose. Later on, the prehistoric inhabitant of Lepenski Vir knew enough to select rounded pieces of solid rock and the hardest minerals for his building needs, tools, jewelry, and weapons. Not only that, he used attractive red and reddish - brown porphyrite, marble, crystalline limestone, and sandstone for fences, plates, tables, shrines, and tombs. We also mention the magnificent Pyramids of Egypt, about 3,500 years old, which were built of compact limestones and Nubian sandstones, with obelisks and columns carved in granular and porphyroid granites, sphins of basalt and gabbros, etc.

Non - metallic mineral raw materials, above all geological building materials, today represent the product we extract in greatest volume from the bosom of the Earth. After energy raw materials, they are the second most significant form of mineral wealth. The majority of non - metallic mineral raw materials have enormous reserves, and for some (building stone, for example) there is no need to estimate them quantitatively, since they are very plentiful in the Earth's crust. This means that the future looks optimistic where the majority of non - metallic raw materials are concerned, and the tendency to use them as replacements for rare metals is increasingly pronounced.

Non - metallic mineral raw materials are characterized by the enormous diversity of rocks and minerals. The group of main raw materials includes the following: *limestone, crushed rock, rock for block making, sand, gravel, phosphate rocks, clays, sulfur, potassium salts, and gypsum.* However, there is a whole series of economically valuable non - metallic materials apart from the most important ones mentioned. The group of second - rank raw materials is made up of: *abrasives, asbestos, asphalt, barite, benthonite, borates, bromine, diamonds, diatomite, feldspar, fluorite, garnets, graphite, magnesite, mica, olivine, perlite, pumice, quartz, sodium, talc, vermiculite, and zeolites.*

The application of all the listed raw materials is quite varied. Two groups can be distinguished on the basis of the area of their application: 1. agrochemical and chemical raw materials (fertilizers and raw materials for the chemical industry); and 2. geological building materials.

Agrochemical and Chemical Raw Materials. *Deposits of raw materials for production of fertilizers* are unquestionably among the most necessary of resources. As the source of the world's food reserves, plants require nitrogen, phosphorus, potassium, calcium, and sulfur, which they extract from the soil. The role of fertilizers is to correct the content of those very often deficient elements. This need is especially pronounced in the poor soils of Brazil, Australia, and an increasing number of other countries throughout the world. The need for fertilizers worldwide is increasing at an accelerating rate. To be more precise, it doubles every ten years.

Phosphorus, potassium, and nitrogen fertilizers are the most important.

The most widespread mineral of phosphorus - apatite - is poorly soluble, and very complex chemical reactions are employed to obtain *superphosphate.* Syenites of the Kola Peninsula (Russia) and carbonatites of Brazil and the Republic of South Africa are rich in apatite. About 82% of the world's needs are satisfied by apatite of *marine sedimentary deposits,* 16% by apatite of *volcanic magmatic rocks,* and 2% by apatite of guano. In the USA, for example, phosphate - enriched deposits were formed in a shallow marine basin encompassing parts of the present - day states of Idaho, Nevada, Utah, Colorado, Wyoming, and Montana with an area of more than 160,000 km^2.

Soluble minerals of *potassium* are also used as fertilizers. They occur exclusively in marine evaporites formed by accumulation of salt during evaporation of seawater. Evaporites are in fact the main source of *halite* (kitchen salt), gypsum, and potassium salts. In the USA alone, evaporitic salts are distributed over an area of more than 160,000 km^2 (in the states of New Mexico, Texas, Oklahoma, and Kansas), not counting the much larger potassium salt deposits. Basins in the Canadian province of Saskatchewan represent enormous reserves elsewhere in the world, together with deposits on territory of the former Soviet Union, South America, and Southeast Asia.

Nitrogen is extracted from the atmosphere and introduced into the soil in the form of various soluble nitrates, although the natural nitrates of Chilean deposits are also used in smaller amounts.

Sulfur is predominantly obtained from evaporites following secondary enrichment (USA, Mexico) or from sulfuric volcanic gases (Japan). Sulfur production attained a level of 52 million tons in 1982. In relation to needs, sulfur reserves are limited, but alternative sources of sulfur are being discovered in the meantime.

Halite (NaCl) stands out among the diverse non - metallic materials used as *raw materials for the chemical industry*. Enormous reserves of that salt are found in seawater, as well as in marine evaporites. Salt domes are known in many regions of the world (in Europe, South America, the former Soviet Union, and the Middle East), but are especially concentrated in the coastal zone of the Gulf of Mexico.

Geological Building Materials. *The more important natural stone materials* include the following: 1. *building stone*, above all limestones and granites, which in the USA account for 70% of total production (1,2 million tons in 1982); 2. *crushed rock*, primarily of limestone and dolomitic composition, whose production in the USA alone comprised 800 million tons in 1982 (when worldwide production was 2.5 billion tons); and 3. *sand* and *gravel*, used especially in roadwork and mixing with cement (consumption in the USA was 835 million tons in 1982).
The following products of rock processing stand out among others:

1. *Cement*, made from suitable mixtures of dolomites or sandstones with marls and clay, with a worldwide production of the order of 889 million tons in 1982;
2. *Alabaster*, obtained by heating gypsum (just in the USA, gypsum - bearing rocks cover 10% of the country's territory);
3. *Clays* for various uses in ceramics, obtained predominantly from large surface digs;
4. *Glass*, a product with fast - growing needs obtained by melting of rocks or minerals, especially quartz;
5. *Asbestos*, the fibrous variety of serpentine minerals (former USSR, Canada), used widely up until the 80's of the last century, but noticeably less so after that on account of its carcinogenic properties.

Energy and Mineral Raw Materials of Yugoslavia

On the territory of Yugoslavia, magmatic, metamorphic, and sedimentary processes over the course of geological time were especially favorable for formation of different kinds and types of rocks and deposits of mineral raw materials, resources needed to improve the economic and the cultural life of the population. This natural wealth has been exploited for a relatively long period of time: for about 5000 years in the cases of copper, gold, and silver. The period of Roman rule was characterized by significant intensification of exploration and exploitation of iron, silver, lead, and copper ores, as well as mercury[48]. The Kopaonik region and the mountains

---

[48] The Suplja Stena mercury mine in the vicinity of Mt. Avala is one of the oldest mines in the world, as is indicated by archaeological evidence found in the remains of a prehistoric settlement from the Younger Iron Age in the village of Vinca and pieces of ceramics from this settlement found in the mine.

Rudnik, Kosmaj, Avala, and Brskovo were well - known for mining activity, and traces of old mines have also been preserved to the present day in the vicinity of Krupanj, Majdanpek, Rudna Glava, and other places. In those days, the wealth of people and states was valued from the amount of precious metals, gold, and silver in their possession. At the time of the great migration of people with the decline of Byzantium, mining production - which up to then had been highly developed - declined noticeably until establishment of the Serbian Medieval State. Thus, after a long period of stagnation, mining attained a high level of development during the 13th and 14th Centuries, with ores of silver, lead, copper, and iron being extracted in the greatest amounts. The advance of the Turks and gradual establishment of their rule at the end of the 14th and beginning of the 15th Century marked the beginning of a new stage in organization of economic activity and thereby in mining as well, albeit on a minor scale. Such a state of affairs was maintained practically until the first decades of the 20th Century, when Serbia gained limited independence. Mining activity then flourished again: already in 1837 the first coal mine in Serbia was opened in Miliva, and it was followed by many other mines. Mines of ferrous and non - ferrous metals were opened somewhat later, and their exploitation increased at a rapid pace.

Energy Raw Materials. Energy raw materials are present in all of their main forms, but with distinct dominance of coal.

1. *Coal* is represented mainly by lignite (95% of all available reserves). The great lignite deposits of Yugoslavia are characterized by favorable tectonic structure, considerable depth of the beds, and the possibility of surface exploitation. Among such deposits is the Kosovo Basin, with an area of about 10.5 km^2 and estimated reserves of 11.7 billion tons, one of the largest deposits in Europe. The Kostolac, Kolubara, Metohija, Mazgoski, and Kovin Basins can also be singled out;
2. *Oil* and *gas* are predominantly concentrated in Vojvodina (in The Banat region above all), where oil has been produced since 1956 and gas since 1952. Total reserves of oil in Serbia are estimated at 9.5 million tons, while gas reserves are estimated at 20 billion m^3.
3. *Oil shales* extend from Bolevac to Pirot, and also occur in the Babusnica, Vranje, Aleksinac, Krusevac - Cacak, and Valjevo - Mionica Basins. With the highest percentage of kerogen (18 - 24%), the Aleksinac deposit is the most significant from the economic standpoint;
4. *Uranium* deposits and occurrences have been discovered in several places. Uranium was found first in 1947 at the confluence of the streams Dobroticka Reka and Vojnicka Voda south of Prokuplje and later in a granite massif in the vicinity of Kalna (in the Stara Planina Mountains of Eastern Serbia). Explorations have shown that uranium - bearing granites, volcanites, Paleozoic schists, and some Tertiary sediments occur in Sumadija. Elsewhere, uranium - bearing trachytes have been found south

of Gnjilane, uranium - bearing granites on Mt. Cer, etc. Uranium content in ore (pitchblende) ranges from 50 to more than 2,500 ppm. It is characteristic that the Janja and Bukulja granites have a greater clarke of uranium (with reserves of the order of 3,000 t of U_3O_8) than in other granites in Yugoslavia.

Metallic Raw Materials. The territory of Yugoslavia is characterized by many deposits and occurrences of ferrous and non - ferrous metals. The following four metallogenic provinces can be singled out (Fig. 2.31.): 1. The *Dacian Province*, of limited area, with sedimentary deposits of gold and clay; 2. the *Carpatho - Balkan Province*, with several lithogenic zones and characteristic presence of deposits of copper ore, uranium, gold, and iron; 3. the *Serbo - Macedonian Province*, with many endogenous deposits of lead, zinc, and antimony, as well as exogenous deposits of iron, nickel, magnesium, bauxite, and boron; and 4. the *Dinaric Province*, with deposits of chromium, iron, bauxite, lead, and zinc.

About 30 deposits and occurrences of *iron* have been registered on the territory of Serbia. Because they have small reserves and ore of low quality, there is not a single active mine of iron ore today. The situation is similar with *manganese*. *Chromium* reserves are exclusively associated with peridotite massifs, among which the richest in this metal is the Djakovo Massif (with more than 25 smaller deposits), whose reserves have been for the most part exhausted. Among deposits of *molybdenum*, Mackatica represents the largest deposit of this metal in Southeastern Europe. *Copper* is one of the most economically important mineral raw materials on the territory of Yugoslavia, and it occurs in 25 deposits concentrated in the Bor - Majdanpek and Ridanj - Krepoljina zones. Of similar economic importance are *lead* and *zinc*, deposits of which are likewise significantly represented. There are today 13 active mines of lead - zinc (polymetallic) ore. These mines are concentrated in the Serbo - Macedonian Metallogenic Province. Ore of *aluminum* (bauxite) is present in limited reserves associated with limestone terrains in the Dinaric Province. *Antimony* reserves are numerous in Southwest Serbia, i.e., part of the Dinaric Province. The distribution of occurrences and deposits of *gold* is limited to the eastern half of Serbia. *Mercury* ore (cinnebarite) occurs in several places.

Non - Metallic Raw Materials. Deposits of non - metals are many and diverse. Among 50 kinds represented, the following raw materials are undoubtedly the most important economically: 1. *Magnesites*, formed in peridotites of the Serbo - Macedonian Province, with production of 253,000 t (in 1990); 2. *dolomite*, with many deposits; 3. *phosphorite*, whose main deposit is the Lisina deposit near the town of Bosiljgrad; 4. *cement marl*, a component of several geological formations, with significant reserves; 5. deposits and occurrences of *asbestos*, associated with ultrabasites of the Vardar Geotectonic Zone (Strugari, Korlace, Borance, and Stave); 6. *limestones*, among the best represented of non - metallic raw materials, with wide distribution in the Dinarides and Carpatho - Balkanides; 7. *quartzites* and *quartz sand*; 8. *clay* deposits, which are numerous in Serbia (especially widespread are ceramic and brick clays); and 9. Occurrences and deposits of *fluorite*, with significant reserves in the vicinity of Krupanj.

Of special interest to us are various ecological non - metals to be discussed later. Here we single out *medicinal non - metals (peloids)*, fine - grained unbound sediments or broken - up and pulverized solid rocks that in combination with water form a "batter" used in balneotherapy. Rocks with granulation of under 2 mm (clays, carbonaceous clays, oil shale, chalk, tuffites, etc.) are used, or solid rocks that are broken up to achieve the appropriate granulation (bentonite clays, limestones, volcanites and tuffs, crystalline schists). Like various other ecological non - metals, the given raw materials are very well represented on the territory of Serbia (Tomanec et al., 2001).

Mineral Raw Materials - Risks and Positive Effects

In many areas of exploitation and processing of mineral raw materials (including energy sources), the environment is disturbed to a greater or lesser extent, often with serious consequences to public health. However, it is unrealistic to expect significant reduction in the scale of such activity. Except for products of agriculture, fisheries, and forestry, practically everything that humans use in the modern world - oil, gas, coal, metals, non - metals, mineral fertilizers - comes from the Earth.

In contrast to agricultural, forest, and water resources, the majority of mineral resources are unrenewable, and they are exhausted in the process of exploitation. On the other hand, the expenditures involved in building constantly growing cities rise considerably due to long - distance transport of geological building materials. The global problem of permanent or sustainable development of mankind will unquestionably depend largely on the management of mineral resources, i.e., protection of raw materials from exhaustion and irrational exploitation.

At the present time, the shortage of energy throughout the world as a whole is not acute, even in the case of oil, not counting political aspects of the problem and international disputes, which especially in the last decades have led to devastation of territories, injury and death of people, and pollution of the environment. Without doubt, we need to begin using ecologically more suitable new sources as soon as possible, so as to avoid restrictions in energy use that would entail an appreciable decline in the standard of living.

Environmental problems remain to be resolved. Use of some new forms of energy resources prompts debates as to adverse consequences, above all to human health. For example, public protests occasioned by the serious consequences of possible breakdowns in nuclear power plants have halted the construction of new units in certain places in the USA and other countries. In utilization of coal, on the other hand, many difficulties arise due to release into the atmosphere of oxidized sulfur during coal burning, with unfavorable consequences such as smog and acid rain[49].

---

[49] Another problem very present in the world is that of underground combustion of coal, with numerous consequences to the environment. The given phenomenon is especially present in China, USA, Australia, and India. In Northern China, for example, under-

For this reason, gasification of coal and production of gasoline from coal increasingly suggest themselves as possible solutions to the problem of effective utilization of the world's enormous coal reserves with minimal damage to the environment. Waste disposal represents another big problem throughout the world. All of this will be more closely considered at a later time in the text on anthropogenic factors. Risks and positive effects in areas of mineral deposits under natural conditions are treated below.

The service to civilization that certain microbes performed by causing coal to be formed from plant remains was enormous, but they also created the source of great disasters, namely *marsh gas* or *methane*. That gas killed thousands of miners in Couriere (France), and in many other coal mines throughout the world. On the other hand, enormous potential resources of methane - interesting from the industrial and balneological standpoints - are dissolved in many water - bearing horizons of the Earth's crust. The best - known zone of anomalous pressures is found in the United States, along the coastline of Texas and Louisiana. In this zone (which is 100 km wide and several hundreds of kilometers long), sedimentary rocks contain layers with dissolved methane in hot brines at very high pressures.

Pollution of the environment by *ecologically toxic chemical elements* (primarily mercury, cadmium, lead, arsenic, chromium, copper, nickel, fluorine, and sulfur) represents an even more important problem in evaluating the influence of mining production and planning protective measures. In addition to this, there are also problems linked with dispersion aureoles as natural sources of pollution in regions of mineral deposits. There is no need to emphasize particularly the fact that occurrences and deposits (and thereby dispersion aureoles) are associated with certain geological (metallogenic) formations, as is the case, for example, with the extensive diabase - chert formation in Serbia or the sedimento - volcanogenic formation in the wider vicinity of Konjic in Bosnia - Herzegovina.

In contrast to many other pollutants, chemical elements do not get involved in processes of self - purification: their concentration in the course of migration is only diluted. Moreover, chemical elements (and above all the most toxic heavy metals) take part in all types of migration and biological circulation, which inevitably leads to pollution of vitally important media, namely water, air, and food. Numerous studies have shown the ability of elements to accumulate in living organisms, with toxic influence on many of their systems, such influence resulting in the

---

ground coal fires are present over an area about 5,000 km long and 750 km wide (Fig. 2.34.) (X. M. Yhang, C. J. S. Cassells, and J. L. van Genderen, 1999).

incidence of specific diseases, weakening of immunobiological reactions, and aggravation of overall nonspecific ailments (especially of the allergic type). Typical further consequences of the negative action of toxic chemical elements on living organisms include, above all, disturbance of the reproductive function, decline of bioproductivity, and mutagenic and carcinogenic effects. For all these reasons, we shall dwell on natural geochemical anomalies and components of the environment linked with natural processes of migration and dispersion of substances from deposits. This is especially important because in regions of mining activity, plans are usually made for the creation of a developed economic infrastructure, including settlements, agricultural concerns, water - engineering projects, and recreation zones.

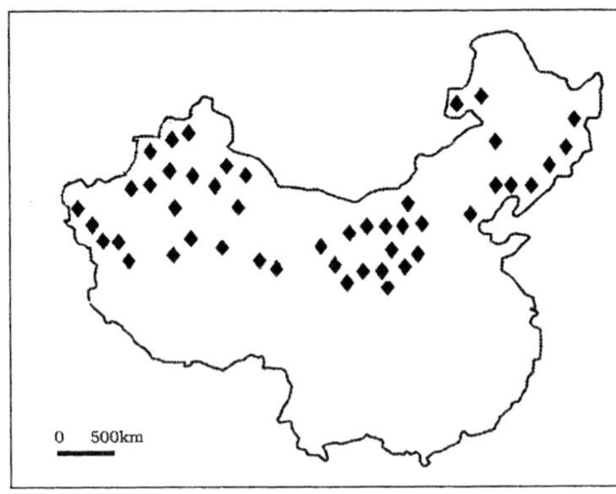

Fig. 2.34. Map of locations with distribution of fires in subsurface deposits in Northern China.

A field with increased content of ore elements and elements accompanying them is formed in rocks, water, plants, and the atmosphere around every ore body. This is the so - called *dispersion aureole* of the deposit. The rocks surrounding ore bodies contain a fairly high concentration of the chemical elements that mark the *primary aureole*, which is formed at the same time as the deposit. The dimensions of primary aureoles and quantity of their chemical elements usually exceed parameters of the ore bodies themselves (Fig. 2.35.). Their composition includes both the main elements of the deposit and a whole series of accessory elements - mercury, lead, zinc, copper, cadmium, arsenic, silver, fluorine, bismuth, cobalt, indium, selenium, tellurium, iron, and sulfur - currently held to be ecologically toxic and hygienically dangerous.

Disturbance of the deposit sends elements into biogenic, physico - chemical, and mechanical migration. For this reason, the soil in valleys and on slopes or in streams near the deposit contains greater quantities of indicator elements than in the more distant surroundings (Fig. 2.35.). The extent of such *secondary dispersion aureoles* in loose rocks and soil is measured in hundreds and thousands of meters.

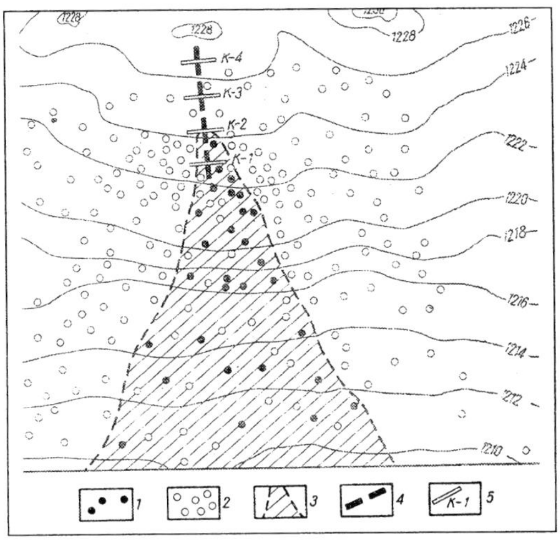

Fig. 2.35. One of typical examples of secondary dispersion aureoles in the area of wolframite deposit. 1 - quartzite particles with wolframite; 2 - granites particles with quartzite; 3 - aureoles of scattering of mining particles; 4 - position of ore vein; 5 - exploratory trenches.

Underground and surface water in the near or distant surroundings of ore bodies also contains greater amounts of indicator elements than at more remote locations. The dimensions of *hydrogeochemical (water) aureoles* likewise attain hundreds and thousands of meters. When plants take root in an ore body or a dispersion aureole and absorb indicator elements from them, a *biogeochemical aureole* is formed in the plants. With elevated content of methane, carbon dioxide, radon, mercury vapor, and other indicator gases in air of the soil, *gas aureoles* are recorded around oil and gas deposits, deposits of radioactive ores, coal - bearing and mercury deposits, and deposits of other kinds.

By determining the content of indicator and accessory elements in rocks, soil, water, plants, and the atmosphere, it is possible to establish the extent and chemical composition of dispersion aureoles. It is much easier to find ores when an aureole is discovered. It is also easier to establish the risk to public health in given regions, which is of greater interest to us.

Water aureoles of ore deposits as a rule are of a multi - component nature. In complex study of the Chu - Il' ore zone - one of the important metal - bearing regions of Kazakhstan - investigators isolated and described more than 130 hydrogeochemical anomalies with high content

(150 - 170 mg/l) of mercury, copper, zinc, lead, molybdenum, aluminum, titanium, cobalt, vanadium, niobium, fluorine, etc. Another characteristic of water aureoles is that their content of macro - and microelements often is not constant in time, but instead depends on the level of precipitation.

In a ***zone of sulfide deposits with oxidative conditions***, oxidation of sulfide occurs in addition to other reactions of biochemical transformation (oxidation). Thus, in groundwater near sulfide ore bodies, acidic sulfate water with high content of heavy metals and formation of secondary minerals can occupy large areas. According to S. L. Shvertsev (1973), maximal metal concentrations in such water can attain the following values (g/l): Cu, 45.6; Zn, 12.0; Co, 3.6; As, 0.4; and Ni, 30. Due to combined action with the surrounding natural environment, water dispersion aureoles undergo differentiation, with the result that it is possible to speak of zonality of their composition. In arid regions, the chemical composition of water aureoles depends to a significant extent on the influence of processes of continental salinization, with formation of soda, sulfate, and chloride water of high mineralization and elevated content of arsenic, iodine, molybdenum, selenium, uranium, etc.

One of the most important problems linked with mine water of sulfide deposits is the problem of protecting the environment from pollution, above all pollution of water resources. In the Borska and Kriveljska Rivers and lower course of the Timok, for example, the flora and fauna have been completely destroyed under the influence of "*blue*" copper - bearing water. The content of sulfate in such water varies from 1.24 to 2.78 g/l, copper content from 0.5 to 1.7 g/l, and iron content from 2.5 to 3.0 g/l.

In any event, such regions must be the subject of careful geochemical and hydrogeological investigation, since they carry potential risks to human health, whether through water or through soil (food). The negative and positive influence of macro - and microelements on human health will be discussed in greater detail in a special section of the book.

In neutral and weakly alkaline environments in the vicinity of sulfide deposits, conditions of formation of the chemical composition of groundwater are sharply altered, and its enrichment with heavy metals occurs for the most part at the expense of electrochemical dissolution.

Water in ***zones of silicate***, ***carbonate***, and ***oxide mineralization*** with its dispersion aureole differs significantly from that of sulfide deposits. Enrichment of such water with heavy metals in zones with oxidative conditions essentially occurs through the action of physical and chemical weathering and radioactive processes. With respect to the solubility of ore minerals and degree of groundwater enrichment with the corresponding metals, non - sulfide deposits can be arbitrarily divided into three groups: 1. *boron deposits*; 2. *deposits of silicates, carbonates, and oxides of various metals*; and 3. *deposits of radioactive elements*. The content of *boron* in water of its deposits can very often attain 100 mg/l, and water aureoles are also characterized by high content of accessory elements, namely fluorine, copper, arsenic, mercury, etc. Increased concentrations of *aluminum*, especially in compounds with organic components, are peculiar to regions of bauxite deposits under conditions of a humid climate. *Deposits of the second group* are characterized by lower concentrations of elements in water of ore bodies and dispersion aureoles, while water of *uranium* deposits under oxidative geochemical conditions has a very unique composition. The highest concentrations of uranium, radium, radon, and accessory elements (arsenic, fluorine, lead, copper, etc.) are linked with water of metallo - riftogenic and hydrothermal uranium deposits. Moving toward the peripheral zone of a water aureole, only the content of uranium and sometimes that of radon and highly migrative accessory salts are above background levels.

A very rich spectrum of components is formed in water aureoles of ***petroleum deposits***. Together with increase in total gas saturation of water following contours of the deposit and the degree of its enrichment with methane homologs, biogenic nitrogen, organic acids, benzene, toluene, phenol, and other organic components, anomalies are also recorded here with respect to sulfur, heavy metals, pure phosphorus, radio - and carbon isotopes, and nitrogen.

Significant enrichment of groundwater in the vicinity of ***salt deposits*** is primarily dictated by the high solubility of salts, especially chloride salts. Thus, mineralization of water as a rule comprises about 300 g/l near halite deposits, which are very widespread. When magnesium salts occur in salt deposits, magnesium ions can prevail over sodium ions. Water of this type has high mineralization (above 200 g/l) and a significant quantity of microcomponents. Groundwaters of potassium deposits are in fact brines with high concentrations of chlorine, bromine, calcium, sodium, magnesium, potassium and various microcomponents. Apart from very significant application in industrial production of salt and soda and use in spa and recreational tourism, highly mineralized saline waters represent a threat to the environment: saline springs and watercourses destroy the surrounding plant and animal life and indirectly act adversely on human health. Moreover, due to melting of salt deposits, empty spaces are created and terrain subsidence with consequent material damage occurs in zones of direct exploitation (for example, near Tuzla in Bosnia - Herzegovina). For all that, we must not forget the great importance of salt in human life, especially since it is possible through preventive exploration and regionalization to isolate zones of occurrence of unfavorable elements linked with mineral raw materials in general.

Radioactive Elements in Rocks, Soil, and Groundwater

Pitchblende was probably known already in the Bronze Age as a mineral of unknown composition to miners of cassiterite in Cornwall (Southwest England), where these two metals occur together. It is quite certain that this black ore came to man's attention already in 1512, when exploitation of silver began in the now famous mine of St. Joachimsthal (currently Jahimova in the Czech Republic). Finally, the German apothecary Martin Heinrich Klaproth established 210 years ago that the black resinous ore in question is a mineral of a new element, which he named after the planet Uranus (discovered in 1781). With the discovery of radioactivity 104 years ago, the new element of unusual behavior entered the annals of the so - called *atomic era*, marking it above all as material for a heretofore unseen super weapon and an excellent source of energy.

Since the beginning of the 50's of the last century, we have been witness to the greatest search for some ore in the history of mankind. Thus, apart from Jahimova and Cornwall, the Urgeirica Mine of uranium (radium) was opened around 1912 in Portugal. The following year, the Shinkolobwe deposit was discovered in what was then the Belgian Congo (Zaire), while the Eldorado deposit was discovered in 1931 on the shores of Great Bear Lake in Northern Canada. Mass construction of nuclear power plants starting in the 1960's led to spectacular discoveries of major uranium deposits, and already at the outset of the next decade the Cigar Lake deposit was discovered in Canada. This deposit is the unsurpassed world record holder with more than 10% average uranium content and more than 10,000 t of reserves of this metal[50,51]. Deposits and reserves of uranium were discussed in greater detail in the chapter on energy raw materials.

Characteristics of Natural Radionuclides

Radioactive elements and ionizing radiation are present everywhere in space and on our planet as well. Very high - energy particles reach the Earth's surface from the depths of space in the form of *cosmic radiation* and *electromagnetic radiation* of varying energy from the stars. There is also the *radioactive radiation* of radionuclides dispersed in the Earth's

---

[50] Already in 1980 there were 186 nuclear power plants using U_{235} in the world, and these plants satisfied 7.6% of all energy needs (as much as 45.5% in the USA and 28% in Europe).

[51] It is interesting that a phenomenon observed at the Oklo deposit in Gabon suggested the possibility of nuclear reactors. To be specific, it was found that losses of U_{235} in ore are a consequence of consumption in the process of fission, which here took place in a natural nuclear reactor. It is believed that the circumstances that made a chain process of fission possible in Oklo do not represent an exceptional case in the Earth's past.

crust. Radioactive elements from rocks enter water and the atmosphere as a result of migration and through them gain admission to the biosphere and the forms of life in it. Contributions to radioactivity of the atmosphere are also made by radionuclides that are formed as a result of human activity (whether in processes of fission or by irradiation of stable isotopes in nuclear reactors or irradiation of high - energy particles) and which today unquestionably represent the most numerous and diversified sources of radiation in the world.

The abundance of radiation in the Earth's remote past was considerably greater than today. To put it more precisely, amounts of U_{238} in the Earth's crust have due to decay been reduced since its formation by exactly half, while amounts of U_{235} have shrunk approximately fivefold. Thus, the natural radioactive elements that we encounter today represent only a part of the abundant radionuclides that once existed, but which have disappeared due to radioactive decay. Their half - lives were short in relation to the Earth's lifetime.

By *radioactivity* we mean instability of the nucleus, while *radioactive decay* refers to the transformations undergone by radioactive isotopes. The main types of radioactive decay are *alpha (α) decay, beta (β) decay,* and *gamma (γ) decay. Alpha rays* are positively charged helium ions, and this radiation occurs mainly in the heaviest radionuclides. It has the greatest strength of ionization, but passes poorly through different materials. Although the range of action of alpha radiation is limited, radioisotopes that emit alpha particles penetrate the organism and cause serious damage to it. *Beta particles* represent elementary particles carrying a negative (more rarely positive) charge, i.e., electrons. They are more penetrative than alpha particles. Because they have lower ionizing ability, these particles are not as dangerous as alpha particles to living cells. *Gamma radiation* is photonic energy emission, and gamma rays are identical in nature to *X - rays,* but of different origin. In contrast to α - particles, gamma rays can penetrate soil and rocks to a depth of the order of a meter, as well as the whole human body. The biological damage caused by gamma radiation is very similar to that caused by X - rays.

There are three main series of radioactive decay: *the uranium - radium* (U_{238} - Ra_{226}) series; *the uranium - actinium* (U_{235} - Ac_{227}) series; and *the thorium* (Th_{232}) series. The final (stable) members of the series are lead 206, 207, and 208. Apart from the indicated series, there exist in nature many other radionuclides that ert very little influence on the overall background of radiation. The most significant among them is *potassium - 40* because it makes large contribution to total gamma ionizing radiation on the surface of the Earth due to its relatively high content (2%) in the Earth's crust.

About 63 radionuclides occur in nature. On the basis of their half - lives, they can be divided into two groups: *long - lived* (U, Th, Re, Rb, V, K, etc.); and *short - lived.* The long - lived category includes radionuclides whose half - life exceeds a hundred million years and which remain present in nature for billions of years (the so - called *primary radionuclides*: U_{235}, U_{238}, Th_{232}, Rb_{87}, K_{40}, etc.). The half - life of short - lived radionuclides amounts to only several days, minutes, or seconds (Ra_{223}, Rn, Po, Bi, Ti, etc.), or several years in exceptional cases. Let us dwell on the more important radionuclides (radioisotopes).

Natural uranium consists of three isotopes (U_{238}, U_{234}, U_{235}). It exists exclusively in the four - valent state in the UO_2 and UO_3 oxide (uranate) form (uraninite and pitchblende being mixtures of these two oxides) or as the uranyl complex UO_2^{2+}, of which a large number of minerals are built. Apart from uraninite, minerals of the best - represented isotope of uranium (U_{238}) are also known, namely torbernite and carnotite. It is held today that uranium in soil is found in two main forms, inert and mobile. The *inert form* of uranium has no significance whatever for radioecology and biotechnology, even though more than 97% of total uranium in the environment belongs to this form. The *mobile form* of uranium - although it represents only 0.1 to 3% - is of great biological and ecological significance, since through the process of ionic alteration it can pass from the soil into plants or penetrate plants in the form of complex compounds with organic acids secreted by the root system. The solubility of many uranium compounds in water is fairly high, with the result that almost all groundwater is to a certain extent radioactive. Uranium in solution circulates only in the form of uranyl comples and can be adsorbed by different materials (zeolites, clays, natural colloids).

The most widespread isotope of *radium* Ra_{226} is a product of decay of U_{238} (half - life: 1622 years). Another isotope of this element is Ra_{228}, a product of decay of thorium Th_{232} (half - life: 67 years). Radium 226 in nature is found in varying concentrations in virtually all kinds of rocks and soils that contain natural uranium, occurring in the greatest amounts in the mineral uraninite. Radium Ra_{228} is the only isotope of this element that is a beta - emitter. In nature it is found in soil and water in a ratio of 1:1 with Ra_{226}. However, because of the different half - lives of these two radionuclides, Ra_{228} disappears faster from organisms, so that in older persons this ratio is displaced in favor of Ra_{226}.

Radon is a radioactive inert gas formed by emanation of Ra_{226}. It is an invisible gas without taste and odor that is seven and a half times heavier than air. It has 12 short - lived isotopes. The most widespread of these isotopes is Rn_{222} with a half - life of 3.8 days. The remaining isotopes of radon have a very short half - life (several seconds) and are of no radioecological significance. $Radon_{222}$ is found in rocks, soil, and groundwater and air. It migrates readily along fissures and faults in the Earth's crust, often over great distances from the place of its formation. Elevated concentrations of this gas are associated with deposits of uranium and phosphorus that have high radium content, areas with greatly cracked granitoids, and certain thermal springs. Radon is considered to be a very important source of natural radiation.

Among ***cosmogenic radionuclides***, the best - known are *radioactive carbon* C_{14} and *tritium* H_3, while the other radionuclides of this group are without radioecological significance. In the phase of life of organisms, carbon and organic substances are found in virtual equilibrium with carbon from the atmosphere, i.e., the ratio between radioactive carbon C_{14} and stable carbon is the same as in the atmosphere. After the death of organisms, the equilibrium of C_{14} is no longer maintained and its concentration decreases at a rate of 50% every 5,700 years.

This regularity makes it possible to ascertain the age of organic matter by determining the presence of C_{14}. With a half - life of 12.26 years, the radioactive isotope of hydrogen tritium H_3 is continuously formed in the atmosphere by the interaction of cosmic radiation with nuclei of nitrogen and hydrogen atoms. Up until the period of more intensive nuclear tests in the 50's of the 20th Century, the content of H_3 in water ranged from 222 to 888 mBq/l, only to increase five - to tenfold and more in the course of nuclear testing (for example, a level of as much as 2,937 tritium units was measured in Ottawa in 1954).

Among ***radionuclides of artificial origin*** (formed during nuclear reactions in production of atomic energy, during tests of nuclear weapons, etc.), the most interesting are *strontium* Sr_{90} and *cesium* Cs_{137}. With about a thousand other known artificial radioisotopes, they represent sources of radioactive pollution that are dangerous to human health. On the other hand, owing to the possibility of reliably monitoring them by measuring radiation, they have found very diverse applications in many branches of medicine, industry, science, and other domains of human endeavor.

Concentration of Natural Radionuclides in Rocks and Soil

Table 2.12. presents the content of uranium and thorium in different rocks and some magmatic rocks of former Yugoslavia. Perceptible variations in the content of those two radionuclides are also present in acidic or intermediate alkaline rocks, but uranium and thorium are in the same constant ratio of 1:3.5.

Table 2.13. gives average content of uranium and thorium in magmatic and sedimentary rocks. It follows from the table that granites and granitoid rocks are richest in these elements, but their share is also appreciable in sedimentary rocks. Minerals of the feldspar and mica group are of greatest significance for the concentration of these elements in granitoids. Since these are very widespread minerals, considerable concentrations of uranium can be released from them in spite of relatively low uranium content. Considerable concentrations of uranium in granitoids are linked with zircon and monazite. Monazite is regularly found in the alluvial sediment of rivers passing through regions built of rocks containing it (sediment of the Pek River, for instance)[52].

The concentration of uranium in soil is very different in different parts of the world (Table 2.14.), ranging somewhat more often in the interval of from 7.4 to 37.7 mBq/g, although there are regions in the world (Peru, for example) where it attains 76.2 mBq/km^2.

Thorium is found in soil in concentrations ranging from 8.1 to 33 ppm. Its main minerals - thorianite and thorite - readily pass from granites and pegmatites into debris. As the main economically significant

---

[52] As for uranium, it is interesting to note that the bulk of ore is obtained from deposits in sandstones and conglomerates, the rest from uranium-bearing phosphates, schists, granites, and granodiorites.

carrier of thorium, monazite is an accessory mineral considerably more frequent than the two minerals mentioned, and much more of it is present in debris, especially in sediments of marine beaches. The insolubility of thorium and its very low specific activity cause thorium to be found in quite small amounts in the biological world, i.e., they account for the relatively slight radioecological significance of this radionuclide.

TABLE 2.12.
Contents of U and Th in some magmatic rocks of former Yugoslavia. 1- Marlovo; 2 - Kratovo; 3 - Timok - Pek; 4 - Gnjilane; 5 - Cer; 6 - Bukulja; 7 - Janja; 8 - Bujanovac.

Ppm	1	2	3	4	5	6	7	8
U	9.7	5.7	1.3	11.7	4.5	4.3	6.4	1.9
Th	33	25	4.6	49	13	15	29	4.5
Th/U	3.4	4.4	3.5	4.2	2.9	3.2	4.5	2.4

TABLE 2.13.
Average concentrations in U, Th, Pb in magmatic and sedimentary rocks (ppm).

Type of rock	U	Th	Pb	Th/U
Granite with low Ca	3.0	17.0	19.0	5.7
Granite with high Ca	3.0	8.5	15.0	2.8
Syenite	3.0	13.0	12.0	4.3
Basalt	1.0	4.0	6.0	4.0
Ultramafic rocks	0.001	0.004	1.0	4.0
Clay schists	3.7	12.0	20.0	3.3
Sandstones	0.45	1.7	7.0	3.8
Carbonates	2.2	1.7	9.0	0.77
Deep - sea clay	1.3	7.0	80.0	5.4

TABLE 2.14.
Uranium concentration in soil (B.Petrovic, R.Mitrovic, 1991).

Area	μg U/g
New Zealand	60
USA	0.2 - 47
France	1.2 - 37.6
Gabone	1 - 100
Poland	0.5 - 12.5
Former USSR	0.2 - 13
Japan	0.7 - 1.9
Serbia	0.08 - 5.9

Radionuclide Concentration in Water

In nature, all surface and groundwater contains radioactive elements (except thorium) in measurable quantities, i.e., it is radioactive to one extent or another. However, in the case of uranium behavior in the presence of water under exogenous conditions, we are dealing with very dynamic and often reversible processes of dissolution, transport, fixation, and renewed dissolution, the chemical nature of which has not yet been elucidated. These processes accordingly are hard to foresee, and predictions of places where they might occur are most often dubious. It is therefore understandable that uranium concentration in natural water fluctuates within a very wide interval, from 0.05 to 10 ppb. The average content of uranium in water is approximately 1.5 ppb. It is usually more than 200 ppb in groundwater that circulates through uranium - rich rocks. In the United States, for example, the concentration of Ra_{226} varies within limits of from 370 mBq/m^3 (Portland, Oregon) to 2.14 Bq/m^3 (Joliet, Illinois). According to E.M. Durrance (1986), uranium content comprised 0.07 - 7 ppb (0.5 ppb on average) in 100 tested Russian watercourses; 0.1 ppb on average in American rivers; 1 - 10 ppb in surface water of uranium - bearing terrains; 1 - 150 ppb in groundwater of such terrains; and 15 - 400 ppb in water inside uranium deposits.

Radon Concentration in Rocks, Soil, and Groundwater

Radon is the heaviest of all inert gases, but like all of them it is exceptionally capable of migration. Of gases most often present in the Earth's crust, only CO_2 is more soluble than radon. For this reason, radon is often found in water, which is the main agent of distant transport of this gas. It is especially present in groundwater and (under special conditions) can be transported as far as many kilometers. The indicated high ability to migrate enables radon to move relatively easily and rapidly through rocks impermeable to water (schists for instance) and emanate from rocks into surrounding water or air. In tectonic zones, all gas formed from radium can be released from rocks (the coefficient of emanation is equal to 100%). Apart from groundwater, the main channels or conduits of radon transport are usually fault fissures[53]. It should be kept in mind that the existence of most radon hydrochemical anomalies in the vicinity of tectonic zones is determined not only by migration of radon from greater depths, but also by the fact that the structure of such zones makes it possible for formations of uranium ore to come into contact with groundwater and be degraded by it.

Background concentrations of radon in the air of soil and in groundwater are of a predominantly local nature and most often depend on the rock substrate. Thus, for example, T. Tadolini and M. Spizzico (1998) cite experimental results from the karst region of Apulia in Italy,

---

[53] In the presence of pre-seismic tensions, fault fissures open and radon concentrations increase, which is why this gas is taken as one of the possible indicators (predictors) of earthquakes.

where they analyzed the ratio of radium in terra rossa (as the primary source of this gas) and radon 222 in groundwater (Fig. 2.36.).

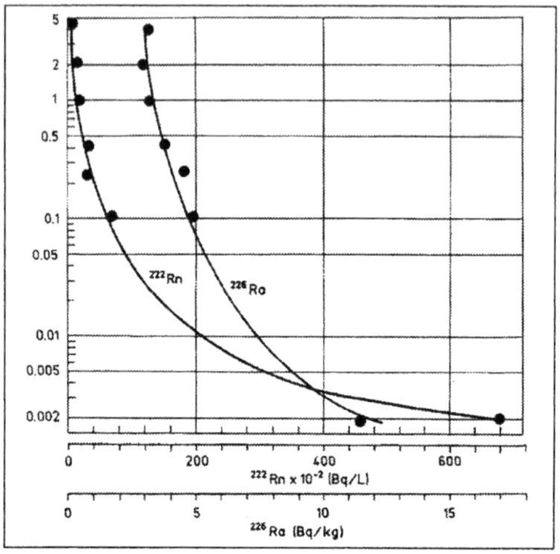

Fig. 2.36. Relation between grain - size and Ra_{226} activity in footwall of terra rossa and appropriate activities of Rn_{222} in water (T. Tadolini, M. Spizzico, 1998).

Radon content in groundwater varies from 3.7 to 1.11×10^3 Bq/l, the average approaching 74 Bq/l or somewhat exceeding it. The greatest concentration of this gas for different regions varies from 1.65×10^3 to 14.8×10^3 Bq/l. Groundwater with Rn concentration above 1.85×10^2 Bq/l is considered radioactive, while radon waters of high concentration contain more than 7.4×10^3 Bq/l. One of the extreme concentrations of this gas in water (20×10^3 Bq/l) was measured in the vicinity of Skopje in FYROM (Former Yugoslav Republic of Macedonia), and average content in water of springs in the immediate surroundings was of the order of 40 Bq/l. For the deep tectonic zones of the Caucasus, Ural, and Tien - Shan Mountains, S. L. Shvartsev et al. (1982) reported that radon concentration varies within an interval of from 133.2 to 277.5×10^3 Bq/l.

Natural Radionuclides and Risks

In their paper "**Two centuries since discovery of the chemical element uranium**" (1989), V. Omaljev and A. Antonovic wrote as follows: "*Uranium today is tasked with a heavy burden of guilt for current and future radioactive pollution of the planet. Especially worrying are uncertainties regarding possible (uncontrolled) mutagenic changes in living organisms, man above all. Systematic biological and medical investigations of this question for practical purposes began only after the atomic bombing of Japan, and a period of 45 years is too short to see the long - term risks.*" The authors ask *whether this means the end of the rule of uranium.*

Together with economic and other global questions, safety is certainly a pressing concern, and most people hold it to be the most alarming problem of the present day. In the event that nuclear power plants prove to be faulty for any reason and radioactive substances are emanated into the environment or if nuclear tests are carried out and nuclear weapons used on a massive scale, the consequences could be catastrophic. *At the present level of civilization, the modern world is increasingly destructive, and the consequences are unfortunately moving ever closer to a state of catastrophe.* We note only what was done recently to the population and environment of Iraq, Yugoslavia, and Bosnia and Herzegovina by tens of tons of depleted uranium used in the making of projectiles.

The anthropogenic component of the biosphere's radiation background is what has raised this global question. However, even after eventual cessation of the rule of uranium, there would remain a group of natural radionuclides with all potential risks to human health[54]. According to the results of studies performed to date, natural radiation of the lithosphere (biosphere) is responsible for one fourth of the total *basic radiation load* of all living beings. To be specific, the role of natural radionuclides and their radiation as sources of radiation risk to the biocenosis has been present since the beginning of life on Earth. Thus, natural radionuclides have through radiomutations not only affected the evolution of living organisms, but also influenced the dynamics of equilibrium of ecological systems.

In contrast to the large doses of ionizing radiation produced predominantly by fabricated radionuclides, it is fairly logical to maintain that in the case of so - called *small doses*, i.e., doses that do not cause clinical symptoms but which ert distinct bionegative action, natural sources of radioactive radiation assume the leading role. It is known that opinions among radiobiologists are greatly divided about the effect of small doses. Still, many authors hold that a slight increase of absorbed radiation raises the probability of cataracts, increases the incidence of radiation illness, contributes to the formation of tumors, shortens life expectancy, and slows fetal development in the mother's uterus. Following the engagement of world experts on an international project during the period of 1980 - 1984, it was concluded (among other things) that: 1) ***small doses of ionizing radiation ert distinct bionegative action*** (boldface mine - M.K.); 2) *the bionegative action of small doses of ionizing radiation conforms to a linear dependence with no threshold*; 3) *since it is impossible to isolate small doses of ionizing radiation as one of a large number of carcinogenic factors, it is necessary to prevent increase in the level of all of the indicated*

[54] According to propositions of the International Committee for Radiation Protection published in 1976, *risk* means the probability that some organism will experience a certain harmful effect as a result of irradiation.

factors; and 4) *we need to monitor constantly the radiation load of the population through the whole food chain, including cultivated plants, livestock, and foods of plant and animal origin.*

Medical geology can make a great contribution in establishing natural preconditions for occurrence of small doses of ionizing radiation in rocks, soil, and water, and in planning measures designed to lower the level of carcinogenic factors. This is especially true because the content of radioactive elements in the natural environment is not always the same, and organisms inhabiting a given region are exposed to radioactive radiation of varying strength: greater in some cases, lesser in others. We shall therefore dwell in the text to follow on certain geological factors that deserve special attention.

The first and very important step in investigating a region of interest is to isolate ***uranium - bearing rocks and formations*** in it. We have already discussed the kinds of rocks (such as acidic granites) that are potential carriers of radionuclides. However, it was not stressed that uranium - bearing geological formations can potentially encompass enormous areas and thereby increase the natural radiation load of a large number of people. Thus, schists rich in organic substances and uraninite (for example, the Chittanunga formation in the states of Alabama and Kentucky, covering an area of the order of 100 thousand km^2) occur in many regions of the world. In the United States, the Phosphoria formation (phosphate deposits with 0.003 - 0.03% uranium) and other formations on the Colorado Plateau and in Eastern Utah, Northeast Arizona, and Northwest New Mexico encompass an even greater area. The Francevillien formation in Gabon covers an area of about 35 thousand km^2. When it is taken into account that the Oklo deposit or phenomenon (a natural nuclear reactor) is found in that formation, it is easy to grasp the potential danger from constant reactions of fission in that region. Large amounts of thorium are found in sands of marine beaches in Southern India and (to some extent) Australia. By conducting regional geomedical investigation (mapping) of the natural radionuclides in rocks and soil in an area of interest to us, we are able to isolate *radioecogeological ranges* (regions) in which the biocenosis (including human beings) carries a certain radiation load. It is understandable that some regions of the world are well - known for high content of radioactive elements (the state of Kerala in India, the La Plata Plateau in Brazil, the wider region of the Oklo uranium mine in Gabon, etc.).

That man has been unable to adapt to natural radiation fields is shown by the results of international research indicating that regions with elevated content of natural radionuclides in rocks and soil are characterized by slow population increase, increase in the number of birth defects, increase in the number of organic diseases, and increased mortality (especially among children). Complex studies on the question of existence or non - existence of *radiogeochemical endemias* remain to be carried out. In their book "**Radiation Hygiene in Biotechnology**" (1991), B. Petrovic and R. Mitrovic state that there is research indicating decisive significance of radiogeochemical endemias, but also work that negates them. For example, English investigators maintain that elevated concentration of radon Rn_{222} in drinking water in the county of Devon (Great Britain) represents the main cause of more frequent incidence of malignant diseases among the population. Similar conclusions were reached by American investigators who carried out radioepidemiological studies on more than a million people in 111 cities in the states of Iowa and Illinois: they assert that considerably elevated concentrations of the indicated radionuclide in drinking water above the average level lead to significant differences in mortality from malignant diseases of the bones (which are manifested especially in persons more than 30 years old). Also, Soviet investigators reported results of measuring natural radioactivity in the bones of persons who died of leukemia and ones who died of various injuries: the total activity of gamma - emitting radionuclides was more than two times greater in bones of those who died of leukemia than in bones of persons who died of traumatic causes. From the results of his own investigations and reports of others, the French scientist Pincet concludes that there exists a significant correlation between the level of the radiation background in the biosphere and mortality due to malignant diseases. It follows from all that has been said that familiarity with the radiogeological regionalization of a certain area is very important in order to establish the radiation load of the population in it: doctors need to pay special attention to the most dangerous radioecogeological ranges.

Depending on the rocks from which it was formed and content of the clay component, ***soil*** can be radioactively contaminated to a greater or lesser extent. Natural radionuclides accumulated in the soil are incorporated metabolically into plants and through contaminated food find their way to the organism of man and animals. The given danger to living beings understandably is especially expressed within radioecogeological ranges marked by an elevated background of natural radioactive elements. The problem is further complicated by the presence of strontium, cesium, and other artificial products of *long - lived* radionuclides, which will be discussed in a special chapter.

Uranium can be relatively easily transported, concentrated, and carried away from its primary deposit, depending on geological (hydrogeological) conditions. A very important role in this is played by **groundwater**, since uranium from the moment of its oxidation and dissolution in water moves almost freely through the Earth's crust (Fig. 2.37.). Uranium by itself is not necessarily all that dangerous, but the products of its radioactive decay not only threaten nature with radioactive radiation, they are also toxic as elements[55]. The role of aqueous solutions is therefore of primary significance for the creation of chemical dispersion aureoles of uranium and its decomposition in the surface clay layer of river sediments. We have in mind here *anomalous zones*, with uranium content of up to 200 ppb and more (as much as 400 ppb and sometimes ten times higher in mine waters). Under favorable hydrogeological conditions (for example, the presence of extended fault zones), uranium can be transported over very great distances, which must be kept in mind. Thus, for example, the origin of high uranium content (50 - 100 ppb) in one spring in Colorado is linked with rocks at least 100 km away. For all of these reasons, in hydrogeological investigations of a region, special attention must be paid to the content of radioisotopes in drinking water and water used for production of food, so as to prevent them from entering the human organism in this way.

Pollution of the environment by radioactive elements is certainly at the very top of the list of all problems that have ever existed and called for protection of nature and mankind. The complexity and danger of pollution of nature by radioactive elements have forced man to study in the minutest detail processes involving radioactive elements. Natural and artificial radionuclides introduced into the organism by ingestion or inhalation is distributed to individual organs in keeping with metabolism of the radionuclide itself and sensitivity of the organ to radiation. All of this is taken into account in estimating size of the contribution of radionuclides from drinking water, food products, and other sources of radiation to total irradiation of the population.

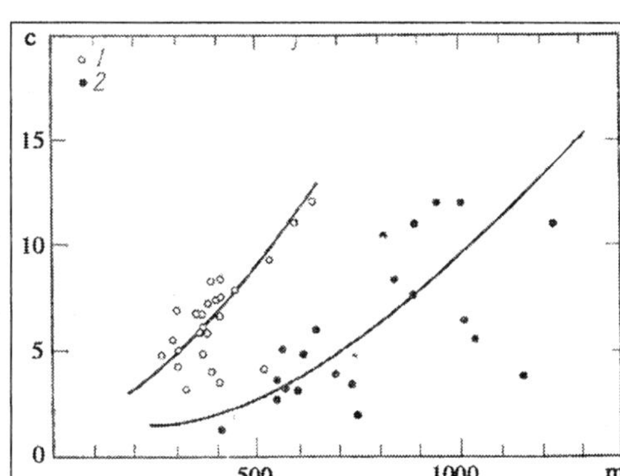

Fig. 2.37. Ratio of uranium concentration **c** (µg/l) and water mineralization **m** (mg/l) from Ogallala formation in Texas (USA)(S.N.Davis, R.J.M.de Wiest, 1966). 1- samples with HCO_3, over 50% of all anions; 2 - samples with HCO_3 is less than 50% of all anions.

---

[55] It is interesting to note that when the most widespread U_{238} isotope of uranium is present in water, the danger of chemical toxicity causing diseases of the liver is greater than the danger from radiation.

Radiation that from the place of its generation enters inhabited regions or regions not directly connected with production or utilization of nuclear energy should not exceed 0.5 *sieverts (Sv)*[56] per person annually. In regions with widespread radioactive minerals, the dose of background irradiation a man receives can be four times greater than the given limit, which must have greater or lesser harmful effects on human health[57]. If some short-term radiation is involved, more considerable changes in the cardiovascular system can set in at a dose greater than 150 - 200 mSv, while a dose exceeding 700 - 1,000 mSv is usually fatal. The term *maximal permissible concentration* is used in practice, although it is highly questionable whether there exists a limit below which safety is guaranteed. Even if a maximal permissible dose of radiation is found, it remains for us to establish the action of individual radioisotopes, especially inasmuch as they emit different kinds of radiation and at different rates. Moreover, some elements have a tendency to accumulate in different parts of the human organism. Radioisotopes of such elements cause concentrated biological damage. For example, plutonium, radium, and strontium accumulate in the bones; iodine accumulates in the thyroid gland; and lead and water - soluble isotopes accumulate in the kidneys. The lungs are the critical organ in the case of radionuclides in undissolved form.

Differences in the biological danger caused by different radioisotopes are generally great. For example, if we compare Ra_{228} and H_3 (both with a half - life of several years and emission of low - energy beta rays), we see that the time of their residence in the human organism is different. A large part of H_3 disappears from the organism after several weeks, whereas R_{228} remains in the bones forever. Several other specificities in behavior of the more important radioisotopes are examined below.

The most widespread **isotope of radium Ra_{226}** is the most toxic of all inorganic components and has very strong carcinogenic action. Its biological half - life is quite long, 45 years. Its chemical behavior is similar to that of calcium. By means of resorption from the soil, it readily enters plants and later reaches man by way of animals and food of animal and plant origin. Although the average concentration of this element introduced with food is 111 - 185 mBq/day, American investigators established great variability of its concentration in different kinds of food. For example, the very high activity and content of R_{226} in filberts as a natural phenomenon has occupied the attention of many investigators.

---

56 The *sievert (Sv)* is the biological equivalent of the *roentgen* and corresponds to the value of absorption by the body of any ionizing radiation equivalent in its activity to absorption by man of one roentgen of X- or gamma radiation.

57 It is known that the biological consequences of ionizing radiation can be divided into somatic consequences, which are manifested on the irradiated organism itself; and genetic consequences, which result from damage to the cell's genetic material and can appear as a reaction in the organism's progeny, often several generations later.

According to B. Petrovic and R. Mitrovic (1991), many workers have investigated the content of Ra_{226} in the human organism and attempted to determine the radiation load of this radionuclide. It has been established that in inhabitants of Europe and Central America, its average amount in bones is 0.37 - 0.55 mBq/g of ash (which for the average man corresponds to an activity of 1.11 - 1.48 mBq). This figure is somewhat higher in regions of North America and Asia.

The presence of **radon** in water is not dangerous to human health. However, radon in the gaseous state erts undesirable influence. Radon is fid in the lungs as the critical organ. The real danger of radon is from its short - lived products bismuth and polonium. Radon and its products emit predominantly alpha rays, which have a very limited range, but are energetically quite strong and within a diameter of less than 1 mm cause heavy damage to the lungs. The risk of cancer is directly dependent on the concentration of this gas in the space where man spends his time and is especially high for miners working in shafts in uranium deposits.

The positive influence of radon and unknown doses of alpha rays from products of its breakdown on the human organism is widely used in balneology. Radioactive (radon) waters alter the blood picture, lower blood pressure, affect certain allergic diseases, act on functioning of the central and vegetative nervous system, stimulate many compensatory - adaptive responses of the organism, etc.

Radioactive Elements in Rocks, Soil, and Groundwater on the Territory of Yugoslavia

In Yugoslavia, the first occurrences of uranium were discovered in 1947 at the confluence of the streams Dobroticka Reka and Vojnicka Voda south of Prokuplje (M. Ristic, 1965). The uranium mineral autunite (*autonite*) was found in crystalline schists in a pegmatite vein there. After this discovery, the Geological Institute in Belgrade (Geoinstitut - Beograd) conducted intensive investigation of uranium mineralization on the entire territory of former Yugoslavia. Already in 1947, the first measurements were performed in Permian sandstones in the vicinity of the peak Zirovski Vrh about 30 km west of Ljubljana. It was subsequently established that this is in practical terms the most interesting uranium deposit not just in the former Yugoslavia, but in Southeastern Europe as well, and its exploitation was launched. It also turned out that there are many other occurrences of uranium, including uranium - bearing coals near Sinj in Dalmatia and Rasa in Istria; uranium - bearing veins of Zletovska Reka east of Kratovo and south of Prilep; in rhyolites near Radovise and Strumica; in the granites of Bukulja and the Stara Planina Mountains; occurrences in the Golija region and along the upper course of the Pcinja River; in granites of the stream Slatinska Reka; occurrences on the southern slopes of Mt. Cer; and in trachytes south of Gnjilane. In different mineral associations and genetic types, uranium mineralization is found in virtually all metallogenic provinces of the given territory. Let us dwell on interesting localities on the territory of Serbia and Montenegro, where *the Carpatho - Balkanides, the Vardar Zone,* and the southern half of *the*

Serbo - Macedonian Mass can be singled out as regions with many occurrences.

1. A large number of occurrences and deposits of uranium are found in the region of the ***Carpatho - Balkanides***. The area of the *Stara Planina* Mountains has been evaluated as especially promising due to elevated content of uranium in granitoids and in surface and groundwater. Granitoid massifs on western slopes in the central part of these mountains are characterized by interesting mineralization. The Janje ore field extends over an area of about 30 km^2 (S. Gertik, 1978). Uranium content in ore bodies here is exceptionally high, about 300 g/t (300 ppm). Investigation of up to 30 km of mine corridors was carried out in the exploratory phase. During trial exploitation from 1960 to 1966, the amount of uranium ore extracted from the deposits in Gabrovnici and Mezdreji comprised 90 thousand tons. Also known in the Stara Planina Mountains are uranium mineralizations in the vicinity of Inovska Reka and Tumba near Kalna, with average content of this element in the range of 300 - 700 g/t (300 - 700 ppm).
   The attention of investigators has been attracted as well by a formation of Permian red sandstones developed on an area of about 800 km^2. Systematic radiometric investigations were carried out in the *Zaplanje region* south of the Nisava River, where about 130 anomalies with increased radioactivity were isolated. A significant number of these anomalies are secondary dispersion aureoles. The more important occurrences include Gadzin Han, where uranium is found in Permian sandstones and conglomerates (with up to 200 g/t of uranium and gamma radioactivity of up to 1,500 nR/h); occurrences in sandstones of Neogene age in Donje Prisiane (up to 300 g/t of uranium) and in Gornje and Donje Dragovlje; and occurrences in sandstones with coals along the Mlaka River (more than 800 g/t in coals) (S. Visic, 1969). Several tens of radioactivity anomalies have been registered in black Paleozoic schists on an area of about 100 km^2 in the neighborhood of Mts. *Kucaj* and *Beljanica*, where more than 60% of surface and groundwater samples had higher than permissible radium concentration and some were in good measure contaminated by radon.
2. Apart from the Kratovo - Zletovo volcanogenic region in Macedonia, the *Pcinja region* can be singled out within the ***Serbo - Macedonian Mass*** as an area with a considerable number of uranium mineralizations. High levels of uranium, thorium, and potassium have been established in the Slatinska Reka granites (on slopes of Mt. Kukavica). Marked by anomalies with elevated uranium content are granitoids of the *Surdulica Massif*; the pegmatites of *Vidojevica* and *Dobrotic* near Prokuplje; and the *Grbko* gneisses near Stalac (up to 260 g/t of uranium). Characterized by anomalies with elevated thorium content are monazites from granitoids in the vicinity of *Stalacki Visovi* (2,300 ppm of uranium and 50 thousand ppm of thorium); sphene from the granodiorites of *Surdulica* (up to 380 ppm of thorium); gneisses of the *Poljno* locality near Krusevac; etc.

3. Radiometric measurements in Sumadija have shown that granites, volcanites, Paleozoic schists, and some lacustrine sediments of Tertiary basins are uranium - bearing. The situation is similar at many localities in the remainder of the ***Vardar Geotectonic Zone***, from Mt. Cer in the northwest through Mt. Golija and Brzec to Gnjilane in the south. The *Bukulja granitic massif*, with an area of about 40 km^2, has been extensively investigated. Up to 10 m thick and more than 1 km long, the Paun Stena and Cigankulja veins have been isolated within that massif. Also of interest in the Sumadija region are the *Brajkovac granitic massif* and the volcanites of *Medvednjak*, *Rudnik*, and *Borac*, with high average content of uranium and thorium. High uranium content is especially characteristic of volcanites in the western part of *Mt. Golija* (the region of Muhovo) and is also recorded in surface and groundwater there, as well as in clayey alluvial sediments (10 - 75 g/t). On the western slopes of *Mt. Cer* (Iverak near Dragnice), uranium is linked with lignite residues in lacustrine sediments of Neogene age. On the opposite side of this geotectonic unit, trachytes near *Stublovaca* south of Gnjilane contain appreciably more thorium than uranium.

From what has been said above, it is possible to discern regions or geological formations with enhanced uranium - bearing mineralization on the territory of Yugoslavia. These areas with increased risks to human health are appreciably expanded as a result of transport of natural radionuclides by surface and groundwater running from regions with mined uranium. Thus, for example, springs formed in fault zones in the region of the hydrochemical radon anomaly near Mt. Kitka south of Skopje (FYROM) are marked by exceptionally high content of this isotope (up to 19,640 Bq/l), even though they are fairly far removed from the anomaly. Or in the case of an anomaly in the granitoids of Mt. Cer, the spring Majur - Slatina in 1978 had radon content of the order of 3,600 Bq/l.

In order to illustrate the dangers of radioactive contaminants, let us dwell on waters of terrains that have been investigated. In the eastern part of the *Bukulja granitic massif*, from which water gravitates toward the Malo Jezero lake (a source of water supply for the town of Arandjelovac), all water analyses showed radon content of more than 10 Bq/l, 55% of the samples contained more than 100 Bq/l, and some samples had even as much as 462 Bq/l of radon. In the southern part of the massif, around the lake Garasko Jezero (formed above the western part of the Paun Stena uranium deposit), radon content in half of the samples of groundwater was 120 - 126 Bq/l. In the domain of the Cigankulja deposit in the massif's western part, close to 80% of 64 samples of groundwater contained more than 7 Bq/l of radon, 11 samples contained more than 100 Bq/l, and even levels of 500 to 1000 Bq/l were recorded in 8 samples. In view of the fact that this region is fairly densely populated, the cited data emphasize the need for geological (geomedical) studies and appropriate protective measures[58]. Elsewhere, underground and surface water in the region of black schists on the mountains *Kucaj* and *Beljanica* predominantly contains radium in amounts greater than permissible and is also in good measure contaminated with radon. In the *Stara Planina Mountains*, several occurrences and deposits of uranium have been investigated in the Dojkinci - Brlog - Jelovica - Lokve region southeast of Kalna. It was established that levels of radioactive elements (U, Ra, and Rn) in groundwater are elevated (U up to 40.9 μg/l, Ra up to 0.294 Bq/l, and radon up to 63.2 Bq/l). Of the many water samples tested, about 38% had increased content of uranium, while about 34% were with increased content of radium. Radioactivity anomalies were isolated northwest of Dojkinci (Dojkinci - Paleski Vrh), in the Brlog - Jelovica region, and north of Ponor (J. Kovacevic, 1999). Due to high radon content, the *"Skolska Cesma" Spring* and water from travertine around the spa Niska Banja are extremely radioactive (about 766 Bq/l). The *Studenica Mineral Spring* near Usce also has significant radioactivity (407 Bq/l).

Where groundwater is concerned, it can be concluded from the health standpoint that the results of radiohydrogeological investigation of the indicated large areas are alarming, especially in the case of the densely populated Bukulja massif (B. Ilic and J. Kovacevic, 1996).

How do things stand with the radionuclide concentration in crops, food, and the organisms of domestic animals and man? In attempting to answer this question, we rely on the book "**Radiation Hygiene in Biotechnology**" (B. Petrovic and R. Milivojevic, 1991), which provides abundant data on the radiation load of biocenoses on the territory of Serbia[59]. Primarily due to lithologic heterogeneity, the activity of Ra_{226} in soil of this territory is highly variable, ranging from 18.5 mBq/kg to 10.5 Bq/kg. For all crops tested (except onions), activity of Ra_{226} in plants is linearly dependent on its activity in soil. It is therefore understandable that activity of this element in crops grown in different parts of Serbia is highly variable. As for soil, it is in general appreciably greater on untilled land than on tilled land. Table 2.15. gives the average activity of this isotope in several crops. Like uranium, Ra_{226} accumulates best in onions, while in beans its concentration can vary greatly.

---

[58] Omaljev (1996) cites two cases where farm families live above fairly intensive anomalies: 1) the Kostovic house in the settlement of Ljuljac on the southern slopes of Mt. Golija is located in the middle of a strong radon anomaly posing high risks to humans; and 2) composed of the volcanic rocks of Borac, the entire hill above the Jakovljevic house has elevated radioactivity, and rocks from local quarries are used in construction of farmhouses.

[59] As a part of the carbonate Dinarides, the territory of Montenegro is characterized by a far lower background of natural radioactivity.

TABLE 2.15.
Activity of Ra_{226} in plant cultures at the territory of Serbia
(B. Petrovic, R. Mitrovic, 1991).

Type of plant culture	mBqRa_{226}/kg of fresh sample
Bean	7.4 - 303.4
Wheat	3.7 - 162.8
Corn	7.4 - 111
Cabbage	7.4 - 88.8
Potato	7.4 - 7.7
Onion	18.5 - 980

Macro- and Microelements in Rocks, Soil, and Water: Biological and Medical Role and Geochemical Risks

Selective accumulation of microelements in organisms has been of exceptionally great significance throughout the entire evolution of life. The role of metals was especially conspicuous here. To be specific, it is known that the first organisms that arose in the hydrosphere 3.4 billion years ago were surrounded by a solution of different metals that must have affected their metabolism. Many ions were absorbed into the tissue of living cells and participated in biochemical reactions. Metals played a large part in the evolution of redox processes of the cell during different periods of transition from anaerobiosis to aerobiosis (M. Jaredic and J. Vucetic, 1982).

Today it is known that about 60 elements are found in the human organism, 36 of which are metals. With the improvement of analytical methods, it becomes possible to detect new metals in organisms, even ones that are present in extremely small quantities.

Both metals and all elements of interest from the biological and medical standpoints will be treated in the text to follow. They include above all the so - called *essential macroelements*, without which plant and animal life would not be able to survive: C, H, O, N, P, S, Ca, Na, K, Mg, and Cl (Fig. 2.38.)[60]. In addition to the listed macroelements, life also

[60] According to E. J. Underwood (1981), 15 microelements (Fe, I, Cu, Mn, Zn, Co, Mo, Se, Cr, Ni, Sn, Si, V, As, and F) are essential for animals and man. Their *essential nature* is based on the following criteria: 1) they must be present in living matter; 2) they must be capable of interacting with living systems; and 3) their deficit in nutrition must lead to a reduction of biological function, which can be prevented or restored by introducing physiological amounts of the microelements in question. Of those listed, the last six are ones with a relatively recently discovered role, and their practical significance in human and animal nutrition has not yet been defined. The etiology of certain endemic diseases

requires the presence of *essential microelements*: Co, Cu, Mn, Mo, Se, Zn, B, Fe, I, and F, as well as (to some extent) Ni, Li, Si, V, Sr, Sn, and Cr (Fig. 2.39.). Opposed to these elements are *non - essential* and *harmful elements*: Cd, Hg, Pb, Ti, Be, Ba, Al, As, Sb, Bi, U, Th, and others that threaten life, even though the certain biochemical and biological role of some of them (As, Hg, etc.) cannot be ignored. The differentiation of elements into essential and harmful categories is a consequence of long evolutionary selection and development of unsuccessful and successful (resistant) species, i.e., ones that do not survive and ones that do.

Whereas macroelements constitute the bulk of living material, microelements are present in the organism in insignificant quantities of the order of 10^{-1} to 10^{-12}%[61]. Nevertheless, the biological importance of many microelements is very great. Deficits of some of them lead to diseases of animals and plants, and the disease symptoms can be successfully removed simply by adding the missing ingredient to the diet. Deficiencies of microelements in food and water can also cause serious pathological processes in humans and animals.

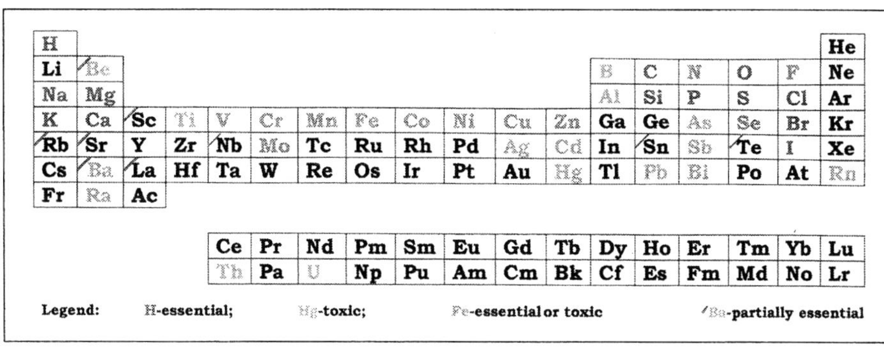

Fig. 2.38. Essential (black) and toxic (grey) elements.

The general dependence of geochemical composition of the *soil* on rocks in the substrate was stressed earlier. Excesses of heavy metals generally occur on soils formed from minerals rich in heavy metals.

---

and some "diseases of civilization" (such as cardiac ischemia, diabetes, osteoporosis, and many malignant diseases) is not yet known, but multifactor analyses have shown that diet is very important. For this reason, it is impossible to exclude a potential role for microelements recently found to be "essential," as well as ones whose role is yet to be established (Z. Maksimovic, 1985). *All the indications are that medical geology will make a very great contribution in this field of scientific research.*

[61] Relatively large amounts of essential macronutrients (above 100 mg every day) are needed for human life.

Among other things, soils formed on sulfate, copper, and silicate minerals or on serpentinites are characterized by elevated content of nickel and chromium, while ones formed on sedimentary and metamorphic rocks are characterized by a high share of cadmium. We are especially interested in sedimentary rocks, since they account for approximately 75% of all lithological products on the Earth's surface. The concentration of microelements in these rocks depends on minerals and adsorptive characteristics of the basic rock and the concentration of metals in the water in which sediments were formed. For example, clays are characterized by high concentrations of many elements, which is a consequence of their ability to adsorb metal ions. Concentration range of metals found in agricultural soil in different parts of the world is noticeable. A more detailed treatment of the geochemical origin of individual microelements and their chemical behavior in soil is given in the book "**Heavy Metals in the Environment**" (Ed. by R. Kastori, 1997).

Land that as a rule is highly populated and farmed, *river valley sediments* represent the hypsometrically lowest base for concentration of transported products of rock weathering and soil erosion, as well as for accumulation of water. The appreciable concentration of heavy metals formed in this way in the overlying clay horizon of alluvial sediment can threaten human health[62]. For example, systematic measuring of heavy metal content in the clay fraction of Danube river sediment in Germany revealed pronounced anomalies of cadmium (up to 45 ppm), lead (100 - 200 ppm) and zinc (500 ppm) (U. Fortsner and D. T. W. Wittmann, 1981). Anomalous levels of mercury in the sediment of this river course have been established both in Germany and in Austria and Hungary. Downstream, the problem of concentration of metals and other toxic elements in Danube sediments appeared immediately after construction of the Iron Gate Dam, when changes occurred in the entire aquatic system (Z. Maksimovic and M. Rsumovic, 1988). Just the question of the form of binding of heavy metals in aquatic sediments is today very pressing.

The content of individual microelements in *groundwater*, mineral and thermal waters above all, was likewise considered in a special chapter. It was stressed that such waters are characterized by considerable diversity of composition, depending on the rocks in which they move. However, exceptionally high concentrations of microelements sometimes occur, for example, in the waters of fumarole springs in regions of New Zealand with current volcanism (11,350 mg/kg of Fe^{2+}, 2,018 mg/kg of Al^{3+}, and 871 mg/kg of F); in waters of regions where saline solutions occur (up to 3,000

---

[62] The geography of endemic nephropathy is known to be almost exclusively linked with alluvial valleys of stream courses on the territory of Serbia, Republic Srpska, Romania, and Bulgaria. However, the main cause of that disease has not yet been established, although it is logical to assume that geological factors play a decisive role.

- 6,000 mg/l of Br, up to 2,000 - 6,000 mg/l of Sr, up to 500 - 700 mg/l of Fe, and up to 100 - 700 mg/l of Mn); and in brines of fault zones (with anomalously high content of Zn, Pb, and Cu).

The application of scientific principles in study of vital practical problems showed that a relationship exists between the supply of microelements and health of humans and animals, and enabled man to conquer such illnesses as pernicious anemia and Kashin disease (Z. Maksimovic, 1985). However, much remains to be done in attempting to correct imbalances of microelements in man, especially in growing children (Fig. 2.39.) The risks that lead to abnormal mental and physical development of children and to the appearance of certain chronic diseases later in life have been poorly defined, except in the case of exposure of children to high concentrations of lead (W. Mertz, 1981). Significant in this connection is the discovery of a high percentage of anomalies in the kidneys, ureters, and urinary bladders of children in families with endemic nephropathy, which indicates the action of unknown harmful factors in their development (N. Calic - Perisic et al., 1981).

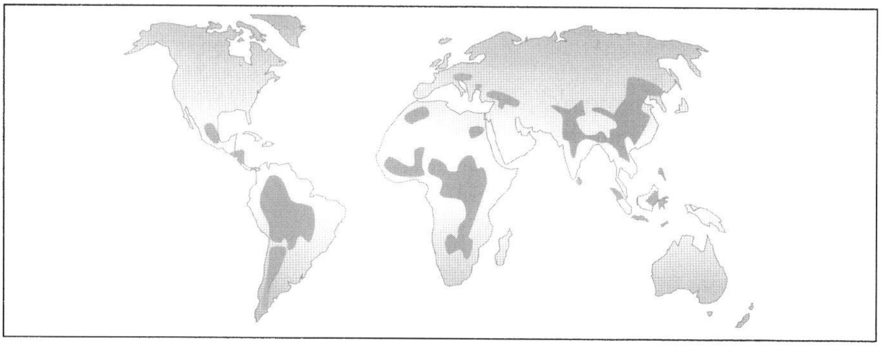

Fig. 2.39. Areas of iodine deficiency in the Earth (Dunn and van der Haar, 1999).

Role of Elements in Plant Nutrition

Plant nutrition has been a subject of interest and study since the earliest times. The first investigations were aimed at discovering factors that determine soil fertility and substances of which plants are made. Our current knowledge in this area is a result of the work of scientists from the oldest times to the present day. Knowledge about the role of individual elements in plant nutrition has expanded and deepened noticeably in the last decades, especially owing to the application of modern physico - chemical methods in the area of plant physiology. During that period, investigators have increasingly studied the role of individual elements in vital processes of plants and the nature of morphological and anatomical changes caused by their deficiency or excess. In writing the following brief discussion of the given material, we relied on the work "**The Role of Elements in Plant Nutrition**" (R. Kastori, 1983).

Elements entering the composition of plants do not have the same significance: some are essential, without which plants cannot complete their life cycle; others behave in a stimulative way; and one group of elements (especially non - essential heavy metals) in higher concentrations ert very toxic action on plants. They affect the vital processes, anatomical and morphological structure, chemical composition, yield, and distribution of plants.

Plants play an important role in the circulation of heavy metals in nature. Heavy metals for the most part enter the food chain through plants. For this reason, knowledge of the ecology and mechanisms of accumulation, distribution, and metabolism of heavy metals in plants is of great ecological, scientific, and practical significance. On the other hand, the creation of varieties and hybrids characterized by lower accumulation of heavy metals would greatly reduce the chances of their entering the food chain.

Essential microelements are found in the dry matter of plants in far greater amounts than toxic metals and other microelements. Plants could not normally complete their life cycle without them. However, they differ significantly among themselves with respect to physico - chemical characteristics and their role in physiological and biochemical processes of plants. Whereas nitrogen, phosphorus, and potassium often limit the organic production of plants, carbon, oxygen, and hydrogen rarely limit the yield of cultivated plants.

The role of 19 of the most significant elements in plant nutrition will be considered below. These elements can be relegated to one of the following four groups: 1) *essential macroelements* (N, P, S, K, Ca, and Mg); 2) *essential microelements* (Fe, Cu, Mn, Zn, B, Mo, and Co); 3) certain *microelements that can ert physiological action* (Ni, F, and Li); and 4) *useful elements* (Na, Si, and Cl).

Essential Macroelements. *Nitrogen.* Nitrogen enters the composition of many compounds important for the vital processes of plants, compounds such as proteins, nucleic acids, nucleotides, chlorophyll, etc. Owing to this, its role in physiological processes of plants is significant and multifold. Its high mobility and capacity for transport are properties that enable nitrogen to be rapidly and repeatedly involved in processes of synthesis and breakdown of organic compounds, whereby the importance of nitrogen in circulation of plant substances becomes all the greater.

An adequate amount of nitrogen in the nutrient substrate is needed for normal and harmonic growth and development of plants. Insufficient nitrogen nutrition very quickly causes specific changes in the metabolism and

morphological and anatomical structure of plants: growth is reduced, tips and edges of leaves die, the stem shortens, the root system lengthens, fruits shrink in number and size, and their quality deteriorates. Symptoms of nitrogen deficiency are partly specific for individual species and even varieties of plants.

Like nitrogen deficiency, a surplus of nitrogen has an unfavorable effect on plant productivity. Excessive nitrogen nutrition promotes growth of vegetative organs and bushiness of plants, increase of stooling in grain, shortening and thickening of the roots, retardation of ripening, decline of plant resistance to disease and drought, etc.

Use of greater doses of nitrogen in certain vegetable plants can cause accumulation of nitrates, which with longer standing are reduced to nitrites. Nitrites act very unfavorably on human and animal health. Examples of poisoning with serious consequences are known in the literature, especially poisoning by nitrate - rich spinach (*spinach meth - hemoglobinemia*).

Phosphorus. Phosphorus is a constitutional element because it enters the composition of important organic compounds. Many significant physiologo - biochemical reactions in living organisms cannot proceed without the participation of different phosphorus compounds. Phosphorus deficiency is manifested in cessation of growth, formation of tender and unresisting stems, and decrease in the yield and quality of production. Contrary to nitrogen surplus, an excess of phosphorus very rarely occurs in plants because its ions are quickly bound in soil.

Sulfur. Sulfur also belongs to the group of essential constitutional elements. Its role in plant metabolism is multifold: in addition to structural and catalytic functions, sulfur is involved in maintaining equilibrium of redox processes in the cell, is linked with the chain of electron transport, etc. Symptoms of sulfur deficiency are rarely observed on plants in nature. Plants in industrial regions and near larger cities usually are adequately supplied with sulfur due to the increased share of SO_2 in the atmosphere. Symptoms of rare occurrences of sulfur deficiency have much in common with those of nitrogen deficiency, and it is for this reason sometimes difficult to distinguish them. On the other hand, the surplus of SO_2 in the air has lately been posing a constantly increasing threat to the plant world.

Potassium. An alkaline metal, potassium is considered to be a *functional element.* Owing to its role in specific activation, i.e., as a co - factor of enzymes, potassium directly or indirectly affects many physiologo - biochemical processes, as a result of which it is of exceptionally great significance in the life of plants. Due to the multiple role of potassium in vital processes, a deficiency of this cation causes far - reaching changes in metabolism and thereby in the growth and development of plants. A potassium surplus can cause effects that are indirectly unfavorable.

Calcium. Calcium is one of the most widespread elements in nature. It plays a multiple role in vital processes of plants. Although it does not enter the composition of organic compounds, it plays a very important part in plant metabolism. It affects physico - chemical characteristics of protoplasm, stability of cell membranes, activity of certain enzymes, etc. Calcium deficiency can affect plants directly (if plants do not contain this element in sufficient quantity) and indirectly (if its content in the soil is insufficient). Calcium deficiency in nature has been observed to date in lettuce, cabbage, peanuts, potatoes, flax, tomatoes, and apples. The first symptoms are manifested on the youngest organs, the plants take on a bushy appearance, they lag significantly in growth, and signs of chlorosis and necrosis are soon observed on the leaves. Calcium deficiency has recently occurred with increasing frequency on apple fruits.

The presence of a greater amount of calcium in soil can affect the accessibility of certain soil microelements (B, Mn, Fe, Cu, and Zn), with *chlorosis* in plants as a possible consequence. Some fruits (peach, bird cherry, etc.) and certain decorative plants are especially sensitive to chlorosis caused by a surplus of calcium.

Magnesium. This earth - alkaline metal is very widespread in nature. It is very similar to calcium in chemical properties. However, its role in plant metabolism is significantly different. In greater concentrations, magnesium is much more toxic than other essential macroelements. Symptoms of deficiency of this element were first noticed in regions where potassium fertilizers were used in increased amounts for a number of years. They were subsequently observed on fields irrigated with water containing large quantities of sodium. An inverse correlation also exists between manganese content and magnesium content.

Both deficiencies and surpluses of magnesium can cause morphological, anatomical, and physiologo - biochemical changes in plants. Surpluses of this element occur rarely in nature, primarily on soils formed from dolomites and serpentines.

Essential Microelements. It has long been observed that iron, zinc, manganese, copper, boron, molybdenum, and cobalt are needed in addition to the described essential macroelements for normal growth and development of plants. Discovery of the importance of these microelements in the life of plants made it possible to explain a whole series of phenomena leading to abnormal anatomical and morphological changes in plants. Thanks to the further development of science, many physiologo - biochemical processes have been discovered in which microelements play a significant part. Moreover, it is increasingly possible to detect in nature latent and acute signs of deficiency of essential microelements.

Iron. Iron is among the elements needed by plants. Its polyvalence and ability to form chelate comples are two significant characteristics that account for the important activity of iron in many physiologo - biochemical processes of plants. Symptoms of iron deficiency are today widely disseminated, especially in fruits and grapevine. Registered in different regions of Yugoslavia, *Fe - chlorosis* leads to significant decrease of the yield.

Iron surpluses occur rarely in nature, and for this reason there are very few data pertaining to them in the literature.

Copper. Copper belongs to the group of elements whose role in plant metabolism is predominantly catalytic. Owing to the fact that it enters the composition of enzymes, copper indirectly or directly affects many processes important for plant life. Copper deficiency in plants is an occurrence more widespread than is often supposed. It has been observed in many countries throughout the world, and many soils in Western Europe are deficient in copper. Dangerous diseases of livestock such as *chronic diarrhea, enzootic ataxia*, etc. occur on pastures whose vegetation does not contain enough copper. Copper surpluses also have an adverse effect, not only on plants, but through them on the state of health of man and animals as well.

Manganese. This is an important regulator of redox processes in plants. Manganese activates a number of enzymes, including many that regulate very important processes. On the basis of its role in plant metabolism, manganese is counted as one of the heavy metals most significant in the life of plants. Its deficiency occurs predominantly on lime and alkaline soils. Inadequate provisioning of plants with this element causes characteristic morphological, anatomical, and cytological changes. Especially sensitive to diseases caused by manganese deficiency (*dry spot, chlorosis*) are oats, potato, fodder and sugar beet, red beet, bird cherry, citrus fruits, and apples.

Symptoms of manganese surplus in plants under natural conditions are more widespread than might be imagined. On soils with high manganese content, low pH, and a high reducing potential, plants can accumulate manganese to such an extent that it becomes toxic for them.

Zinc. Zinc in small quantities is needed by higher and lower plants, as well as by man and animals. Due to the multiple role of zinc in plant life, its deficiency causes considerable changes, both in metabolism and in anatomical structure. Occurrence of zinc deficiency in perennial plants is very common throughout the world, and in Yugoslavia is especially pronounced on alluvial soils, chernozem, and loess. Large concentrations of zinc (as in the case of other metals) act toxically on plants, and the resistance of individual plant species to surpluses of this metal is extremely variable.

Boron. Boron is needed by higher plants, and its deficiency or surplus - especially in dicots - causes significant physiological and morphological changes. Toxic action of boron in crop plants can be expected when the content of this element (dissolved in boiling water) in soil exceeds 5 ppm.

Molybdenum. The significance of molybdenum and need for it in plant metabolism have been established in many experiments. Deficiency of this element usually occurs on acidic soils. Symptoms of molybdenum deficiency have been most thoroughly studied in cauliflower, since this plant is especially sensitive to it. The resistance of individual plant species to molybdenum surplus varies. High content of this element in fodder can have an unfavorable effect on the state of health of animals, causing molybdenosis in them.

Cobalt. Cobalt is a regular ingredient of plant cells, but its role in vital processes is not well enough known. It enters the composition of vitamin B_{12}, and for this reason its content in plants is of great significance for human and animal health. In normal circumstances, cobalt never accumulates in plant cells in great enough measure to be toxic for animals and people.

Microelements That Can Ert Physiological Action. In addition to the seven essential microelements, plants also contain 14 other elements whose individual concentration in dry matter usually does not exceed 2 ppm, but which can affect synthesis of organic substances in a favorable way and take part in physiologo - biochemical processes. Special attention has been devoted more recently to study of the harmful (toxic) action of certain microelements. Of the group of toxic elements, the influence of nickel, fluorine, and lithium on vital processes of plants has been studied in greatest detail.

Nickel. Nickel is chemically and physiologically similar to cobalt. Its role in physiological processes of plants is multiple. On the practical side, the influence of this metal on the yield of cultivated plants is very significant. Certain authors unjustifiably ignore the undoubted favorable action of nickel on plant growth and development, placing it in the group of elements with predominantly harmful action. This is especially wrong because many essential heavy metals - copper or cobalt for instance - at high concentrations are no less toxic than nickel.

Fluorine. At certain concentrations, fluorine can significantly affect the overall metabolism of plants and thereby influence their growth and development. Like many other elements, fluorine in higher concentrations can have a very unfavorable effect (causing *necrosis* and other conditions). Negative influence in some cases is manifested at low concentrations, due to the cumulative action of fluorides.

Lithium. Lithium affects certain physiologo - biochemical processes of plants and in some cases can even stimulate their growth and development. Higher concentrations of this element can cause morphological changes in both vegetative and reproductive organs of plants. Citrus plants are especially sensitive to higher concentrations of lithium.

Useful Elements. These are elements that are not needed by plants, but which in certain concentrations can stimulate their growth and development. They include above all sodium, silicon, and chlorine.

Sodium. The presence of this element can favorably affect growth and development of sugar beet, alfalfa, and other plants, and it can even be essential for certain halophytes. Sodium deficiency occurs very rarely in nature. Unfavorable action of its surplus occurs more often, especially on salinized soils, where it can be toxic.

Silicon. Favorable action of silicon on plant growth and development has been demonstrated in many studies carried out to date. Since it is in nature the second most widespread metal in the Earth's crust, symptoms of silicon deficiency have been recorded in few places, primarily fields of rice, oats, onions, and beans. Toxic action of this element has not yet been recorded.

Chlorine. The role of this element in vital processes of plants is still not well enough known, but it is a fact that chlorine deficiency causes various morphological and physiological changes in plants. On the other hand, it erts toxic action if its concentration in dry matter attains 0.5 to 1.8%.

Role of Elements in Nutrition of Animals and Humans

Continuous exchange of substances with the surrounding environment is a common characteristic of all living organisms regardless of the degree of their development. This exchange is especially characteristic or quantitatively more pronounced in higher forms of living organisms, a category that includes man and animals, particularly ones raised for production of milk, meat, wool, etc. In the healthy organism of animals raised for production of something, milk for example, anabolic processes prevail over catabolic ones and primarily depend on the chemical composition of plants. The chemical composition of animals and man cannot be studied apart from that of soil, water, plants, food, and the surrounding medium, i.e., the atmosphere (Fig. 2.40.). This means that nutrition of animals and man, like all responses of the living organism to natural and artificial factors of the environment, must be studied within the framework of a system that includes *rocks, water, soil, air, plants, animals, and man,* with whom the circle is closed. Studying overall relations of the system's components enables us to model many vital processes, which should ensure a long and healthy life.

Systematic study of the chemical composition of upper layers of the Earth's crust - as well as that of plants, animals, man, sea water, and the atmosphere - has shown that the former erts definite influence on the latter. Moreover, certain regularity exists in the distribution of chemical elements between the Earth's crust and the biosphere. It is logical that the chemical composition of all living organisms is directly correlated with that of food and water, i.e., with that of the environment in which they live. Of 89 natural chemical elements in the Mendeleyev periodic system, about 45 have been identified in the living organisms of animals, plants, and humans, and a biochemical role has been established for 27 chemical elements. These are the so - called *bioelements.*

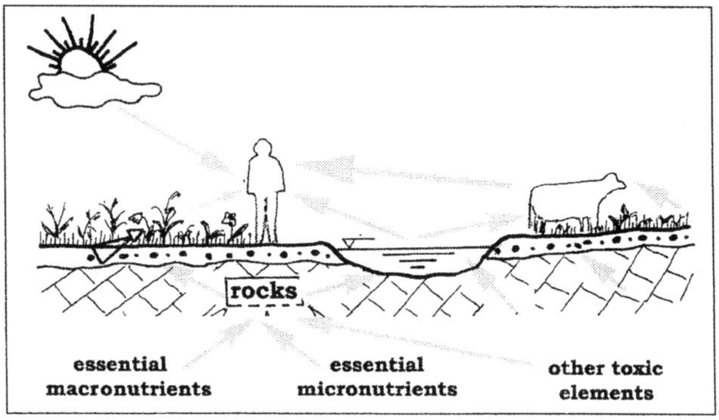

Fig. 2.40. Transport of natural trace elements within the *system rocks - soil - plants - animals - humans.*

In growth and development of animals and man, macro - and microelements are needed for mineralization of growing tissues, formation of the fetus in pregnant females, egg formation in egg - laying hens, wool growth in sheep, hair growth in different domestic animals (cattle, goats, and pigs), renewal of tissues, etc. It is therefore very important that they be systematically introduced into the organism for maintenance of life, health, and productivity.

Medical science started to take an interest in microelements and their role in pathological processes only since the beginning of the 20th Century (L. R. Nozdrjuhina, 1977), even though Vernadskii much earlier than that called attention to their enormous role in the biochemical processes of organisms. Studying geochemical processes in the Earth's crust, Vernadskii discovered the part played by organisms in these processes and stressed the interconnection between them and the evolution of living creatures.

On the whole, macro - and microelements perform numerous and very diverse functions in all organisms. Their biochemical characteristics and thereby their biological and physical roles depend on the position they occupy in the periodic system of Mendeleyev (Fig. 2.38.). With the exception of molybdenum, tin, and iodine, most of the elements of interest to us have a low ordinal number (below 34) and belong to the first to fourth periods of the system. On the basis of the periodic system and periodic law, it is possible to predict with high probability which chemical elements can have an essential role in metabolic processes. For example, the position of aluminum and silicon in the periodic system and their properties suggest that they could be involved in the building of complex biologically active compounds, whereas lithium and beryllium could be expected to play a structural role in enzymes. It is also noticeable that with the transition from light to heavy metals of the same sub - group - for example, the *zinc - cadmium - mercury* sub - group - the toxicity of an element increases at the same time as its quantity in the organism decreases (Fig. 2.41.). In all of this, it should be kept in mind that much also depends on the nature and kind of living organism (plant, animal, or human) and its physiological state.

Within the organism, macro - and microelements are unevenly distributed among organs and tissues. However, a constant reversible dynamic equilibrium exists between them in the healthy organism. The majority of microbioelements are found in greatest concentration in the liver, and the liver is for this reason considered to be the functional depot of these elements in the organism (D. Kolarski, 1995). Due to their great mass, bone and muscle tissues are storage places for elements such as calcium and phosphorus. Also, large amounts of some other elements are present in certain tissues: iodine in the thyroid gland, manganese and chromium in the hypophysis, zinc in the pancreas, etc.

Microbioelements can ert specific and non - specific influence on metabolic processes. Examples of specific influence are the connection between zinc metabolism in pancreatic tissue and biosynthesis of insulin; influence of iodine on the function of the thyroid gland; etc. On the other hand, copper, cobalt, fluorine, and some other microelements ert non - specific influence on the function of this gland.

Different mineral ingredients act in combination with vitamins, enzymes, and proteins, jointly stimulating and thereby regulating the course of complex biochemical reactions. Hence it follows that a deficit of any mineral or microelement involved will cause the given set of functions to be disturbed. Instead of unimpaired occurrence of the described processes, a serious breakdown arises in the work of certain organs or the organism as a whole (Fig. 2.38.). A state of disease sets in due to the deficiency of some element, whereas a surplus of it - if not toxic - can still be harmful.

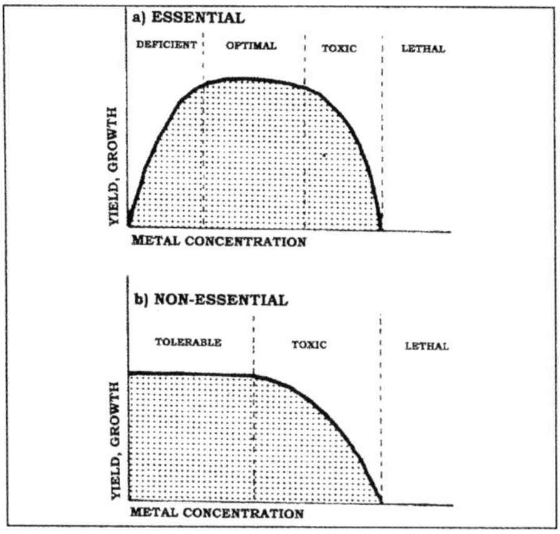

Fig. 2.41. Deficiency and surplus consumption of essential and non - essential elements in traces (C.B. Dissanayake, R. Chandrajith, 1999.).

As we have seen, the basic functions of macro - and microelements are very diverse and numerous, depending on the nature of the element, the kind of compound, and its structure. The biochemical and biological functions of macrobioelements (Ca, Mg, Na, Cl, K, P, and S), microbioelements (Mn, Fe, Co, Cu, Zn, Se, Mo, I, and F), and other microelements (Si, V, Cr, Ni, As, Sr, Cd, Sn, Hg, Pb and Ta) are discussed below, together with the effects of their deficiencies and surpluses. For this discussion, we rely on the book of D. M. Kolarski "**Fundamentals of Nutrition of Domestic Animals**" (1995) and other works.

Macrobioelements. *Calcium.* Calcium has an important role in the biosphere and represents a typomorphic element for many natural environments. A natural environment is considered favorable if it is rich in calcium. A calcium deficit in some environment reduces fertility of the soil, renders it acidic, and affects living organisms. For this reason, combating calcium deficiency represents a very important means of increasing the biological productivity of soil and improving public health.

Calcium is the most abundant metal in the human organism, making up about 1.5% of the weight of an adult person. The bulk of calcium is found in the skeleton. Mineralization of the organic matrix of bone tissue is one of the key roles of this macroelement, but it is also unusually important for normal functioning of nerves, muscles, and the heart. A portion of calcium is found in body fluids in the form of calcium carbonate. Ionic calcium has many functions: it is needed for maintenance of blood vessel tonus, contraction of myocardial muscles, and transmission of nervous impulses. Moreover, calcium takes part in virtually all phases of blood coagulation and is absolutely necessary for secretion of gastric juice, activation of some enzymatic reactions, and secretion of certain hormones. It promotes absorption of iron in the intestines, alleviates insomnia, generally stabilizes nervous and psychic functions, and is important during pregnancy and the period of nursing...

A deficiency or surplus of calcium in the diet of humans and animals is manifested in characteristic biochemical changes and clinical symptoms: rickets, curvature of the spine and ribs, loss of appetite, lameness, osteomalacia (softening of the bones due to their demineralization), osteoporosis, etc. In the case of a more prolonged deficit of calcium, "leaching" of this element from the bones causes increased absorption of lead from polluted environments and its incorporation into bones and teeth. If calcium is present in excess, its absorption from the lumen of the small intestine increases, which causes increase of calcium concentration in the blood or *hypercalcemia*. The recommended daily dose of calcium in humans is 600 - 1,200 mg/l, depending on a person's age and physiological state.

Magnesium. Magnesium is an element well represented in the Earth's crust. Its content in soil depends in large measure on the parent rock and its mineral and chemical composition. Magnesium cations are the fourth best represented cations in the organism of man. Together with calcium and phosphorus, it forms the foundation of bone tissue, and 60 - 70% of all magnesium in the human organism enters the composition of bones. It is a biologically essential element not only for humans, but for animals and plants as well. Magnesium plays an exceptionally important role in human metabolism, taking part in the activation of about 300 enzymes. It is vitally important for unimpaired functioning of nerves and muscles. Magnesium contributes to health of the cardiovascular system and serves to prevent cardiac stress. It slows the aging process, prevents clogging of the blood vessels, and helps to lower, i.e., normalize, the level of cholesterol in the blood. Together with calcium, magnesium acts as a natural calming agent, assists in combating stress, and helps to overcome periods of depression. Intercellular deficiency of magnesium is a significant etiological factor in the occurrence of hypertension. Deficiency of this important element in human nutrition is being increasingly studied because it can be linked with the pathogenesis of cancer or heart disease and stroke in regions with soft water.

In animals, magnesium deficiency in the food they consume is manifested in very diverse symptoms. In calves, these symptoms include decrease of magnesium concentration in the blood, mobilization and consumption of magnesium from the bones, and typical tetany and death. In egg - laying hens, magnesium deficiency results in hypomagnesaemia and reduction of lying, while in adult animals out to pasture it gives rise to hypomagnesaemia known as *grass tetany.*

Sodium and chlorine. The need for salt or sodium chloride has been manifested since man and animals started to feed on plants, which are for the most part poor in sodium and chlorine. Salt loss is especially pronounced in animals and humans in warm regions, where significant amounts are lost through sweating.

These two elements are linked in metabolic processes. Salt deficiency in consumed food is manifested in loss of appetite, slowed growth, and decrease in the amount of milk in mammals. If animals do not take in salt over longer periods of time, their body weight decreases significantly and they experience hair loss, exhaustion, etc. On the other hand, symptoms of strong poisoning can be manifested in animals if they consume large quantities of salt.

In living tissue of humans, sodium is found almost exclusively in the form of the Na^+ ion. Reduction of intake of this element into the organism is recommended for all persons, especially individuals suffering from hypertension. Consumption of large amounts of sodium directly affects (lowers) calcium content in the organism and influences the occurrence of osteoporosis and kidney disease.

Potassium. Together with sodium and chlorine, potassium belongs to the group of macrobioelements that (in contrast to phosphorus, calcium, and magnesium) are found mainly in fluids and soft tissues of the animal organism. Owing to its electron structure, potassium is one of the most mobile elements in nature. It is an essential microelement for plants, animals, and man. The biochemical and biological role of this cation is linked with that of sodium. They maintain the normal balance of water, osmotic and acid - base equilibrium, and equilibrium of metabolic processes in the cell, especially metabolism of carbohydrates. Potassium promotes cell growth and (together with calcium) plays a significant part in regulation of neuromuscular activity.

Deficiency of potassium in food, especially in ruminants, is very rarely manifested under normal feeding conditions. Large amounts of this element in fodder mis for calves can ert toxic action.

In humans, it has been demonstrated that potassium salts lower blood pressure. Potassium deficiency causes a whole series of disturbances in the organism, such as disturbances in work of the heart muscle, kidneys, digestive organs, respiratory organs, etc. *Hypokalemia* occurs with reduction of potassium intake, movement of potassium from the extracellular into the intracellular fluid, and increased loss of body fluids. A great deficiency of potassium can cause death. A surplus of this element more rarely leads to *hyperkalemia.*

Phosphorus. Like calcium, this essential macrobioelement is very widespread in nature. These two elements are antagonists and have much in common in metabolic processes. Among other things, they are directly linked with the same organic matter in bones.

The biochemical and biological function of phosphorus is essential to life, but adequate attention is not paid to this bioelement because it is present in nearly all foods. Apart from belonging to the group of basic biogenic elements, phosphorus has many functions in the organism, more than any other bioelement, for there are almost no significant metabolic reactions in which phosphorus is not involved.

In animals, phosphorus is among other things needed for growth, development, and formation of the skeleton; synthesis of milk components, i.e., production of milk; and production of meat, wool, eggs, etc. The lack of phosphorus in consumed food or a very pronounced deficiency of this element can be a cause of death in the first days or weeks of life. If an animal consumes fodder with a moderate phosphorus deficiency, it leads to the occurrence of so - called low - phosphate forms of rickets, which are manifested in slow growth. If *hypophosphatonemia* lasts for a longer period of time, growth stops and bone mineralization is disturbed.

In man, phosphorus is vitally important for normal functioning of the kidneys, proper work of the heart, and normal structure of the bones and teeth. It promotes growth and maintenance of the body in good condition, reduces arthritic pain, and is needed for transmission of nervous impulses. Vitamin D and calcium are extremely important for the proper functioning of phosphorus. Increased phosphorus intake can lead to *hyperphosphatemia*, usually in children.

Sulfur. Together with carbon, hydrogen, oxygen, nitrogen, and phosphorus, this essential macrobioelement belongs to the group of basic organic elements of which all living organisms are made. The biochemical and biological functions of sulfur (or compounds containing sulfur) are very diverse and numerous, since organic compounds with sulfur manifest their biochemical function through characteristics of the electron structure of this element, namely variable valence, ease of oxidation and reduction, and ability to form complex compounds with certain macro - and microbioelements, respiratory pigments, and enzymes with very significant biological roles.

The main limiting sulfur compound in consumed food is methionine. A partial or complete lack of methionine slows or stops growth and development of animals that have not yet attained morphofunctional maturity and lowers the productivity of animals raised for meat, milk, and wool. Large amounts of inorganic sulfur in the form of flowers of sulfur or sulfates in chickens and pigs retard the growth process and lead to rickets and gastroenteritis.

In humans, sulfur is especially important for healthy hair, skin, and nails. It helps maintain the oxygen equilibrium needed for proper brain functioning; assists the liver in secretion of bile; acts (together with vitamins of the B complex) on basal metabolism of the body; and forms part of the amino acids needed for tissue building.

Microbioelements. *Manganese.* Manganese is among the chemical elements essential for plants and animals. In man, it is present in the heart, liver, and kidneys. The biochemical and biological role of Mn is very significant because this microbioelement activates many enzymes of great importance for maintenance of the life and productivity of animals. Deficiency of Mn in consumed food is manifested in various symptoms, since it is directly or indirectly involved in the metabolism of basic nutrients with structural, energy, and biocatalytic roles in the organism. Symptoms of deficiency of this microbioelement are as follows: retardation of growth and development; irregularities in bone formation; reduced productivity; frequent nervous disorientation; etc.

In humans, manganese promotes activation of enzymes needed for proper utilization of biotin, vitamin B_1, and vitamin C in the organism; is needed for production of thyroxin, the main hormone of the thyroid gland, and for normal bone structure; plays a crucial role in digestion and utilization of food; and is important for reproduction and normal functioning of the central nervous system. A shortage of manganese results in ataxia (loss of muscle coordination). When greater amounts of this element are introduced into the organism, it acts as a poison, causing the disease known as *manganese madness*.

Iron. Iron is a necessary component of all plants and animals, including man. The biochemical and biological function of iron is very significant for maintenance of life and for reproduction. Iron plays an important part in transport of oxygen from the lungs to tissues, in transport of carbon dioxide from cells to the lungs, and in the process of respiration in the cell. Its function is maintained by many enzymes that contain it, without which there would be no life on Earth.

Manifested in retardation of growth, hypochromic anemia as a result of inadequate synthesis of hemoglobin is the main symptom of iron deficiency in all domestic animals.

For human beings, iron is a vitally important oligoelement. It is needed for formation of red blood cells (hemoglobin), myoglobin (the red pigment in muscles), and certain enzymes. Iron promotes growth, fosters resistance to diseases, and prevents fatigue and anemia. The amount of iron in the blood is increased in some pathological disturbances, but reduced in relation to the healthy organism in others. The monthly loss of Fe in women is twice as great as in men. It has been established that iron and calcium are most often deficient in the diet of women in America.

Cobalt. Cobalt is an essential chemical element as an integral part of vitamin B_{12}, although it takes part in biochemical processes in the form of a free ion. It is needed for growth of all living organisms and is of crucial significance for the red blood cells. Deficiency of cobalt in the food of animals is manifested in different symptoms because it takes part in many biochemical processes important for maintenance of animal life and productivity. In cattle and sheep, signs of cobalt deficit take the form of appetite and weight loss, exhaustion, and anemia. This form of anemia in humans is known as *pernicious anemia.* Toxicosis due to excess of Co occurs mainly in chronic form, and its symptoms depend on the amount of Co introduced. In humans, inhalation of Co leads to incidence of bronchial asthma, and exposure to Co salts has an especially unfavorable effect on the skin, thyroid gland, lungs, heart, and bone marrow.

Copper. Copper is an essential microelement for plants, animals, and man. It is a structural component of many enzymes significant for growth and development of animals, and for maintenance of their life and productivity. The biological significance of copper was first established experimentally in 1928 on rats fed only milk; the animals developed anemia, which was later alleviated by the action of copper from liver extract. Symptoms of dietary Cu deficiency (*hypocuprosis*) include anemia, impairment of growth and development, depigmentation of wool and hair, disturbances in synthesis of bone tissue (manifested in bone fractures), and degenerative changes of the blood vessels.

Among animals that continuously take in excessive amounts of Cu, the element accumulates in tissues, primarily those of the liver, with damage to its function. In the case of chronic toxicosis of animals, a longer period of time (several weeks or months) is needed for signs of poisoning to appear. Copper toxicosis rarely occurs in humans because copper and its salts introduced *per os* cause vomiting. It has been established that the spinal fluid contains twice as much Cu in persons suffering from *Alzheimer's disease,* which causes destruction of nerve cells in the brain.

Increase of copper content accompanies certain diseases, while others are characterized by its decrease. The organism loses the capacity to retain copper in the early stages of cancerogenesis, and copper is not involved in metabolism in the later periods. Copper in the blood is not bound with proteins in malignant forms of cancer. Accumulation of copper in the organism occurs in acute and chronic hepatitis and in cirrhosis of the liver. *Hypercupremia* is also characteristic of epilepsy.

Zinc. Like copper, zinc is an essential microelement for plants, animals, and man, taking part in many physiological reactions. Zinc behaves like a traffic policeman, directing and controlling the flow of processes in the organism and regulating enzyme systems and cells. The liver is the main organ in which metabolism of zinc takes place. This microelement is found in all cells and is also stored in the bones and in the prostate and testes among organs. Its metabolic and physiological role consists of activation of certain enzymes and participation in the formation of many metalloenzymes.

Zinc deficiency has an especially negative effect on fast - growing cells and tissues, and there are many possible symptoms of it. A shortage of zinc can result in hypertrophy of the prostate, atherosclerosis, and hypogonadism.

Zinc contributes decisively to resistance of the organism to all ailments, including malignant diseases. It also prevents decline in working capacity of the brain.

Symptoms of the rare instances of zinc poisoning in humans are similar to those of poisoning with other heavy metals. As with copper, extremely large amounts of introduced zinc salts are rejected by vomiting.

Selenium. The content of this metalloid is relatively significant in some 40 minerals, and soil of volcanic regions or arid and semi - arid areas (where it is not washed by precipitation) can sometimes have a high percentage of selenium. Few elements have had a history as interesting as that of selenium in study of the interrelationship of rocks, soil, plants, animals, and man. This element was for a long time considered toxic, until its addition to food was shown to prevent the occurrence of *Keshan disease* (endemic cardiomyopathy), which is characteristic of extensive regions in China with soil especially poor in selenium.

Selenium deficiency causes illness in cattle, sheep, goats, horses, pigs, and poultry. They most often suffer from the so - called white muscle disease or nutritive muscular dystrophy. The disease is manifested as weakness of the skeletal muscles, myocardial weakness, or a combination of the two.

Investigations have shown that Se plays three important roles in humans: 1) it is an anticancerogen; 2) it improves immunity; and 3) it protects the blood vessels. In addition to being an anticancerogen, Se also erts antiviral action. Organic compounds of selenium protect the organism against all types of ionizing radiation, including the radioactive kind. An unusually important role of selenium is its role in protection against poisoning by heavy metals (Cd, Hg, and Pb). Deficiency of selenium and other antioxidants is today held to be responsible for oxidative stress, decline of immunity, and occurrence of different diseases. Together with this, it considered to be a microelement essential in preventing the incidence of muscular dystrophy.

Selenium deficiency causes illness that goes by the general name of *selenosis*, while a surplus of selenium in consumed food (as has already been stated) erts toxic action. Poisoned animals experience blindness, salivation, a certain degree of paralysis, grinding of the teeth, respiratory disturbances, and death. Recent research indicates that development of the AIDS virus in humans can be caused by a deficit of selenium (E. W. Taylor et al., 1994, 1995; B. M. Dworkin, 1994; etc.).

Molybdenum. Molybdenum began to be studied more intensively from the time when it was established that a surplus of this microelement in animals causes changes that were subsequently referred to as *molybdenosis*. Molybdenosis occurs in many parts of the world and results from the presence of greater concentrations of molybdenum in livestock fodder obtained from natural sources or from intensive addition of Mo for the purpose of stimulating plant growth.

Molybdenum represents an essential element for animals, but the organism's need for it is minimal. Its biochemical function in the animal organism is manifested through enzymes of which Mo is a structural component. If all other conditions are met, Mo deficiency is rarely expressed, since the requirements of animals are satisfied through normal feeding. When present in greater quantities, molybdenum becomes toxic, especially for ruminants. Toxicity in the presence of a molybdenum surplus is manifested in the guise of exhaustion, anemia, depigmentation of the hair, and pain in the joints. In chronic cases, metabolism of the bones is disturbed, which leads to *osteoporosis*.

In man, molybdenum assists in the metabolism of carbohydrates and fats. It is a vital part of the enzyme responsible for utilization of iron. It helps prevent anemia and improves the general wellbeing of the organism.

Iodine. The amount of this essential microbioelement in terrestrial plants is correlated with the amount of it in the soil. Geochemical provinces in which water and food are poor in iodine as a consequence of iodine content in the soil (for example, mountainous regions of Bosnia, Serbia, and Slovenia, as well as the region of the Alps) are referred to as being goitrogenic.

Iodine - linked diseases are especially widespread in countries of the tropical zone of South Asia (Bangladesh, Myanmar, Indonesia, Nepal, Sri Lanka, and India) and are particularly prevalent in Vietnam (Fig. 2.39.). According to Dissanayake and Chandrajith (1999), 1.6 billion people and 50 million children are at risk from mental retardation and brain damage due to iodine deficits in food and water, and 100 thousand cretins are born every year. For this reason, the geochemistry of iodine in rocks, soil, water, sea, and air in relation to the incidence of iodine - deficiency diseases today represents one of the most interesting fields of research (Dissanayake and Chandrajith, 1999).

Metabolism of iodine is directly linked with synthesis and metabolism of thyroid hormones in the animal organism. Its deficiency in consumed food and drinking water lowers the concentration of iodine in the thyroid gland, with consequent decline in synthesis of thyroxin and triiodothyronine in the thyroid gland, which becomes enlarged in *endemic goiter*. On the other hand, large amounts of iodine in food for domestic animals are poisonous.

In man, two thirds of the body's iodine is found in the thyroid gland. Because the thyroid gland controls metabolism and iodine erts strong influence on its work, a deficiency of this mineral can result in slowing of psychic reactions, weight increase, and lack of energy. *Goiter* (*hypothyroidism*) occurs due to shortage of iodine.

Fluorine. Fluorine is present in the form of compounds (fluorides), which are widespread in rocks, but unequally distributed (Fig. 2.6.). Fluoride concentration in soil increases with depth. In soils of the USA, for example, fluoride content is 20 - 50 mg/kg at depths of up to 7.5 cm, but 20 - 1,620 mg/kg at depths of 0 - 30 cm. In regions rich in fluorine - containing minerals, groundwater can contain up to 10 mg/l of fluorides and more (for example, certain springs of water used for drinking purposes in Spain contain more than 20 mg/l of this ion). Fluorine belongs to the group of nutritive substances that have cumulative properties. It is introduced into the organism with food and water. As an essential element, it is found in the greatest amounts in the teeth and bones of animals and man, and traces of it are present in the thyroid gland and skin. Of total fluoride content, about 96% is stored in bone tissue.

It is estimated that in recent years more than 43 million people in China have suffered from dental fluorosis, which is mainly caused by the presence of fluorides in drinking water. This is particularly characteristic of groundwater. Many countries such as India, Sri Lanka, Ghana, and Tanzania have very pronounced incidence of dental and (in some cases) osteal fluorosis (Dissanayake and Chandrajith, 1999).

It has been demonstrated that the presence of fluorine in drinking water in concentrations from 1 to 1.2 μg/l significantly reduces the percentage of dental caries. In fact, it has been incontrovertibly established that this halogenic element erts protective action against dental caries, is essential for optimal growth and fertility of animals, and is very significant for maintenance of hardness of their skeletons. The teeth and bones (skeleton) of animals are damaged if the water they drink contains large amounts of fluorine over long periods of time (especially during growth). Fluorotoxicosis, so - called *chronic fluorosis*, generally is of a chronic nature. Fluorosis of the bones (*osteofluorosis*) is manifested in pain and stiffness in the joints, calcification of tendons and ligaments, and occurrence of *exostosis* and lameness. It later leads to *cachexia*, loss of appetite, slowing of growth, weight loss, and finally death.

In regard to its dissemination, course, complications, and consequences, *dental caries* represents the most massive and frequent non - infectious disease of man, one that is accompanied by significant socio - medical problems. Since 1945, fluoridation of drinking water has in many countries of the world been used or favored as a method of mass prevention of dental caries, and the results of studies in 20 countries show that the incidence of this disease was reduced on average by 30 - 90% as a result of it. Similar effects are achieved by fluoridation of salt, fluoridation of milk is also practiced, and tablets containing fluorine are sometimes used.

The first symptom of chronic fluorine poisoning is *dental fluorosis* or diseased dental enamel, which occurs as a consequence of consumption of water with fluoride content of more than 2 mg/l over a period of years. *Osteal fluorosis* is characteristic of geographic regions with extremely high fluoride content in drinking water. *Endemic fluorosis* occurs in regions where drinking water has fluoride content higher than 6 mg/l. *Osteosclerosis* occurs in 10% of adults if the daily intake of fluoride with water is more than 8 mg/l, and *paralytic fluorosis* occurs at concentrations of 20 - 80 mg/l or higher.

Other Microelements. Silicon, vanadium, chromium, nickel, arsenic, strontium, cadmium, and tin are elements that are found in the organism of animals and man, but whose essential nature has not been confirmed with certainty. In addition to those indicated, many other chemical elements are also present in animals and humans, depending on the geochemical composition and other factors of the environment in which they live. Some - like mercury, lead, and thallium - represent elements with highly toxic action.

Silicon. Silicon is among the microelements essential for growth and development of higher animals and man. In the animal organism and in man, the greatest amounts of silicon are found in connective tissue of the aorta, tendons, and trachea, and in other connective tissues.

Silicon is essential for growth and development of the skeleton of higher animals and man. Animals fed the same food but without Si are significantly smaller and finer - boned, many organs are atrophied in them, legs and crests are pale, and the skeleton is stunted owing to considerable slowing of its development. On the other hand, Si in large quantities is toxic.

Vanadium. Experimental data and studies with vanadium in the diet of rats and chickens confirmed that this microelement is essential for the tested animals. Vanadium is believed to be essential for humans also, since the organism of an adult man contains about 3 mg of it. It is a relatively toxic element.

Chromium. Extremely widespread in nature, chromium is present in soil, water, plants, and food, albeit in very unequal concentrations. It is an essential element for animals because its presence is needed for normal metabolism of carbohydrates and lipids. In humans, an increased need for Cr is felt where there is a reduction in tolerance for glucose. In contrast to other microelements, the level of Cr in human tissues declines with age, and the organism loses the capacity for normal metabolism of glucose, a condition known as *senile diabetes.* Chronic exposure of humans to dust of chromium compounds is correlated with increased incidence of lung cancer.

Nickel. This essential element is needed for growth and for resorption of iron. Nickel is non - toxic for the most part and is poorly absorbed from the lumen of digestive organs. It has been established that the amount of Ni in the human organism increases in certain pathological states, for example in patients with cancer, myocardial infarct, and thyrotoxicosis. Due to deficiency of vitamin B_{12}, Ni content is reduced in the blood of persons suffering from cirrhosis of the liver and chronic uremia.

Arsenic. Arsenic is found in nearly all soils, its content depending on geological composition of the terrain. It exists in minerals of many rocks, and its content in clayey media can be as much as 15 mg/kg and more. In slightly mineralized water, the concentration of this element ranges from less than 1 to 50 µg/l, more rarely to 200 µg/l (2,000 µg/l in Bad Dirkheimer Maxquelle and even 12,000 µg/l in the capped horizon in Eastern Wisconsin).

The toxicological role of arsenic is today amplified primarily by the anthropogenic factor (pollution). Of all elements after lead, it represents the greatest toxicological risk for domestic animals and humans. Cases of arsenic poisoning have been recorded in Southern Taiwan, Chile, Argentina, Mexico, China, West Bangal (India), and Bangladesh, and the governments of several countries have assigned priority status to research dealing with the influence of arsenic on health. In regard to toxicological characteristics, arsenic belongs to the category of poisons that accumulate in the organism. Chronic arsenic poisoning is manifested in loss of appetite, disturbances in the gastrointestinal tract, and weight loss. Skin diseases - including pigmentation, hyperkeratosis, and skin cancer - are the most typical symptoms of chronic exposure to arsenic through drinking water. Others include renal, gastrointestinal, neurological, hematological, cardiovascular, and respiratory symptoms. It has been established that small doses of As in the air increase mortality from lung cancer, and that a correlation exists between exposure to arsenic and cancer of the kidney, bones, and large intestine.

On the other hand, symptoms of arsenic deficiency in consumed food are evident in growth and many functions of the organism in domestic animals and include increased juvenile mortality. They also include changes in the content of bioelements in many tissues and organs of pigs and goats, as well as newborn piglets and kids.

Strontium. Strontium is found in the form of stable nuclides in the Earth's crust. In the organism of man and animals, it is concentrated in the bones and partially replaces calcium. Metabolism of strontium is very similar, but not identical, to that of calcium. Absorption, transport, and excretion of calcium are more efficient than in the case of strontium. It has been experimentally established that large amounts of Sr in consumed food negatively affect mineralization of the bones, even if adequate amounts of vitamin D are present in the food.

Cadmium. Cadmium is very widespread in nature, but considerable amounts of this microelement constantly enter air, water, and soil as a result of anthropogenic activity. In the human organism, it is stored primarily in the liver, kidneys, salivary glands, testes, and pancreas. Metabolism of Cd is characterized by defective work of the control mechanism for maintenance of homeostasis in the organism. With a half - life ranging from 12 to 30 years, cadmium is retained for a very long time in humans.

Many data suggest that Cd can be an essential element, but the amounts have not been established. The chemical reactions of Zn and Cd and their metabolic pathways are similar. However, while Zn is important as an essential microelement, Cd is better known for toxicity and metabolic antagonism to Zn and other elements. Introduction of a greater amount of Cd into the organism of man and animals leads to a wide range of metabolic changes. Severe toxicosis at high doses of Cd in humans is accompanied by serious, acute, and often fatal inflammations of the lungs. Hypertension or lung cancer (especially characteristic of persons who work in cadmium smelteries) can also occur.

Tin. It has been experimentally demonstrated that tin is essential for growth of animals and mammals, but the optimal and suboptimal amounts of this microelement have not yet been established. It is needed for proper growth of bones and connective tissue.

Mercury. After arsenic, mercury is the most important microelement in regard to toxicological significance. With its organic and inorganic compounds, mercury has a very wide range of applications. This being the case, mercury pollution of the biosphere takes many forms, with the possibility of acute and chronic poisoning by means of ingestion or inhalation. Animals are exposed to toxic action of Hg ions in air, soil, water, and food. Cases of professional, accidental, and medicinal poisoning have been known to occur in humans.

Lead. Lead is not an essential element of the living organism. It is considered to be one of the main pollutants of the environment and represents the most important cause of accidental toxicosis in domestic animals. Lead is a systematic poison that damages different tissues. With respect to its course, lead poisoning can be acute and chronic. Chronic professional lead poisoning is especially prevalent in humans. Along with cadmium and arsenic, lead is one of the leading carcinogenic elements.

Thallium. Thallium represents a toxic substance for all species of animals. The mechanism whereby this element erts toxic action is not known, but is presumed to involve interference with the metabolism of sulfur - containing compounds through binding with them, which happens in nearly all tissues in the organism.

Microelements in Rocks, Soil, and Water on the Territory of Yugoslavia

Great differences in the content of individual microelements on the territory of Yugoslavia are a consequence of its diverse lithological composition and wealth of geological formations. The first information about the spatial distribution of microelements was obtained by geochemists, who since the 1950's have been engaged in study of their content in different geological environments, river sediments, soil, water, and vegetation for the purpose of finding concealed ore deposits. These investigations have enabled them to single out geochemical anomalies with elevated metal content and on a broader scale made it possible to isolate geochemical regions and geochemical provinces.

Two geochemical provinces stand out clearly in Yugoslavia: a Tertiary *lead* geochemical province, which encompasses the region of the *Inner Dinarides* and the *Serbo - Macedonian Mass*; and a Cretaceous - Tertiary *copper* geochemical province in the Yugoslav part of the *Carpatho - Balkanides.* These geochemical units and regions with ultrabasites are described below according to Z. Maksimovic and M. Rsumovic (1988).

The **lead geochemical province** includes significant deposits of lead and zinc, deposits of antimony ore, and a large number of smaller occurrences of these metals, as well as some deposits and occurrences of molybdenum, arsenic, mercury, and thallium. Genetically linked with these deposits, Tertiary magmatic and volcanic rocks are considerably enriched with many microelements, especially lead and zinc, but also antimony, bismuth, cadmium, and thallium. A large number of molybdenite - pyrite mineralizations occur in Mackatica and the wider region of the Surdulica eruptive massif. The rocks themselves sometimes exhibit molybdenum content as much as 25 times greater than average content in the Earth's crust, and this to some extent also applies to soil, water, and vegetation.

Numerous antimony mineralizations (sometimes accompanied by mercury) occur in rocks and soil of Western Serbia, between Koviljaca and Krupanj. Since volcanic rocks are widely disseminated within the lead geochemical province, elevated levels of heavy metals can be expected in soil and vegetation of large areas.

The **copper geochemical province** in Eastern Serbia extends across the Timok eruptive regions into the mountains of Bulgaria. It is marked by significant deposits of copper and a large number of mineralizations and occurrences of this metal. Genetically linked with deposits of copper ore, volcanic rocks occupy a large area in Eastern Serbia. The copper is accompanied by molybdenum, gold, selenium, and tellurium. The rocks are very poor in lead and (especially) chromium, which is one of their characteristics.

Within the framework of the Dinarides of Western Serbia and Western Bosnia (Bosnia and Herzegovina), there are large masses of ultrabasic rocks and serpentinites, which cover an area of more than 5,200 km^2. Soils developed on these rocks are considerably enriched with Mg, Fe, Cr, Ni, and Co. Because these rocks are extremely poor in other components, it is possible to expect elevated levels of Cr, Ni, and Co in water and food, but very low content of a whole series of biologically important microelements[63]. The ultrabasic rocks and serpentinites are carriers of chrysotile asbestos deposits (Korlace and Strugari) and many dispersed mineralizations and occurrences, which by contaminating air and water with minute hanging particles of asbestos can have carcinogenic effects.

We note that results of measuring the concentration of selenium in rocks, soil, food, and humans have shown that our country is deficient in this element[64]. Especially low average content is recorded in metamorphic schists (28 ppm) and Cretaceous - Tertiary formations (about 17 ppm) - rocks that are very widespread in Serbia - and in soils formed from them. The situation is similar with the loess formation in Vojvodina and along the southern rim of the *Pannonian Basin* (in which the content of selenium ranges from 5 to 50 ppm), a circumstance that is reflected in the concentration of this element in agricultural soil there (chernozem). Selenium levels lower than values average for the world has also been established in soils of lands along the Morava River and on the plain of the Zeta

---

[63] Studying the geochemical characteristics of alpinotypic ultrabasites of Serbia and Bosnia, Z. Maksimovic (1975) cites the following content of trace elements in peridotites: Cr, 260-5000 ppm (1722 ppm on average); Ti, 10-1250 ppm (341 ppm on average); V, 10-210 ppm (69 ppm on average); Ni, 750-4000 ppm (2008 ppm on average); Co 62-275 ppm (120 ppm on average); Mn, 300-1400 ppm (944 ppm on average); and Hg, 0.014 ppm on average.

[64] This does not mean that anomalies of the opposite sign do not exist. For example, galenites of the Rudnik deposit near the town of Gornji Milanovac contain as much as 87 g/t of selenium, and high levels of this element are present in phosphates around Bosiljgrad, Susevo, Kriva Lakovica, etc.

River. Such levels of this important element in cereals are reflected in the low selenium status in the human population of Yugoslavia. In view of the pronounced deficit of selenium, the possibility has been considered that its deficiency is one of the main etiological factors in endemic nephropathy (Z. Maksimovic, 1987), and increase in mortality from malignant diseases as a consequence also seems certain.

It is well - known that the territory of Serbia is characterized by many occurrences of chemically diverse types of slightly mineralized, mineral, and thermal waters. The natural medicinal properties of mineral and thermal waters have been exploited for a long time. However, the uncontrolled use of many of these waters doubtlessly entails negative consequences for public health, and it is one of the tasks of medical geology to divide a territory into regions based on the risk or positive influence of waters in them and make recommendations for the use of these waters. The importance of such scientific research is illustrated below by several examples of extreme concentrations of certain elements and other parameters.

1. A unique hydrological phenomenon in Serbia, the water of "*Djavolja varos*" ("Devil's City") on Radan Mountain in fact represents a mixture of dilute sulfuric and hydrochloric acid, a real poison for the human and animal organism. The water is characterized by extremely high potassium content (180 mg/l), content of metals (Al, Fe, Cu, etc.) above 1.5 g/l, and sulfur content of 2.63 g/l. It is also characterized by extreme acidity (pH 1.5) and high mineralization (15 g/l).
2. Saline (fossil) mineral waters in Serbia occur for the most part in the region of the former Pannonian Sea (Vojvodina) and its bays (Lestani, Ritopek, Valjevo, Mionica, etc.). For example, the mineral water in *Ovca* has high content of chlorine (9760 mg/l), sodium (6250 mg/l), HBO_2 (88 mg/l), strontium (29 mg/l), iodine (8.5 mg/l), bromine (28 mg/l), barium (19 mg/l), Al_2O_3 (7.0 mg/l), etc. Water from a borehole in Lestani is also similar to seawater (with mineralization of 18 - 20 g/l, bromine content of 45 mg/l, and iodine content of 5.0 mg/l).
3. Many other occurrences of groundwater are likewise not for oral use. We note several of them: a) the spring "*Suva Cesma 2*" near Prokuplje, which has very high concentrations of SiO_2 (730 mg/l), sodium (up to 1896 mg/l), Al_2O_3 (16 mg/l), and Ti (1.7 mg/l); b) as many as 56 mineral water occurrences in Serbia with fluorine content greater than 2 mg/l (the Yugoslav standard for drinking water is 1 mg/l), especially in waters of the *Serbo - Macedonian Mass* and northern part of Vojvodina, where it attains a maximal value of 25 mg/l; c) arsenic waters of the peak Crni Vrh northwest of Bor from an exploited deposit of copper ore have As content of up to 4.8 mg/l; and d) levels of H_2S in waters of the Mataruska Banja spa and Slatina are high (21 and 15 mg/l, respectively).

CHAPTER 3

Other Natural Factors

CLIMATIC FACTORS

Climate, or the physical state of the aerial environment, is one of the leading natural factors from the standpoint of human health. As Hippocrates said already in 400 B.C.: "*The human organism behaves differently in different seasons. Some people feel better in winter, others in summer. Dry weather is healthier and less dangerous than rainy weather.*" Beyond those general observations, we have today the developed scientific disciplines of *medical climatology* (human bioclimatology) and *medical meteorology*, which study the relationship between the organism and long - term or short - term weather conditions and their direct connection with health and disease.

It can be arbitrarily postulated that a temperate climate represents the norm, and extreme or even moderate deviations of heat or cold and dryness or wetness significantly affect the physiological functions of man and his health. Climate affects the capacity or inclination of human beings for physical activity at different temperatures, reduces salt levels in warm climates, and increases the desire for caloric (fatty) food in cold lands. There are so many different examples that they cannot be enumerated at this time. It is our aim to give a concise review showing the role of this natural factor on human health. For the following discussion, we draw on the work "**Balneoclimatology**" by T. Jovanovic et al. (1977) and other literature.

In the first place, the weather situations to which man is exposed can be simple or complex, the latter having a much more pronounced negative biotropic effect on the human organism than the former. Strong heat waves and cold spells are *simple weather situations*, which also include sudden changes of barometric pressure, wind, fog, global solar radiation, precipitation, and atmospheric ionization. *Complex weather situations* occur more rarely. They usually are frontal in origin and represent a combination or complex of simple weather situations. So - called *enhanced weather situations* occur even more rarely. They occur when strong air pollution - smog - is superimposed on the complex of meteorological factors. In order to make it easier to study biotropisms of the human organism in response to weather situations, there are a number of classifications of weather states giving rise to certain biotropisms.

Meteorological events on the whole take place in the troposphere, i.e., the envelope of air up to an altitude of 12 - 14 km, and this layer of the atmosphere is popularly called the "*weather kitchen.*" Weather represents the real state of values of meteorological elements and occurrences at a given point in time. For this reason, weather is a concept that is defined by numerical values of meteorological elements and symbols of meteorological occurrences. In contrast to weather, climate represents the set of weather phenomena or atmospheric occurrences characterizing the average physical state of the atmosphere above some region.

Meteorological elements include the following things: solar radiation, terrestrial radiation, day length, horizontal visibility, temperature, pressure, humidity, air streaming, evaporation, snow cover height, and height of precipitation. *Meteorological phenomena* include fog, clouds, rain, snow, hail, sleet, frost, and ice. Greater non - periodic, unexpected, and sudden fluctuations in the values of meteorological elements are of special interest for meteorobiology.

The different climates that occur on the Earth bear witness to the controlling influence of several climatic factors: 1) geographic latitude; 2) zones of constant low and high pressure; 3) prevailing winds; 4) storms and interacting air masses; 5) disposal of land and sea; 6) distance from the sea; 7) barriers in the form of mountain ranges; 8) ocean currents; and 9) elevation above sea level. Accordingly, certain large areas of the Earth have a definite climate, which can be polar, tropical, continental, Mediterranean, subtropical, etc. Bioclimatologists, biogeographers, doctors, and other specialists deal with specific complex questions of human life and acclimatization in polar regions, the damp tropics, arid zones, high mountains, and regions with a maritime climate.

With respect to biological effects, the following four meteorological comples can be singled out: thermal, hygric, actinic, and aeric. The *thermal complex* (air temperature, movement of air, and thermal radiation) gives rise to thermoregulatory responses of the organism. The *hygric complex* (concentration of water vapor in the air and air pressure) hinders the thermoregulatory process and imposes considerably greater demands on the thermoregulatory apparatus of man. The *actinic complex* includes solar radiation, and the visible and ultraviolet parts of this radiation are of special significance. The *aeric complex* consists of air pressure and partial pressures of the individual gases and admixtures of which air is composed.

In addition to studying the numerous physiological reactions of the human organism to the indicated comples, meteorophysiology also monitors physiological reactions brought about by the short - term, intensive, and predominantly dynamic influence of sudden weather changes. Most physiological reactions of the organism occur in response to changes in the surrounding conditions, and their intensity depends on the rate of such changes. We list below several factors and indicate how the organism reacts to them:

1. *Air temperature.* Among other things, the depth and rate of respiration, rate of blood circulation, and provisioning of cells and tissues with oxygen depend on air temperature, as do characteristics of the metabolism of carbohydrates, salts, and fats and work of the muscles;
2. *Relative humidity.* Increased temperature is easier to withstand in the presence of drier air. However, if relative humidity is lower than 20%, then evaporation from the surface of the mucosa lining the respiratory pathways is so great that it begins to dry out. In deserts and mountains, this causes an unpleasant sensation of dryness in the throat and nose, cracking of the lips, and decline in their protective role as factors preventing dust and microbes from entering the organism. On the other hand, high air humidity in cold weather leads to colds and frostbite of the hands, feet, ears, and cheeks due to the prolonged cooling of the organism. Increased humidity in cold weather usually causes aggravation and recurrence of rheumatism and bronchial asthma. This is the case, for example, in the Eastern United States, where the number of acute attacks of rheumatic fever attains a maximum at the end of winter and beginning of spring (Fig. 3.1.);

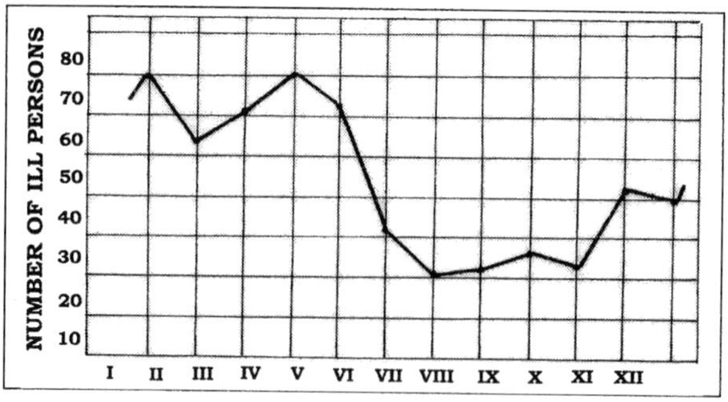

Fig. 3.1. Seasonal variation of number of person's disease of recurrence of rheumatism.

3. *Atmospheric pressure.* Statistics confirm that the number of hypertonic crises is twice as great on days with low atmospheric pressure than on

days with normal values of it. Especially hard to bear is the combination of reduced pressure with high air humidity. It has been established that a decrease in atmospheric pressure by only 10 millibars with a change of weather can cause vascular catastrophes;

4. *Wind.* The influence of wind on man can be either positive or negative, depending on a number of things. For example, the "bura" (a dry cold intermittent wind that blows along the Adriatic seacoast) acts positively on the human organism: sick persons feel better, and there is improvement in the subjective sense of wellbeing of hypochondriacs and asthmatics. On the other hand, the "jugo" (a damp coastal wind) evokes effects opposite to those induced by the "bura." Similarly, the hot dry intermittent winds known as "fens" have a harmful effect on the human organism, especially the psyche.

Adaptation of the organism to changes in the environment is frequently hindered, depending both on the conditions in which it occurs and on the state of adaptability of the people concerned. The mechanism of adaptation is complex and has not been fully clarified.

If a state of equilibrium is not achieved between adaptive responses and changes in the environment, disturbances arise in the organism that fall in the domain of *meteoropathology*, i.e., *meteorotropic diseases.* To be specific, weather conditions can induce the onset of certain diseases, aggravate the state of sick persons, or cause complications of diseases present in the human organism. *Meteorotropisms of certain diseases and states* are discussed below.

Cerebrovascular diseases. Yoshiyuki Ohno established that cerebral hemorrhages can occur with increase of air temperature above 25°C, decline of relative atmospheric humidity below 40%, and decrease of barometric pressure below 1,010 mb. These complications are also often recorded in the course of some physical activity during the cold period of the year. Apoplexy most often occurs after a sudden drop of barometric pressure, with an increase of air temperature, and under conditions of strong wind and cloudiness.

Infarct and *coronary disease.* The greatest incidence of infarcts is recorded in the cold period of the year. Sudden cooling has an unfavorable effect on patients with coronary disease. Decrease of air temperature coupled with increased cloudiness and fog also has an unfavorable biotropic effect on patients with angina pectoris.

Ulcers. Ulcers represent a typical case of poor adaptation. Perforations occur most often during the winter in Western Scotland, but during the summer in Northern Scotland. It is a fact that patients with ulcers respond to sudden changes in the values of many meteorological factors, strong winds especially.

Mental diseases. Psychic patients are subject to meteorological influences. Patients with the clinical pattern of schizophrenia react to the advance of hot air. Many authors attribute a negative biotropic effect to atmospheric ions.

Chronic bronchitis and *bronchial asthma.* Patients experience a worsening of their condition and discomfort with increase of atmospheric humidity and under cloudy conditions. With a decrease of temperature by more than 10°C in a short period of time, asthmatic attacks occur in all patients. Reactions are recorded during passage of meteorological fronts.

Rheumatic diseases. It has been known for a fairly long time that persons who suffer from various rheumatic diseases are "living barometers" and react to sudden weather changes. Discomfort arises with greater fluctuations of temperature and humidity and after a sudden drop of barometric pressure. Atmospheric ions also play a definite role in these reactions.

Traffic accidents. There are many causes of traffic accidents. Apart from factors that make driving conditions worse (fog, slippery roads, heavy rain, snow, and strong wind); there is also the influence of sudden weather changes on the human organism. To be specific, weather changes that occur under unstable cyclonic conditions give rise to exhaustion, apathy, and slowing of reactions in humans. If the driver also suffers from some chronic disease, then significant reduction of driving ability and general concentration is amplified.

As well as risks to the human organism, *meteorological* and *climatic factors* can also ert positive influence. For example, the aim of climatic therapy of asthma and bronchitis patients is to remove the sufferer temporarily from a polluted environment and induce as good and lasting remission as possible by means of exposure to favorable climatic conditions in conjunction with other methods of treatment.

HYDROLOGICAL FACTORS

Water has occupied a position of high priority in the traditions of all peoples and cultures. From time immemorial, man has known that he is inseparably linked with water and can survive only where water also survives. It is therefore understandable that water has erted appreciably greater influence on the development of civilization than all other natural resources, influence such that the entire history of mankind can today be described in terms of our needs for water. Earlier civilizations arose exclusively around great river courses: the Tigris and Euphrates in Mesopotamia, the Nile in Egypt, the Indus in India, and the Yangtze in China. The river - oriented Egyptians constructed complex water - engineering systems, irrigation systems especially, already around 3200 B.C. Thus, for example, the historian Herodotus records that the first pharaoh of Egypt, Pharaoh Min, built a dam on the Nile that altered the river's course. The historian was amazed by the artificial lake, which in his words extended for a length of 450 miles and was almost as long as the whole seacoast of Egypt.

Man marked his holidays and his holy places by regulating water courses and improving springs. He was grateful to the gods for such a valuable gift and feared that the same gods might take it away in punishment for his sins. With the advent of the Industrial Revolution and increasing preoccupation with material values, man removed water from its mythical throne and started to behave toward it as toward every other natural resource. Probably because he believed in the inexhaustible abundance of water on the planet, he subordinated water to uncontrolled industrial development, using it and polluting it without mercy. Only a few decades of such behavior were needed for mankind to be confronted with the *global water problem*, i.e., the *problem of clean drinking water.* Ensuring a supply of clean drinking water and protecting it from pollution and greedy exploitation have thus become one of the most pressing problems facing humanity, and water is today the primary strategic raw material.

Water Resources of the Planet

The distribution of the total volume of water on the Earth is evident. It can be seen that 96.5% (1,338 million km^3) of the world's water resources falls on salt water, i.e., water of the oceans and seas. Only 35 million km^3 or about 2.53% belongs to the category of *low - mineral water*, which is of the greatest interest to mankind. Of this amount of low - mineral water, virtually 70% - or 24 million km^3 to be more precise - is accumulated in *ice* of the Arctic and Antarctica and in glaciers of Asia, North America, and (to some extent) Europe. About 30% of low - mineral water occurs in the form of *groundwater* in water - permeable rocks and soil as very important temporary reservoirs of atmospheric precipitation. Only 0.006% of all low - mineral water falls on terrestrial surface water resources, viz., lakes, reservoirs, swamps, and riverbeds.

Atmospheric precipitation represents the main input element or element of water renewal on the Earth[65]. It can be stated that a layer with average thickness of close to 0.9 m falls annually on the Earth's surface. Renewal of the reserve separates water from practically all other raw materials in nature, which is to it's and man's advantage. Unquestionably, attention must also be called to negative aspects of the situation, namely geographic unevenness in the distribution of precipitation (Fig. 3.2.) and inequality of its regime in terms of time.

---

[65] As Solomon observed, all of the rivers run into the sea, yet the sea is not full, even though the rivers flow eternally. Not even the greatest minds, from Aristotle (384-322 B.C.) to Kepler (1571-1630 A.D.), were able to explain this riddle. Measuring of atmospheric precipitation and river drainage in the 17th Century led to formulation of the concept of the *hydrological cycle*. This concept embraces movement of moisture from the sea into the atmosphere, thence to land in the form of precipitation, and then through drainage back into the sea.

A significant portion of precipitation falling on land (approximately 62%) is returned to the atmosphere by means of evapotranspiration. The remainder of precipitation, in the form of surface and underground drainage, is available for use (Fig. 3.3.). Many geographers, hydrologists, and meteorologists have worked out the water balance of the Planet, and the obtained values do not differ very much (by up to 10%). It is calculated that the inhabited part of dry land annually receives an average of 834 mm of precipitation, of which 294 mm is drained or infiltrated into the ground, while 540 mm is evaporated. Water drained on the surface represents the total flow of all rivers, which according to M. M. L'vovich (1974) annually comprises 38,830 km^3 or 1,232,000 m^3/s of water (44,500 km^3 or 1,413,000 m^3/s according to some authors).

Fig. 3.2. Average distribution of precipitation in the Earth (average values of precipitation height, in cm).

As for individual continents, *South America* has the most intensive water balance (D. Dukic, 1984). The values of all elements (precipitation, evaporation, and drainage) are higher there, and South America is considered to be the wettest continent. The outflow of the Amazon alone is 2.23 times greater than the total outflow of all the river courses in Europe. By way of contrast, *Australia* is the driest continent. The average annual level of precipitation there is 2.2 times lower and the drainage coefficient 3.36 times less than in South America. With an area of 2.68 million km^2, the western plateau of Australia has no surface drainage whatsoever. *North America* has the lowest level of precipitation and the same drainage coefficient as Australia: 0.31. Thus, the average total outflow of all the rivers of North America is less than that of the Amazon, and only seven rivers have an average annual outflow of more than 100 km^3 or 3,170 m^3/s. *Africa* lies largely in the subtropics, and there are vast deserts north and south of the equator. The equatorial zone is where more than 85% of the continent's total annual outflow is formed. It is also characteristic that enormous reserves of groundwater are present in the wider area of the lower course of the Nile, the middle course of the Niger, and the lower course of the Senegal. A large part of *Asia* (about 30% of its territory) is represented by desert regions, but in spite of this the average drainage coefficient of the largest continent is high: 0.40. Asia has as many as 10 river courses with average annual outflow of more than 350 km^3 or 11,100 m^3/s. Almost all of *Europe* is under the strong influence of damp air masses from the Atlantic Ocean. For this reason, the average coefficient of water drainage is the greatest among all the continents: 0.43. Still, only nine European rivers have an average annual outflow of more than 50 km^3 or 1 - 500 m^3/s, among them being the Volga (with 255 km^3 or 8,150 m^3/s) and the Danube (with 201 km^3 or 6,370 m^3/s).

Fig. 3.3. Skradinski buk on the river Krka, Croatia.

Rivers are a significant link in the planetary circulation of water. For example, the Amazon has a drainage area of 7.1 million km^2 and is 80 - 150 km wide at its mouth. Its yellow water can be seen in the ocean at a distance of 300 km from the coast. The Nile is the longest river in the world (6,671 km). Egypt's "Gift from God," it has a number of dams and hydroelectric power plants (the Aswan High Dam has a capacity of 130 billion m^3 of water), as well as irrigation canals with a total length of 25,000 km. Larger river courses are navigable over a great part of their length, and some (the Mississippi and Yangtze) are prone to catastrophic flooding.

Lake water also plays a significant part in human life. About 280 thousand km^3 of water is accumulated in **lakes**. The Caspian Sea is the largest lake in the world (with an area of 371 thousand km^2), Lake Baikal the deepest (1,620 m). Lake Horpa in Tibet lies at an elevation of 4,400 m above sea level, while the Dead Sea in Palestine is 392 m below sea level. The chemical composition and degree of mineralization of lake (and river) water is closely dependent on the climatic zone in which it occurs. For example, in regions with a humid climate, lake water usually is weakly mineralized (up to 0.3 - 0.5 g/l) and characterized for the most part by hydrocarbonate magnesium - calcium composition. Widespread in arid zones are saline and brine lakes of varied composition (chloride - sodium, etc.) with significant content of lithium, boron, potassium, rubidium, bromine, strontium, and iodine.

Together with rivers and lakes, enormous areas of land (about 11%) are occupied by **glaciers** (about 15 million km^2 in Antarctica and 1.8 million km^2 on the island of Greenland). Average thickness of the covering glaciers is about 1,500 m (up to 4,000 m in the central part of Antarctica). The Antarctic ice cover represents the largest reservoir of low - mineral water on Earth.

Especially significant for the life of humans and their economic activity, **groundwater** accumulated in permeable rocks and soil on land has already been treated in a separate chapter.

More than 2/3 of the Earth's surface is occupied by **oceans**: the Pacific, Atlantic, Indian, and Arctic Oceans (193.1 thousand km^2). Average depth of the oceans is about 4 km, and their total quantity of water amounts to 1,370 million km^3. The many island archipelagos in the Pacific Ocean occupy an area of about 3.6 million km^2 and represent the most important geographic characteristic of this dominant body of water on the planet. Water salinity usually varies within limits of 3.2 - 3.7%, average salinity at the surface being 3.5%. In seas bordering oceans and land - locked seas, which occupy about a tenth of the World Ocean's area and possess an abundance of water (37 million km^3), salinity ranges from 0.5 - 0.8 to 4.0 - 4.2%. All chemical elements and their isotopes are recorded in water of the oceans. The total content of metals in ocean water is enormous, for example 6 x 10^6 t of gold and 0.5 x 10^9 t of uranium. Kitchen salt (NaCl), magnesium, and bromine are today extracted from seawater. Owing to its enormous amounts of water and the composition of this water, the World Ocean is of exceptionally great significance for life on Earth.

Water and Health

The significance of water for life of plants, animals, and man is very great. Water is as essential a factor as air for human existence in the natural environment, and it makes up approximately the same percentage of the human organism as in composition of the biosphere: about 70%. In picturesque terms, it can be said that the human organism is essentially an aqueous solution in which all processes of metabolism take place. It is therefore easy to grasp that in order to satisfy the organism's needs, obtaining adequate amounts of water from the natural environment represents a basic biological requirement. Thus, under conditions of the Central European climate, an adult man must take almost three liters of water daily to be able to live; in a desert, on the other hand, he needs as much as 12 l a day. Because of its function, water cannot be replaced by anything else: it is a constant participant in the biochemical processes that transpire in the human organism.

In discussing the more important aspects of water's significance, we rely for the most part on works of V. Jorga (1999) and Cebedzic (1999).

The *significance of water for health* has been known for a very long time. Already Hippocrates recommended bathing as the most effective measure of protection against many diseases. That the ancient Romans well knew the significance of water for health can be seen from the famous Roman aqueducts, public baths, and spas that they built and improved throughout the Roman Empire.

The health significance and role of water is reflected in the fact that a supply of hygienically good water in adequate amounts improves living conditions, raises the level of public health, and reduces mortality (primarily from infectious diseases). In this way, it directly increases the length and quality of life. Classical medical science and practice and current medical postulates based on them indicate unambiguously that the supply of drinking water is closely correlated with the danger of infectious intestinal diseases.

Rest and *active recreation* are important for healthy people. They promote much needed increase in general resistance of the organism, strengthen the body's overall musculature, improve functions of the cardio - respiratory system, and raise the level of general immunity as an expression of the organism's natural strength and ability to defend itself from disease.

Many scientific studies in the area of medical hygiene and modern sports medicine emphasize the great positive effect of different methods and procedures employed in *utilization of water for therapeutic purposes.* Recognition of the health, esthetic, and sport and recreation value of water is constantly growing in keeping with the development of hygiene consciousness of the population and improvement of general health and physical culture.

The *physiological significance of water* is exceptionally great and the role of water so diverse because all vital processes in the organism are linked with its presence. The presence of water in cells ensures all activities of cell metabolism. In the extracellular space, it makes possible metabolic exchange of cells in tissues and exchange between blood and tissue fluids. Water represents a universal solvent for most organic and inorganic substances and the medium for chemical and physico - chemical reactions linked with commerce of matter and energy in the organism. Nutrients are transported with water. It maintains the normal structure of all tissues in the organism and eliminates the end products of metabolism. Water has a significant if not primary role in maintaining constant body temperature and regulating the course of other vital processes.

The human organism finds it difficult to withstand a *water shortage*. Loss of water amounting to 2 - 3% of body weight already gives rise to the need for re - establishment of the water - mineral balance. Water shortage in the organism increases osmolarity of the intestinal contents and threatens equilibrium of the processes of absorption and secretion of water and fluids from the lumen to the blood vessels. This causes thirst, which of all human instincts is the hardest to deny. If water loss is not compensated in time, all health parameters deteriorate due to disturbance of physiological processes, and death occurs with loss of 10 - 20% of total fluids.

The *hygienic significance of water* is manifested above all in its role in maintaining personal and general hygiene. The hygienic and other every-day needs for water require considerably greater amounts of it than are needed to satisfy basic physiological demands. If consumption of water is limited, the difficulty of maintaining general and personal hygiene unfavorably affects the general health situation and leads to the appearance of lice, supporating and fungal diseases of the skin, intestinal ailments, and other infectious and parasitic diseases.

Water's significance in transmission and spreading of infectious diseases was known already in ancient times. Hippocrates recommended that only boiled water be used. Herodotus wrote that the Persian king rs carried boiled water in barrels with silver bands for the use of his army (the first description anywhere of water disinfection).

The *epidemiological role of water*, i.e., the role of water in transmission and spreading of infectious diseases, today is well - known. We note several infectious diseases below.

Cholera has been known since the time of Hippocrates, when it was given that name. It first occurred in the form of epidemics, but in 1817 appeared in Asia as a pandemic that lasted until 1823. In the course of a second pandemic (1826 - 1837), this dangerous disease embraced all of Europe and was carried by immigrants to the American continent. From that time until 1923, four more pandemics were registered in which hundreds of thousands of people contracted the disease. Imported cases of cholera appeared in Europe in 1989 and 1994. High resistance of the agent (bacteria of the species *Vibrio cholere*) is the reason for continued existence of endemic foci. The homeland of cholera is India, where the deltas of the great rivers Ganges and Brahmaputra represent the main endemic foci. Discovery of the agent of cholera is linked with the name of Robert Koch.

Typhoid fever is a severely acute infectious disease caused by bacteria of the species *Salmonella typhi*. Great epidemics occur in settlements on large rivers and are most often caused by consumption of water contaminated during floods. *Paratyphoid fever* is similar to typhoid fever and is caused by the bacteria *Salmonella paratyphi* A, B and C.

Bacterial dysentery is caused by bacteria of the genus *Shigella*. Contaminated drinking water is significant in the spreading of this acute infectious disease. Dysentery is widespread throughout the whole world, especially in warm regions with monsoon rains.

The agent of so - called *Legionnaire's disease* (*Legionella pneumophila*) is a very resistant species of bacteria that survives in water for up to a year.

Poliomyelitis (infantile paralysis) is a disease caused by polioviruses. Polioviruses are spread via contaminated objects of general consumption, food, and especially water that are virologically not in order. With introduction and use of the Sabin vaccine, the number of polio sufferers has declined since the 1960's, and it is reckoned that the disease will soon be eradicated.

Coxsakie viruses are the agents of a whole spectrum of diseases such as vascular stomatitis, meningitis, infections of the upper respiratory pathways, myocarditis, etc. The virus is transmitted with contaminated food or drinking water and objects of general use.

Hepatitis A is another viral disease, one that occurs often in children of school age as a result of poor hygiene. The disease in most cases is contracted by consumption of water contaminated with the virus. A clinical picture more serious than that hepatitis A is characteristic of *hepatitis E*, whose endemic focus is in Algeria.

River valleys are especially rich in agents of infectious diseases. As a rule, flatland relief is predisposed to homogeneity of the soil cover, uniformity of vegetation, and a more or less uniform structure of the biocenosis. Lakes with surrounding forests, temporary river courses (*wades*) in desert regions, reservoirs, and irrigation canals can also have great epidemiological significance. An instructive example of this is the rapid multiplication of mosquitoes and other disease - carrying insects that occurred along the 800 - km - long Kara - Kum Canal, which was built between the Amu - Darya River and the city of Ashkhabad in Turkmenia for irrigation of the sandy desert terrain there.

We have already considered the action of geochemical factors under natural conditions, above all the role played by microelements of groundwater and soil in the occurrence of a number of diseases. In arid regions, where the negative action of these factors is amplified by high temperature of the environment, it is possible to single out (among others) regions with increased incidence of fluorosis and goiter as a consequence of elevated fluorine content or iodine deficiency in the water. According to certain authors, cancer of the esophagus in arid conditions is linked with (among other things) the shortage of good drinking water and use of desalinized water lacking the optimal combination of minerals and microelements.

We see that water is exceptionally important and irreplaceable for two reasons: it is a physiologically essential substance on the one hand, and it can directly or indirectly threaten human life and health on the other. A supply of hygienically good water is an imperative for modern life and a decent standard of living, and it can serve as the best synthetic measure of the prosperity and wellbeing of a population.

In economically developed countries, an adequate amount of good - quality water has been ensured by water - supply systems since the beginning of the 19th Century. Owing to this, the further occurrence and spread of cholera and later typhoid fever was checked already in the course of that century. Unfortunately, these results were spoiled by constantly increasing pollution of water sources in the first half of the 20th Century, and various epidemics have occurred more and more frequently.

The picture is even more unfavorable in developing countries, which are home to 2/3 of the world's population. On an enormous territory, only a third of the urban population is supplied with water from public water mains. If the total population is taken into account, the situation is worse still.

According to data of the World Health Organization (WHO), about 1.3 billion people (predominantly residents of developing countries) in the middle of the 1980's had no communal water supply and 1.8 billion had no system of sewage disposal. The World Health Organization also reports that 80% of all diseases in the world arise from unhygienic water: about 400 million people suffer from gastroenteritis, 160 million from malaria, 30 million from onchocerciasis, and 200 million from schistosomiasis.

BIOLOGICAL FACTORS

The Earth was formed 4.6 billion years ago, and the first rudiments of life (the first biological forms) on it appeared in waters of the World Ocean before more than 3 billion years. Thus, life that arose in the ocean started to develop in exceptionally reductive (anaerobic) conditions. Already about 1 billion years ago, intensive development of aerobic life began in the marine environment, and a new stage in evolution of the biosphere started before about 400 million years with the emergence of life on dry land. Luxurient vegetation and animals then made their appearance.

A million years ago, another miraculous event occurred in the Earth's history, namely the appearance of man. Man became the fruit of evolution of the living world, the crowning achievement of the biosphere, a thinking being fated to deliberate about his life and perhaps about life of the planet as well (Z. Jovicic, 1997).

The concept of the *biosphere* was introduced in the literature in 1875 by the Viennese geologist Eduard Suess, who used it in a monograph without any interpretation[66]. This term began to be employed in 1911 by V. I. Vernadskii, who in a work entitled "The Biosphere" (1926) explained its essence and geological (geochemical) significance. In Vernadskii's interpretation, the biosphere is an entirely real geological shell inhabited by animals and plant life that determine geochemical characteristics of processes within the shell itself. The biosphere is a truly complex notion, since it signifies the range of life, whose boundaries embrace the troposphere, all of the hydrosphere (including the entire World Ocean), and a considerable part of the lithosphere (to a depth of the order of 3 kilometers).

Discussion of Biological Factors

According to Vernadskii, there are no chemical forces on the Earth's surface that are more constant and more powerful in their final consequences than living organisms taken as a whole. The total effect of the geochemical activity of organisms throughout the entire course of our planet's geological history is very great. For example, A. I. Perelman (1975) states that production of organic substances over the last billion years has been ten times greater than total mass of the Earth's crust, regardless of the fact that the mass of organic substances of the biosphere as living material is constant and comprises only 0.1% of the mass of mineral matter of the Earth's crust. Material of the biosphere is impregnated with mineral material of the Earth's crust, so that today it is often difficult to separate what is mineral material and what is organogenic material of biological origin. The enormous task of destruction of dead organic matter, a kind of unique cleaning of the environment, has been performed predominantly by microorganisms. Mineral compounds formed during decomposition of organic matter are to some extent reabsorbed by plants, and chemical elements once more become part of the bodies of organisms, to be mineralized again in the process of respiration and especially after the death of plants and animals. In this way, *biological circulation of atoms* occurs in nature. Whether tundra, taiga, steppe, or some other kind of landscape arises depends on the amount of organic matter formed in a given place on Earth, the composition of that matter, and the rate at which it decomposes.

---

[66]According to A. L. Yanshin (1986), the term biosphere was first used by J. B. Lamarck at the beginning of the 19th Century.

The ocean represents a special area of the biosphere, since water cannot be separated from life, or life from water (O. Gazenko and V. Makarov, 1987). Thus, the greatest production of living mass on Earth is in the oceans, and it is logical to treat them as a very promising food resource. Here microscopic algae synthesize organic matter by means of photosynthesis and bind solar energy to it. Their production serves as food for planktonic animals, which are eaten by more developed organisms, i.e., fish and a number of farther removed consumers including whales, birds, and finally man. The "unconsumed" planktonic mass decomposes and is again transformed into inorganic matter that will be revitalized by algae.

A crucial characteristic of living matter is its ability to perform *photosynthesis*, a unique chemical reaction widespread on the Earth's surface by which organic matter is synthesized from carbon dioxide and water with the aid of solar energy. This reaction is accompanied by evolution of oxygen[67]. The amount of free oxygen in the present - day atmosphere comprises about 280,000 billion tons, and all that oxygen—the product of life on Earth—is a result of photosynthesis. A second important component of air, nitrogen, is created by the activity of microorganisms. On the other hand, green plants have absorbed carbon dioxide from the air over a long period of geological time, with the result that CO_2 content in the present - day atmosphere is low (0.03%). The energy of sunlight needed to decompose water into oxygen and hydrogen is absorbed by plants and preserved in their organic substances. In this way, green plants act as unique accumulators of energy, which animals utilize in nutrition, since they are themselves incapable of creating organic matter from simple mineral bodies (CO_2 and H_2O).

Through their activity, living organisms largely determine the chemical characteristics of landscape components (water, soil, and air). For example, oxygen, carbon dioxide, and organic acids are products of the metabolism of organisms, and a significant part of the calcium, sulfur, magnesium, and other elements in water enters it as a result of decomposition of the dead remains of organisms. Living matter affects the chemical properties of soil in no less measure.

The living world is very rich and diversified[68]. It is estimated that the Earth is inhabited by about 1.5 million different forms of life, which differ among themselves in their relationship with and dependence on the environment, as well as with respect to their development cycles. The

---

[67] Plants also require compounds of nitrogen, phosphorus, and many microelements. They can grow and reproduce as long as they don't use up all existing phosphates or nitrates, after which the process of photosynthesis ceases.

[68] Biodiversity or biological diversity has been increasingly discussed of late in scientific circles and wider.

plant kingdom includes about 500 thousand species, the animal kingdom close to a million (I. Radovic and R. Mandic, 1998). Bacteria alone number up to six thousand species, which are for the most part invisible to the naked eye (there are exceptions). While some bacteria cannot survive without oxygen (aerobic species), others live only in media without oxygen (anaerobic species). Disease - causing (pathogenic) species exist among bacteria. In spite of this, plant life is of vital significance to man, and it is logical to devote special attention to plants, which constitute one of the two groups of living organisms (Z. Jovicic, 1997).

Very diverse living communities - *biocenoses* - exist on the Earth[69]. Every such community possesses certain characteristics. The specific characteristics of biocenoses are determined above all by their habitats and by local climate. Typical are large communities: biocenoses of the deciduous forest kind and other types (tropical rain forest, steppe, meadow, desert, tundra, etc.). Plant and animal life is very scarce in zones of permanent ice and snow immediately around the poles. In the spacious areas of arctic tundra just below the zone of the North Pole, biocenoses are characterized by low shrubby vegetation, mosses, and lichens. The zone of the taiga (Canadian, European, and Siberian) is a zone of coniferous forests, which farther south pass over into regions of deciduous forests (Western and Central Europe, East Asia, and Southeastern North America). Various communities of evergreen forests and shrubs are developed in warm subtropical regions, as in the case of Mediterranean regions, while tropical and equatorial vegetation appears farther south.

The plant kingdom has an irreplaceable role in circulation of matter in nature. Thanks to photosynthesis, plants create a special kind of energy - biomass - from inorganic material. About 350 billion tons of oxygen is produced annually on the Earth. Because of oxygen and assimilation of carbon dioxide, the plant kingdom is said to represent the source of life on Earth, and the biological interdependence between plants and man is a special dimension of the biosphere. Plants have other important functions in human life, above all in nutrition, but also in dress and as raw material for industrial processing, including use in the pharmaceutical industry (in the manufacture of medicines). The animal kingdom likewise is greatly dependent on plants, both as food and because plant communities (forests, pastures) provide habitats for animals (Z. Jovicic, 1997).

Forests represent a unique form of plant community, a valuable natural resource of inestimable significance for mankind. They occupy a third of the world's land area. Of the continents, the Americas and Asia

---

[69] On the territory of Yugoslavia, we encounter practically all of the biomes characteristic of Europe or five of the Planet's total of 12 terrestrial biomes. We note that biomes are biocenoses with mutual similarities.

are the richest in forests, while Europe and Australia are the poorest in this natural treasure. Equatorial and tropical forests stand out with respect to the wealth of species per hectare, number of levels, and height of trees, and they are also of planetary significance as oxygen factories.

The animal kingdom is much diversified, with different divisions. The most widespread types of animal life are unicellular organisms. Amoebas, foraminifera's, radiolarians, and other microorganisms are very widely distributed, but much less so than poriferans, annelids, and arthropods or considerably more organized types such as reptiles, fish, birds, and mammals. It is interesting to note that about 8,600 species of birds exist on the Earth; they live in all meridians and parallels, on rivers and seashores, on high mountains, and in settlements (Z. Jovicic, 1997). Fish make up a goodly part of the human diet, as do many species of domestic animals such as cattle, pigs, sheep, and poultry. Humanoid monkeys live predominantly in tropical regions of Asia and Africa.

Different regions of plant and especially animal life are formed due to the action of biotic and abiotic factors. Although these regions do not coincide in space, interdependence and interconnection between plant and animal life is manifested in each of them.

Six *floristic regions* are singled out on the Earth's surface: Holarctic, Paleotropical, Neotropical, Australian, Kaplan, and Antarctic. The largest of them is the Holarctic floristic region, which encompasses more than half of the globe `and includes temperate and polar zones of the Northern Hemisphere.

There are seven *faunistic regions*: African, Indo - Malaysian, Holarctic, Neotropical, Australian, Polynesian, and Antarctic. In this case also, the Holarctic region occupies the greatest area: Europe, Africa with the Sahara, and a large part of Asia and North America. However, the African and Indo - Malaysian regions have the richest faunas.

According to evolutionists, **man** belongs to the zoological group of primates, whose evolution lasted 50 - 60 million years. The evolution of man proceeded in five phases: Australopithecus, Pithecanthropus, Neanderthal, Homo sapiens, and modern man. As the oldest *fossil man*, Australopithecus (from the end of the Tertiary) has been discovered in South Africa and China, as well as on Java. The Paleolithic (Würm glaciation) began with Homo sapiens, who was succeeded in alluvium by modern man (Homo sapiens recent).

Three separate races - Mongoloid, Europoid, and Negro - Australoid (Equatorial) - developed in the course of human evolution. The *Mongoloid race* numbers about 1.5 billion people and is geographically fairly compact, inhabiting areas of East and Southeast Asia. The *Negroid* or *Equatorial* ("black") *race* inhabits areas of Africa south of the Sahara and considerably smaller parts of Australia and Southeast Asia. With more than 2 billion people, the *Europoid* (or "white") *race* inhabits Europe, North America, India, North Africa, and other parts of the world. The origin of races is a question that has not been definitively settled, but it is fairly certain that natural factors (especially climatic ones) erted decisive influence.

The history of human influence on nature is divided by I. Radovic and R. Mandic (1998) into four ecological phases: the *Initial Phase*, the *Early Agricultural Phase*, the *Early Urban Phase*, and the *Modern Industrial Phase*.

1. In the *Initial Phase*, the human population probably differed hardly at all in the extent and nature of its ecological activity from populations of other omnivorous mammals. It is possible that the use of fire, beginning more than half a million years ago, caused the first qualitative changes in relations between man and his surroundings.
2. In the *Early Agricultural Phase* (beginning about 12 thousand years ago), man domesticated animals and plants and developed farming technology. Despite all the changes brought about by agriculture, human activities still did not ert any more significant influence the environment, above all meaning natural biogeochemical cycles of the biosphere such as the cycles of carbon, nitrogen, and phosphorus.
3. The *Early Urban Phase* (beginning 5 thousand years ago) brought about a series of fundamental changes in the organization of human society, primarily an increase in the number of people on a limited space and dependence on food produced by surrounding farmers. Due to the uniformity of food, specific diseases such as rickets, scurvy, beriberi, and pellagra appeared and began to spread. However, as in the previous phases, natural biogeochemical cycles remained untouched, and the biosphere as a whole was still essentially in a state of dynamic equilibrium.
4. The *Modern Industrial* or *Technological Phase* (beginning 150 - 200 years ago) introduced many changes in human activities and has erted influence on the biosphere completely out of proportion to duration of the phase. Man's destructiveness will be discussed in a special chapter. Here we note his massive and irreversible destruction of life, which has become a major global biological problem. This can be illustrated by some numerical data: the number of species irreversibly lost every year has climbed to a fantastic figure of 27,000. If extinctions proceed at the present rate, 20% of today's species could disappear in the next 30 years. "*In its scale, this can be compared only with the natural disaster that occurred 65 million years ago, when dinosaurs disappeared from the face of the Earth*" (Radovic and Mandic, 1997).

Plants, Animals, and Health

In certain biogeochemical provinces, all floral and faunal components of the environment are characterized by specific features of chemical element composition. Inseparably linked with this is the concept of biogeochemical microelement endernias, which can occur when soil, water, air, and food of plant and animal origin exhibit an excess or deficiency of microelements

(I, Mn, Co, F) or more complicated forms of imbalance, especially anomalous ratios of their concentrations or an imbalance of micro - and macroelements (P and Mn for instance). For this reason, mechanisms governing the development of certain biogeochemical endemias can be understood only through multiple study of the interconnection and interdependence of links (*rocks— soil—water—plants—animals—man*) in the complex chain of biogeochemical systems (A. A. Keller, O. P. Shchepin, and A. V. Chaklin, 1993). Here both higher and lower plants can be biological indicators of potential dangers to animals (including man) inherent in the biogeochemical situation of a region under study.

The influence of biological factors on human health can be negative or positive. Among these factors of the biosphere, especially dangerous from the standpoint of health are disease agents, animals that are reservoirs or carriers of diseases from their natural foci, and poisonous animals and plants. These things can be the causes of a number of dangerous diseases of humans and animals. Diseases of man caused by contact with poisonous animals or products of their vital activity represent a little studied chapter of geographic pathology. In the work "**Introduction to Geographic Pathology**," A. P. Avtsin (1972) proposes a working classification of local and general pathological processes caused by some form of contact between man and certain poisonous and provisionally pathogenic (nonpoisonous) animals. He also cites Voronov's classification of infections and invasions (A. G. Voronov, 1965).

Many representatives of the **plant kingdom** can possess toxic properties. Such properties are predominantly characteristic of certain parts of plants - fruits, leaves, bark, roots, juice, seeds, etc. For example, it is little known that the very widespread decorative plant black locust possesses poisonous roots and bark. Another very widespread decorative plant, oleander is also poisonous. Opium poppy contains many toxic (and medicinal!) substances. Cases of poisoning by several species of mushrooms are well - known... It is understandable that substances isolated from poisonous plants are often used for purposes of pharmacology and toxicology.

According to the estimates of investigators, there are in nature about a thousand species of plants that under the influence of mechanical processing or specific enzymes are capable of releasing *hydrogen cyanide*, a very poisonous gas which paralyzes breathing. These are the so - called *cyanogenic plants*, and the toxic substances they contain are cyanogenic glycosides. For example, the leaves of wild cherry in the United States can produce 200 mg of hydrogen cyanide per 100 g of mass, and it is estimated that only about 125 mg of such leaves is enough to kill an animal weighing 50 kg (about 150 mg is needed to kill a man of average weight). Sudden death of domestic animals for no apparent reason was most often caused by hydrogen cyanide released when the animals consumed greater amounts of green mass.

An enormous number of poisonous plants (blind - your - eyes, candelabrum tree, poison ivy, etc.) have been described in the tropics. Contact with these plants can cause fatal poisoning, itching, skin rash, facial swelling, elevated body temperature, headache, etc. On the other hand, many species of useful medicinal plants have been registered in the rich flora of tropical and subtropical regions [70]. For example, the sandalwood tree is used to treat gonorrhea in India, Indonesia, and many other countries; hydnocarpus oil and preparations made from it are used in India, Burma, Thailand, and other countries to treat leprosy, psoriasis, and other skin diseases; croton leaves are used in medicine to treat snakebites... More than 300 species of tropical plants can be used as medicinal raw materials.

Russian scientists have shown that all plants release into the atmosphere especially volatile organic substances called *phytoncides*. Phytoncides ert physiological action on the organism. Many of them possess the ability to destroy microorganisms, including ones that are pathogenic. Certain phytoncides kill microbic agents of disease in animals. It is believed that our organism requires phytoncides, and that their deficiency causes certain lung diseases. To be specific, the existence of phytoncides explains why the air of forests, fields, and pastures has a favorable effect on weak and diseased lungs[71].

The highly diversified **animal kingdom** erts great influence, both positive and negative, on human health. Thus, for example, there are few birds that are not useful as sanitation agents, since they rid fields of various pests such as insects and rodents. On the other hand, many animal species of mountain regions ensure the existence or facilitate the spread of different parasitic and transmissible diseases such as plague, malaria, fevers, etc.

The negative role of animals is especially pronounced in tropical regions. Many representatives of the animal kingdom serve as reservoirs of the agents of infectious diseases: monkeys in the case of yellow fever; rodents and others in the case of plague; mosquitoes of the genera *Anopheles*, *Aedes*, and *Culex* in the case of malaria; etc. Among arachnids, whip scorpions are extremely dangerous; the Malay centipede is the most

---

[70] The plant life of mountain massifs is rich in resources of vitamin-containing and medicinal herbs. We mention dog rose, hawthorn, and walnut. The medicinal properties of plants are more pronounced at higher elevations.

[71] "Oxygen factories" are lost with the disappearance of forests. Especially great attention is being devoted to forested zones around urban settlements because suburban forests absorb the carbon dioxide exhaled by man and produced by industry and traffic in great amounts.

dangerous of centipedes; and the most dangerous of poisonous snakes are the black mamba, Egyptian cobra, Gabon viper, etc. Poisonous animals are also very widespread in tropical waters: false coral in the Caribbean Sea and the Pacific and Indian Oceans; the sea wasp and hiropsalmus among medusas; etc. The bite of *Hapalochlaena maculosus* (the most dangerous Australian octopus) can kill a man. The group of poisonous fish numbers about 350 species, most of which live in the Pacific and Indian Oceans...

Food and Health

According to I. Coulier (1986), five of the 12 main causes of death in the United States - atherosclerosis, hypertension, infarct, diabetes, and cirrhosis of the liver - are directly or indirectly linked with nutrition. Thus, about 40% of all of the main causes of death have a link with food. Dietary delusions of all kinds, consumption of risky food, and experimentation with their own bodies have been with people from ancient times to the present day. Whether by reason of ignorance or increasingly due a shortage of food that is safe from the standpoint of health, a smaller part of the world's population feeds itself correctly.

The starvation of hundreds of millions of people on the Planet today can also be attributed to the unjust distribution of food. In addition to this, the race to increase profits has led to growing use of chemical agents, with the result that food is of constantly declining quality and often dangerous to health. Degraded natural conditions for food production are characteristic of agriculturally developed regions such as Western Europe and North America.

The scientific approach to nutrition began in the 18th Century, when Lavoisier clarified the nature of respiration on the basis of heat generation in the living organism. According to Lavoisier, respiration is an oxidative process, and heat liberated in the organism is also a consequence of oxidative processes. After Lavoisier, van Helmout scientifically formulated the theory of *digestion*, i.e., decomposition of food in the digestive organs. Soon to follow were scientific studies on enzymes in the digestive organs and cells, which laid the foundations of modern biochemistry. Improvement of methods in chemistry, physics, physical chemistry, physiology, microbiology, and many other sciences has made a great contribution to further development of the science of nutrition.

Nutrition has three basic functions in the organism. The *first function* is to supply the organism with energy, the *second* is to provide it with plastic or building substances (proteins above all, carbohydrates to a lesser extent), and the *third* is to supply it with biologically active substances needed for regulation of metabolic or vital activities. Thus, the science of nutrition is concerned with processes involved in the metabolism of different kinds of nutritive substances (energetic, structural, and biocatalytic) and their interaction in the maintenance of life.

Maintenance of biological functions of the organism requires constant replacement of consumed substances by means of renewed intake of food and water. The organism needs water, proteins, carbohydrates, lipids (fats), vitamins, minerals, and microelements. Of these substances, proteins, lipids, carbohydrates, and vitamins belong to the category of organic nutrients, while water, minerals, and macro - and microelements are inorganic nutrients. Starvation or malnutrition sets in if the organism does not receive the indicated substances or if it receives them in inadequate amounts. Overnutrition or obesity occurs if the organism takes in more than it needs. It has been calculated that mortality in humans increases by as much as 20% with a 10% increase of body weight and by even more than 40% with a weight increase of 15 - 20%.

Proteins are basic complex organic nutrients that have energetic, structural, and regulatory - catalytic roles in the organism of man and all animals. They constitute the structural and functional basis of every living organism, and the survival of life depends on their activity. In other words, proteins are the most widespread complex organic molecules in the cell. Linked with their activity in the living organism are all manifestations of life: growth, development, active regulation and functioning of the organism, sensitivity, irritability, contraction, adaptability, transmission of inherited traits, etc.

Proteins in foods of plant and animal origin are very similar in elementary chemical composition. However, foods differ significantly in the amount of protein they contain: those of plant origin contain 15 to 18% raw protein, whereas ones of animal origin contain 57 to 84%.

Carbohydrates are basic complex nutrients that have a very important role as a donor of energy in the nutrition of man and nearly all animals. They are divided into sugars (monosaccharides and oligosaccharides) and polysaccharides (starch and cellulose). It is a fortunate circumstance that the organism can convert carbohydrates to fats and partially substitute the former for the latter in food.

Lipids (fats) are basic complex organic nutrients that have energetic, structural, and protective roles in the living organism. About two thirds of the total energy needed by the cell (or the living organism) is satisfied by energy arising from the oxidation of fats. It is calculated that the daily meals of an adult individual should contain at least 70 g of fats. During the summer, this amount can be somewhat reduced, but never below a limit of 40 g; of this amount, 25 g should be of animal origin.

Vitamins are organic compounds of varying chemical structure and molecular weight. They are basic nutrients among biologically active substances. These biocatalysts are essential in small amounts for the health of man and animals. They are needed for health maintenance, growth, development, sustenance of life, etc. A deficiency or surplus of vitamins in food can cause specific and various disturbances in metabolic processes (*hypo* - and *hypervitaminosis*, respectively). About 30 vitamins are known, and nearly all of them have been isolated in crystalline form.

Minerals and *microelements* are an inseparable part of our food. Sodium chloride or kitchen salt is the only mineral that is used in natural form. Other minerals form part of other food components. Minerals are lost every day during cooking of food: 50 - 95% when cooking is done in water, 25 - 59% in oil, 20 - 60% in fat, and 19 - 54% in butter. Loss of minerals also occurs in the human organism through urination and perspiration, so it is necessary to replenish them. Thus, for example, a man on average loses the following amounts of minerals every day: 7 g of sodium, 4.75 g of potassium, 1.25 g of calcium, 1.63 g of phosphorus, 0.5 g of magnesium, 0.05 g of iron, 0.05 g of sulfur, etc. It is believed that of the total number of fatal diseases, microelements are involved in various ways in 70% of the cases. Copper, iron, zinc, manganese, molybdenum, cobalt, iodine, fluorine, selenium, and some other elements are carriers of vital functions of the organism because they participate in the enzyme system, without which necessary biochemical processes could not take place. The role of macro - and microelements in the organism of man and animals was discussed in greater detail in the previous chapter.

According to D.B. Jellife (1967), dietary disturbances represent a pathological state caused by a relative or absolute deficiency or excess of one or several necessary food components. The author divides the given dietary disturbances into four groups: a) malnutrition; b) specific forms of alimentary deficiencies; c) obesity; and d) imbalance.

Food risks are understood to mean the probability that consumption of some food will result in harmful action and cause damage to the organism. Unfortunately, they are very much present today in the world, both in developing countries and in developed ones.

According to a list published in 1973 by the Federal Drug Administration (FDA) of the National Academies of Sciences of the USA, first place among food risks was occupied by naturally present microorganisms and pathogenic microorganisms. In second place was malnutrition, which was followed by soil pollutants, natural toxicants in food, pesticide residues, and (finally) additives (I. Coulier, 1986).

1. It is practically impossible to ensure absolute security from *toxigenic and pathogenic microorganisms* present in food. First on the list of toxigenic agents are bacteria of the genus *Salmonella*, which in food or water cause serious poisoning of humans: salmonollosis. During the period of 1972 - 1974, this agent in the United States alone caused a total of 60 thousand cases of food poisoning in humans, including 12 that were fatal. Serious risks are also posed by mycotoxins, which can contaminate entire harvests of corn, cotton, and other crops[72].

---

[72] Among substances of biological origin, *mycotoxins* are today considered to be the most significant. Such mycotoxins are formed as a result of metabolic processes of molds multiplying on products of plant and animal origin rich in proteins and carbohydrates. Toxic substances produced by molds aroused special interest in the world only relatively recently, with appearance of the first more serious disturbances caused by consumption

2. One eighth of the world's population today suffers from *malnutrition*. Data of the World Health Organization indicate that between 30 and 40 million people throughout the world die annually as a result of starvation or malnutrition. Scurvy, rickets, beriberi, goiter, pellagra, and other chronic deficiency diseases in great measure affect health of the Earth's population[73].
3. *Contaminants from the soil* that enter food through plants and drinking water include heavy metals (lead, mercury, and cadmium), arsenic and selenium, halogenic elements (iodine and chlorine), and other components;
4. *Pesticide residues* are strictly controlled, but there still exists a risk of toxic action from food contaminated with them;
5. The risk from food *additives* is slight, but nevertheless cannot be completely excluded.

Investigations of cholesterol and other components have shown the extremely great influence of food on human health. There are many who feel that man dies more often because of food than due to disease.

The group of *alimentary diseases* depends on a number of factors: natural, anthropogenic, socio - economic, and ethnic. Extensive information about the systematic malnutrition of large groups of inhabitants of Latin America was collected some time ago in the now classical book "**The Geography of Hunger**" (J. de Castro, 1954). Problems

---

of food contaminated by these agents. To be more precise, the disturbances called *mycotoxicoses* have been known for a long time, but not by that name. In feudal Europe, this disease resulted in thousands of deaths caused by introduction of rye and other grains infected by the mold *Claviceps purpurea* into the organism. Alimentary toxic aleukia was a second mycotoxicosis that attacked humans with tragic consequences. It occurred in Russia in regions where the population ate moldy grain after its overwintering in fields. This disease attracted the attention of experts only in the 60's of the last century, following the deaths of turkeys, ducks, pheasants, and partridges on poultry farms in England and significant losses of a similar kind in Kenya, Uganda, Thailand, and the Philippines. It was established that fungi in corn meal was the cause of death. The isolated toxin was given the name aflatotoxin, and the disease caused by its action was called *aflatotoxicosis* (D. Markovic et al., 1996).
Real danger from micotoxins is hidden in long - term consumption through contaminated food. That is why it is possible formation of carcinogenic, mutagenic and other changes within cells, but also liver malfunctions and other organs.

[73] The cause of starvation and malnutrition lies above all in agro-business, i.e., concentration of political and economic power over agricultural production and trade. In addition to this, the petrochemical industry employs all means to prolong the mass use of chemical agents in agriculture, and a newly created branch of economic activity - production of hybrid seeds - has become a big part of agro-business. The first test of growing from artificial seeds was performed in 1987, and a decade later (at the height of the *transgenetic revolution*) genetically altered agricultural crops were being cultivated on 15 million hectars throughout the world. *It must be asked how all the altered genetic composition of plants and animals will affect the global food chain and above all the health of man and his gene material.*

of protein deficiency and forced uniformity of nutrition affect many regions of developing countries in other parts of the world as well. The very pronounced protein deficiency in the tropics has resulted in extremely high mortality rates among children (30 - 40% in Egypt and nearly 100% in the Belgian Congo). At the same time, a number of diseases are linked with overweight (obesity) of patients.

In the field of nutrition, there have been no scientific or other polemics as numerous as those pertaining to *cholesterol*, above all as to the pathological level of this fatty ingredient of human cells and cell membranes that is introduced into the organism with animal tissues, eggs, cheese, seafood, etc. Worthy of note are the results of 10 - year studies at 12 American research centers, which can be summarized as follows:

- Heart diseases are closely linked with the level of cholesterol in the blood serum;
- Lowering the percentage of cholesterol in the blood significantly reduces the risk of heart attacks (infarcts).

In 1983 - the year in which the results were published - a million Americans suffered a heart attack, and half of them did not survive it. We note that Americans consume as much as 60% more cholesterol in food than is recommended by the American Heart Association and three times more than the Japanese consume (Fig. 3.4.). As the nation that eats the fattiest food, the Finns have the highest cholesterol levels in the blood and the greatest number of heart diseases in the world.

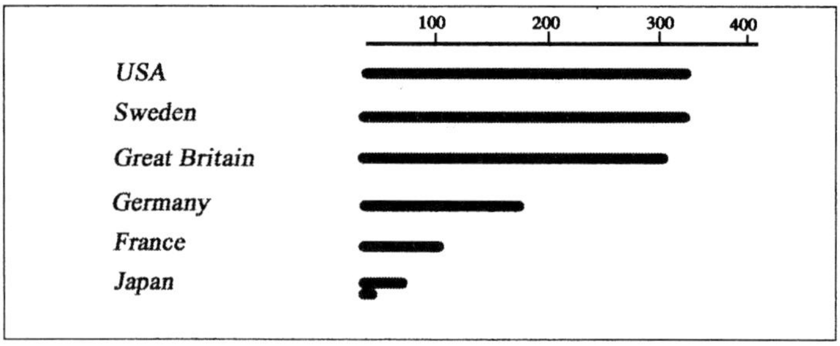

Fig. 3.4. Mortality caused by heart attack per 100,000 persons.

Discovery of the causal link between clogging of the arteries (atherosclerosis) and cholesterol in food was anticipated by the Russian pathologist N. N. Anichkov, who already in 1913 succeeded in artificially creating a fatty deposit on walls of the arteries of experimental mice. A direct correlation between the frequency of atherosclerosis and the nature of nutrition of the population was demonstrated later, although many doctors believe that cholesterol is only one, possibly the most important, of numerous causes of cardiovascular diseases.

In addition to cholesterol as a component of food, the long chain of causes also includes obesity, hypertension, smoking, stress, and inactivity. The American pathologist R. Minick explained the situation as follows: "*Like members of some orchestra, they all play a part in the occurrence of heart diseases.*"

Many indicators support the assertion that modern eating habits, especially in the most highly developed countries, have led to a real explosion of *obesity* and *diabetes*, the diseases of the 20th Century. It is estimated that 15 - 40% of the adult population of developed countries suffers from obesity, as a consequence above all of increased caloric intake (Fig. 3.5.). During the 90's of the past century, obesity took on the character of an epidemic among Americans, and in 1994 71% of the population belonged to the category of *overweight*. Diabetes today ranks third as a cause of death in the United States, and more than 90% of the 5 million American diabetics are overweight (I. Coulier, 1986).

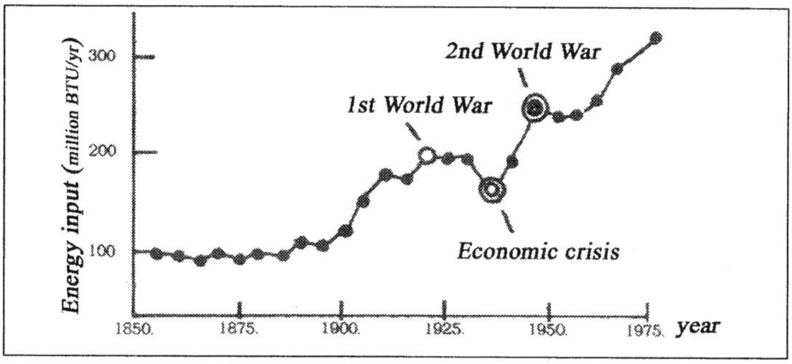

Fig. 3.5. Energy consumption by humans in USA from 1850 to 1975.

High blood pressure (hypertension) affects more than 20% of the world's population. Although the true cause is not known in most cases, there is solid scientific evidence indicating a key role for sodium ions in regulation of blood pressure, and limited salt intake is prescribed for hypertensives.

Microelement deficiency in food is manifested in a whole spectrum of damage to the organism. Contributing to this are various modern technological processes and constantly increasing use of refined food, which deprive man of all traces of minerals.

During the past decades, man has found himself confronted with yet another problem where nutrition is concerned, namely the ever increasing *content of radionuclides* in the food chain. The danger is magnified by the fact that the organisms of some animals accumulate radionuclides in their bodies, whose consumption in the form of meat and other products leads to a dramatic increase in the effects of exposure of the human organism to radioactive radiation.

Study of the relationship between *nutrition* and the incidence of *malignant diseases* has confirmed that some components of food increase the risk of sickness, while others reduce it. Thus, toxic and carcinogenic action has been established for luteoscirin and cyclochlorotin (products of different species of rice molds). The risk from aflatoxin B - 1 is extremely great, since it is the strongest known carcinogen in food. Benziperen causes cancer of the esophagus and intestine when it is ingested in food and lung cancer when inhaled in smoke. Elevated content of selenium in wheat or corn can cause cirrhosis of the liver and tumor in humans. Cancer of the large intestine is linked with excessive consumption of meat. Esophageal cancer can be caused by vitamin deficiency in food and its uniformity (predominance of fats and carbohydrates). On the other hand, vitamin A protects the organism from different types of cancer... It must be stressed in conclusion that while food is only one of the causes of cancer, it is unquestionably very important[74].

---

[74] Keller, Shchepin, and Chaklin (1993) assert that dietary and nutritional irregularities make up about 35% of the total number of factors responsible for occurrence of malignant diseases. In the book of I. Coulier (1986), estimates of a group of experts indicate that 40% of cancers in men and 60% in women are linked with food. More comprehensive statistical processing of available data with detection of pertinent correlations will make it possible to establish the share of food-related causes in the incidence of various diseases.

CHAPTER 4

Anthropogenic Factors

In 1953 some rural inhabitants of the island of Kyushu in Japan started to exhibit unusual symptoms: deafness and blindness were observed in many cases, and every fourth sufferer died. All those who died had eaten fish contaminated with mercury. It was discovered that a year earlier a factory began using mercury compounds in a technological process whose waste waters were disposed of in a nearby bay of the sea. The disease did not reoccur after the factory stopped getting rid of toxic waste in this way. Unfortunately, similar and more complex cases of anthropogenic action on the environment and human health have been recorded throughout nearly the whole of our planet.

The studies of many authors indicate that the causes of occurrence and wide dissemination of diseases such as cardiovascular pathology, malignant neoplasms, trauma, genetic anomalies, etc., are often linked with pollution and unfavorable environmental conditions, urbanization, and negative consequences of scientific - technological progress, which are characteristics of our reality. Polluting substances like pesticides, heavy metals, petroleum and its derivatives, and others set off a whole series of chain reactions in the environment, and their presence in it therefore cannot go unremarked.

Unfortunately, examples of severe foreseeable tragic effects of destructive action on the environment are becoming increasingly more numerous. To put it more precisely, the influence of man on the environment has acquired a planetary dimension, and the negative effects have become alarming. Here it is useful to cite the illustrative approach of I. Radovic and R. Mandic (1998): "*If we reduce the period of the Earth's existence to the length of human life, then every 100 million years of life on the planet would be equal to one year, and the planet on that time scale would now be 46 years old. Bacteria (or the first prokaryotic forms of life) would in that case be 35 years old, while the first unicellular eukaryotic forms of life would be about 15 years old. On that scale, dinosaurs disappeared before approximately eight months, and man appeared only seven days ago. The more intensive process of destruction and pollution of the environment started just three seconds ago. What are a mere three seconds in relation to the whole 46 years of life? Perhaps it is enough time for man to destroy both himself and his own civilization, but not to destroy nature itself taken as a whole*" (Radovic and Mandic, 1998).

The conditions of life on Earth have been altered by negative environmental effects caused by increase in the concentration of carbon dioxide and global warming of the climate; destruction of the ozone envelope; the advent of acid rain, a source of great damage in many regions of the world; wasteful exploitation of natural resources; the action of heavy metals, radionuclides, and pesticides in water, soil, plants, and animals; cutting down of forests; expansion of erosive and desertification processes; reduction of biodiversity as manifested in the daily disappearance of some 80 biological species; etc. The organism of man and other living beings on the Planet has for the most part exhausted its capacity for adaptation and adjusts with increasing difficulty to rapid changes in the environment. Influence has been especially strong on the physiogical integrity and genetic structure of the living organism. Analyzing mutagens and breakdowns of genetic information, N. P. Dubinin reached the following conclusion already in 1975: "*On the whole, it can be stated that whereas much has been said up to now and is still being said about the threatening* ***ecological catastrophe****, today we face the problem of preventing a* ***genetic catastrophe***." According to Keller, Shchepin, and Chaklin (1993), three interrelated tasks are without doubt very pressing today: 1) *protection of the natural environment from further pollution*; 2) *protection of human health from adverse effects of environmental contamination*; and 3) *renewal and healing of the damaged environment.*

The literature on the many anthropogenic factors and causes of environmental pollution and degradation is very voluminous. Our treatment of these factors (in contrast to natural ones) will understandably be compressed. This does not mean that greater attention need not be paid to these problems (especially the response of the geological environment to anthropogenic factors) in future geomedical research. *In the newly arisen conditions, geological factors unquestionably play an important role through their defense mechanisms and ability to dictate both the process of pollution and effective measures for sanation of the environment.*

PROCESSES OF POLLUTION AND DESTRUCTION OF THE GEOLOGICAL ENVIRONMENT

Man as a geological force appeared relatively recently. However, the very short geological time of his action has not prevented man from destroying and altering many things in both the biosphere and the lithosphere. Among other things, he has extracted and processed the ores of ferrous and rare metals and dispensed enormous masses of potassium, phosphorus, and other compounds on fields throughout the globe. Anthropogenic redistribution of substances of the lithosphere, hydrosphere, and atmosphere has had increasingly negative consequences.

We note that life for billions of years avoided a whole series of elements with toxic and carcinogenic properties, whereas those elements today have already polluted a significant part of the geological environment and accumulated in agricultural soil, groundwater, and plants in amounts tens, hundreds, and thousands of times higher than maximal permissible concentrations. Moreover, unique anthropogenic (secondary) deposits of chemical elements have been formed in the vicinity of great industrial concerns, centers of mining, and thermal electrical power plants, and a circumstance to which we cannot remain indifferent.

Viewed geochemically, the consequences of human activity on the geological environment (surroundings) are both visible and manifold. According to M. Babovic (1992), changes associated with *destruction, transport, accumulation, and lithification can be distinguished.* Man takes part in *destruction* of the geological environment by digging, building, releasing various fluids, carrying out conscious interventions (tilling of agricultural land), etc. Through *transport* of rock masses, he has altered the natural relief and promoted processes of downfall (landslides and rockfalls). Changes of the geological environment caused by artificial *accumulation* take on significant dimensions in regions with hydrotechnological projects, where newly deposited sediments have been formed[75]. In a work published in 1978, F. V. Kotlov speaks of *anthropogenic lithogenesis* and singles out seven genetic comples of newly created formations (landfills, sedimentary deposits, embankments, etc.).

The given anthropogenic changes of the geological environment occur in the context of certain economic activities: agriculture, forestry, industry, construction, and mining.

From the standpoint of environment protection, geochemical changes of agricultural soil are the most important of all anthropogenic changes.

---

[75] In the large Iron Gate (Djerdap) Reservoir, for example, the rate of sediment deposition in the past two decades on average comprised about 1 m annually. The sediment has very high concentrations of toxic elements (M. Perisic, 1997).

Such changes will be considered later on in the text. In addition to this, significant changes of the geological environment occur with the destruction of forests and other vegetation. It is estimated that the area of forested land on our planet has shrunk by two thirds just in the course of human history. Among other things, this has resulted in unchecked development of erosive processes, destruction of the soil cover, and expansion of deserts. Soil erosion has led to complete loss of fertility of more than half of all arable land in the world. Just between 1960 and 1970, about 6 million km^2 of fertile soil was transformed into desert in the region of the South Sahara[76]. This resulted in a million deaths and creation of approximately as many invalids due to inadequate nutrition.

Every day anthropogenic activity represents an increasingly important factor in the inception and development of landslides and other exogenous geological processes of all kinds. Human activity not only alters the appearance of the Earth's surface, but also causes significant changes in the upper part of the Earth's crust that in scale and consequences can be compared with geological processes. Significant activation of landslides and rockfalls has been known to occur during construction and exploitation of large reservoirs, road building, and other kinds of construction work.

Forms of destruction similar to those caused by construction projects are present in strip mining (enormous excavations on deposits of coal and copper, underground chambers, etc.). More will be said about this in the text to follow. We mention the exhaustibility of geological wealth as one of the most important problems facing mankind. The collapse of geological energy potentials could cause a great break in trends of human development. Pollution of drinking water and degradation of agricultural land are things that could be truly fatal for the human race.

The devaluation of geological wealth caused by urbanization probably occurs much more often than we know, since ore concealed by settlements is later hard to find. In conclusion, we mention dumping grounds of communal and industrial waste, which already result in considerable degradation of the geological environment.

---

[76] The scale of destruction can be illustrated by the phenomenon of "yellow" rain (snow) in regions of Northern Europe very far away. In just one episode of "yellow" snow, about 50,000 t of dust raised by sandstorms in the Sahara fell on the Scandinavian Peninsula. During the period 1989-1990, occurrences of elevated heavy metal concentrations were observed at Norwegian monitoring stations in episodes involving arrival of air masses from North Africa.

ARTIFICIAL PHYSICAL FIELDS - RISKS AND POSITIVE INFLUENCE

Artificial (Anthropogenic) Electromagnetic Fields

Of extraterrestrial origin, natural electromagnetic fields have different frequencies, behavior, and intensity, and all living beings are completely adapted to them. By way of contrast, artificial fields have greater coherence, a higher level of energy, and more stable frequency and duration. For this reason, they ert stronger influence on the biosphere. On the basis of frequency, electromagnetic fields are divided into electromagnetic fields (if their frequency is greater than commercial frequency - 50 Hz in Europe, 60 Hz in the United States) or electric and magnetic fields (if their frequency is less than the indicated values).

The influence of radio and TV transmitters is most frequently examined. Children are very sensitive to increased electromagnetic (EM) radiation and become irritable, cry, and feel the need to move from the place where they are. It has been established by measuring that horses, dogs, and humans react negatively to places with increased radiation, while cats, bees, and wasps seek out just such places. The sensitivity of organisms to the action of strong fields is attributable to the presence of elementary magnetite (Fe_3O_4) in them. Every living cell has its own EM frequencies and the human brain virates at the same frequency as the Earth. That is why all living beings respond differently to EM radiation. The result of harmful action is manifested in blood vessel shrinkage and greater tissue acidity, which facilitates penetration of heavy metals and microbes. Heavy metals attract EM waves in such a way that changes of cell gene structure occur after a period of several years, which leads to the development of chronic and malignant diseases (J. Kunosic, 2001). Although the organism struggles to re - establish the original state of equilibrium in its cells, defense mechanisms weaken in time and the person sickens.

The consequences include increased mortality from cancer (J. Matousek, 1987) and diseases of the nervous, reproductive, and cardiovascular systems with symptoms such as tachycardia, arrhythmia, reduction of arterial tension accompanied by decrease in voltage of the QRS complex on the ECG, slowing of circulation in veins of the internal organs, decrease of cholinesterase content in the serum, increase of histamine content, and changes of blood composition. These symptoms lead subsequently to headache, sweating, exhaustion, insomnia, dizziness, impaired memory, libido loss, muscle pain, asthma, increased frequency of chromosome aberrations (transmission of gene mutations to the progeny), etc. Also, personnel engaged in maintaining telecommunications and radar systems are prone to the "*microwave syndrome*" (with headaches, nervousness, and back pain reported by 60% of those tested; and conjunctivitis and fatigue by 40%) (D. Backovic, 1998). It should be mentioned here that artificial fields are used in medicine for therapeutic and diagnostic purposes (for electro - stimulation of the brain, heart, or muscles, as well as for destruction of tumors in oncology).

Artificial electric fields are associated with:

- *Electrostatic fields.* Like natural fields, they ert negative (unfavorable), positive (favorable), or neutral influence on the organism. The ideal ratio in living space is 60% negative ions (mainly oxygen ions) and 40% positive ions. Synthetic fibers in the textile industry lead to the disappearance of strong electric fields, which is reflected in worker depression and fatigue, as well as in changes of blood pressure. Artificial materials used in the building industry (synthetic paints and lacquers, artificial flooring, wallpapers, etc.) also affect ion content. These materials absorb negative ions, become charged, and create artificial fields with predominantly positive ions. This results in total weakening of the organism's immunity (R. Terzic, 1998); and
- *Long - distance power lines.* The phenomenon of so - called "*electromagnetic smog*" occurs here. The influence of such smog must not be underestimated and is manifested in various symptoms (disturbances of the central nervous system, insomnia, exhaustion, changes of pulse and blood pressure, etc.). Whereas fields of more than 1000 V/m are formed in the immediate proximity of electrical generators, motors, transformers, and distributive networks, the fields of household installations are in the range of 10 - 100 V/m (V. Stojanovic, 1998).

Artificial magnetic fields are created by industrial activity. In contrast to an electric field, a magnetic field passes easily through house walls, almost all metals, and tissues of the human body. It is a task for geophysicists to establish the existence and characteristics of such fields, on the basis of which they can ascertain any correlation between mortality of the population and artificial fields. According to the standards of Eastern European countries, it is permissible for the human body and head to be exposed to a magnetic field with a strength of 0.03 T and a gradient of 0.05 - 0.2 T/m, while the breasts and hands can be exposed to a field with a strength of 0.07 T and a gradient of 0.1 - 0.2 T/m. Fields in the range of 0.2 - 0.4 mT arise during use of household appliances.

It has also been established that an artificial field of terrestrial origin with a frequency of f = 0.01 - 5 Hz causes increase in frequency of the pulse and changes in the psycho - physiological state of humans (exhaustion above all), while a magnetic field with a frequency of f = 0.2 Hz and variable strength (5×10^5 to 1×10^6 nT) prolongs the reaction time of drivers. It should be noted that in the case of mobile telephones, the depth to which radiation penetrates the human body comprises 10 - 100 mm, whereas penetration is only 1 mm in the case of radar installations (whose frequency is as high as 10 GHz).

It is interesting to note here that researchers during the past 10 years have been intensively conducting experiments in which artificially created vertical geomagnetic field components are used to promote faster germination of the seeds of various cereals and improve their yields (E. Comino et al., 1995).

According to N. Trifunovic (1998), malignant diseases, myocardial infarct, and cerebral accidents (strokes) are caused in part by anomalous increase of the magnetic - electromagnetic (M - EM) field (the sum of the Earth's natural magnetic field and the artificial field consisting of induced and remnant magnetizations) in living and working space. The greater frequency of cancer in industrially developed countries and urban environments is attributed to the influence of electrification and automation in enclosed space (where man spends most of his time), which causes a significant increase in normal values of the M - EM field. Anomalous increases of these fields cause accumulation in tumor tissue of electrophiles (known in medicine to be risk factors, various substances that exist in neoplasms are electrophiles with increased ionic conductivity). In addition to inducing their formation, magnetic fields of anomalously high intensity thereby represent an environment favorable for the development of neoplasms.

Artificial ionizing radiation causes erythema of the skin and dermatitis, atrophy of skin glands, hyperkeratosis, and tumors (Dj. Sofrenovic, 1993).

Artificial non - ionizing radiation arises from different technological sources (B. Vulevic, 2000).

Ultraviolet (UV) radiation is emitted by lamps with inert gases and hydrogen lamps, halogen and fluorescent lamps, and lasers. Most exposed to it are welders, physiotherapists, cosmeticians (including ones who work with quartz lamps), printers, laboratory and medical personnel, etc. The consequences of excessive exposure to this kind of radiation have already been presented in the section of the book dealing with natural UV radiation.

Infrared (IR) radiation is emitted by hot and red - hot sources (industrial furnaces, apparatuses for gas welding, heaters, burner rooms), sources with electric discharge (arc lamps and arc - welding equipment), and lasers. Most exposed to IR radiation are blast furnace workers, workers involved in glass production, paper and cellulose factory workers, and workers employed in the textile, wood, and chemical industries. Physiotherapists are also at considerable risk. Harmful consequences of IR action were described in the chapter on natural IR radiation.

Radiofrequent (RF) radiation comes from modern - day apparatuses used in industry, medicine, research, etc. (TV, radios, radar devices, radio - navigation equipment, radio - telemetric devices, etc.). For this reason, most threatened by the given kind of radiation are persons who work around industrial RF heaters, industrial microwave (MW) furnaces, and installations for microwave and shortwave (SW) diathermia in medicine. Harmful action is registered as effects of overheating, with increase of body temperaturc, hyperthermia, and burns. The most sensitive are organs with weak circulation and weaker local thermoregulatory mechanisms (the lens of the eye, testes, etc.). Apart from thermal effects, there are also nonthermal ones: action on CNS, changes of the EEG, changes in synthesis and transduction of DNA, and altered flow of Ca ions from the brain cells.

Electromagnetic fields of extremely low frequencies (ELF) are created around aboveground and underground lines of high - tension (more than 35 kV), high - voltage apparatuses and installations, industrial machines and installations that operate on electricity, and electrical devices used in medicine, households, scientific research, etc. Workers who repair and maintain long - distance power lines are exposed to such fields, as are persons who work in or around substations, induction furnaces for smelting of metals, large electromotors, and transformers. Also at risk are installers and electricians who work on medium - and low - voltage circuits, radio and telecommunications installers, electrical engineers, etc.

The population is exposed to electric and magnetic fields of circuit frequencies that arise in the vicinity of transformers or around the electrical circuiting of houses, as well from household electrical appliances and light sources. Particularly threatened are persons who live near substations, long - distance power lines, and generators. The consequences of harmful action of ELF radiation include carcinogenesis, altered reproductive capacity (and disturbances in pregnancy), and neuro - behavioral disorders (electrical hypersensitivity). According to some authors, brain tumors (neuroblasts), leukemia, and breast cancer can occur during exposure to this kind of radiation.

Artificial Radioactive Fields

In America during the 1960's, elevated concentrations of radon were established in apartments built of materials made from the waste of uranium mines, as well as in buildings near dumps of highly radioactive waste. Investigations were initiated on the concentration of radon in enclosed spaces and its harmful action on human health. The pertinent data are presented in Table 4.1. (I. Petrovic, 1998). It was established that strength of the absorbed dose of gamma radiation is on average twice as great indoors as it is outdoors. Also, radiation is more intensive in zones with a colder climate. It is weakened by rain, snow, and increased atmospheric pressure. The concentration of radon is twice as great at night than by day, the minimum occurring at noon.

TABLE 4.1.
Radon concentration in rooms and their harmful effect on health.

Organ of human body	Professional workers		Citizens
	Quaternary	Annually	Annually
Whole body exposed to radiation, sexual glands, marrow	30	50	5
Skin, bones, thyroid gland	150	300	30
Arm, legs, knees	400	750	75
Other organs, tissues	80	150	15

The main sources of radon in enclosed spaces are the soil under and around the building, the materials that went into its construction, and the water used in it. Table 4.2. presents values of specific activity of Ra_{226}, Th_{232}, and K_{40} in different kinds of specially prepared samples of standard building materials.

External radiation of the population and penetration of the food chain occur as a result of contamination of the soil with radionuclides. Monitoring of radiation of both natural and artificial origin is resorted to for the purpose of implementing adequate protective measures.

Unfortunately, construction measures designed to reduce the energy needed to maintain buildings contribute to intensive concentration of radon. For example, measurements conducted in Sweden showed that the concentration of radon inside apartment buildings was as much as 5000 times greater than the level in the outside air. Resulting from energy - saving measures including overuse of thermoinsulation, this is precisely the reason why 14,000 new cases of cancer (primarily lung cancer due to accumulation of radon isotopes) and leukemia are recorded every year in Sweden.

Many human activities tend to strengthen radiation. They include mining (in the case of radon); production of energy from fossil fuels (accumulation and dispersion of radioactive elements with ash); setting off nuclear explosions; operation of nuclear power plants; application of mineral fertilizers (K and U in phosphates); use of radioactive isotopes in industry and medicine; etc.

Artificial Seismic Fields

Artificial seismic fields are formed as a result of chemical and nuclear explosions, mining activity, and seismic research. Their manifestations are the same as in the case of weaker earthquakes, and they have already been studied for some time by investigators in the field of engineering geophysics. Artificial earthquakes are known to have been induced by construction of hydroelectric structures (for example, the Kariba Dam in Central Africa), major chemical accidents, nuclear explosions, mining activity, or seismic research.

Instead of a **conclusion**, it remains only to be noted that the field of *ecological geophysics* constitutes the youngest branch of geophysics, one that began to be considered important only in recent years. The discipline known as sustainable civil engineering has developed rapidly in counterpoint to the kind civil engineering practiced up to now (which has been responsible for great environmental disturbances due to use of extensive land areas, excessive consumption of raw materials and energy, production of large quantities of waste materials, and degradation of the natural environment). For all these considerations, it can be concluded that research in the area of ecological geophysics is capable of making a significant contribution (S. Komatina, 1998). Naturally, this presumes a more elastic (and responsible) approach to problem solving and awareness of the fact that *experts of different profiles must be simultaneously involved in every task if we hope to achieve the goal of sustainable development.*

POLLUTION OF AGRICULTURAL SOIL

Due to the presence of humins and clays, soil can bind very high concentrations of harmful materials or elements. The distribution of elements is itself determined primarily by the ion - altering characteristics of soils, their adsorptive properties, and the presence of independent compounds of certain harmful components. Transport of harmful substances through soil is accomplished mainly by water and partly by air currents (winds) carrying with them dust or aerosols enriched with these substances.

Soil possesses the capacity for biological self - purification: decomposition of fallen detritus and its mineralization occur in the soil. In the final analysis, soil compensates at its own expense for lost mineral substances. If any component important for mineralization processes is lost due to overburdening of the soil, then the mechanisms of its self - purification are inevitably disturbed and complete degradation occurs as a result. Conversely, creation of optimal conditions for self - purification of the soil promotes preservation of ecological equilibrium and conditions for the existence of all living organisms, including man.

TABLE 4.2.
Minimal, average and maximal values of Ra_{226}, Th_{232} and K_{40} specific activity in samples of various building materials (G. Pantelic, I. Petrovic, 2000).

Type and number of samples		Minimal activity	Average activity	Maximal activity
		Ra_{226}		
Block	6	29.0	50.7	125
Cement	14	6.3	53.2	119
Brick	17	8.5	48.0	167
Gypsum	11	9.1	278	
Granite	83		64.4	213
Stone	26		44.2	236
Lime	11		8.3	17.2
Marble	26		44.7	299
Ash + Slag	6	62.2	194	323
Sand	10	6.8	12.8	25.7
Gravel	6	7.6	10.8	21.3
		Th_{232}		
Block	6	9.9	40.2	59.1
Cement	14	1.4	22.6	41.1
Brick	17	8.9	62.0	163
Gypsum	11		13.9	25.6
Granite	83		112	432
Stone	26		56.6	306
Lime	11		1.2	3.8
Marble	26		63.0	297
Ash + Slag	16		53.2	118
Sand	10	6.2	14.9	24.1
Gravel	6	8.6	13.0	18.6
		K_{40}		
Block	6	83	411	632
Cement	14	24.9	209	438
Brick	17	100	650	903
Gypsum	11		10.7	27.6
Granite	83	1.3	1058	2067
Stone	26	6.1	746	1616
Lime	11		11.4	28.8
Marble	26		868	1620
Ash + Slag	16	54.3	284	416
Sand	10	100	308	515
Gravel	6	100	259	421

On the other hand, high - intensity food production requires that plants be protected from a number of pests, diseases, and weeds. For this reason, utilization of agricultural soil is today almost unthinkable without the use of *pesticides* (insecticides, fungicides, herbicides, etc.), substances toxic even in small doses[77]. Herbicides have gained widespread application in the process of chemization of agriculture, and their manufacture has increased appreciably throughout the world. Man has even used them as a biological weapon. In one of their aggressions (the aggression against Vietnam), the Americans "succeeded" in completely destroying with the aid of herbicides the rich vegetation and animal life on vast areas of jungle in Vietnam, contaminated the country's soil with them, and threatened the lives of its people. The ecological consequences of this imprudent human action were disastrous, with long - term negative effects.

Contamination of vital food products by pesticides represents an ever - present problem. There have been described in the literature nearly 3 thousand cases of non - professional poisoning occurring after consumption of plants treated with organochlorine and organophosphorus preparations, as well as ones based on arsenic and mercury. Among mercury preparations, fungicides are widely used for seed protection. Their use caused poisoning on a large scale in Iraq at the end of 1971 and beginning of 1972, when 6530 persons were poisoned and 459 died. The clinical picture of this poisoning - which was caused by Mexican seed wheat treated with an organomercury fungicide - was characterized by the following symptoms: paresthesia, ataxia, reduction of coordination, loss of vision to the point of blindness, dysartria, and hearing loss. Similar mass poisoning occurred on a state farm in Ghana in May of 1967.

Tests have shown that among pesticides, such well - known preparations as DDT, tifios, and other organochlorine preparations have the most toxic and dangerous ecological properties. The insecticide properties of DDT were discovered before the Second World War, and enormous amounts (about 4.5 million tons) of this very toxic compound were used over the next two decades on fields, in forests, and even in living space throughout the world. Owing to the fact that it possesses exceptional stability and a remarkable capacity for migration, DDT has been recorded over vast areas of the biosphere: in the organisms of penguins in Antarctica, in seals of Lake Baikal, and even in mother's milk. A known characteristic of DDT is its tendency to accumulate in fatty tissue of the human organism, where it acts as a carcinogenic agent.

Fertilizers applied to agricultural soil contain the following four harmful ingredients: nitrogen compounds (nitrates, nitrites, ammonia, and organic nitrogen), sulfates, chlorides, and phosphates. In addition to the

[77] Experiments have also demonstrated carcinogenic effects in certain kinds of pesticides.

ingredients listed, heavy metals are also present in fertilizers. A constant increase of nitrate content in both soil and groundwater has been recorded in most agricultural regions over the course of the past quarter century. For example, this is the case with nearly all of Northwest and Western France. Nitrates are undesirable in food and drinking water because they cause decline in the level of hemoglobin in the blood, which is fatal for the progeny. Moreover, they promote formation of carcinogenic nitrosamines, which represent a danger to living organisms at all stages of development. Phosphates in soil bind various microelements, and uranium (one of the most undesirable of pollutants) can also be introduced into soils.

Although the level of *heavy metals* in most agricultural soils is not yet high enough to cause acute problems of toxicity, their heightened presence in certain regions has been increasingly often reflected in mass poisonings of the population. Heavy metals are introduced into agricultural soil in some of the following ways: 1) with chemical protective agents; 2) through application of fertilizers; 3) with waste water (industrial waste water, sewage, and water from livestock farms) used for irrigation purposes; and 4) with emissions from the air. As time passes, pollution of soil by chemical elements is becoming an increasingly pressing problem. This refers primarily to ecologically dangerous elements capable of accumulating in organisms with definite consequences to them, elements such as mercury, arsenic, cadmium, lead, chromium, nickel, zinc, antimony, fluorine, strontium, and some others. Due to their ability to accumulate in the organism, heavy metals can be toxic even in weak doses. For example, arsenic accumulating in the human organism can produce a carcinogenic effect on the skin. Mercury, on the other hand, causes disturbances of the gastro - intestinal tract, nervous disorders, and kidney damage. Heavy metals in soil are treated in greater detail in the book "**Heavy Metals in the Environment**" by D. Bogdanovic, M. Ubavic, and V. Hadzic (1997).

WATER POLLUTION

By pollution of water, we mean deterioration of its quality under the influence of chemical components, heat, or bacteria to an extent that it unfavorably affects (without necessarily creating a threat to human health) water use in everyday life, for agricultural and communal purposes, and in industry. Surface water and water in shallow horizons are most predisposed to pollution. Such water is directly exposed to the unfavorable influence of various anthropogenic factors, above all waste water and contaminated atmospheric water. The question of its protection is therefore especially pressing (M. Komatina, 1990).

The reasons for water pollution are to be found in the intensive urbanization and industrialization that have characterized the last two centuries of development of human society. The main pollutants are industrial waste water and communal sewage; pollutants linked with agriculture; ones arising in mining regions; and petroleum and petroleum products.

Among all groups of pollutants, a leading role can unquestionably be assigned to *industrial waste water*, primarily because of the great quantity of such water and high content of toxic components in it. Dregs of such metals as lead, mercury, cadmium, vanadium, nickel, zinc, cobalt, molybdenum, manganese, etc., on entering water even in insignificant quantities become toxic. Active ions of the compounds of these elements can cause serious damage to the health of persons who drink water containing them. In the United States, for example, more than 100 thousand sectors have been recorded where industrial waste water and communal sewage are disposed of on the Earth's surface, and there are 50 thousand registered dumps of toxic chemical waste, 75% of them being located in regions with important water resources. Industrial waste water is rich in various inorganic and organic compounds. Moreover, it can contain radioactive substances or be hot. Much of this water is highly toxic and highly mineralized. Experience indicates that in cases of larger basins of waste water disposal, the dispersion aureole of pollutants in groundwater can measure tens of square kilometers.

Marked by a number of polluting components (various microorganisms, nitrogen compounds, organic acids, chlorides, etc.), fecal matter represents the basic form of waste in urban centers. However, the presence of bacterial pollutants constitutes the main characteristic of *communal sewage*.

Water pollution in agricultural regions is caused by pesticides, fertilizers, drainage from livestock farms, waste from poultry processing plants, etc. (Fig. 4.1.). Especially dangerous is pollution caused by pesticides, followed by contamination with nitrates and heavy metals from fertilizers. Mineralized water represents a potential pollutant of groundwater on territories with soil improvement.

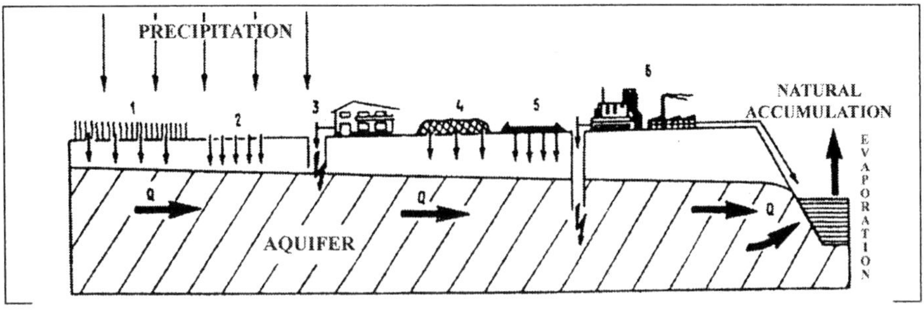

Fig. 4.1. Sources of groundwater contamination caused by agricultural activity. *Fertilizers and pesticides:* 1. during irrigation; 2. during fertilizing; *Waste waters*: 3. Municipal; 4. Industrial; *Disposals*: 5. Hard; 6. Liquid.

Exploitation of *large mineral deposits* often leads to deterioration of water quality on the wider area around mines and strip mines. Large amounts of water are generally evacuated during drainage of the deposit, and this disturbs the regime of groundwater and upsets the hydrochemical equilibrium.

Large foci of pollution by *petroleum and petroleum products* are linked with regions where oil wells and refineries are located. It is estimated that total losses of this liquid form of energy raw material constitute about 2% of world production, which represents a serious threat to underground and surface water. The problem is complicated by disasters occurring during transport of oil, such as happened, for example, in 1967, when about 100 thousand tons of oil leaked into the sea from the tanker Torrey Canyon, completely contaminating beaches on the coasts of England and Normandy and destroying fish life in the region. Not to mention the even greater damage to human health and the natural environment caused by the bombing of Serbian oil refineries and petrochemical installations by the United States and its Western European allies during the spring of 1999.

The listed pollutants cause the following *basic forms of water pollution*: chemical (inorganic and organic), hydrocarbon, biological (bacteriological), radioactive, and thermal. *Chemical pollution* of water is associated primarily with industrial waste water, escape of technological solutions, chemization of agriculture, etc. Such polluted water can have high content of heavy metals, aromatic hydrocarbons, phenols, detergents, cyanide, sulfates, chlorides, phosphates, etc. *Hydrocarbon pollution* is characterized by increased content of petroleum hydrocarbons of varying composition, which become very toxic as soon as their content exceeds the limits of perceptible smell and taste. *Biological (bacteriological) pollution* is caused by various organisms: viruses, bacteria, algae, etc. The most dangerous consequences (epidemics) are caused by pathogenic bacteria and viruses[78]. *Radioactive pollution* occurs when uranium, radium, strontium, cesium, tritium, or some other radioactive elements enter the atmosphere and reach the Earth's surface after nuclear tests or with certain kinds of industrial waste. *Heat (thermal) pollution* is manifested in a change of temperature under the influence of technogenic factors, for example during evacuation of water from thermoelectric power plants.

From the standpoint of influence, it is significant whether the introduced pollution is:

- *biodegradable,*
- *bioresistant,* or
- *toxic.*

When it enters water, *biodegradable pollution* undergoes intensive processes of biological decomposition, the strongest action here being erted by microorganisms. *Bioresistant pollution* in principle presents less of a problem if it is biologically inactive, inert inorganic material forming sludge that is a suitable environment for creation of benthos and development of biological processes in it. *Toxic pollution* of inorganic (heavy metals)

---

[78]Pathogenic bacteria transported by water are one of the main causes of illness and death in many countries, especially developing ones.

or organic (pesticides, etc.) nature represents the most harmful kind of bioresistant pollution.

The risk to health from harmful and dangerous chemical substances present in drinking water is most often defined as the probability that an adverse effect on health will appear after exposure to such substances. Depending on their effects on health, chemical polluting substances are divided into ones that are:

- *toxic (acute and chronic),*
- *carcinogenic,*
- *teratogenic,* and
- *mutagenic.*

Metabolic processes transpire mainly in the liver and to some extent in the spleen and kidneys, but the bulk of toxic substances are deposited in the bones.

In the matter of transport and interaction of microelements (especially heavy metals), *river courses* represent the most complex aquatic systems. We take the Rhein in the 1970's as an example of the extent to which a river can be overburdened with such elements. As a result of chemization of agriculture, this river received annually about 800 kg of mercury, 30,000 kg of arsenic, and 900 kg of herbicides. With waste water of industrial concerns on its banks, the river took in 11,000 t of zinc, 2,000 t of chromium, 1,400 t of copper, 1,850 t of iron, 500 t of nickel, 2.2 million t of sulfates, 400 thousand t of nitrates, 1,200 t of chromic acid, and 200 t of cadmium. Navigation also contributed to heavy pollution of the river, which received about 75 thousand t of petroleum and petroleum products annually.

In comparison with rivers, the degree of pollution of *groundwater* - which in many regions represents the only source of drinking water - is appreciably lower. That this is the case is primarily due to its better natural protection. For example, artesian water - bearing horizons are generally well protected from sources of pollution because they are covered by a thick clay overlayer. In the case of an alluvial medium, apart from possible lateral influence of a polluted surface stream, sectors with a thin clay overlayer can be directly threatened from the surface. A karst water - bearing medium is undoubtedly the most sensitive because the risk of infiltration of pollutants through ponors and karst collectors is ever present, while the capacity for self - purification is perceptibly lower due to the great speed of underground currents.

AIR POLLUTION

From the time of the Industrial Revolution to the present day, air in the vicinity of settlements, industrial concerns, and mining basins has differed to a constantly growing extent from air of the natural environment. To be specific, the atmospheric envelope in the indicated places is increasingly becoming a dump for the products of combustion of all kinds of coals, oil, and oil derivatives, as well as for many synthetic and other compounds evaporated into the air. According to certain data, 240 billion tons of the *gas of life* was destroyed in the past 100 years, in exchange for which the atmosphere received 360 billion tons of toxic carbon dioxide, 53 million tons of nitric oxide, 200 - 250 million tons of dust, and 120 million tons of ash. In this way, life in many cities has for decades been growing less and less bearable.

We call to mind the smog that engulfed London from the 5th to 9th of December in 1952, when the hospitals overflowed with patients suffering from respiratory and cardiac ailments and there were 4 thousand deaths, which was not the only time something like this happened in the city. Or Los Angeles, also a *city of smog*, where the incidence of lung cancer increased by 40% in areas with high content of carcinogenic hydrocarbons (20% in other parts of town). Later, the center of smog - induced catastrophes shifted to the densely populated industrial regions of Japan, where disastrous situations occurred in 1968 in Osaka and in 1970 in Tokyo.

Skipping over global changes, we note that the biological effect of air pollution can be local or regional. In the case of *local action*, the quantity of toxic ingredients is such that it causes rapid local sickening of the respiratory pathways and lungs. Thus, for example, dangerous poisoning by SO_2 from an industrial concern occurred in 817 citizens of Yokohama (Japan) in January of 1972. A *regional effect* is manifested, for example, on the territory of England, where mortality from chronic bronchitis - which comprises 30 thousand deaths a year - exceeds the mortality from tuberculosis and lung cancer combined and is ten times higher than in agricultural countries.

According to the data of H. Heimann, about 1.5 million t of SO_2 enters the air every year from burning of 32 million tons of coal in New York alone. This results in an increase in the incidence not only of bronchitis and other lung ailments, but of heart diseases as well[79]. The extent of air pollution with hydrogen sulfide is high in a good many cities of the USA

---

[79] The biological consequences of pollution are also very evident in the case of plant life. It has been demonstrated that the process of pollution, even at low concentrations of SO_2, can lead to serious physiological changes in plants. Apart from its harmful effect on vegetation, sulfur dioxide erts destructive influence on stone and concrete buildings, as well as cultural and historical monuments.

and Europe, while air pollution with SO_2 (*acid rain*) leads to increase in acidity of the soil.

With the development of *metallurgy*, a number of problems have arisen in the area of air hygiene. In the process of ore processing, dust rich in metal particles enters the air as a side product. As a result of both its quantity and quality, such dust can cause lasting deterioration of health due to its direct influence on the organs of respiration. In non - ferrous metallurgy, for example, dust from copper smelteries is the most dangerous, since extremely poisonous volatile substances can be formed from released elements under the influence of atmospheric humidity. The situation is similar in regions with lead and zinc smelteries.

The majority of undesirable ingredients act on the process of metabolism. Some of them enter the bloodstream through the lungs and become involved in metabolism in that way. Others (mainly ionizing radiation) do this without going through the lungs. A negative effect also results from the fact that a *cap* of dust and smoke over cities retains the biologically most useful ultraviolet rays, which can lead to various metabolic disturbances and the onset of sickness (for example, D avitaminosis). Toxic substances entering the organism through the lungs and bloodstream can cause serious metabolic disturbances more dangerous than when they are introduced with food because they bypass the liver, in which initial detoxication takes place.

Automobile traffic represents the main source of harmful emissions in developed countries today. It is estimated that in the United States at the beginning of the 1980's, 94 million tons of carbon monoxide entered the air, three fourths of it having been produced by automobile traffic. This presents a special danger to children and the sick (persons suffering from insufficient circulation of blood in the brain or myocardial infarct, as well as patients with enhanced thyroid function and anemia). Unfortunately, exhaust fumes contain increasing amounts of carcinogenic 3 - 4 benzopyrene. This accounts for the direct relationship between increase of lung cancer in inhabitants of large cities and growth of automobile traffic, which has been demonstrated by a number of investigators (T. Bakacs, 1977). Of elements present in the air, molybdenum, arsenic, zinc, vanadium, and cadmium contribute to the occurrence lung cancer in males and bronchitis in persons of both ses. In many countries, a special problem is presented by high concentrations of lead (from high - octane gasoline), and investigators have established increase in the amount of this metal to harmful levels in the blood of urban drivers. According to B. P. Dark (1970), the content of lead in tissues and organs of Americans was elevated 120 - fold.

Radioactive air pollution will be discussed later in the text.

ARTIFICIAL RADIOACTIVE POLLUTION

The first diseases that that we know today were caused by radioactive radiation were lung diseases. They were caused by internal sources of radiation, radioactive substances that people introduced by inhaling radioactive dust. To be specific, data from the year 1500 indicate that a large number of miners in a European mine died from a then unknown lung disease. At the beginning of the last century, it was concluded that they died of lung cancer: the cancer was caused by radiation of uranium and products of its decay that the miners inhaled while extracting cobalt ore (P. Bojovic, 1981).

The modern *atomic era*, meanwhile, has been marked by the sudden mass appearance of various anthropogenic sources of radiation. We now have nuclear weapons, various nuclear installations, radioactive isotopes, nuclear machines in laboratories and clinics, and along with them radiation accidents of greater and smaller dimensions that have increasingly affected man, animals, and plants.

Contamination of the environment by radioactive materials has become one of the most significant problems of modern civilization. The technological revolution brought about by application of nuclear energy for peaceful purposes is indisputably leading to increase in radioactive contamination of the biosphere, in spite of great advances in the matter of environment protection that have been made in this field. The dangers of nuclear weapons use and accidents occurring at nuclear installations make this problem even more pressing.

Artificial (fabricated) radionuclides enter the biosphere for the most part in the following three ways:

- As a consequence of radioactive fallout after experimental nuclear explosions;
- Through escape of radioactive substances due to breakdown of nuclear reactors or other nuclear installations; and
- Due to use of radioactive isotopes for medical, technological, and other purposes.

Regarded from the radioecological standpoint, *radioactive fallout* must be considered the most significant form of contamination of the biosphere, since it is rapidly dispersed over extremely great areas of our planet following nuclear explosions.

The first mass pollution of the atmosphere occurred in 1945 at the time of the atomic bombings of Hiroshima and Nagasaki in Japan, which caused radiation sickness on a large scale (with severe immediate or chronic consequences). Nuclear tests have been performed without interruption from 1945 to the present day. Altogether, 1,418 tests of nuclear weapons were carried out during the period of 1945 - 1983, the culmination occurring in 1961 - 1962 (Fig. 4.2.). The given tests released enormous quantities of fission products into the biosphere. Still, the share of radioactive fallout in overall irradiation of the population has declined since 1962, when nuclear explosions in the atmosphere were almost completely stopped.

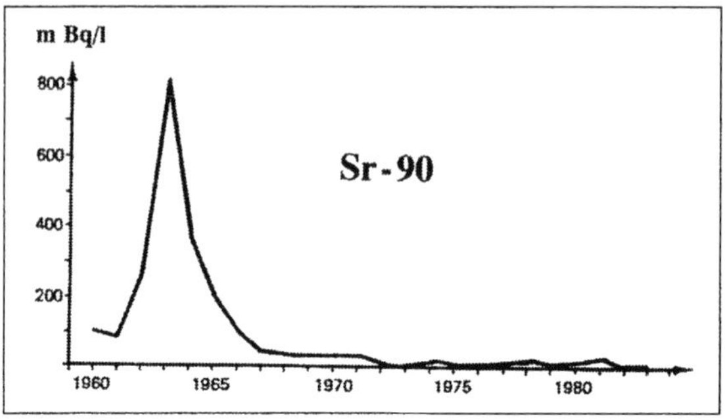

Fig. 4.2. Graph of Sr_{90} local distribution in atmosphere for 1960 - 1985 period (B. Petrovic, R. Mitrovic, 1991).

How great can be the *radiation risk to the population* from breakdowns at nuclear installations for peaceful use of atomic energy is demonstrated by the accident that occurred at the "Lenin" Nuclear Power Plant in Chernobyl (Ukraine) on 26 April 1986. It turned out that released radionuclides can pollute large areas, with consequent accumulation of radionuclides such as I_{131}, Cs_{134}, and Cs_{137} in the biosphere[80]. A great many nuclear accidents of varying degrees of gravity have been recorded to date at nuclear installations throughout the world.

Of the more than 200 radionuclides formed during nuclear fission, Sr_{89}, Sr_{90}, Cs_{134}, Cs_{137}, Cs_{144}, Pu_{239}, Ru_{103}, Ru_{106}, Ba_{140}, I_{131}, and Zn_{65} can be singled out as being biologically significant. We shall dwell on radioactive strontium (Sr_{89} and Sr_{90}), radioactive cesium (Cs_{134} and Cs_{137}), and the iodine radionuclide I_{131}, and the plutonium radionuclide P_{239}. These radionuclides did not exist in nature before atomic tests. During the past half century, they have acquired a special role in the radioactive contamination of drinking water and foods of plant and animal origin.

---

[80] During operation of a nuclear power plant, about 99% of radioactive material remains in the fuel elements in the form of radioactive waste. The disposal of such accumulated waste and radioactive isotopes formed in laboratories and at medical institutions for the time being presents a problem with no easy solution.

1. The *radionuclides* Sr_{89} and Sr_{90} have a significant fission return (about 5.5%). Their main biological specificity is that they behave in the organism like calcium, Sr ions being deposited in the bone system. The bones of young individuals bind 5 - 10 times more Sr than do the bones of adults due to the organism's great need for calcium during the period of bone growth, a fact that must be remembered in planning the diet of children. The biological half - life of Sr - 90 in an adult man is 1000 days (spongy tissue) or 1500 days (compact tissue);
2. Unlike Cs_{134}, the *radionuclide* Cs_{137} has a high fission return (about 6.2%) (Fig. 4.3.). The main biological specificity of these two radionuclides is that they behave in the organism like potassium and are found in every cell of the organism. The biological half - life of Cs_{137} is 10 - 70 days in animals (depending on the species of animal) and 10 - 110 days in man (depending on age of the organism);
3. The *radionuclide* I_{131} has a high fission return (about 7%). It enters the organism of humans and animals primarily with lettuce and other vegetables radioactively contaminated by atmospheric precipitation. The physical half - life of I_{131} is 8 days and comprises 80 - 120 days in an adult man. Deposition of I_{131} in the human thyroid gland depends on age of the organism and is greatest in newborn infants;
4. Among *radionuclides of plutonium*, Pu_{239} is best represented in the biosphere, although even it is present in relatively small amounts. However, because it has a very long physical (24,100 years) and biological (200 years in the human skeleton) half - life, the biological significance of this radionuclide is enhanced. On account of their exceptionally strong carcinogenic action, radionuclides of plutonium present one of the greatest threats to survival of man on Earth.

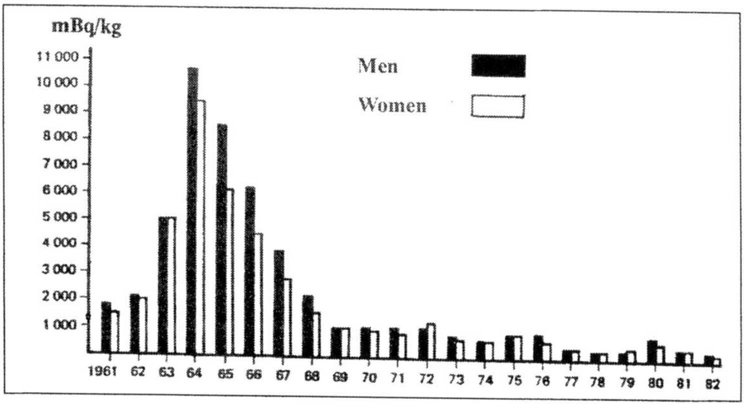

Fig. 4.3. Level of Cs_{137} activity in human beings during 1961 - 1982 periods.

Research on the biological action of ionizing radiation was initiated already during the first decades of the last century, at which time its harmful influence on the human organism was established. Interest in such research grew especially in the 1950's, when nuclear energy started to be used for peaceful purposes in different fields. Together with radiation as a cause of a number of human diseases, accumulation of strontium - 90, cesium - 137, and other newly created isotopes in various minerals, water, soil, animals, and plants was seen to be a problem of great significance in connection with pollution of the Earth's biosphere with radioactive substances. For man, radionuclides in the food chain are especially dangerous. We have in mind here the phenomenon of radionuclide concentration in animal organisms, especially in certain organs and tissues[81]. For example, cows that consume large amounts of contaminated grass concentrate radionuclides in the thyroid gland or in the milk. A similar effect is achieved in fish and shellfish after passage of a large amount of contaminated water through their bodies.

As a result of experiments on animals and study of the consequences of radiation in persons who survived the atomic explosions in Hiroshima and Nagasaki and survivors of nuclear accidents, one - time exposures to lethal doses causing *acute radiation sickness* have been investigated the most thoroughly. Bone - marrow, intestinal, tomic, and cerebral forms of acute radiation sickness can be distinguished, depending on the sensitivity of cells and tissues and the dose of irradiation (D. Markovic, S. Djarmati, I. Grzetic, and D. Veselinovic, 1996). The biological effects of radiation have been treated more extensively by J. Martinovic (1981).

Chronic radiation sickness occurs following multiple or long - term external irradiation or long - term internal irradiation (when radioactive substances are introduced with food or water over a long period of time). Some of the radioactive material entering the organism is eliminated from it, but some of it is absorbed and accumulated in organs and tissues (isotopes of iodine in the thyroid gland, strontium - 90 and radium in the bones). Mucous of the intestines, liver, thyroid gland, and other organs is sensitive to strong irradiation.

Depending on the dose of radiation, effects that do not appear at once, but instead set in some time after irradiation can be: 1) *somatic* (malignant tumors); and 2) *genetic* (birth defects and disturbances that are transmitted to future generations). Moderate and high doses of radiation for the most part induce somatic effects, while low doses induce genetic effects.

---

[81] Norwegian investigators in 1988 recorded a level of Cs_{137} activity in venisen of the order of 90,000 Bq/kg (K. Hove and A. Ekern, 1988). The deer fed exclusively on forest vegetation and fruits with a level of contamination considerably higher than that of arable land. In Yugoslavia, activity of the indicated radionuclide in meat of these ruminents comprised something under 100 Bq/kg just after the Chernobil accident (S. Stankovic et al., 1993).

Pathological changes in different tissues and organs can occur with the passage of time in organisms recovering from acute injuries caused by radiation, as well as after chronic exposure to it. There is not the slightest doubt as to the carcinogenic effect of ionizing radiation. This effect was observed in man already during the first days of radiology and was subsequently confirmed by many examples in humans and by experiments on animals. On the basis of these data, it is believed that radiation can induce any type of cancer.

The occurrence of malignant tumors in different organs and tissues largely depends on the specific ability of a radionuclide to be stored in a certain organ or tissue.

Leucosis occupy an important place in the pathology of tumorous forms with delayed action. Leukemia appears many months after the action of leukomogenic factors, but pre - leukemic cells are found in the bone marrow soon after irradiation. During the period of 1948 - 1950, for example, five times as many people died of leukemia in Hiroshima than in other regions of Japan.

POLLUTION AND DISTURBANCE OF THE ENVIRONMENT CAUSED BY ORE EXPLOITATION AND PROCESSING

Mining stretches back into distant pre - history and already in the Middle Ages constituted part of the foundation of every state. In the course of its long history, mining started with a phase of ore and rock gathering. The technology for exploitation of only rich deposits of metals and non - metals near the surface was developed in the next phase. With accelerated industrialization and far greater needs for energy raw materials in addition to metals and non - metals, the transition was made to strip mining and underground mining on a massive scale (with breaking of ores and accessory rocks). For all practical purposes, it was in this phase that man embarked on massive degradation of the geological environment. The depth of some mines in Africa and India attains 4,000 m, while great strip mines of coal can be more than 1,000 m deep. The enormous "hunger" for more and more inexpensive new raw materials can be illustrated by the fact that 35 to 40 tons of ore and slag per capita is excavated (detached from the geological environment) annually throughout the planet. Of that quantity, only about 2% represents useful material.

Together with the indicated problems caused by disturbance of the geological environment, problems of environmental pollution are equal in rank. Such pollution can be associated with surface processing of coal deposits; the presence of sulfur in ores, coal, and oil; drilling for oil at sea and its transport; fluorine loss at aluminum plants; radioactive waste; etc.

Potentially, any deposit can be a powerful complex source of pollution. The reason for this is that elevated concentrations of a wide assortment of toxic chemical elements are inevitably present in extracted and processed ores, as well as in slag. The complex nature of a deposit as a source of pollution also stems from diversity of the natural environment in which it is found. Negative effects in the form of toxic chemical accumulation can be seen in all vitally important resources: water, air, soil, and food. As numerous data indicate, this produces negative biological reactions in all living organisms, including man (increase in occurrence of general and specific diseases, decline of biological productivity, far - reaching mutagenic consequences, and finally death).

During the course of exploitation and processing of mineral and energy raw materials, the lithosphere, atmosphere, biosphere, and hydrosphere are all disturbed to a greater or lesser extent. In the case of *strip mines*, there are many visible disturbances:

- *Changes in morphological properties of the terrain*;
- *Changes in the rate of exogenous processes* (landslides, rockfalls, gulley erosion in the slag, etc.);
- *Changes in hydrological and hydrogeological properties of the terrain* (patterns of drainage, pollution of surface streams and groundwater, thermal pollution of water, etc.);
- *Changes of flora and fauna* in the case of non - ferrous metallurgy, such changes being noticeable over a radius of up to 100 km (elevated content of trace elements in vegetation, fruits, vegetables, and other agricultural crops, disappearance or migration of plants and animals, etc.);
- *Changes in the atmosphere* (mechanical pollution by different kinds of dust, chemical pollution by gases and fumes, etc.); and
- *Changes in the living and working conditions of man* (decline of work capability, health problems, shortening of life expectancy, etc.).

Operation of *underground mines* is accompanied by terrain subsidence, appearance of cracks, stability problems, and formation of slag on the surface. Very serious problems arise when water is pumped and evacuated from many shafts and abandoned underground chambers, operations that often result in pollution of surface and groundwater. Special problems are presented by the difficult conditions of underground work, with many attending risks (methane formation, pit fires, etc.). Mining is plagued by incurable professional diseases, work injuries, and shortening of the miners' life expectancy.

Strip mining of coal on the territory of Appalachia in the United States transformed a picturesque natural setting into a desert with gaping holes, slag heaps, and infertile barren earth. Masses of materials are thrown out in the process of extraction and refinement of ore. Some of it is washed by rain into rivers, where it has a serious effect on the organic life in them and their water regime. It is an especially difficult task to evacuate polluted mine water without polluting surface and groundwater (A. Allison and D. Palmer, 1980). The situation is similar in many other countries of the world. It is a fact that large strip mines of coal with dumps and installations for its preparation are located near settlements and in flatlands, where they ert additional influence on the environment, especially disturbing agroecosystems and the groundwater regime.

Methods of gasification are being increasingly used in an attempt to resolve the problem of effective exploitation of the world's enormous coal reserves with minimal damage to the environment[82].

Mining basins with exploitation of metallic mineral raw materials and their metalworks are great polluters of air, water, soil, and plant and animal life, above all with numerous toxic elements[83]. For example, the unrestored strip mine on Mt. Kopaonik has become a permanent source of mineral particles rich in heavy metals. Wind action has led to increase in the concentration of heavy metals in the soil, the following levels (mg/kg) having been recorded at Suvo Rudiste (map reference 1,960): Fe, 208,000; Mn, 3700; Cu, 340; Zn, 240; Pb, 210; Cr, 67; and Cd, 2 (Z. Vukmirovic and M. Curic, 1993). In the vicinity of the Trepca lead smeltery near the town of Kosovska Mitrovica - a major source of environmental pollution - the following levels of heavy metals were found in soil (in mg/kg): Pb, 4000 - 7500; Zn, 2000 - 7000; Ni, 4 - 14; and Co, 0.2 - 1.2 (Z. Vukmirovic, 1997). As for metalworks, roasting of ore minerals results in the release of significant amounts of sulfur dioxide, a poisonous gas that has a harmful effect on plants and animals. Fluorine is released in the technological process employed in aluminum plants, and it also erts harmful influence on vegetation, plowed fields, and domestic animals.

The ecological action of *secondary geochemical anomalies* created by man is defined by the levels of chemical elements in living organisms (i.e., biogeochemical linkage in the environment - organism system) and the

---

[82] World coal reserves are so great that it can be stated with certainty that man will be able to count on stable production and consumption of coal for the next 200 years. Inasmuch as more than 4 billion tons of this fossil fuel is extracted from the bowels of the Earth every year, it is clear that the use of *clean coal technologies* (*CCT*) needs to be expanded and combined with recultivation of exhausted strip mines in order to protect the environment.

[83] *Transgressive technogenic diseases* of a chemical nature are linked with movement of chemical pollutants of the environment over great distances by means of atmospheric or hydrological circulation (A. P. Avtsin et al., 1983). A classical example of transgressive phenomena in the biosphere is the case of *acid rain*, which is caused by massive emission of sulfur and nitric oxides and other toxic ingredients (including trace elements) into the atmosphere. The sources of pollution are industrial concerns in the United States, England, France, Germany, and other industrially developed countries.

biological consequences of that phenomenon. Research on the biogeochemical and biological consequences of environmental pollution by chemical elements has been carried out with increasing intensity over the past quarter century, mainly through experimental study of the response of living organisms to elevated concentrations of chemical elements entering the organism in various ways (with food, air, or water). These studies have resulted in the discovery of a broad and diversified scale of negative consequences, expressed both in the form of unhealthy reactions specific for a certain chemical element and in the form of reactions of a non - specific nature, when the action of elements contributes to increase in the frequency of deviations from normal physiological development of the organism or increase of overall illness (Fig. 4.4.). This is illustrated by the following examples:

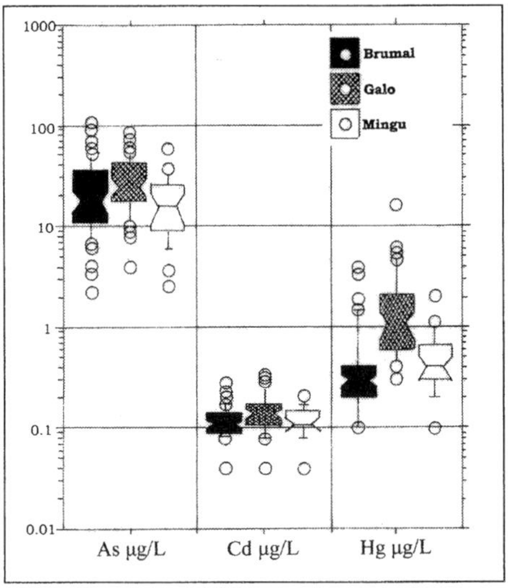

Fig. 4.4. Concentration of arsenic, cadmium and mercury in children urine in settlements: Brumal, Galo and Mingu (J. Matschullat et al., 2000).

1. First registered in the middle of the 20th Century, the widely known *Minamata sickness (mercuriosis)* was caused by water pollution with industrial waste water rich in mercury. The clinical picture of poisoning of many persons was complex, with significant morphological changes in the central nervous system (Keller, Shchepin, and Chaklin, 1993);
2. Also recorded for the first time in Japan during the 1950's, *itai - itai disease* resulted from increased intake of cadmium into the organism with rice. The rice fields responsible were irrigated from a river that received cadmium from a nearby mine. This disease basically involves specific affliction of the kidneys (*cadmium nephropathy*) and is characterized by severe damage to the skeleton;

3. Gold fever is especially present in Amazonia, where about million gold rinsers handle mercury in order to separate the precious metal. The local population was confronted with a serious threat of intoxication in this way. Investigations showed that the average level of mercury as a poison was equal to that recorded in developed countries with high consumption of fish (Japan, Australia, and the West Coast of the United States). For example, analysis of skin of the inhabitants established that mercury content was highest (6.6 μg/g) in Amero Indians, and in one village the poison was found in more than 80% of children and 60% of adults. A similar situation prevails in Guyana;
4. According to E.B. Davies (1983), there are numerous data indicating that a correlation exists between the distribution of lead in soil and the incidence of *atherosclerosis* and *caries* in regions with polymetal mines. Moreover, near the dump of an old mine in West Wales (with the following levels of metals in soils, in mg/kg: lead, 1,760 - 14,910; copper, 32 - 42; and cadmium, 1.4 - 2.9), animals died of lead poisoning, and elevated Pb content was recorded in the blood of children;
5. Increasingly present problems of *industrial beryIliosis* have been found to cause mass sickness among workers and also affect children in neighboring houses. It is most often manifested in the form of lung sickness, for which no specific methods of treatment have been discovered to date.

Thermal electric power plants and *installations for coke production* as a rule are heavy polluters of air, water, and soil. They are above all the greatest anthropogenic sources of heavy metals. The most negative consequences arise due to simultaneous emission of oxides of sulfur and nitrogen, which in the presence of atmospheric water vapor are transformed into strong acids. For example, soil acidity caused by the smoke from many coke furnaces using coal with high sulfur content destroyed vegetation and intensified gulley erosion on hill slopes in the state of Pennsylvania in the United States (A.D. Howard and I. Remson, 1978). Together with pollution of the atmosphere by smoke fumes, surface and groundwaters are also contaminated. Thus, in the neighborhood of Yugoslav thermal electric power plants that use brown coal containing a series of dangerous elements, from 12 to 14% of the primary mass of coal remains in the form of ash following combustion. With the exception of mercury, some of the arsenic and cadmium, etc., the majority of elements remain in the solid phase (in ash and smoke sediment). It was established by S. Matic et al. (1993) that smoke sediment from the Morava thermal electric power plant near the town of Svilajnac contains the following elements (in mg/kg): Cr, 48.7; Mn, 497, Fe, 69,700; Co, 7.5; Ni, 43.6; Cu, 65.2; As, 39.0; Cd, 2.0; and Pb, 59.0. Ash in the firebox itself was estimated to contain 4.7 to 10.9 mg/kg of uranium (soluble in water).

COMPLEX HARMFUL INFLUENCES OF URBANIZATION ON THE NATURAL ENVIRONMENT

The first alarming signs of environmental pollution coincide with the process of urbanization, which is with the Industrial Revolution. Although the first cities were founded in Mesopotamia 5 - 6 thousand years ago, there was no urbanization on a more significant scale in any country of the world until the middle of the 19th Century. Thus, for example, in 1800 only 2.4% of the population lived in cities, and only four cities had a population of more than a million. There were 19 such cities at the beginning of the 20th Century and 141 in 1965. By the end of the year 2000, more than half of the world's population lived in urban environments. City dwellers today constitute 90% of the total population in Holland, Belgium, the Ruhr, Silesia, etc., and we are witness to the formation of enormous cities - megapolises - in a number of industrially developed countries.

Limited space with constant growth in the number of inhabitants per unit of area and consequent increase in the amount of communal waste has contributed to ever greater pollution of the environment. On entering the environment, waste participates in biological circulation, weakens the capacity for self - purification, and causes breakdown of ecological equilibrium. The soil loses its capacity for self - purification, and the biological value of groundwater deteriorates. Polluted drinking water in cities became a source of serious intestinal diseases in the past, with the result that inadequate communal hygiene in Europe at the end of the 18th Century gave rise to the famous era of great destructive epidemics (typhus, plague, smallpox, and - at the beginning of the 19th Century - cholera as well) (T. Bakacs, 1980).

Industrialization was a significant factor in stimulating the rate of environmental pollution in the 19th Century. Pollution of the environment took on a new character: whereas it had previously been a local phenomenon, pollution by toxic substances and various end products came to be present on an alarming scale with the development of industry[84]. The amount of atmospheric dust is hundreds of times greater and that of exhaust fumes 5 to 25 times greater in cities than in the country. Atmospheric pollution in cities absorbs about 20% of sunlight and more than

---

[84] Increase in the rate of industrialization, development of new technological processes, and improvement of the standard of living have caused rapid growth in the amount of industrial and communal waste. The task of sanitary and controlled waste disposal has thereby become a major problem, and waste control is now an imperative of the modern world. It should be noted that environmental concerns were also characteristic of pre-Christian history. Of great interest here is the city dump that existed in the Sumerian city of Ur (about 2500 B.C.). This dump is a treasury of significant finds that enable us to trace the city's development and history over the course of three centuries.

50% when the sun is low in the sky. All of this affects public health by increasing the number of respiratory, cardiovascular, malignant, and other diseases.

The number of diseases increases precipitously in times of smog. Automobile exhaust fumes are the main cause of an especially dangerous type of smog, so - called *photochemical* or *Los Angeles* smog[85]. On the streets of Paris alone, more than 1.3 million automobiles move daily on an area of 105 km^2. More than 4 million automobiles in Los Angeles release about 1100 t of nitric oxide, the concentration of ozone is 100 - 200 times greater than in clean air, various organic compounds are formed (toxic formaldehydes and organic peroxides), and several hundred chemical compounds enter the air. Generally speaking, air takes on an unpleasant smell during photochemical smog, visibility worsens, the mucousa in humans is irritated, and the condition of sufferers from lung and other diseases deteriorates. Domestic animals and plants also suffer.

The constant *street noise* of city traffic has a harmful effect on hearing. Moreover, noise causes tension, a feeling of fear, and insomnia. It also frequently contributes to serious disorders of the nervous system, somatic changes, and stress - linked diseases such as hypertension, impaired secretory work of the stomach and intestines, hormonal disturbances (diabetes for instance), etc. (T. Bakacs, 1980). The following pathological states (after Bakacs) can arise due to strong stress influences:

- Gastritises and enteritises, gastric and duodenal ulcers;
- Cardiovascular diseases, including coronary disease and hypertension;
- Diabetes and certain other metabolic diseases of hormonal etiology, disturbances of thyroid function, etc.;
- General irritability and nervousness, all the way to mental illness.

Apart from noise, constant and ever growing nervous tension (caused by the fast pace of life) and industrial and traffic - induced air pollution represent important pathogenic and etiological factors characteristic of large cities. A typical disease of urban areas is hypertension, in whose pathogenesis nervous strain plays a significant part. In addition, stress factors are increasingly blamed for the incidence of coronary illnesses, which among members of certain professions (intellectual workers, doctors, managers, etc.) have acquired the nature of professional diseases.

[85] Growth of traffic has caused an enormous increase in the number of road accidents. According to data from the late 1980's, the total number of persons injured in traffic accidents in West Germany attained an average of 1.7 million annually. In countries with more highly developed automobile and rail traffic, road accidents ranked third to fourth among all causes of death and first in the case of children up to five years old.

We note in conclusion other professional diseases (according to A. P. Avtsin, 1972):

- Cancer of the urinary bladder in workers who come into contact with products of aniline and hydroquinones;
- Cancer of the respiratory pathways in employees of electrolysis works and persons involved in production of nickel, such cancer being characterized by long - term development; and
- Cancer among chimney sweeps in England.

CONSEQUENCES OF THE ACTION OF POLLUTANTS ON MAN

As we have seen, anthropogenic influence on living organisms - humans, animals, and plants - is diverse and for the most part entails negative consequences. The following key consequences of pollutant action are singled out in the book "**Physico - chemical Basis of Environment Protection**" (D. Markovic et al., 1996):

- Threat to life on the Planet, probably enhanced by possible decrease of oxygen in the atmosphere;
- Threat to the survival of individual species of living organisms and indirectly to that of other organisms dependent on them;
- Genetic changes of living species, especially man, that can lead to his demise as an intelligent and the currently reigning species;
- Hindering of the existence and threat to the survival of individuals and the groups to which they belong; and
- Direct and rapid negative action on individuals or narrower groupations.

All of the indicated consequences are a result of the combination of a large number of greater or smaller changes on the Earth that have acted to disturb the unity of organisms and the environment.

The action of chemical pollutants depends on a number of factors such as *physico - chemical and toxicological characteristics, particle size, concentration, time of action (exposure), and the manner of entry into the organism.* Where the human organism is concerned, such action can be classified as *irritating, fibrogenic, allergic, dermal, toxic, mutagenic, carcinogenic, and embryotoxic.*

- Many substances irritate the eyes, nose, upper respiratory pathways, lungs, or skin, and their action can be expressed in varying degrees;
- Many forms of dust (coal dust, asbestos dust) cause fibrosis, the advance of which leads to pneumoconiosis, a serious disease that every year kills several thousand people throughout the world;

- A polluted environment contains many allergens that are taken into the organism by breathing, with food products, or through contact;
- Skin diseases can be caused by a whole series of substances such as acids or bases, solvents, or detergents;
- Entry of poisons into the organism causes poisoning (intoxication) with possible acute, subacute, and chronic forms of action;
- Mutations are chemical (molecular) changes of genes that are inherited (passed on to the next generation), and a whole series of more conspicuous mutations are known today to have occurred in plants, animals, and man;
- Two kinds of agents ert carcinogenic action: *endogenous carcinogens* (certain hormones, vitamin D_2, cholesterol, and bile acids); and *exogenous carcinogens* (chemical substances; physical agents; ionizing and non - ionizing radiation; biological agents such as viruses, bacteria, and parasites; and combinations of the indicated carcinogens);
- During pregnancy, chemical substances can cause various disturbances in development of the fetus, two types of effects - *teratogenic* and *embryotoxic* - being distinguished here.

In order to prevent or suppress the harm done by chemical agents on living organisms, it is necessary to have a closer knowledge of the modes and mechanisms of unfavorable action and ascertain the connection between the amount and concentration of a substance and the adverse effects it produces[86]. Some mechanisms of harmful action are considered below.

Mechanisms of Toxic Action. Classical toxicology deals with the effects of chemicals on humans and domestic animals under conditions of their direct exposure. Experiments have shown that poisoning requires interaction between poisons and the organism, i.e., their entry into the organism, poisoning occurring only when the dose of a harmful substance attains a value at which deleterious effects appear. Depending on the amount of a given substance, it can be indifferent, medicinal, or poisonous in its action on the organism. For example, the well - known poison arsenic is used as a medicine in small doses, while many medicines become poisons with a significant increase of dose.

Poisons enter the organism by means of inhalation and ingestion or through the skin. On penetrating a living cell, poisons alter physico - chemical characteristics of the cytoplasm, destroy the membranes of organelles, change the reaction of the cell medium, and disturb the conditions needed for normal functioning of cell proteins. Enzymes - cell biocatalysts - are especially sensitive to the action of poi-

[86] *Risk estimation* is a basic element in the complex plan of human health protection on an area under examination. This involves estimation of influence erted by environmental health factors (factors classed as very significant and most significant from the standpoint of health). Estimation of the level of morbidity and mortality under the influence of anthropogenic factors is carried out according to the concept of *environmental health impact assessment (EHIA).*

sons. Poisoning of some enzyme that takes part in an important metabolic process in the final analysis has a lethal outcome. Poisons act as enzyme inhibitors, either *general* (salts of heavy metals such as Ag, Cu, Hg, and Pb) or *specific* (cyanides, H_2S, sulfides, azides, and CO, which act on metals).

Entry of a poison into the organism evokes defensive responses that attempt to limit the poison's toxic action. Such responses include elimination of the foreign substance in unchanged form from the organism; deposition (storage) of the foreign substance in tissues of the organism; and decomposition of the poison to simpler substances that are eliminated or involved in general processes of metabolism. The most widespread response of every organism to an introduced foreign substance is its decomposition, processes of biotransformation occurring primarily in the liver, but also in the gastro - intestinal tract, kidneys, and fatty tissue.

Ecotoxic Action. When a chemical substance causes damage to the ecosystem, we say that it erts ecotoxic action. A chemical can alter the total quantity of living organisms (biomass) of the ecosystem or change the ratio of individual elements of biomass. Bioinhibitory action (resulting in decline of reproduction and growth and increase of mortality) and biostimulatory action (eutrophication) can be equally harmful to the ecosystem. In ecotoxicology, any action that introduces changes in functioning and structure of the ecosystem is important.

In estimating the relationship between the real concentration of a pollutant and its effect, we need to take into account the substance's chemical persistence, mobility, and bioconcentration (bioaccumulation). Persistence refers to long - term retention of toxic substances in the environment. Here it is very important to establish whether the substance remains in its original form or undergoes processes of degradation. Degradation of a substance in the environment can be accomplished chemically, photochemically, and biologically, usually with formation of inorganic substances as end products. It should be stressed that there are a large number of compounds which retain unaltered high toxicity over a longer period of time (compounds of heavy metals and metalloids, certain organic compounds).

Combined Action of Pollutants. Inasmuch as the environment simultaneously houses pollutants of very complex composition with different mechanisms of action, the combined action of such substances represents a central problem of ecotoxicology. The problem is complicated by their combination with climatic, biological, and geological factors.

Depending on whether the effect is strengthened or weakened during simultaneous action of poisons, the concepts of *synergism* and *antagonism* were introduced at an early stage in study of the combined action of toxic substances. More recently, the term synergism has been used to express the state where the effect of combined action of toxic substances surpasses the sum of effects of each individual poison. The term additive action is used if the given effect is equal to this sum, while antagonism refers to the case where it is weaker than it.

Natural factors that act in combination with harmful substances include the following:

- *High air temperature*, which usually accelerates development of the toxic process and increases sensitivity of the organism to the action of toxic substances, while toxic substances lower resistance of the organism to overheating;

- *High atmospheric humidity*, which significantly alters the irritating and general toxic action of substances that are readily hydrolyzed; and
- *Reduced air temperature*, which strengthens the toxic action of pollutants, a circumstance attributable to alteration of conditioned refles and disturbance of the body's thermoregulation.

Basic Mechanisms of Mutagenesis. Man is subject to the same laws of inheritance as every other organism on Earth, and mutations can arise in him also. Thus, it is known that 46 chromosomes are found in the nucleus of cells of a normal man, although there are individuals with 47. Such people have organic damage and are usually characterized by reduced mental capacity. According to data of the World Health Organization, 20% of childhood diseases and 15 - 20% of deaths among children are linked with birth defects (physical, mental, and psychic) resulting from somatic or generative (germinal) mutations, i.e., damage to genetic material.

Numerous data indicate that the genetic material of man has recently undergone increasing damage, the number of harmful mutations in the human population growing at an ever faster rate. According to data presented at a meeting of the Soviet Academy of Sciences in 1989, the genetic danger to humanity is rising and becoming alarming. It is believed that genetic processes in somatic cells cause (among other things) malignant tumors, which are responsible for the deaths of a fifth of the population. If the rate of gene mutations under the influence of environmental pollution continues to increase, the biological consequences to man will be catastrophic when the frequency of mutations is doubled.

Research results show that many chemical pollutants are harmful agents, since they affect gene structure and function and cause somatic and generative mutations. Cancer, autoimmune aggressions, degenerative diseases (physical, mental, and psychic), accelerated aging, etc., are held to be caused by somatic mutations, while various congenital diseases, many birth defects, and early fetal death are believed to result from generative mutations.

Causes of Malignant Diseases. Professional cancer has been known since 1775, when P. Pott described cancer of the scrotum in young chimney sweeps that pulled themselves through factory chimneys in order to clean them more thoroughly and thereby came into direct contact with considerable amounts of soot. Although many similar cases were recorded subsequently, it was only in 1933 that benzo(a)pyrene was identified as the most significant carcinogenic component in soot. Cancer in Scottish miners employed in exploitation of oil shales was described by A. Scott, while cancer of the urinary bladder was recorded in 1895 among workers employed in the chemical industry in Germany.

According to data of the World Heath Organization, environmental pollutants are the cause of more than 70% of cancers in humans. A direct dependence has been established with certainty for lung cancer, which occurs more frequently in the inhabitants of polluted cities than in country dwellers. Also, changes of benzo(a)pyrene concentration in the air always affect mortality from lung cancer.

The number of carcinogenic substances is relatively great, although there are no precise data as to the final figure. The list of carcinogens published by the International Agency for Research on Cancer (IARC) includes 783 substances.

Chemical carcinogens are divided into two groups, *inorganic* (certain heavy metals and their compounds, minerals, and radionuclides) and *organic* (polycyclic aromatic hydrocarbons, alkylating agents, aromatic nitrogen compounds, and various other organic compounds). The greatest number of chemical carcinogens belongs to the group of polycyclic aromatic hydrocarbons, which induce tumors locally after one month to two years. A second significant group is composed of alkylating compounds (ethylene, ethylmethane sulfate), which unite readily with proteins, DNA, and RNA, causing changes in their composition that alter genetic material of the chromosomes.

ROLE OF ECOLOGICAL MINERAL SUBSTANCES IN ENVIRONMENT PROTECTION

Many mineral raw materials and rocks in general - in the raw state or after a certain degree of processing - play a significant part in environment protection. For this reason, such raw materials can be called ecological raw materials.

Most significant from the standpoint of environment protection are various non - metals: zeolites, different kinds of clays and silicates, glauconites, granites, serpentinites, dolomites, limestones (lime) and marbles, etc. They are being increasingly used as natural sorbents to remove suspended particles or dissolved substances from industrial waste water (a major pollutant of water resources and soil) and for other purposes. Based on the paper of R. Tomanec, D. Milovanovic, and D. Vucinic (2001), a brief survey of the uses of such raw materials is given below.

Among different kinds of clays, it is possible to single out bentonite clays, kaolinite, palygorskite clays, and sepiolite. Use of *bentonite clays* is based on the fact that they possess a high capacity for cation exchange and can be effectively employed for selective adsorption of certain ions of heavy metals (Cu, Zn, Ni, and Cd for instance) from waste (contaminated) water. They are also used to remove alkaloids and insecticides from fruit extracts, in production of fodder for livestock, in recultivation of degraded soils, etc. More effective as an anion exchanger, *kaolinite* is also used for purification of water. *Palygorskite clays* are used for melioration and improvement of the structure of sandy soils, as an excellent adsorbent in the oil and gas industry, as a fixative of lead accumulated in soil near highways, etc. A complex mixture composed of several minerals - bentonite, zeolite, perlite, and vermiculite - in addition to water - soluble fluorides is used to remove a whole spectrum of heavy metals present in solutions for preparation of mineral raw materials or in mine waters.

Employed in practice for a long time now, *quartz sand* is one of the mineral raw materials used most frequently for relatively inexpensive filtration of water. Of siliceous sedimentary rocks, we mention *diatomite*, *opoka*, and *trepele*, which are excellent adsorbents and substances for filtration. Diatomite is also used to remove radioactive isotopes from contaminated soil, as a carrier of microelements, for improvement of soil quality, etc.

Zeolites constitute an extremely important group of ecological mineral raw materials, since they possess the capacity for ion exchange and exhibit adsorptive characteristics. They are used in many ways: in treatment of relatively low - radioactive waste and contaminated water; to remove heavy metals from waste; to eliminate Fe, Mn, and Sr from groundwater; for recultivation of soil; etc.

Vermiculites are being employed to an increasing extent in the field of environment protection. They have been used to neutralize Cs_{137} of radioactive waste water; to remove heavy metals (Pb, Zn, Cu, and Cd) and ammonium ions from surface water; and as the bed of different filters for purification of mine, industrial, and other kinds of contaminated water.

Owing to its adsorptive properties and ion - exchange capacity, *glauconite* (glauconitic sand) has many applications in the domain of environment protection. When covered with a layer of glauconitic sand 0.5 m thick, recultivated soils in areas of abandoned strip mines give a very high yield of various crops. Other possible applications are for purification of radioactive and other waste water; to improve soil quality (through improvement of aeration, maintenance of soil humidity, and adsorption of pesticides from the soil); to increase the bioproductivity of aquatic basins (by stimulating accumulation of phytoplankton and increasing the productivity of fish in them); etc.

Serpentinites represent a cheap and widely disseminated mineral raw material. In the first place, finely ground serpentinite is extremely useful as an adsorbent of anion collectors such as xanthates and oleates (which are very frequent and dangerous pollutants of river courses in zones with industrial installations for preparation of mineral raw materials). It can also be used for purification of mine water, etc.

It is felt that *garnets* will have a significant role to play in the field of water purification in the near future.

Transformed into MgO, *magnesite* represents a very useful product in treatment of contaminated water of different origin (for removal of heavy metals from dilute solutions and elimination of algae, organic substances, phosphorus and ammonia, boron, and other pollutants).

As the main product of limestone, *lime* is used in many areas of environment protection: for treatment of drinking water, wastes, and sludge; to neutralize certain kinds of industrial waste; and for desulphurization of gases from thermal electric power plants. Use of lime in the form of calcium oxide in farming for calcification of agricultural soil is today practiced on a scale smaller than might be expected.

Due to limitation of space, it is impossible to present other ecological mineral raw materials such as *barite*, *gypsum*, and *peat* (useful for blotting of waste water from livestock farms, as an adsorbent in the case of water and soil polluted by petroleum and petroleum derivatives, etc.), *peat vivianite*, *tremolite* (a substitute for asbestos), *dawsonite*, etc.

The wide range of possible uses of ecological mineral raw materials or repair of degraded land or environment protection has not yet received the attention it deserves.

CHAPTER 5

Socio - Economic Factors and their Influence on Human Health and Health Protection

In addition to natural factors, socio - economic conditions play an important part in formation of the environment and its influence on human health. Their role is especially great on territories with a developed economy, where they can have a crucial effect on human health. Generally speaking, socio - economic conditions or the conditions in which an individual lives influence his exposure to disease and determine the level of his resistance. By socio - economic conditions, we mean residence in town or in the country, housing, the temperature and humidity at home and at work, living space, hygienic conditions, nutrition, and all the circumstances of poverty or plenty.

Poor living conditions and overcrowding of cities can become a significant and even dominant cause of infection of the upper respiratory organs by viruses or pneumococci for instance. According to J.R. Paul (1966), the consequences of wars and the circumstances accompanying them create *microclimates* in which starvation, tuberculosis, and typhus epidemics rage[87]. Not only does the microclimate affect the host, it also clearly acts on the parasite and its vectors as well.

Medical geography has dealt with the influence of socio - economic factors on human health for a fairly long time. In planning medico - geographical research on a concrete territory, these factors must certainly be taken into account. We therefore recall on the present occasion the

---

[87] By microclimates, Paul means the social climates of different peoples and nations, including their way of life and customs, local circumstances, housing conditions, and places of work. Such climates can be very different on relatively small areas, where rich and poor live in luxury or in squalor. Ryle (1948) applied the term *social pathology* to the influence of microclimates on disease or mortality.

procedural approach presented in the "**Handbook of Medical Geography**" of Keller, Shchepin, and Chaklin (1993). We start from Shoshin's 1962 division of significant socio - economic factors into two categories: 1) *economic - geographic conditions*; and 2) *medical - sanitary conditions*. The following elements are examined in the framework of economic - geographic conditions:

- The system of government and administrative division of a territory;
- Its population;
- Settlements;
- Industry and agriculture; and
- Transportation and means of communication.

Medical - sanitary conditions include the following elements:

- The level and structure of illness of the population;
- The system of organization and state of local health protection; and
- The system and state of veterinary protection.

ECONOMIC - GEOGRAPHIC CONDITIONS

The *system of government* represents a decisive condition in protection of public health. The organization of that protection depends on the *administrative division* of a territory. Local organs of government play an important part in protection of public health and treatment of patients, as well as in the implementation of environment protection measures.

Demographic data on the *population* of a studied territory represent an important source of information needed for a medico - geographic evaluation of the region and for planning of preventive, sanitary, and health measures. Among the most important elements that must be established are the numerical status, composition, and density of the population; its distribution and content by social and economic groups; literacy and education of the population; its age and sex composition; international and internal migration patterns; mortality and its causes; and average life expectancy.

Settlements constitute a significant element in the socio - economic conditions of a territory. It is possible to distinguish two main kinds of settlements, urban and rural, which differ in the functional structure and size of the permanent population. The nature of a settlement is of great significance in planning measures of protection from adverse environmental influence and especially from anthropogenic pollutants, as well as in attempting to mitigate the consequences of natural disasters in both the settlement under consideration and other settlements of the same administrative region. For maintenance of public health, it is particularly important to ensure an adequate amount of good - quality drinking water, removal of sewage, cleanness of atmospheric air, and healthy conditions at places of work and in enclosed space.

The nature of *industrial and agricultural production* determines the incidence of many professional diseases. Such diseases can be caused by the use of various poisons, radioisotopes, and lasers in production, as well as by pollution of the air in production halls, strong noise, vibrations, etc. These characteristics of industrial and agricultural production not only demand that arrangements be made for treatment of the respective diseases, but also necessitate the development and implementation of special preventive measures.

The degree of development of the *transportation infrastructure*, especially the road network, fairly often dictates the system of medical assistance and patient treatment. Knowledge of the state of the road network in a region is of great significance in mitigating the consequences of natural disasters and in dealing with mass sanitary crises. The *means of communication* - telephone, telegraph, radio, and other ways of communication - are important for the management of health protection in a given region.

The examined elements of economic conditions must be analyzed in carrying out a medico - geographic study of any region. Practice has shown that demographic information, the water supply and waste disposal systems, and the nature of industrial and agricultural production are of crucial significance from the standpoint of public health and organization of health protection.

MEDICAL - SANITARY CONDITIONS

Study of the *incidence of disease* in a population carried out in the framework of medico - geographic research is not the same as study of the geography of disease (nosogeography). In the former instance, data are collected on the structure of illness of the population, the frequency of certain diseases, their seasonal occurrence, and other questions. The aim of nosogeography, on the other hand, is to discover regularities governing formation of the ranges of selected human diseases and ascertain the causes dictating their occurrence and dynamics.

The general structure of non - infectious diseases among members of a local population to a considerable extent reflects the influence of natural conditions and characteristics of life and work of the population. Knowledge of it enables us to judge the possible dynamics of these diseases if necessary preventive measures are not taken in time. The diseases in question here include cardiovascular ailments, for example, diseases of the digestive organs, etc.

Medico - geographic evaluation of the sickness of a population is given on the basis of collection and analysis of the following data: a) the general structure of incidence of diseases among members of the local population (with permanent and temporary residence); b) the dynamics of infectious diseases and characteristics of major epidemics; c) the seasons of occurrence and distribution of selected infectious and non - infectious diseases; and d) characteristics of the course of these diseases. These data can be found in the existing documentation on examinations of patients and causes of death compiled by all medical institutions in the city, region, province, and country as a whole. Each of them has its own source of information and starting statistical base.

Very important additional data can be obtained by examining the results of exhaustive periodic surveys of the kind generally conducted during years of census taking.

Together with information about their diseases, an awareness of physical development of the inhabitants of a region is also needed. The state of physical development largely reflects medico - geographic conditions of the territory under consideration. The following traits make up the complex on whose basis physical development is judged: 1) *anthropometric traits* (body height and weight, rib cage volume, head volume, shoulder breadth, forearm length, tibia and femur length); 2) *physiometric traits* (lung capacity, fist muscle strength); and 3) *somatoscopic traits* (constitution, musculature development, state of nourishment, form of the rib cage and legs, foot shape, expression of secondary sexual symptoms).

Information about the *system of organization and state of local health protection organs* is needed to determine their ability to carry out therapeutic, prophylactic, sanitary, and counter - epidemic measures. To evaluate this information, we need to be familiar with the general principles of organization and structure of the staff of public health institutions. Also needed here is knowledge about the following things: availability of medical personnel to the population; the system of training and educational status of the indicated personnel; the availability of emergency medical assistance to the population; the distribution of medical institutions (hospitals, polyclinics, sanatoria, etc.) throughout the territory; their basic characteristics; the distribution and characteristics of basic prophylactic, sanitary, and epidemiological institutions; the system of preparation and distribution of medicines; the distribution of medical supply storage centers; and the location of larger pharmacies.

We need to acquaint ourselves with the *veterinary service* of a region in order to gain information about the distribution of zoonoses among domestic and wild animals and determine the possibility of involving local veterinary institutions in the tasks of preserving public health and developing health protection measures.

In analyzing local veterinary institutions and their personnel, it is necessary to clarify the following things: the existence and location of institutions; the capacity of such institutions; the existence and qualifications of veterinary workers; and the places where they are stationed.

ROLE OF THE WAY OF LIFE AND HARMFUL HABITS IN FORMATION OF NOSOGEOGRAPHIC CHARACTERISTICS OF A TERRITORY

Medico - geographers have investigated many regions of the world with different groups of inhabitants. Among other things, they have established that influence of the environment on public health depends on the degree and nature of its action, on the possibility of adaptation of the inhabitants to the given environment, and to a great extent on their way of life. Of special significance here are the habits and customs of people living in different parts of the planet. Without taking into account data on the way of life, habits, and customs of the people, it is impossible to estimate the level of their health.

For example, ***drug addiction*** represents a major problem at the present time. In the beginning, drug addicts and addiction were mainly found in those countries and regions where plants with the corresponding properties (poppies, hemp) grew in abundance. Heroin use was customary there, and this harmful habit was widely known. During the 19th Century, Great Britain and several other European colonial states began to show an interest in the markets of Southeast Asian countries with the aim of purchasing opium. The market expanded at a relatively rapid rate in this way and the number of addicts increased daily, with the result that in the 1920's 22 million addicts were registered just in the United States and 1 million in France.

Study of the geography of ***smoking*** represents an important page in the history of medical geography. The negative role of smoking is generally known, but its contribution to pathology in one region or another often is not taken into account. It must be kept in mind that the number of smokers in the world is enormous. In 1980, for example, there were 18 million smokers (or 40% of the adult population) in Great Britain and 50 million (or 37% of the total population) in the United States.

Mortality among smokers is high. This harmful habit contributes to the incidence of many diseases, above all lung cancer. Statistics indicate that persons who for 25 years smoked 40 or more cigarettes daily contracted lung cancer 52 times more often than did individuals who during the same period of time smoked less than 10 a day. Smoking leads to other diseases as well:

- Myocardial infarct is present to a far greater extent among smokers than among non - smokers;
- In 33% of cases tested, smoking causes diseases of the respiratory organs;

- In many regions of Asia and South America, cancer of the oral cavity is very common as a consequence of the custom of smoking cigarettes with the burning end in the mouth.

Alcoholism is also a matter of great concern today. For example, alcoholics make up 1.6% of the adult population of Switzerland. In France - where the given figure is 1.4% - 16.7 l of pure spirits or 152 l of wine is produced annually for every inhabitant of the country (including women, the elderly, and children). Alcoholism ranks third as a cause of death there, after heart diseases and cancer, and 70 of accidents are linked with alcohol consumption.

Drunkenness and alcoholism not only create enormous economic damage, but also cause nervous and psychic diseases, cirrhosis and cancer of the liver, cancer of the esophagus, and heart disturbances. The British Medical Association reported that 20% of all beds in English hospitals are occupied by patients with ailments caused by excessive alcohol consumption.

In judging the state of public health of a given region, it is important to know the extent of distribution of drug addiction, alcoholism, smoking, and certain other harmful habits. In the past two decades, meanwhile, the problem of *AIDS* has also become increasingly present. It is today expressed in alarming form in many countries, especially among homosexuals and other persons with sexual abnormalities. Already in 1989, the AIDS epidemic embraced as many as 144 of the world's countries. It too has thus become a medico - geographic problem, one similar to that of venereal diseases. Knowledge of the geography of AIDS and medical studies of human migrations will be of assistance in the struggle with this disease.

SOCIO - ECOLOGICAL FOCI OF HUMAN DISEASES

Although they have brought many blessings and comforts to human life, urbanization and *scientific - technological progress* have altered the environment and influenced the level, structure, and boundaries of distribution of many diseases. While medicine has suppressed infectious and parasitic diseases formerly widespread among the world's population, illnesses linked with exogenous factors such as disturbance and dangerous pollution of the environment are becoming increasingly conspicuous in the structure of disease incidence. Characteristic of modern society, this tendency has embraced a growing number of industrially developed regions and in the last decades has not even by - passed rural regions. We have already spoken of diverse anthropogenic factors as causes of many *diseases of civilization and urbanization.*

The reason for the relatively wide dissemination of modern *pathologies of civilization and urbanization* lies less in absolute increase in the number of cases of such diseases than in the fact that conditions of human work and life give rise to unfavorable situations in the environment that promote the appearance and spread of those diseases. Thus, situations arise in the new conditions of medicine where a pathological process appears whose true ecological cause and pathogenic nature are difficult to ascertain. Only with great tardiness have people begun to realize the damage to nature and human health that can be done by unreasonable and excessive use of modern technology and the great advances in chemistry.

Under current conditions, scientific - technological progress with its social consequences has created for medical science many new theoretical problems and tasks requiring involvement of the whole system of health protection. The circumstances prevailing in rural settlements can be taken as a case in point. With improvement of the living conditions of country dwellers and availability of specialized medical assistance to the population, the danger to human health and life from serious infectious diseases has disappeared. However, diseases formerly not encountered in the country are being recorded with increasing frequency. In Armenia, for example, studies have shown that congenital anomalies in newborn infants are perceptibly on the rise in rural enclaves. Among 1,095 analyzed families with congenital and genetic anomalies from 36 rural regions of Armenia, only 14.4% of cases of such anomalies could be linked with and attributed to family - related hereditary factors. In the great majority of cases (85.6%), the anomalies were not inherited. Complex investigations indicated that they had a social basis and were linked with agricultural activity rather than local natural conditions. To be more precise, genetic diseases have been recorded with increasing frequency in rural regions due to comprehensive chemization of agricultural production, especially in villages with high pollution of the environment by toxic chemicals.

The given example and many other communications enable us to conclude that we are dealing with *socio - ecological foci of human diseases.* This means that practical health protection requires socio - ecological regionalization and mapping of a territory with a corresponding estimate of the extent of favorable or unfavorable influence of this factor on the state of public health. The medico - geographic approach and cartographic presentation of information could be very convincing and of great practical usefulness here.

CHAPTER 6

The Geography of Disease (Nosogeography)

Throughout the many centuries of human history, people have died most often of hunger, war, and (especially) infection. Culminating in dirty and densely populated cities, conditions for spreading of infections were created with the formation of settlements. Change can be brought about through improvement of living conditions, especially by ensuring a supply of bacteriologically sound drinking water and by introducing safe ways of waste disposal. Even greater influence on decline of mortality from previously fatal infections was erted by great medical discoveries made for the most part in the first half of the last century, above all by the devising and widespread application of vaccines and serums, insecticides, and (particularly) antibiotics.

Changes in the structure of illness and mortality of the population and prolongation of life expectancy in the last decades have put non - infectious somatic ailments in the group of leading diseases. Scientific forecasting on questions of health protection has now become impossible without detailed research on diseases of the heart and blood vessels (infarct, etc.), malignant tumors, and cerebrovascular diseases (stroke, etc.)[88]. Such research provides new data on the etiology and pathogenesis of these diseases and can open up new possibilities in developing programs of prevention and organized treatment of diseases that take into account the geographical and geological characteristics of their distribution.

---

[88] In Serbia (as in the Western world), the main causes of death are as follows: cardiovascular diseases (infarct, etc.), 40%; malignant tumors, 15-20%; cerebrovascular diseases (stroke, etc.), 15%; inadequately defined symptoms and states, 7-8%; chronic lung diseases, 5%; and injuries and poisonings, 5%. It follows that more than half of all deaths by illness result from diseases of the blood vessels, especially those of the heart and brain (Z. Radovanovic et al., 1995).

There can be no doubt as to the need for application of medico - geographical methods in the field of *non - infective epidemiology*, whose theoretical foundations are unquestionably weaker than in the case of classical *infective epidemiology*. A complex approach to research is obligatory, inasmuch as the relationship between man and the environment is very complex and environmental influence on human health and life expectancy is extremely diverse. It thus becomes imperative for investigators to elaborate the geological and geographical foundations of a territory of interest to them. In other words, they need to evaluate the chemical composition of rocks, soil, and water in it; ascertain the level of radiation; determine the nature of climatic and meteorological factors; establish characteristics of food and the manner of food preparation; evaluate social and hygienic conditions; determine the presence of anthropogenic pollutants; etc. Of great importance here are information on cases of disease, demographic data on the population, and data on the economic state of the territory.

A more comprehensive treatment of the geography of disease is given by Keller, Shchepin, and Chaklin (1993). We shall dwell briefly on the most widespread diseases of the present day: diseases of the heart and blood vessels, malignant tumors, infectious and parasitic diseases, and some diseases whose causes most likely originate in the geological environment.

NON - INFECTIOUS SOMATIC DISEASES

Diseases of the Heart and Blood Vessels

Among diseases of the heart and blood vessels, the most important are heart disease (affecting blood vessels of the cardiac muscles) and cerebrovascular disturbances, i.e., damage to blood vessels of the brain, which most often are a consequence of hypertension (high blood pressure) as the third - ranked pathological state in this group of diseases. Also of great epidemiological significance are rheumatic disease of the heart valves, congenital heart defects, and cardiomyopathy (diseases of the cardiac muscle itself).

Diseases of the heart and blood vessels are considered to be the tax (albeit not unavoidable) man pays for development of his civilization. Taken together, those diseases in many countries are the cause of every other death that occurs. However, considerable differences exist within the group of diseases as a whole. There are many more cases of cardiomyopathy in developing countries, whereas coronary and hypertensive heart diseases - together with cerebrovascular disturbances (stroke above all) - are dominant in developed countries.

The leading risk factors include the following: 1) ingestion of cholesterol and other animal fats as the cause of *atherosclerosis* (accumulation of fatty substances on the walls of arteries) in conjunction with factors that accelerate the process itself (smoking, high blood pressure, exposure to psychic stress); 2) high consumption of kitchen salt, obesity, and hereditary factors as causes of *high blood pressure* (hypertension); and 3) increased consumption of sugar and fats as the cause of the sudden expansion of *diabetes* and *obesity*, which are very important risk factors where cardiovascular diseases are concerned. The relationship between the incidence of cardiovascular diseases and hardness of water represents one of the most intriguing, but still undefined, aspects of the problem (Bernardi et al., 1995).

In study of the geography of heart and blood vessel diseases, a number of important data have been obtained indicating the significance of diet along with climatic and meteorological factors in the occurrence primarily of infarct, atherosclerosis, and ischemic heart diseases. For example, mortality from myocardial infarct is highest in regions where saturated fats and refined sugars are heavily present in the food and where the drinking water contains little calcium and magnesium; in temperate and high latitudes, deaths caused by coronary diseases occur most frequently during the cold season of the year; a sudden drop of atmospheric pressure can lead to cardiac disasters; and so on. Held in especially high esteem are the studies of A. Keys (1957, 1963) on the link between development of the process of atherosclerosis and excessive consumption of animal fat in different countries of the world: these studies enabled the author to establish that a correlation exists between the two values. An international project whose work was published under the title "**The Geographic Pathology of Atherosclerosis**" (1968) dealt with atherosclerosis, one of the mass diseases of mankind. Stressed among other things was the dependence of atherosclerosis on the total number of calories in food (which is linked with fat intake) and the level of cholesterol in the blood serum. According to the authors of the project, geographic differences (in occurrence of atherosclerosis) are more significant than differences of race and sex.

Geographic characteristics of the distribution of heart and blood vessel diseases are discussed in greater detail in the book of Keller, Shchepin, and Chaklin (1993).

Malignant Tumors

Perhaps the first observation of a professional carcinogen was that made by P. Pott, who already in 1775 described the susceptibility of chimney sweeps to tumors of the scrotum. The beginning of the epidemiological way of thinking about malignant tumors (cancer) dates from a much later time with the work of R. Stern, who at the outset of 1842 gathered statistical data on mortality from cancer in the city of Verona. The geographic distribution of cancer was a subject that attracted increasing attention in the second half of the 19th Century and in 1915 Hoffman wrote a book on the rates of mortality and geographic pathology of cancer throughout the world. This was followed by a large number of urban investigations.

In the last half century, there have been many studies in the world devoted to the geography of malignant tumors. Registration of patients in many countries has made it possible to achieve a more comprehensive approach in studying these serious diseases and mortality caused by them. This has resulted in the discovery of local (regional) characteristics of their distribution.

The latest research indicates that cancer is (after cardiovascular diseases) the greatest killer in many industrially developed countries, and that it originates for the most part from environmental causes. Investigations for now are based on the hypothesis that the environment can be assumed to affect the incidence of cancer until it is proven otherwise. This means that geochemistry plays an important role in the etiology of this disease (Dissanayake and Changrajith, 1999). A good example is contamination of drinking water with nitrogenous substances (human and animal waste, nitrogen - containing fertilizers, etc.) in developing countries, where diseases caused by excessive nitrate content (cancer of the stomach and esophagus, metahemoglobinemia) are prevalent. Since penetration of the food chain by these chemical ingredients is for the most part a geochemical process, *medical geochemistry of cancer* has developed into a special area of research (Dissanayake and Werasooriya, 1987).

About five million people in the world die every year of malignant tumors, which means that approximately every tenth person to die was a victim of cancer. The number of new cases is almost twice as great, even not counting very early forms of certain tumors and slowly developing types of malignant tumors of the skin. More than half of all malignant tumors occur in developed countries, despite the fact that not even a third of the world's population lives in them (Z. Radovanovic et al., 1995). This can be attributed in part to the considerably longer life expectancy of the population of more developed countries, due to which more favorable conditions exist for tumors to act on the organism (to develop in them), and in part to other factors (chemical, radioactive, and other pollutants of air, soil, water, and food; viruses; smoking).

As for the location of malignant processes, the stomach is most often affected (close to 700 thousand new patients throughout the world annually), followed by the lungs (600 thousand), breasts (550 thousand), large intestine (more than 500 thousand), cervix (460 thousand), etc. Malignant tumors of the lungs and stomach are most frequent in men, while breast and cervical cancers are most common in women. Of the many possible factors leading to the occurrence of malignant tumors, diet and tobacco smoking are held to be responsible in the greatest number of cases (35 and 30%, respectively).

Characteristic of most of the 20th Century, the trend toward increased mortality from malignant tumors is especially noticeable in more highly developed countries. These diseases rank second among all illnesses of the present day in the majority of civilized countries: the USA, Canada, the greater part of Europe, Australia, and New Zealand.

A significant difference is present in the distribution of various kinds of malignant tumors in different countries. The greatest affliction among males is recorded in Uruguay, Scotland, and Belgium, while the greatest affliction among females occurs in Uruguay, Denmark, and Hungary. Appreciable differences are also evident in the frequency of distribution of individual forms of these diseases by regions. Thus, the frequency of incidence of liver cancer varies a thousandfold, being highest in blacks of the Bantu tribe in South Africa and lowest in the inhabitants of many regions of Canada, the USA, the Scandinavian countries, and the European part of Russia. The indices for cancer of the esophagus are also quite variable, the highest values being recorded in several countries of southeastern Africa, the lowest ones in Holland, Canada, and countries bordering the Baltic Sea. Hundredfold differences are recorded in the indices for cancer of the sexual organs: the highest values here are recorded in South China, Vietnam, Jamaica, and Uganda, the lowest ones in Israel and several Arab countries (Keller, Shchepin, and Chaklin, 1993).

There are many examples of direct or indirect correlation between the frequency of occurrence of cancer in different parts of the body and environmental conditions (Keller, Shchepin, and Chaklin, 1993). Intensive onco - geographic research is being conducted in the world due to the linkage of groups of malignant tumors with natural geographic and socio - economic factors. Thus, numerous maps of the distribution of different forms of cancer by regions were drawn up in the former Soviet Union, and attempts at onco - geographic regionalization were given in the **"Atlas of Cancer Mortality in Districts of the USA: 1950 - 1969" and the "Atlas of Mortality from Malignant Tumors in Cities and Districts of Japan: 1969 - 1971." The "Atlas of Cancer Mortality in China"** (1981) contains *basic maps* (with data on mortality by districts and by provinces in the case of rarer forms of the disease for the period of 1973 - 1975) and *auxiliary maps* (with data on administrative divisions, population density, climate, soil, etc.). The purpose of the authors was to isolate regions with increased risk of contracting cancer.

Study of the geographical and geological preconditions for occurrence of human diseases as an attribute of territorial comples can be widely used in onco - geographic research. However, a great deal of multi - disciplinary research is still needed in order to solve the *enigma of cancer*, one of the greatest of modern - day riddles. For example, medical geography has not been able to answer why lung cancer ranks first among all forms of cancer in England; stomach cancer in Japan, Russia, and Iceland; cancer of the oral cavity in India; skin cancer in Australia; cancer of the esophagus and liver in certain regions of Africa; and cancer of the urinary bladder in Egypt. *We probably need to learn more about certain natural (geological) and anthropogenic factors as causes of the disease. Perhaps a significant advance could be made using methods of medico - geological (geochemical, eco - geological) research.*

Infectious and Parasitic Diseases

The great *plague of Justinian*, which began in the East in 532 A.D., moved around the Mediterranean over the course of four long periods, each of them lasting 15 years. When the epidemic finally passed, previously inhabited countries resembled a desert. Mortality from the *Black Death* - as plague was often called on account of its severity and fatal outcome in many cases (in which the lungs were frequently attacked and the lips turned blue due to impaired breathing and facial cyanosis) - was extremely high. It is estimated that as much as a fourth of the population of Europe sometimes perished in the waves of greatest expansion of the disease (J. R. Paul, 1966). It was generally agreed that plague originated in the Far East (in China), but nobody had any idea as to the cause or nature of the disease, which attacked rich and poor alike.

For all that, *cholera* is believed to have caused the greatest number of deaths of all infectious diseases. India was long the homeland of this dangerous disease, and its spreading to other countries probably became more frequent after an East Indian firm started trading with Great Britain. Cholera epidemics and pandemics were discussed earlier in the text.

It is difficult to estimate the toll taken by infectious diseases - whether epidemic or endemic - among non - immune or semi - immune populations in the course of the past six or seven centuries. *Smallpox*, *measles*, *scarlet fever*, and *diphtheria* are only some of the diseases that are known to have existed. With introduction of the smallpox or measles virus among the non - immune population of North American Indians, some tribes were virtually wiped out. Two other terrible curses of colonial America were diphtheria and yellow fever, which raged through its colonies in great waves.

For centuries epidemiologists and other investigators tried to find the cause and attempted to classify various infectious diseases. Two significant discoveries were as follows: 1) that the local diet affects the incidence of pellagra; and 2) that infectious diseases can be spread by polluted drinking water. However, the turning point occurred in the era of Pasteur (from 1860 to 1890), when the *germ theory of infectious diseases* was formulated. The whole concept of the causes of infection then underwent a radical change, with stress being placed on microbes and bacteria, and Pasteur and Koch soon thereafter developed methods of coping with these agents (J.R. Paul, 1966).

During the first 50 years of the *bacteriological era,* a great deal of emphasis was placed on the role of parasites as causes of disease, while much less attention was paid to human resistance and the role of the environment in transmission or retention of infectious diseases.

Together with the development of epidemiology, opinions changed both as to the essence of the infectious diseases and the role of geographical and geological factors in formation of the epidemic process. However, those factors never disappeared from the field of vision of doctors engaged in research on causes of the spreading of epidemics. As I.A. Okhotin stated in 1889: "*In order to exist, every form of microorganism requires certain climatic, soil, meteorological, and other conditions, in short has its own geography.*" Today no textbook, manual, or monograph on the special epidemiology of any infectious disease fails to characterize the geographic distribution of that disease, and geographic maps and atlases devoted to the geography of infectious diseases have been published.

The main goal of *epidemiological geography* is to discover objective laws governing the spatial distribution of infectious diseases[89]. At the current stage of development of that discipline, several basic laws can be singled out according to Keller, Shchepin, and Chaklin (1993):

1. *Dispersion and concentration of epidemiological occurrences in space.* This regularity arises from two mutually opposed tendencies in the development of epidemics of infectious diseases: a) the widest possible dissemination among people; and b) adaptation of special nosological forms to certain territories;
2. *Spatial unevenness in the development of epidemiological processes.* This is obvious if we look at a map of the range of any infection (Fig. 6.1.) and is a consequence of many things, above all different biological and socio - economic factors. As an example of the influence of customs of the population, we cite the disease kuru (fatal encephalopathy caused by a slow virus), whose only endemic focus in the world (in Papua New Guinea) owes its existence to the custom of ritual cannibalism, which was until recently practiced by members of one of the local tribes (Fore);
3. *Suitability of the complex of geographical conditions for elements of the epidemiological process.* To be specific, it is known that the existence of a

[89] Proceeding from the premise that medicine is the basic subject and geography an auxiliary concern, A.P. Avtsin (1972) concluded that **geographic epidemiology** is a more correct name.

natural focus of disease is dictated by a certain climate, soil, vegetation, and fauna, and that destruction of the focus occurs when some member crucial to biocenotic interrelations drops out of the picture. In keeping with this regularity, the probability of spreading of malaria in France by agents imported from the tropics is exceptionally low on account of their non - adaptation to local carriers. Also, *Anopheles hyrcanus sinensis* is a carrier of malaria in China, on the Malay Peninsula, and in Eastern India, but does not transmit the disease to other parts of India; and

4. *Zonality and azonality of prerequisites and manifestations of the epidemiological process.* As a law, this is expressed fairly clearly. For example, as we approach the tropics, increases are observed in the number of parasitic, communicable, and infectious diseases; the degree of epidemiological tension; and the index of linkage between foci.

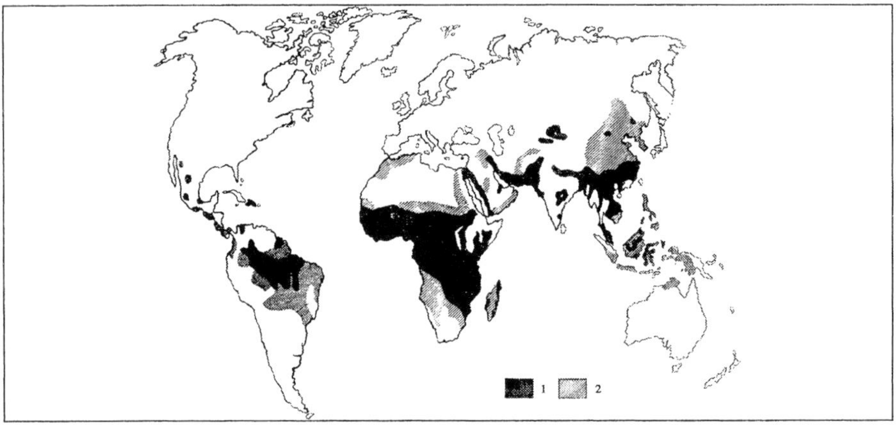

Fig. 6.1. Risks of malaria epidemics. 1 - high and average; 2 - low.

All of these laws influence formation of the epidemiological status of different territories, which determines the world's epidemiological geography (global epidemiology). By the term *epidemiological status*, we mean the sum total of typical (relatively constant) epidemiological occurrences that characterize a concrete territory.

RARE AND STRICTLY LOCALIZED DISEASES

In the history of culture and natural sciences, interest in *strictly localized diseases* remained for a long time in the shadow of *rare diseases* in the wider sense. Increased attention was devoted to those rare diseases that led to a noticeable change in the external appearance of people, to significant deviations in their behavior, and sometimes to completely inexplicable death. Until the advent of the first indications of *scientific teratology*, giants and dwarfs, immensely fat people, and bearded ladies were important attractions for the curious at fairs and carnivals. These unusual examples were generally imported from some distant land where such persons are completely ordinary or constitute a majority of the population there. The history of anthropology and travel contains familiar information about the "long - legged" tall Patagonians, who were exterminated by the colonizers of Argentina; and about the pygmies of Central Africa, who have survived to the present day.

Among the many *rare diseases* with their unusual complications, special attention of investigators has been attracted by illnesses that with greater or lesser justification are classified as *strictly localized* or endemic diseases. The causes of such endemism are sometimes clearly linked with a certain factor of the geographical or geological environment[90].

For specialists in the area of geographic pathology and especially medical geology, so - called *rare* and *strictly localized diseases* are particularly important because there is the real hope of linking their incidence or characteristics with some factor of the geographic or geological environment in the broader sense. This hope is known to have been justified in study of many *biogeochemical endemias*, natural infectious and parasitic diseases of the focal type, as well as certain pathological reactions associated with residence in extreme zones. For support of these notions, we could perhaps dwell on such well studied diseases as endemic goiter or yellow fever, which in no case can be counted among rare diseases.

Our attention will be focused on several *strictly localized diseases*, which are closely linked with the natural environment in which they are systematically registered (Fig. 6.2.). The still puzzling *Balkan endemic nephropathy* is without doubt one of them.

Examples of Rare and Strictly Localized (Endemic) Diseases

Kashin - Beck Disease. This disease was described in 1895 by N. I. Kashin and in 1906 by E. V. Beck, who called it *Osteoarthritis deformans endemica*[91]. Encountered in the Chita and Amur Provinces of the former Soviet Union, it occurs in strictly limited focal areas. Its main symptoms affect the system of the bones and joints, predominantly among younger persons. With the passage of years, sharply expressed deformation of the entire skeleton and especially the joints occurs in natives of the region. The first studies of this disease already indicated that it is endemic. Several hypotheses were suggested to explain its etiology. Among other explanations were the following: a) that the disease is caused by chronic poisoning of the population by spores of the toxin - bearing fungus *Fusarium sporothrycoides*;

---

[90] Our discussion of strictly localized diseases is given for the most part according to the book "**Introduction to Geographic Pathology**" by A.P. Avtsin (1972), with the reservation that we divide the examined factors into geographical and geological categories.

[91] *Hondroosteoarthritis deformans endemica* according to L.F. Kravchenko (1965).

and b) that it is an endemia of biogeochemical origin resulting from deficits of calcium, potassium, and sodium in soil and water in the presence of excessive amounts of strontium and barium.

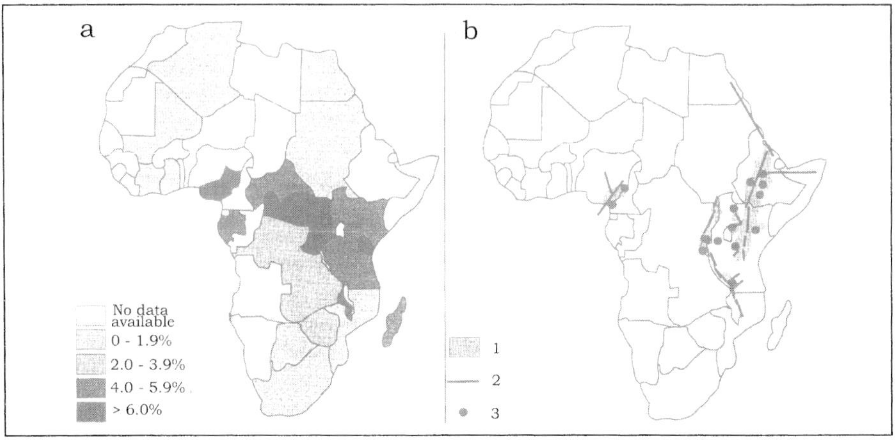

Fig. 6.2. The geographic distribution of Kaposi's sarcoma (**a**) and basalt (**b**) in Africa. 1 - Ultrabasic basalts; 2 - Rift systems; 3 - Area of podoconiosis.

According to Avtsin (1972), it is highly probable that Kashin - Beck disease is a biogeochemical endemia[92]. A disbalance of calcium, strontium, and barium is a very important feature of it. Climatic conditions also play a significant part in occurrence and especially manifestations of the disease.

Kuru sickness. This endemic disease of unclear nature was discovered in high - mountain regions of the eastern half of New Guinea (Papua). It was entirely clear that kuru sickness is disseminated among members of the Fore linguistic group of one of the Melanesian tribes. The sickness affected women primarily, as well as children of both ses, and it invariably culminated in death. The population suffering from this disease lived in forested regions at elevations of from 1500 to 2,000 m. Starting from 1960, the level of mortality in the given zone has declined, with no increase in territory on which the disease is present. Since that time, the whole tribe has in effect been in quarantine.

Kuru is manifested essentially as a wider ailment of the central nervous system, a serious disease with fatal consequences. The cause of this puzzling disease has been the subject of many investigations. Avtsin believes that it cannot be explained in purely geographic terms. The only known environmental factor is the high - mountain relief of the terrain where the disease is prevalent, which obviously promoted natural isolation of part of the Melanesian tribe in question.

[92] The disease was also discovered in the remains of inhabitants buried thousands of years ago, which indicates that similar biogeochemical conditions in which the disease occurred existed in ancient times.

Avtsin therefore concludes that large allowance must also be made for the influence of some other factors of a non - geographic nature.

New facts have been obtained using various research techniques, namely pathohistological, ecological, and experimental - virological methods. It is now clear that important facts of an ecological nature must unquestionably be taken into account in studying the epidemiology of kuru. Most important among them is the practice of ritual cannibalism, in which mainly women and children take part. In addition to other organs, it was the custom to eat the brain of the deceased. *Culinary treatment* of tissues of the deceased was extremely primitive, so that in the opinion of R. Glasse (1967) it was possible for the disease agent to be transmitted in appreciable quantities.

An interesting idea was put forward in 1966 by J. Hotchin. According to this idea, the kuru virus can be transmitted to man through domestic pigs and their parasites. The author did not exclude the possibility that the indicated virus was imported into this high - mountain region, where it encountered a genetically sensitive population and the favorable absence of hygienic conditions.

Vilyui encephalitis. The range of greatest distribution of this disease is confined to riverside localities and embraces a limited area where the Vilyui River empties into the Lena, a region characterized by very low winter temperatures. Here it predominately affects the residents of rural settlements formed on low - lying swampy terrains. The population of this region is mainly engaged in the raising of livestock. Mice and other rodents are very numerous in endemic localities. The disease is registered every year and is therefore of an endemic nature.

Inasmuch as Vilyui encephalitis mainly attacks the working population, it was logical to seek the source of infection in those forms of nature with which the inhabitants come into contact in various ways. The epidemiological significance of the hay gathering period and muskrat trapping season was stressed by A. N. Shapoval (1959).

In 1954 studies were initiated in which attempts were made to isolate the possible agent of the disease. Experiments were carried out by E.S. Sarmanova on white mice. Since that time, it has been assumed that Vilyui encephalitis is an independent viral disease, one that can only be understood by clarifying pathways of transmission of the virus to man and investigating its reservoir in nature. The question of the etiology and epidemiology of this disease remains a subject of discussion.

Buruli pustules. This disease is predominantly distributed on the banks of the Nile in Northern Uganda and on the shores of a number of lakes and rivers (including the Nile and Lake Victoria) in Central Uganda. The disease takes its name from one of the settlements where it has occurred most frequently. Special interest in this disease was shown by G. T. O'Conor and F. Lunn (1966).

The pustules appear primarily on the limbs and very often in the vicinity of joints. Extensive and deep pus fields embracing the muscles are formed in the course of further development of the disease. In some cases, the process of supporation extends even to the bones. MacCallum encountered such a disease in Australia and called it a *mycobacterial infection of man.* Two years later, Belgian investigators in the Congo rediscovered the disease agent and gave it the name

Mycobacterium ulcerans. O'Conor and Lunn indicated the possibility of infection by polluted water and stressed that increase of the disease is dependent on a high water level in rivers and lakes of Central Africa[93].

Biogeochemical Endemias (Microelementoses) of Man

We have already discussed the role of chemical elements in the vital activity and pathology of plants, animals, and man. Characteristics of some biogeochemical provinces or major biogeochemical endemias (microelementoses) of man will now be considered[94]. Here we draw on the book of Keller, Shchepin, and Chaklin among other sources.

Many large regions throughout the globe are known today where the soil and water are characterized by significant deficits of certain microelements or by their increased concentrations, circumstances that have an unfavorable effect on the health of living beings and which lead to the appearance of endemic diseases. It goes without saying that the manifestations of human pathology linked with microelements are themselves also extremely varied. This has served as the basis for isolation of a "*new class of diseases, viz., microelementoses, i.e., diseases in whose etiology the main role is played by a deficiency or excess of microelements in the human organism or their disbalance in the sense of anomalous ratios of micro - and macroelements*" (A. P. Avtsin, A. A. Zhavoronkov, and L. S. Strochkova, 1983).

The term *microelementosis* was introduced in the literature relatively recently, although many diseases and pathological states of the given nature were known previously. Deliberations about the role of deficits or excesses of certain metals and non - metals in the human organism were stimulated by the study of vitamins, as a result of which avitaminoses and hypovitaminoses became classical examples of deficit - induced diseases. It can be said with complete justification that they have now been joined by hypomicroelementoses.

Because they have a pronounced natural or anthropogenically dictated localization, many microelementoses are of great interest from the standpoint of both medical geography and medical geology. As we shall see, they are primarily dependent on geological factors (geochemical factors, to be more precise).

---

[93] It is interesting to note that in all four examples of rare and strictly localized diseases given in Avtsin's book, no mention is made of a possible geological factor in explaining any of them. Among other omissions, nothing is said about the geological (geochemical) composition of the studied region.

[94] With elevated content of Pb, As, F, Hg, Cd, Mn, Ni, and other elements, significant areas are occupied today by technogenic biogeochemical provinces formed around industrial concerns and urban settlements.

According to the classification of Keller et al. (1993), the microelementoses of man belong to four basic groups:

I. *Natural endogenous ones*, congenital (1) and hereditary (2);
II. *Natural exogenous ones*, caused by a deficit (1), excess (2), or disbalance (3) of microelements;
III. *Technogenic ones*, of the industrial - professional (1), neighborhood (2), and transgressive (3) types; and
IV. *Medico - elementoses* (microelementoses associated with intensive therapy).

Natural Endogenous Microelementoses

A microelementosis of the mother is usually responsible for *congential microelementoses*, accompanied by defects in development. The case of zinc deficit can serve as an example of such pathology: when a zinc deficit is present during pregnancy, congenital anomalies are observed in a significant percentage of the progeny. Together with zinc, deficiencies of copper, manganese, iron, iodine, and selenium can also be accompanied by certain anomalies.

Zones of distribution of goiter and cretinism represent the best - studied examples of the influence erted by anomalous biogeochemical situations on the fetus. Large regions in the Himalayas, Southeast Asia, Central Africa, the Andes, Oceania, and the Alps are such zones. Endemic cretinism has practically disappeared in Europe, together with improvement of socio - economic conditions. In many other regions, conditions for the appearance of endemic cretinism - manifested in mental retardation, deaf muteness, blindness, cross - eye, and growth lag - are created where pregnancy occurs in the presence of an iodine deficit.

In contrast to congenital microelementoses, in *hereditary microelementoses* the deficiency, excess, or disbalance of microelements is caused by damage to the genetic apparatus on the chromosome or gene level. *Wilson - Konovalov disease* can serve as an example. In this disease, profound disturbances are recorded in the metabolism of copper and copper - containing proteins in the organism.

Natural Exogenous Microelementoses

Deficits of elements registered in humans and animals can serve as examples of exogenous microelementoses. They include iodine deficits, selenium deficits, iron deficits, and zinc deficits.

Iodine deficit. Endemic goiter occurs not only in mountainous regions, but also on plains in tropical zones of South Asia, Africa, and South America, as well as in the valleys of the great rivers of Siberia and other parts of the globe (Fig. 2.39.). The wide dissemination of this endemia is determined not only by iodine deficit, but also by disbalance of other elements (Mn, Co, Cu, etc.). According to data of the World Health Organization,

200 to 400 million peoples suffer from endemic goiter throughout the world.

Selenium deficit. The problem of the selenium - deficit state in man began to receive attention following publication of a paper on *Keshan disease* in 1979. It was established that Keshan disease (endemic cardiomyopathy) occurs in a wide zone across Chinese territory from the northeastern to southwestern border. Low content of selenium in rocks, soil, and groundwater is recorded in the indicated zone. Administration of selenium to children with their daily meals significantly reduced occurrence of the disease.

The problem of carcinogenesis is another important aspect of the selenium - deficit state. Epidemiological investigations have demonstrated the existence of a pronounced negative correlation between consumption of selenium with food and mortality from cancer of the large and small intestines, mammary gland cancer, cancer of the esophagus, and lung cancer (Schrauzer et al., 1977).

Iron deficit. The iron - deficit state represents the most widespread microelementosis of mankind. The frequency of this microelementosis varies from 20 to 95% in different populations. In India and Central Africa, more than 50% of the population suffers from anemia. Very high indices of iron deficit in the population are registered among inhabitants of the Philippines, Pakistan, and Turkey. Iron - deficit anemias are especially pronounced in regions with extreme living conditions, such as high mountains and the Far North, where *acclimatization iron deficit* occurs.

Zink deficit. The zinc - deficit state was first registered in Iran (A. Prasad et al., 1961), where it occurred in young males in the form of a complex of symptoms: retardation of growth and sexual maturation, testicular atrophy, and other accompanying manifestations. An analogous complex of symptoms was discovered somewhat later in settlements near Cairo in Egypt. In both cases, the diet of patients was characterized by deficiency of animal proteins.

Endemic fluorosis represents a classical example of diseases and syndromes caused by an *excess of certain microelements in the biosphere.* It occurs when too much fluorine enters the organism. We have already discussed this question this and now only mention the fact that foci of endemic fluorosis are found on all continents (Fig. 2.6.). The problem of fluorine excess is of national significance in India, where a so - called *fluorosis belt* exists. The southern part of what was formerly the Soviet Union is characterized by fluorine - rich groundwater, whereas much of the northern part, Siberia, and the Far East represent a gigantic biogeochemical zone of hypofluorosis.

Endemic arsenosis is a disease caused by excessive entry of inorganic forms of arsenic into the organism with water and food. The best - known endemic provinces and foci are in Argentina, China, the USA, Mexico, and Japan. Arsenic - induced skin cancer does not differ histologically from other forms of the disease.

The already mentioned *Kashin - Beck disease* can serve as a typical example of a biogeochemical endemia linked with a disbalance of micro - and macroelements. The etiology of this disease has not yet been established, but great significance is assigned to high content of strontium, copper, molybdenum, and phosphorus in the biosphere, and low concentrations of calcium (disbalance of calcium, strontium, and barium according to Avtsin, 1972).

Technogenic (Anthropogenic) Microelementoses

Professional microelementoses are caused by an excess of certain microelements and their compounds in the zone of production itself. They have been known for a long time to professional pathologists and have been studied fairly thoroughly.

Neighborhood microelementoses (endemias) are diseases and syndromes that develop as a result of environmental pollution in the neighborhood of industrial concerns. Familiar throughout the world, the already discussed Mina Mata disease (mercuriosis) and Itai - Itai sickness (cadmiosis) can serve as examples.

Transgressive technogenic diseases of a chemical nature are linked with the transport of chemical pollutants over great distances as a result of atmospheric processes or water circulation (A. P. Avtsin, 1983). *Acid rain* is a classical example of transgressive occurrences in the biosphere.

It must be kept in mind that the pathology of technogenic pollution with microelements is characterized not so much by sharply expressed severe diseases as by latent forms that often show minimal symptoms. This is the case, for example, with microsaturnism (poisoning by low doses of lead of from 0.6 to 0.07 mg/day, the norm being 0.3 mg/day).

Medico - elementoses

These diseases are encountered practically everywhere in the world, especially in countries with a high level of medical protection. The given paradox is attributable to the fact that modern methods of treatment are becoming increasingly invasive, while the assortment of medicines has grown precipitously. Some medicines can be contaminated by certain microelements, admixtures of aluminum for instance. A survey of this type of microelementoses is given in the book "**Microelementoses of Man**" (A. P. Avtsin et al., 1991).

Urolithiasis

One of the oldest diseases of mankind, *urolithiasis* in its nature represents a broad group of various diseases and syndromes accompanied by formation of stones. Urolithiasis is characterized by very wide distribution throughout the world and is endemic in some regions. In spite of significant research on many questions, this still remains an inadequately studied area of theoretical medicine because the tasks themselves have turned out to be considerably more complex than appeared at first glance. Inclarities are especially present in study of this group of diseases from the standpoint of geographic pathology. *It is logical to hope that medical geology will throw additional light on the nature of most rare and strictly localized diseases, since the basis of their occurrence* (the geological environment with its factors) has not been clearly established. For example, there have been no precise statistical analyses revealing correlations between occurrences of urolithiasis on the one hand and chemical components of drinking water and food from soils of various compositions on the other, and no correlation has been found with the geological (mineral) composition of terrains in general[95].

Zones with greatly increased frequency of stones in the urinary ducts have been discovered in Syria, the southern provinces of China, India, lands bordering the Mediterranean and Caspian Seas, the Volga basin, Sweden, Northern Australia, parts of Madagascar, the southern states of America (especially Florida), the eastern part of Mexico, and Argentina. However, beyond the establishment of clear foci of urolithiasis, not enough progress has been made in clarifying the cause of this medical phenomenon, although many hypotheses have been put forward to explain it. There is a relatively voluminous literature on this question. We shall dwell only on certain points of interest to us.

Earlier authors classified stones in the urinary ducts on the basis of the presumed leading role of anions in stone formation and singled out oxalate, phosphate, carbonate, and urate stones. It was later clarified that cations, especially calcium cations, are responsible in the majority of cases of inorganic stones. In his tabular presentation of diseases that develop with formation of stones in the urinary ducts, Avtsin (1972) indicated that the created stone is often composed of calcium phosphates. It is evident from our examination that the occurrence of urolithiasis is linked to a significant extent with regions having the limestone composition typical of the calcium - bicarbonate type of groundwater (the Mediterranean region or southern provinces of China, for instance). Here it should be kept in mind that magnesium content can be appreciable on such terrains, a circumstance that is reflected in perceptible reduction of calculus formation (E. Trimmer, 1987).

---

[95] Avtsin rightly concludes that study of the geography of urolithiasis solely from the statistical data of urologists must be considered unpromising due to the complexity of the problems involved. It would be far more correct to organize multi-disciplinary expeditions in already registered endemic foci. This would make it possible to study prelithiasis more closely and clarify the factors that cause it. In this sense, experiments with animals are justified. For example, it would be interesting to test the potential lithogenicity of hard water.

Endemic silicon urolithiasis was singled out by V. V. Koval'skii and V. L. Suslikov (1980). According to these authors, the indicated disease is a consequence of human non - adaptation in special biogeochemical provinces to increased intake of silicon into the organism in combination with high content of fluorine, calcium, manganese, nitrates, sulfides, and chlorides in the biosphere. If this view is confirmed, then silicon urolithiasis can join the list of polymicroelementoses. Figure 6.3. presents the distribution of siliceous rocks and endemic urolithiasis on territory of the former Soviet Union. Characterized by decline of phosphorus reabsorption in the kidneys, disturbances of phosphoro - calcium metabolism are recorded in silicon biogeochemical provinces here.

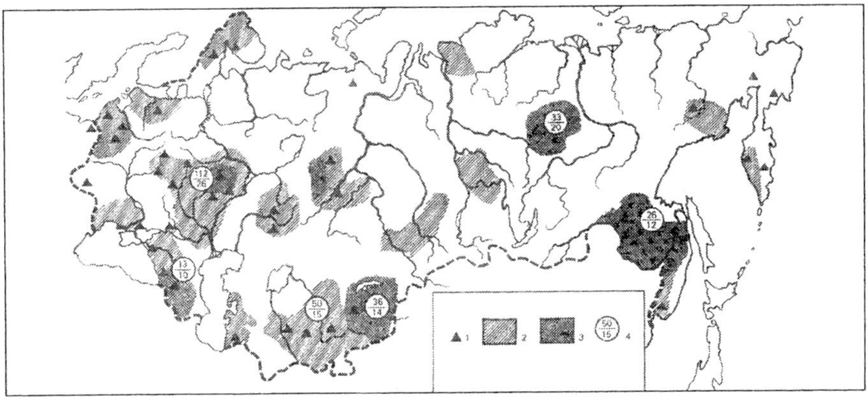

Fig. 6.3. Space distribution of silica rocks and endemic urolithiasis in the territory of former USSR. 1 - exploitated silica deposits; 2 - territories with endemic urolithiasis; 3 - specifically endemic urolithiasis; 4 - marker of public illness.

Questions Connected with the Etiology of Endemic Nephropathy

Up until 1957, nothing was known about Balkan endemic nephropathy as a special kind of chronic nephropathy. The disease was then described for the first time on the basis of medical examinations of inhabitants of the village of Sopic near the Serbian town of Lazarevac (V. Danilovic et al., 1957). By examining church records, Danilovic was able to verify that endemic nephropathy first appeared in this region around 1922[96].

According to S. Strahinjic (1985), endemic nephropathy represents a chronic slowly progressive disease that occurs in endemic foci and endemic regions and has a family - linked nature, unknown etiology, and

[96] In the last two decades, the qualifier Balkan has been used for the nephropathy described in that part of the world to distinguish it from other nephropathies described at the same time in Nigeria, Japan, and elsewhere.

inadequately clarified pathogenesis. It is accompanied by progressive and gradual degradation of the mass of functional nephrons, together with reduction in the anatomical volume of renal tissues, which after long - term evolution of the pathological process leads to terminal insufficiency of the kidneys. Endemic nephropathy is today accepted and described as a special nosological unit.

The existence of endemic foci has been described in all three parts of Serbia (Serbia proper, Vojvodina, and Kosovo), as well as in FYROM, Republic Srpska, Croatia, Bulgaria, and Romania (Z. Radovanovic, 1985).

On the territory of Serbia, the most complete data are available for villages in the valley of the Kolubara River in the vicinity of Lazarevac (Petka, Sopic, and Cvetovac) and the village of Pridvorica (near the town of Lajkovac). The percentage of sufferers here is very high and ranges from 10.23% (Petka) to 6.94% (Pridvorica). The endemic region in the valley of the Southern Morava River embraces villages in the area between Aleksinac and Leskovac (Fig. 6.4.). The following rural areas are conspicuous for the number of patients: Pukovac, Brestovac, Pecenjevci, Moravac, Orlane, and Donje Brijane. The villages with the highest percentage of sufferers are Mezgraja (6.2%), Donja Trnava (2.8%), and Kutles (2.6%). Known endemic settlements in Vojvodina border on the municipalities of Sid and Sremska Mitrovica. The only nephritis focus in Kosovo and Metohija to be described so far in the literature is Vitina with the surrounding settlements. The total number of settlements in Serbia that have been linked with endemic nephropathy comprises 376, which amounts to 6.2% of all settlements in the republic.

In Republic Srpska, the only region under threat is land along the Sava River in the northeast. In 1982 the municipalities of Bosanski Samac and Beeline had the greatest incidence of the disease (0.72 and 0.71%, respectively), which was also significant in Brcko (0.50%).

Foci of endemic nephropathy in Croatia are primarily found on territory of the municipality of Slavonski Brod. The disease has been recorded in 14 villages located in a lowland region with a high level of groundwater.

In Bulgaria, endemic nephropathy occurs on the territory of two districts, Vraca and Mihajlovgrad: in 52 settlements (or 38.5% of all settlements) in the former and four settlements (or 3.0%) in the latter.

The Romanian foci are found in the neighborhood of the Danube on both sides of the Carpathians. The disease occurs on the territory of five districts, with a total of about 40 endemic settlements.

In former Yugoslavia, all of the endemic settlements identified to date are located in close proximity to river courses belonging to the Danube watershed. Those in Serbia are for the most part in floodable regions built of alluvial deposits. Another characteristic feature of endemic nephropathy is their confinements to rural settlements, i.e., patients with this disease as a rule are farmers. The periodic nature of its appearance represents a significant characteristic of the disease.

The causative factor is the crucial question in the case of endemic nephropathy. In order to establish it, living agents (bacteria, viruses, fungi) were tested initially, followed heavy metals. Significant efforts have been devoted to testing the eventual etiological role of drinking water. Experiments have been conducted on animals, heredity has been considered, and immunity aspects have been tested.

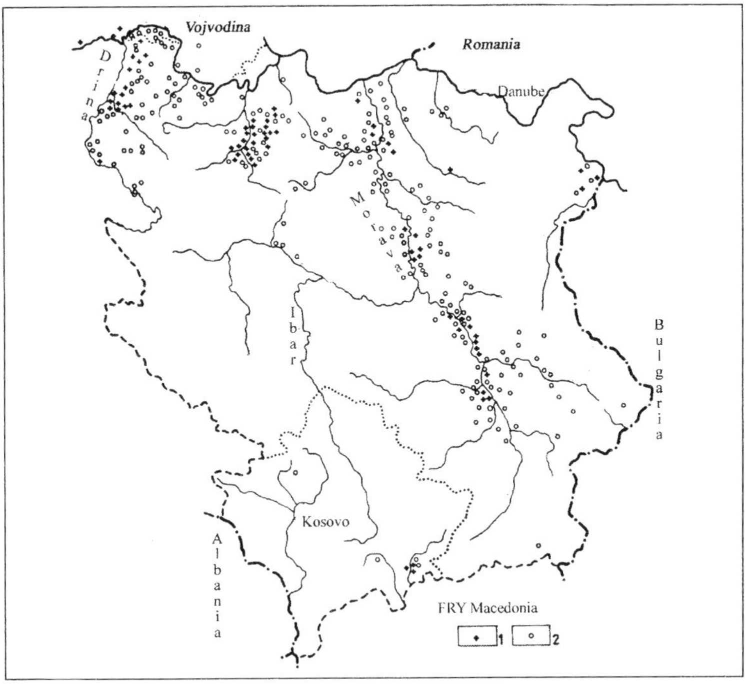

Fig. 6.4. Position of villages with certain and uncertain diseases foci in Serbia
1 - certain foci; 2 - uncertain foci (J. Peric, 1985).

According to Z. Maksimovic (1985), the geographic distribution of endemic nephropathy indicates that this disease is linked with the geochemical environment, as is also the case with other well - known endemic diseases of man, which is sometimes treated as *geochemical diseases*. Most investigators of the etiology of endemic nephropathy have suspected chronic intoxication as a consequence of ingesting subtoxic doses of one or more microelements. In connection with this, lead, nickel, chromium, silicon, uranium, and other radioactive elements have been mentioned as possible causes of this disease, followed by cadmium, boron, beryllium, and nitrates. However, testing of water from wells, rocks, and river sediments in focal regions indicates that practically all of the analyzed microelements are present in amounts significantly below hygienic norms. According to Maksimovic, these data suggest that the possible cause of the disease may lie among chemical elements deficient in water and food of the affected villages. He recommended systematic analysis of rocks, soil, water, and crops in regions threatened by the disease and proposed the compilation of geochemical maps of microelements, primarily ones with anomalously low content.

Hydrogeological factors have been analyzed over the course of many years by J. Peric (1985), who investigated the regions around the villages of Petka and Stubica (in the vicinity of Lazarevac) and Kaniza - Klakar (in the neighborhood of Slavonski Brod). In both endemic and control villages, the electroconductivity of water and its temperature were measured repeatedly; depth to the level of groundwater was determined; and the content of radon, nitrates, boron, and silicon dioxide in water of all dug wells was tested. Also tested were the levels of microelements in water of selected wells. It was concluded that the probability of occurrence of endemic nephropathy increases significantly if the water is mineralized, if polluted water from shallow wells is used for drinking purposes, and when the water has high content of silicon dioxide, boron, and radon. Due to the extremely low rate of real movement of groundwater, in areas where wells are used there is practically no exchange of water during the year in the investigated region, or else its circulation depends on the amount of polluted water. On the other hand, pollution of groundwater by organic substances - which is amplified when flood waves and states of high water occur - is exceptionally great and constantly growing.

Living agents (bacteria, viruses, and fungi) have also been tested as a possible etiological factor in occurrence of the disease. Among other things, it has been observed that fungi of the genus *Aspergillus* are considerably more frequent in materials taken from households where someone suffered from nephropathy or died of it. According to B. Birtasevic et al. (1985), the clinical course of sickness, pathomorphological findings on the kidneys and (especially) significant epidemiological characteristics of the disease suggest an infectious (viral) nature of endemic nephropathy as a disease of the group with natural foci.

We mention in conclusion that V. Momcilovic (1995) investigated the focus of endemic nephropathy in the village of Brestovac near Leskovac. She reached the conclusion that the disease occurs in residential buildings located on privileged hydrogeopathogenic routes of alluvial sediment, which is unlikely in the case of an intergranular water - bearing medium.

Despite the considerable efforts of many Yugoslav and foreign investigators, the causes of Balkan endemic nephropathy (EN) are still unknown even after several decades of multi - disciplinary research. Such a conclusion was reached at the Fourth Symposium on Endemic

Nephropathy (Belgrade, October 1982). At this meeting, Z. Radovanovic (1985) closed the introductory lecture with the following remarks: "*It can be concluded with no shame that we have not yet discovered the cause(s) of EN. The etiology of many other diseases in whose study thousands of experts throughout the whole world are engaged likewise remains unknown. Still, there is a sense of frustration that no basic facts about the epidemiology of EN have been established in the course of 25 years.*"

In our opinion, all that can be said is that a certain geological (hydrogeological, geochemical) environment - an overlying horizon of alluvial sediment in which water probably transmits the unknown agent(s) - plays a part in occurrence and maintenance of the disease. Until we learn the true etiology of endemic nephropathy, it is imperative that water from shallow wells not be used for drinking purposes, and that some new source is found in the threatened villages. Capping of the main (deeper) alluvial aquifer is a rational solution in the given conditions.

CHAPTER 7

Subject and Tasks of Geological and Medical Disciplines in Defining Quality of the Environment and its Influence on Human Health

SUBJECT AND TASKS OF GEOLOGICAL SCIENCES AND DISCIPLINES

Like all of the natural sciences, geology has developed as a result of accumulation of the data of investigators and experiments and their theoretical synthesis. The current level of development of the geosciences is based on the efforts of many scientists (investigators) whose work laid the foundations for our modern understanding of structure and dynamics of the Earth.

The geological sciences are concerned with the composition, structure, and historical development of the Earth's crust; geological processes transpiring in its interior and on its surface; and forces and factors that cause those processes. The geologist primarily investigates the upper parts of the Earth's crust, i.e., those parts that are accessible to direct investigation. In so doing, he studies the kinds, composition, age, and distribution of rocks; the manner of their formation; and their mutual (structural) relations. He relates the rocks he studies to formations in neighboring districts or other (often very distant) regions. In addition to rocks, the subjects of study include various endogenous and exogenous forces, occurrences, and processes, as well as mineral raw materials and groundwater. Today it can be stated that vast regions have been studied and the geological structure of great parts of the continents established. Largely owing to constant demands of the economy, the geological sciences are at the present time developed to such an extent that they are able to resolve practically all problems linked with the environment in which man lives and performs his various activities, namely the geological environment.

Knowledge of geology enables us to gain a deeper understanding of everything surrounding man. It helps us to recognize the beauties of nature; makes it easier to overcome our superstitious terror in the face of powerful, sometimes terrible and destructive, geological occurrences; and enables us to make use of mineral raw materials and groundwater. In addition to all this, such knowledge makes it easier to resolve various ecological problems and helps us to recognize positive and negative factors of the geological environment from the standpoint of human health.

With nearly two hundred separate scientific disciplines, geological science has permeated every pore of contemporary civilization.

It is therefore right to ask whether schools as a whole provide enough knowledge from the domain of geology as a basic natural science of broad cultural significance apart from its practical (economic) importance[97]. Does the problem of inadequate education in general partly conceal frequently inadequate treatment of the environment in the framework of medical geography and nonexistence of medical geology?

The *fundamental geological sciences* are devoted to study of the Earth's structure, origin, and historical development. The list of these sciences includes crystallography, mineralogy, petrology, geochemistry, geophysics, geodynamics, historical geology, paleontology, structural geology, and geomorphology, in addition to their various subdivisions (for example, sedimentology as a part of petrology, paleobotany as a part of paleontology, etc.).

Questions of utilization of the geological environment are the concern of a number of disciplines, namely the science of deposits of mineral raw materials, coal, oil, and gas; hydrogeology; engineering geology; pedology, geo - ecology; etc. Each of these disciplines of *applied geology* has developed its own fundamental base.

Fundamental Geology

In the area of fundamental geological sciences, there are three main directions (branches), each of which studies the Earth's crust from a different aspect:

---

97 The conservation of cultural monuments, for example, is today largely based on knowledge of the petrological, hydrogeological, and other geological characteristics of a given locality (S. Dragutinovic, 2000).

1. *Sciences concerned with composition of the Earth's crust*, often jointly referred to as sciences of the geochemical cycle (crystallography, mineralogy, petrology, geochemistry, etc.);
2. *Sciences dealing with geological processes or dynamics of the Earth* (geodynamics, geophysics, volcanism, seismology, tectonics, etc.); and
3. *Sciences that study the history of development of the Earth* (historical geology or stratigraphy, paleontology).

As the science of crystals, *crystallography* treats their external form and internal structure. Because natural minerals are in most cases crystalline bodies, study of their forms and the laws governing their formation is of great theoretical and practical significance.

Mineralogy is the science of minerals. Minerals - quartz, mica, pyrite, etc. - are unique chemically homogeneous bodies formed as a result of various physico - chemical processes transpiring in the Earth's crust. With properties dictated by their chemical composition, minerals are stable under the given geological conditions. Mineralogy deals with the chemical composition of minerals, their physical properties, and natural processes leading to formation of minerals and their associations. Different combinations of minerals form rocks.

Petrology is the science of rocks, of which the Earth's crust is composed. It treats the properties of rocks, their chemical and mineral composition, relationships between different rocks, changes of rocks with the passage of time, and regularities governing their formation and distribution in the Earth's crust.

Chemical elements and their behavior in the Earth's crust and deeper layers of the planet constitute the subject of *geochemistry*. The geochemistry of processes deals with migration of chemical elements in certain processes transpiring in different parts of the Earth's crust. The geochemistry of systems treats migration of chemical elements in different systems (abiogenic, biogenic, and technogenic) characterized by opposed interconnected processes. The behavior of a certain element in different processes and systems is studied in the geochemistry of elements.

Two major problems of our epoch are closely linked with geochemistry. They are the problem of the environment and the problem of mineral resources. Methods of geochemical research are very important in dealing with these problems. *Environmental geochemistry* can be singled out as a discipline of particularly great interest to us. It deals with the distribution and interrelations of chemical elements and radioactivity in the system composed of rocks, soil, water, air, and living organisms, doing so mainly from the standpoint of human health.

Under modern conditions of diversified geological research, *geophysics* has a special place in both global (regional) and local investigations. In the era of rapid technological progress and high - precision instrumentation, the possibilities of geophysics have become practically unlimited, and it is fair to expect much of geophysical research in the future. The subjects of study of the Earth's physical properties include gravitation and magnetic, seismic, and electrical properties of the lithosphere or parts of the lithosphere.

Geodynamics is the science of processes occurring in the Earth's crust and its superficial zone. Depending on the source of energy, it is possible to distinguish internal (*endogenous*) and external (*exogenous*) processes. Endogenous processes are linked with internal energy and include volcanism, earthquakes, and tectonic movements. Exogenous processes are in the final analysis linked with solar energy. They include the work of wind, surface water, glaciers, seas, etc. Geodynamics includes *geotectonics*, the science of rock formation and deposition, movements of the Earth's crust, and deformations that arise in it. Geotectonics in part borders on *geomorphology*, the science of relief of the Earth's surface, formation of that relief, and the process of its development. *Structural geology* deals with the structural pattern of the lithosphere's surface layer and that of concrete regions, i.e., the mutual disposition and relationships of rock masses in the narrower sense.

Historical geology (stratigraphy) is the science of the Earth's history. It reconstructs the course of geological events from the oldest times to the present moment. The main historical documents used to interpret the history of changes in the planet's surface are rocks and fossil remains of animals and plants in rocks, which constitute the subject of *paleontology*. Preserved in sedimentary rocks formed in the course of the Earth's geological development, the remains of old organisms and traces of life in general represent relatively reliable evidence on whose basis the stratigrapher can determine geological age of the rocks.

Applied Geology

Geology as a science studies our whole planet, but it also treats in greater detail narrower territories of interest from the practical standpoint. The constant demands of practice have brought about a situation in which today we have a number of very well developed disciplines of applied geology.

The *science of mineral deposits* is concerned with research on deposits of various metals, non - metals, and geological building materials. Energy reserves are the subject of *coal geology* and *petroleum* and *gas geology*.

As the science of groundwater, *hydrogeology* treats its origin, movement and evacuation, physico - chemical composition, and relationship to surface water. This branch of applied geology also considers the possibilities of utilization of groundwater for definite purposes and questions of protection of groundwater. There are a number of subdivions of hydrogeology, namely water supply engineering, mining hydrogeology, the hydrogeology and oil and gas deposits, hydrogeothermics, melioration hydrogeology, hydrogeochemistry, protection of groundwater, etc.

Engineering geology is a geological science concerned with properties and dynamics of the thinner superficial part of the geological environment primarily, its rational use and maintenance, and construction and exploitation of civil engineering works. Special attention is paid to the properties of rocks, engineering conditions at locations selected for construction projects, and the carrying capacity and stability of terrains. Existing parallel with engineering geology are *soil mechanics*, a scientific and technological discipline engaged in studying that part of the geological environment built of sands, gravels, and clays; and *bedrock mechanics*, which treats the properties of bedrock as a medium for construction projects.

Pedology is the science of soils. It deals with the physical, chemical, and biological characteristics of different kinds of soil and other factors of the geological environment connected with the needs of agriculture and forestry. *General pedology* is devoted to fundamental scientific research. In addition to it, today we also have *applied pedology*, whose purpose is to establish the suitability of soils for different uses.

SUBJECTS AND TASKS OF MEDICAL SCIENCES AND DISCIPLINES

The history of medicine is extremely long, and man's striving to discover the causes of disease is as old as his efforts to create religion. The transition from supernatural to natural explanations of disease was made already in Ancient Greece, and Hippocrates in that era suggested that elements of the natural environment are direct or indirect causes of disease[98].

Beginning with ancient medicine, the first stage in the development of medical science lasted all the way up to the 19th Century. The year 1800 represents a turning point after which the scientific approach to study of diseases was adopted much more widely. A number of increasingly notable advances in the area of health protection were made at this time, and at the end of the 19th Century the battle against epidemics became incomparably more effective owing to great achievements in the field of microbiology.

During the 20th Century, the highly developed natural sciences - biology, chemistry, and physics - were of enormous significance for the creation of firm theoretical and procedural foundations of the medical sciences. On the other hand, the sudden technological boom especially advances in electronics, automation, and computer technology, made possible accelerated development of new approaches in the diagnosis and treatment of diseases. In this way, modern medicine has become a very complex and involved system of sciences and scientific disciplines.

A considerable part of medical science is of interest from the standpoint of medical geography and geology. That part is treated in the introductory section, where interaction of medico - biological, geographic, and sociological disciplines in solving medico - geographic problems is presented. An idea of the subject and tasks of medical sciences and disciplines can be gained by dwelling on several of them.

---

[98] According to J.R. Paul (1972), this ancient example of observation and conclusion might have laid the foundation of the inductive method of studying diseases. "Had those who came after Hippocrates followed his lead, man's knowledge of diseases might have gained the advantage of several centuries."

Pathology occupies a central position among theoretical medical sciences. It is engaged in studying basic regularities governing the origin, development, and outcome of diseases in the living organism. The main task of *general pathology* is to learn what we can about the main pathological processes, viz., disturbances in metabolism (dystrophies), circulation (dyscyclias), and growth, as well as inflammations, tumors, and developmental anomalies. General pathology also embraces study of the causes and conditions of appearance of pathological processes (general etiology), as well as study of the origin and course of development of disease processes (pathogenesis). In contrast to general pathology, *special pathology* relies on definite criteria to elucidate basic pathological processes in individual organs and organ systems[99].

Pathology represents a very broad area of science, one in which a number of specialized branches have developed. They include, for example, *neuropathology, nephropathology, psychopathology*, and *biogeochemical pathology*. Of special interest to us is *geographic pathology*.

According to A. P. Avtsin (1972), *geographic pathology* represents an independent science engaged in studying regularities governing the origin, spread, and characteristics of the incidence of human diseases in different geographic zones of the Earth. The term *geographic pathology* was introduced in the medical literature by A. Hirsch in 1856. An especially important moment in the development of this science was formation of the International Geographic Pathology Society at the initiative of M. Askanazy in 1929. From then until 1970, the Society held ten conferences dealing with various questions of geographic pathology.

Study of the pathogenesis of human diseases represents an important scientific task of geographic pathology. Geographic pathology does not simply register some pathology of man in a given region; it also attempts to explain the origin and form of occurrence of that pathology on the basis of those data and data obtained in experiments.

Finally, if we analyze endogenous and exogenous pathogenic factors that have been examined by geographic pathologists throughout the long period of their activity, we notice that many of these factors are referable to the inorganic (geological) environment. Such factors include, for example, the geological and geochemical characteristics of rocks, properties of drinking water, biologically significant concentrations of micro - and macroelements in food (dependent on their content in soil and rocks) causing more serious pathological processes in humans and animals, etc. All this and many other considerations mentioned in Avtsin's 1972 book lead us to conclude that it is justifiable to single out a special branch, namely *geological pathology*, within the framework of medical geology.

---

[99] Pathology in the broader sense of the word is not confined to changes in the organism of man and animals (*human* and *veterinary pathology*), but also deals with changes in plant organisms (*plant pathology*).

Physiology is the biological or medical science of vital phenomena and functions of the organism. It studies the functions of cells, cellular organelles, organs and systems, and organisms as a whole[100]. The word *physiology* in its present sense appeared in the 16th Century. In the history of this science, especially great progress was made in the 19th and 20th Centuries, when accelerated development occurred in biochemistry, physiology of the cardiovascular system, respiratory physiology, physiology of the kidneys, and neurophysiology. The latest research in physiology likewise is of a biochemical nature.

General physiology treats general regularities and functions common to all living beings regardless of the degree of their internal organization. Some authors use the term *cellular physiology* in this case, since life is studied predominantly at the cellular level. *Specific physiology* deals with only certain groups of living beings (human physiology, the physiology of domestic animals) or special areas of physiology such as neurophysiology, enzymology, hematology, etc. *Pathological physiology* studies the functions of organisms and their disturbances in states of disease.

Ecological physiology developed in the second half of the 20th Century. It is a branch of physiology engaged in studying the influence of natural environmental factors on formation and performance of physiological functions in the organism. It is logical that this physiological discipline is most directly linked with medical geography and medical geology.

Epidemiology is the medical science dealing with regularities of the epidemiological process and methods of combating infectious diseases of man. As was already mentioned, a turning point in the development of this science occurred in the 1880's, i.e., the era of Pasteur, which was marked by great bacteriological discoveries. A full contribution was later made by *experimental epidemiology*. We also mention *epidemiological geography* and its role in explaining the origin, endemism, incidence, development, and disappearance of various infectious diseases in different parts of the world.

Toxicology is the science that deals with the harmful (toxic) effects of chemical agents on living organisms. Its subject matter includes toxic agents (toxicity), the living organism (as a biosystem that reacts to toxicants), and the toxic effect (the sum of all physiological and biochemical changes). Enormous growth of the chemical industry in the 20th Century was accompanied by rapid development of *professional toxicology*. This discipline has been influenced by modern medical techniques and technology, which enable us in large measure to clarify the processes that occur when a toxin enters the organism. The exceptionally rapid development of biology (especially molecular biology) in the last decade of the 20th Century has made it possible to study the action of toxic substances on the molecular level, which is the subject of *molecular toxicology*. With the task of studying the harmful effects of pollutants on living organisms in the environment, *eco - toxicology* (environmental toxicology) has taken its place in recent years. We note also *aquatic toxicology*, the scientific study of toxic effects of an aquatic environment on its living organisms.

Medical microbiology and **parasitology** are branches of the medical and biological sciences devoted to studying the biological agents of human diseases

---

[100] The *physiology of domestic animals* has been developed parallel with human physiology.

and carriers of the agents of these diseases, as well as their relations with man and the environment in which he lives. These disciplines include clinical microbiology and parasitology, immunology, and sanitary microbiology and parasitology.

As the medical science engaged in studying the influence of environmental factors on human health, working capacity, and life expectancy, **hygiene** elaborates norms, requirements, and various measures designed to achieve the sanitation of settlements and ensure healthy living and working conditions for their people. The branch of hygiene that treats general regularities of environmental influence on human health and the procedures used to study them is called *general hygiene.*

During the earliest stage of its development in the 19th Century, hygiene relied on principles of medicine, physiology, and pathophysiology, which gave rise to the prophylactic concept. The next phase of development was possible only after suppression of the epidemics that carried off hundreds of thousands of lives. The main attention in this phase was transferred to *communal hygiene* and related disciplines, namely the *hygiene of work* and *nutritional hygiene.* In a period of increasing environmental pollution, the main task of the new hygiene is to create conditions in which production does not represent a threat to the environment and does not destroy the optimal ecological equilibrium. In the narrower sense, *environmental hygiene* represents a branch of medical science that elaborates medical aspects of environment protection and scientific ways of preventing unfavorable influence of environmental factors on the population.

Balneoclimatology is a medical discipline that deals with the action of natural physical and chemical factors of the environment on the human organism and the possibility of using them for medical purposes[101]. The natural factors used most frequently are mineral waters, peloids (medicinal mud), carbon dioxide, oxygen partial pressure, heat and cold (mainly in general procedures), and hydrostatic pressure. Depending on which of these factors is used, balneoclimatology is divided into *human bioclimatology* and *balneology.*

Human bioclimatology treats the relationship between the human organism and climate and the direct effect of weather conditions on health and sickness. *Medical meteorology* is a narrower area of bioclimatology dealing with the effect of short - term weather conditions and phenomena on the human organism with respect to a certain disease. On the other hand, study of the effect of long - term weather conditions (climate) on health or sickness is the subject of *medical climatology.*

Balneology deals with the action of balneological factors and the possibility of using them for medical purposes. The list of such factors includes: 1) mineral (medicinal) waters with varying degrees of mineralization, temperature, or radioactivity; 2) peloids of inorganic or organic origin having suitable chemical composition and physico - chemical properties; and 3) therapeutic gases of varying composition, including ones having adequate amounts of radioactive emanations (radon).

---

101 According to the book "**Balneoclimatology**" (Ed. by T. Jovanovic, 1997).

Every medical specialty of a theoretical, clinical, or prophylactic profile has its own *ecological aspects*, whose elaboration is not only desirable, but increasingly necessary. Generally regarded, the ecological approach in medicine involves examination of the interdependence between the human organism and factors of the natural and social environment, and study of environmental influence on the organism. **Medical ecology (human ecology)** - especially ecological physiology as one of its components - has developed rapidly in the last decades.

CHAPTER 8

Methods of Medical Geology

Nowadays, voluminous and different literature in geosciences, medical science, medical geography, biology, chemistry, physics and other scientific disciplines interesting from the aspect of medical geology is present. Thanks to wide scientific, practical and social activity, particularly in USA, Russia, and EU and worldwide, as well as opportunities of modern technology, numerous research methods to solve various theoretical and practical problems are developed within the mentioned sciences. This is why the range of methods and procedures important for medical geology is so wide but, at the same time, made difficult, while the main selection is expected to be performed during the developing stage of the discipline. Namely, during development of medical geology and solving concrete tasks, not only methodological basis will be improved, but also methods of interest will be developed and applied.

On the basis of experience and scopes of medical geography, here is the preliminary choice (review) of the methods. The concept and methodology of research, like also application of methods, depend on certain scientific discipline, in other words - of the problem stated.

Elements (factors) which have to be known within application of the presented methods are discussed in the Chapter 2.

1. Remote sensing. Series of contemporary, fast, cost - effective, reliable exploration procedures is included. Essence of this methodological approach is based on analysis and interpretation of terrestrial, aero - and space records. From the records, abundance of data related to the environment - relief, lithological content and structure of the terrain, but also surface and (indirectly) ground waters, vegetation, etc., is obtained. This is reason why remote sensing is unavoidable integral part of almost all sciences and scientific disciplines dealing with the surface of the Earth's crust and appropriate natural values. Nowadays, the method is widely used for solving various scientific and practical goals - within geological, geomorphologic, geographic and biological exploration, in forestry, soil science, land - use planning, environmental monitoring, etc. Note also successful application of the methods of remote sensing in study of the consequences to the environment caused by underground burning of coal at the territory of the northern China (X.M.Zhang, C.J.S.Cassells, J.L.van Genderen, 1999) (Fig.2.34.).

Long - standing application of stereoscopic analysis of aero - snapshots and accumulated rich experience are reason why the procedure is at the leading place among methods of remote sensing. Wide exploration of the environment by material of cosmic records also presented numerous advantages. Satellite records, however, offer a possibility of study large areas and enable new view on the Earth and appropriate geological, geomorphologic and geographic units.

Possibilities of the remote sensing application are neither completely investigated yet, nor recognized in total.

2. Geomedical mapping. This method is irreplaceable way for regional study of characteristics of the interesting natural environment. In practice, majority of geomedical tasks is not possible to be solved without direct link with terrain.

Field exploration for concrete scientific and practical purposes can be planned successfully on the basis of the studied literature and fund documentation, maps obtained according to processing and interpretation of aero - and satellite images, available geological maps, maps and atlases in domain of medical geography and medicine, etc. Goal in this stage of medico - geological exploration is to collect all necessary data related to various appropriate natural factors which could be of negative or positive effect onto the health of inhabitants, zones of high risk and potential natural foci of infection; to analyze and record: rocks and soil, engineering - geological and geomorphologic characteristics of the terrain; to outline rayons with high radioactivity, geophysical and geochemical anomalies; to register used sources of drinking water; to investigate resources of mineral and thermal waters, as well as rayons interesting from the aspect of balneoclimatology and health tourism; distinguish vegetation as indicator of negative effect of biogeochemical factors; to register medicinal herbs; to locate potential and present contaminants of the environment (mining and industrial structures, agricultural contaminants, landfills, etc.). From the medical viewpoint, in the field, among all, the following activities are carried out: polling of the families and medical checkup of inhabitants of study areas, field dispensary investigation; evaluation of diseases of population (data on disease structure, frequency of distinguished diseases, their seasonal appearance, etc), dynamics of infections and characteristics of significant endemic diseases; data collecting referring to health care organization. Field investigations are used also for sample collecting of rocks, soil, water, plants and food products for laboratory research.

Acquired information are processed and interpreted in the office. Wide - spread use of computers in research makes processing of huge amounts of data possible. In our case, data on medico - geological and medico - geographic characteristics of the study territory, serving not only for outlining characteristic features, but also for their evaluation, are of special interest. By computing processing, necessary data on different medico - geological, that is - medico - geographic indicators are obtained, making producing different specific purpose maps of the investigated territory easier.

3. Methods of fundamental (basic) geology. Geological factors, as mentioned, have significant role in evaluating degree of suitability for living at a natural environment. That is why research starts with analysis of fundamental (regional) geological characteristics and relationships within the study area, using numerous methods of fundamental geology: *mineralogy, sedimentology, petrology, paleontology and stratigraphic methods, as well as methods of geophysics, geodynamics, structural geology, paleogeological analysis,* etc. Result of application of the mentioned methods and geological mapping is *Basic geological map,* particularly important base for medico - geological analysis of some territory. Such map has been published, for example, for the whole territory of the former Yugoslavia (individual sheets covering approx. area of 1,500 km^2), in scale 1:100,000.

During ***mineralogical exploration***, crystal structure is determined by X - ray method, physical properties - by polarization petrographic microscope, and chemical compounds - by chemical and spectrochemical analyses.

For rock investigations (***petrologic investigations***), different methods are applied. Rock composition is defined by field (with the naked eye or magnifying glass) and laboratory (qualitative and quantitative investigations of samples, chemical and other analyses) procedures. In order to define petrologic structure, that is - structural and texture rock characteristics, grain size and uniformity are investigated.

Sedimentological investigations are useful for determining content, features and genesis of sedimentary rocks. By field investigation, lithological content is defined, while more precise results are obtained laboratory analyses - mineral content is defined by microscopic analysis of samples, and chemical content of carbonate and other sediments is evaluated by calcimetry and dolometry. Data regarding grain size and content of some fractures within a clastic rock are obtained by grain - size investigations.

In order to define geochemical characteristics of the environment, detailed data related to chemical content of elements within rocks, soil, water, air, food products, alive organisms and plants are necessary. That is why the whole spectrum of ***geochemical methods*** is applied: *(1) regional geochemical prospection and geochemical mapping; (2) geochemical sampling and lithogeochemical methods; (3) lithochemical methods; (4) radiochemistry* (exploration of radioactive elements in the nature) *and nuclear chemistry* (exploration of chemical point of view of nuclear reactions and investigation of chemical properties of the obtained nuclides); *(5) methods of gas exploration; (6) chemical and physico - chemical analytical methods and technical analysis of samples:* I classic chemical (volumetry, gravity, colorimetry, etc.); II instrumental (polarography, emission spectrography, spectrometry, atomic absorption spectrophotometry, X ray - fluorescent spectrometry, neutron activation analysis); III methods in geochemical prospecting (fast colorimetric methods, simple procedures of emission spectrography, etc.); *(7) unique instrumental methods* (electron and ion microsonde for determining chemical content of minerals).

Modern analytical procedures are completely useful in solving the mentioned tasks, because of very high sensitivity (Fig.8.1.).

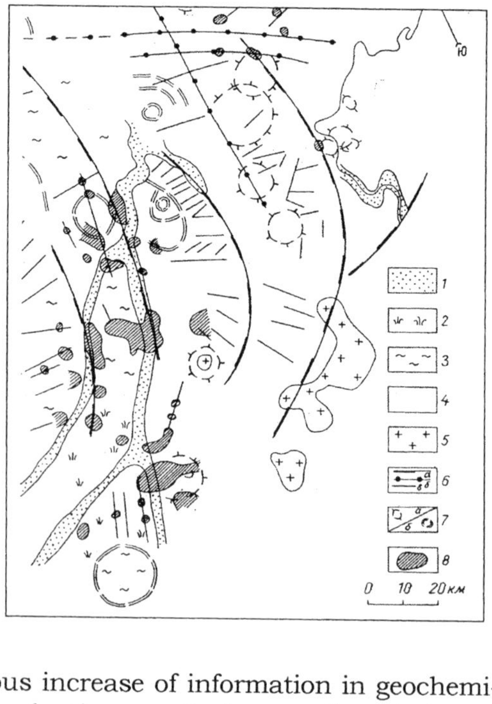

Fig. 8.1. Link between fault and ring structures with hydrogeochemical anomalies. 1 - alluvion; 2,3 - Quaternary clay sediments; 4 - Permian and Triassic sandstones and alevrolites; 5 - granites; 6 - faults; 7 - ring structures; 8 - hydrogeochemical anomalies.

As in other sciences, enormous increase of information in geochemical investigations caused hyper production, particularly noticeable in applied geochemistry. That is reason why highly efficient *information methods* are developed, based on mathematics and computing technique.

By ***paleontological and stratigraphic investigations,*** age and relations among geological formations in the study area are distinguished. Paleontological investigations are based on defining so called characteristic fossils, that is - organisms known by small vertical, and intensive horizontal development. Sediments relationship before important orogenic stages is presented in stratigraphic lines. To determine absolute age of rocks, radioactive isotopes of uranium, thorium, rubidium, potassium, carbon and hydrogen are used.

Various ***geodynamic methods*** are applied to evaluate result of endogenous and exogenous geological forces. Volcanological exploration is particularly used, first of all, in volcanic eruptions prognosis, by monitoring of active volcanoes. Seismographs are instruments for earthquakes registration, while obtained data are useful in seismic regionalization and detailed microseismic zoning of the terrain.

Tectonic investigations exist as inevitable component of practically all geological exploration. Using *methods of neotectonic analysis,* significant material related to the recent tectonic activity (recent uplifting, that is, land sinking) of the territory and active zones (fault structures) is obtained. According to analysis of air - or satellite records, as well as data recorded in the field, all important structural forms: fractures, fault structures, folds, etc., are investigated by *methods of structural geology.*

Formation analysis, as a recent method of geological exploration, is also interesting for medical geology. Subject of exploration are geological formations - occurrences characterized by certain petrologic and geochemical content, metallogenetic, hydrogeological and engineering - geological, and hence, geomedical characteristics.

4. Geophysical methods. In all geophysical methods, procedure of exploration is equal - measuring some physical field at the surface of the terrain and interpretation of measured values in order to collect information regarding distribution of formations with different physical features in the subsurface. For analysis of geological structure (composition and structural fabric, groundwater distribution, ore bodies' deposits, coal, oil and gas deposits, etc.), most frequently used are following quantitative methods: electric methods, seismics, gravity and magnetics. The last decade of the previous century will be famous in the world of geophysics by strengthening of a new discipline of the applied geosciences - ***environmental geophysics*** (S. Komatina, 1996).

Electric methods are based on differences in electric characteristics of investigated geological unit. *Resistivity method* is frequently used, as well as *IP* and *Mise - a - la - masse method.*

Seismic methods are based on analysis of artificially induced elastic vawes propagation. In the practice, *refraction and refraction seismic methods* are used.

Gravity methods are based on different fields of attraction, caused by rock masses. All present methods are classified into two groups: dynamic and static. During dynamic exploration, pendulum or ballistic instruments are used, and during static measuring of gravity relative value gravitymeters.

Geomagnetic methods are useful in defining space distribution of geomagnetic field anomalies, as a result of different rock magnetization. Several types of modern high - resistivity magnetometers are in use. Beside terrestrial, aeromagnetic investigations are also very important, with instruments of higher sensitivity (Fig.8.2.).

5. Geomorphologic methods. Complex geomorphologic analysis includes both qualitative and quantitative analysis of relief. *Quality geomorphologic analysis* means identification of processes and forms. Within the procedure, not only genetic classification of forms is performed, but also determining of their relationship in space and time, and dependence evaluation of the processes intensity and relief on geological structure and the recent tectonic activity of the terrain. *Quantitative geomorphologic analysis* contains numerical parameters of relief through systematic measuring, statistic - mathematical processing, graphic presentation and data interpretation. The first one is carried out by remote sensing methods, and another - by numerous computing techniques.

Main task of ***speleological methods*** is exploration of subsurface karst forms (caves, jamas, ponors) and all other forms into which it is possible to enter, their morphology, speleogenesys and evolution (Fig. 8.3.).

6. Pedological methods. In order to define content, features and quality of soil, various field and laboratory methods are applied: methods of pedo - mapping exploration and sampling; methods for evaluation physical properties of soil; geochemical methods for determining space distribution of trace elements within soil; different chemical and quantitative spectral methods, methods of atom absorption, for distinguishing extraction nutritive substances in soil, radiopedological methods, etc.

7. Hydrogeologic methods. Among geosciences, hydrogeology belongs to disciplines with the fastest development. Reason why dynamics of development and improving extremely numerous and different methods of hydrogeology is, first of all, continual need for drinking, mineral and thermal water. Within the exploration process, methods of fundamental geology, remote sensing, geophysics, hydrometeorology, chemistry, nuclear physics, mathematical modeling and others, are used, as well as sophisticated hydrogeological methods - hydrogeological mapping, tracer methods,

methods of exploratory drilling and testing water intake structures, hydrodynamics, hydrochemistry, monitoring and regime investigations of groundwater methods of groundwater protection, etc. (M.Komatina, 1984, 1986). Thanks to computing processing, need for draft schematization of hydrogeological systems, so often used before expansion of computers, is not necessary any more.

Applied hydrogeology is characterized by different modern approaches of exploration of low - mineralized drinking water, mineral and thermal water; making hydrogeologic basis for drainage of raw mineral deposits; evaluation of hydrogeological conditions and possibility of agricultural soil amelioration. Appropriate methodological procedures are developed within domain of petroleum hydrogeology, engineering hydrogeology, hydrogeothermics, etc. (M. Komatina, 1990, 1995).

Groundwater exploration in discontinual karst medium is very complex task, requiring unique methodological approach. The task is possible to be solved by application of certain geological and hydrogeological methods.

8. Hydrogeochemical methods. Influence of groundwater chemical content and microbiological correctness on human health during the last half a century is studied with particular attention, first of all - because of occurrence of a series of highly - toxic substances, including these in atmosphere (Fig.8.4.). Favorable circumstance is that, thanks to new technical developments and improving numerous laboratory methods of analysis, chemical and hydrogeochemical methods were simultaneously developed. Analytical methods (atomic absorption spectrometry, chromatographic analysis, anode stripping voltmetrics, etc.), as methods of high sensitivity, precision, velocity analysis and automatization, made efficient exploration of micro - and macrocomponents in water possible. Analytical procedure is also useful in easy obtaining of information regarding hundreds of types of organic compounds present in water in micro - and nanogram amounts. Processing and interpretation of obtained results is automatized by computers.

9. Engineering - geology methods. For solving problems in the field of engineering geology (load - bearing capacity of rocks, terrain stability), various field and laboratory exploration works are carried out. The following groups of engineering geological methods are used: (1) *non - geological methods* (climatological, biological, isotope, etc.); (2) *geological methods* (remote sensing, petrologic and mineralogical methods, geomorphologic methods, tectonic investigations, seismological, geophysical and other methods), and (3) methods developed within *engineering geology, soil mechanics, hard rock mechanics and hydrogeology.* In the narrower sense, the following methods of engineering geology are distinguished: engineering - geological mapping, methods of shallow sounding and exploratory drilling, field and laboratory test of physico - mechanical rock characteristics. By detailed exploration, landslides and unstable terrains are treated, with several special ways of exploration, prognosis and sanation.

Evaluation of soil dynamics (instability) and influence of geological conditions on seismic stability of the objects is subject of various investigations of concrete territories. In order to make enough safe foundation for aseismic constructing at wide areas, *seismic regionalization* is performed, while for the areas of limited extent - *seismic zoning.*

10. Nuclear methods. Nowadays, nuclear methods are so developed that it is possible to determine natural content of stable and radioactive isotopes of the elements within different inorganic materials, plants and parts of organisms. That is why they are successfully used in geological, biological, medical and ecological science and practice.

Radiation defining is performed by gas - , liquid and solid detectors. In *exploration of nuclear raw materials,* gamma - detectors are applied, in other words - gamma radiometric method is used. *Gamma radiometric method* is a basis for uranium exploration, particularly during the stage of field reconnaissance, not only from the surface of the terrain, but also from the air. Simultaneously with prospecting from the air, uranium - hydrochemical prospecting is performed, in order to determine uranium content in water and mud sediments. Collected samples are investigated in the laboratory by *gamma - spectrometric analysis.*

Isotope methods are successfully applied in the wide spectrum of tasks within *groundwater exploration.* In practice, the most often used stable isotopes, are deuterium and oxygen - 18, and, among cosmogenic isotopes - tritium and carbon - 14. In the first case, methods of mass - spectrometer measurements are applied, and in the second one - methods of low - phone radioactivity measurements within samples.

11. Biological and biogeochemical methods. As other natural sciences, biology is at the high level of development. During several decades of development, modern molecular biology, for example, has shown that cancerization, teratogenesis and similar occurrences are result of disturbed function of cell genome of the human organism.

Among biological methods of exploration of the natural environment, the following *zoological - parasitological* and *patho - biocenotical,* as well as *biogeochemical methods* have very important role: methods of floristic regional phitogeography (floristic statistics etc.), biotelemetry (remote sensing method), biogeochemical methods and biogeochemical cartography, radiobiological measurements of radiation, ecological ways, approaches of physiological anthropology, bacteriological water and food analyses, physico - chemical investigations for water quality evaluation, microbiological methods for determining water and food quality, methods of biogenetic engineering, population (biological) monitoring of main animal species.

Laboratory analyses are based on analytical methods, showing, for example, that atomic prospecting is ideal method for control of metal traces within biological processes.

12. Methods of bioinorganic chemistry. During the last three decades, new, very important discipline was developed - bioinorganic chemistry. Subject of bioinorganic chemistry is study of reactions at molecular level among some elements, particularly biogenic metals with bioligands (proteins, nucleine acids and their ferments, hormones, vitamins, etc.). This scientific discipline could be developed just when research in the field of inorganic chemistry and biology (particularly molecular biology) reached required level. It is clear to expect that further improving of the most sophisticated physical and chemical exploration methods in domains of biology and chemistry as well as other methods, will make possible that, in the near future, our knowledge on reactions between some elements and bioligands, and generally, on physico - chemical foundations of biological activity of some elements, becomes more voluminous.

Nowadays, in solving scientific and practical tasks, sophisticated physical methods of investigation are used, but also, in less extent, spectroscopy and diffraction methods. Obtained results are used in medicine, biology, farming, environmental protection against heavy metals, etc.

13. Medical methods in a narrower sense. Today, medical staff uses practically unlimited number of different methods and procedures[102]. From the other side, although each of numerous medical sciences, disciplines and specialties possesses own system of knowledge and methodology, overlapping in analysis of the problems, but also in application of some methods, is noticeable. That is why more precise overview of methods is difficult, particularly for some sciences and disciplines. Trend of improving and use of part of methodical procedures within medical sciences is certainly visible. That is, first of all, related to:

a) *Experimental method,* which is still, beside use of electronic microscope and new models, unavoidable way in pathology, epidemiology, physiology and other medical sciences. Modern physiology, for example, is known as extremely experimental science. Tests or experiments are performed on experimental animals or their organs.
b) *Use of genetical principles* in applied medicine (medical genetics) is very important. It can be said that genetics and its role in research of vulnerability of humans to diseases is inevitable, while genetic information is considered in pathology, epidemiology and other scientific fields of medicine.

---

[102] It is realized that, by better basing on real natural and other preconditions (disease carriers), hypertrophied medical science will reject in the future some part of sufficient methodological procedures and drugs, finding more efficient ways to human health preservation (among all, through efficient prophylactic health cure).

c) *Laboratory analytical methods* are widely applied in numerous medical sciences (Fig.4.4.). Atomic absorption methods, for example, are irreplaceable in analysis of pathological and clinical samples of urine, serum and blood, further, in forensic medicine, but ideal in checking traces of metals within biological processes. Chromatography is, from the other side, procedure for separating of substances from the mixture (liquid chromatography, etc.); spectrometric methods are applied in toxicology (spectroscopy in ultraviolet, that is - visible part of spectrum, infrared spectroscopy, etc.); absorption spectroscopy and colorimetric methods - for determining magnesium in clinical biochemistry (serum, urine, and blood plasma), etc.

d) *Development of computing sciences and information technology* is of invaluable importance for medical sciences. That is reason why enormous development in creating new and improving existing exploration methods occurred.

Wide range of methods used in making diagnosis of some diseases is now illustrated by the example of *endemic nephropathy*. In this case, complex exploration is carried out through the following four levels: 1) perlustration of the terrain (conducting a poll, protein checking of urine); 2) field clinical research (blood picture, urine control); 3) policlinic research (laboratory control; methods of evaluation of functional state of kidneys, an X - ray of heart and lungs; intravenous pieolography, radiorhenogram); 4) clinical research (dynamic scintrigraphy, separate kidney clearance, urine electrophoresis, miction chistourethrography, tethrograde urethropieolography, immunological research, kidney biopsy). During the last two decades, radioisotope diagnostics (rhenography, scintigraphy, radioisotope clearance) has shown numerous advantages. *Present diversity of ways to etiology of the disease is partly a consequence of unsuccessful activities in distinguishing appropriate carriers.*

For the sake of discovering pathological changes and monitoring the process, experts in ***pathology*** use autopsy, biopsy and experiments. Based on recent scientific knowledge in the field of microscope optics, biophysics and biochemistry, modern pathology is able to get to the heart of a problem of processes within subcellular structure (subcellular pathology), reaching molecular pathology. Research in pathological institutes, beside autopsy, pathohistologic and bioptical analyses, includes: autoradiography, fluorescent microscopy, histo - chemistry, and, particularly, electronic microscopy. Various epidemiological methods are also applied, as also methods of pathological anatomy, infection pathology and other clinical disciplines (ophthalmology, endocrinology, oncology, etc.), medical statistics and cartography.

Research in ***physiology*** is predominantly of biochemical character for decades. Numerous physiological problems are studied at the level of submicroscope and macromolecular structures. By application of ideas and methods of biophysics and biochemistry, but also introducing new, sophisticated exploration methods and equipment, like electronic microscope, analysis of subcellular and molecular structure of a cell and appropriate compounds is possible. Use of radioactive isotopes significantly contributed to the research. Various modern techniques and methods are applied in study of regulation mechanisms, as well as physiological and biochemical processes of adaptation. Experimental method is still one of basic approaches in physiology.

Infective epidemiology is based on practical tests, experimental method and various abilities of clinical research. Methods of microbiological and parasitological diagnostics are certainly of extraordinary importance among sophisticated methods. By the highest - level precision, serological diagnostics makes high sensitivity and uniqueness in discovering antigens and antibodies or their parts within serum possible. Data processing is performed by biostatistical procedures (distinguishing disease frequency, etc.), and for presentation, different epidemiological graphs are applied.

Medical ecology is characterized by multidisciplinary approach, based on methods of natural sciences, ecological chemistry, biomedicine, medical geography, engineering and technologies, etc. For defining toxic metals and carcinogenic substances, methods of analytical chemistry are applied, and for determining risks of technological procedures - mathematical methods. Effects of toxic substances (alcohol, drugs) on cytological - citochemical, ultrastructural and biochemical features of some tissues and organs are studied. Special attention is paid to radon presence - gas forming the strongest radiation, depositing in bronchi and lungs and causing occurrence of lungs cancer. In this domain, methods of field determination of radon content in soil and air are used and regional and detailed maps of radon risk are made, radon concentration within apartments and offices is measured, as well as epidemiological research method to evaluate risk is applied.

14. Methods of medical geography. Studying general relationships of geographic environment effects on the human health, medical geography deals with factors of geological medium. Numerous methods for evaluation of natural environment are known. That is reason why it is so important for medical geology to use developed methods and available results of medical geography, overlapping each other in scientific and applied domains of exploration.

According to **Handbook of medical geography** (1993), the following methods of medical geography are the most important ones:

- *space - comparative analysis of quantity parameters* (communal feeding; necessary clothes; city and apartment as elements of socio - living environment; work and rest at the open space);
- *medico - geographic evaluation of living environment factors;*
- *medico - geographic cartography* (medico - geographic maps and atlases);
- *modeling in medical geography;*
- *medico - geographical categorization* (taxoning) *of the territories* in unique or hierarchically submitted cells - taxons (zoning);
- *medico - geographical diagnostics;*
- *statistic methods in medical geography;*
- *application of computers in medico - geographic exploration;*
- *Medico - geographic prognosis.*

Information acquired during *expedition visits of the study territory* (medical geographer, biochemist, dentist, radiologist, and pathologist) and *methods developed within basic medical pathology, physiology, epidemiology and climatology* are also of high importance.

In the following text, methods of medical geography which are of interest for our research are presented.

15. Cartographic methods. Importance of medical cartography as one of leading scientific approaches for determining distribution of diseases throughout the world was underlined during fifties of the last century. At that time, prognostic abilities of maps made for epidemic infections, and later for endemic diseases, as geographic map with data referring to distribution of diseases expressing coincidence of certain geographic conditions and studied disease makes possible, were discovered. In the case of making coincidence for several localities, it could be distinguished as general ecological regulation. That is reason why developing of scientific classification of medico - geographical maps, but also way of their making and unification of legends, began. With time, cartography became center uniting almost all methodical procedures, and map was used within medico - geographical exploration beginning from the first stage - acquisition of material, till delivering the report to the customer. Because of continual and complete use of maps, medical geography is different from close medical and biological disciplines.

Now, text referred to methodology of making medico - geographic maps, presented in the **Manual of medical geography** (1993), is cited. Determined goal is to present graphically state and quality of the environment, expressed through: 1) negative influence of geographical factors (natural, social, productive, living, etc.), characteristic for certain geocomples on human health, and 2) presence of health resources, recreation systems and other health stimulators. Because of that, such maps are exclusively of applied nature.

According to the **Manual...**, modern maps are classified into: (1) medico - geographical maps (referring to features of the environment influencing on human health); (2) geographic - medical maps (defining space distribution of diseases at the whole Planet in dependence on some natural and socio - living conditions); (3) maps of health care (as a rule, represent distribution of the network of public health facilities, sanitary and sanatorium facilities, research and other institutes directed to health care. In the following text, the first group of maps will be discussed. According to content (subject matter), within the group, basic (general) medico - geographical maps, maps of medico - geographic evaluation of the environmental, as well as nosogeographic maps are distinguished.

1. ***Basic medico - geographic maps,*** in contrast to other types of maps made on the basis of appropriate thematic maps, are based on data of field exploration and remote sensing. Three kinds of this type are known:
   - *Medico - geographical maps of the environment* (bioclimatic maps, referring to climate influence on the human organism; pedological maps, with pedological soil characteristics of the areas unique from the medico - geographical aspect; geochemical maps, presenting geochemical features of the soil, appearing as biogeochemical endemic diseases; hydrogeochemical maps, defining chemical and other characteristics of groundwater and surface waters and level of their positive or negative influence on the human body; maps presenting features of plants and animals and their influence on human beings, etc.).
   - *Complex (multipurpose) medico - geographical maps* - represent combination of characteristics of natural and production comples of the territory, where integral influence of natural and socio - economical conditions on health of population occurs through qualitative homogeneity and local uniqueness.
2. ***Maps of medico - geographical evaluation of the environment*** are compiled on the basis of thematic maps, made by other experts - geologists, geochemists, pedologists, geobotanists, zoologists, economic geographers, etc. Main advantage of such kind of maps is velocity of their compilation, if a condition that necessary thematic maps are available is fulfilled. Main disadvantage is complete dependence of the maps on credibility of used thematic maps.
   - *Maps of medico - geographic evaluation of the environment* are based on distinguished differences of natural geosystems and effects on health. The following factors are evaluated: climate, soil, geomorphologic forms, surface and ground waters, plants and animals.
   - *Maps of medico - geographical evaluation of socio - economic environment* are made on the basis of exploration of characteristics of production territorial systems and appropriate structural units, level evaluation of the territory organization in accordance with human health. As an example, a map defining consequences of contamina-

tion of the environment in village areas of Armenia, caused by wide application of various toxic preparations in agriculture, is presented.

- *Maps of medico - geographic evaluation of natural - social environment* are referred to relation complex living environment (natural and territorial - production comples) / human health.

3. ***Nosogeographic maps*** outline risky areas with particular disease (nosoareals), their dynamics in time and space. In contrast to basic medico - geographic maps, presenting certain features and states of the environment - assumptions of diseases, subjects of nosogeographic maps are diseases and their distribution.
   - As an example of *maps of geographic distribution of diseases principally caused by natural factors,* can serve maps of biochemicals enemies, maps of allergic diseases or maps of diseases with natural foci.
   - Among *maps of geographic distribution of diseases principally caused by socio - economic factors,* maps of silicosis or maps of professional poisoning are known.
   - In *maps of geographic distribution of diseases with polioethiology,* maps of frequent occurrence of malignant tumors or cardiovascular diseases are classified.
   -

According to the *purpose,* special and universal medico - geographic maps are distinguished.

1. ***Special medico - geographic maps*** are used by certain range of users, for solving concrete tasks.
   - *Cadastre medico - geographic maps* present zoning of the territory according to degree of quality in relation to the state of health in the study area (presence and level of biogeochemical endemics, infective and non - infective diseases, technogenous and municipal contamination of the environment, but also of convenient factors, as well as living and health conditions).
   - *Maps of population health,* as the most reliable integral criterion of effects of environmental features on human body. Basically, three groups of markers are mapped: 1) *demographic;* 2) occurrence of diseases; 3) physical development of distinguished groups of inhabitants.
   - *Sanitary - epidemiological maps* related to sanitary - epidemiologic conditions within geocomples and system of indispensable prophylactic measures for the environment recovery. As an example, classical map made by J. Snow in 1848, in order to outline cholera epidemic in the Soho district (London) is convenient.
   - *Project medico - geographic maps,* made in order to provide medico - geographic information for the needs of planning, use of area for recreation, organizing optimal network of health care institutions.
   - *Operations medico - geographic maps,* applied in solving appropriate medico - geographic tasks (for example, synoptic maps - in order to forecast risky days for weather - sensitive persons).
   - *Recreation medico - geographic maps,* presenting distribution of

zones and places for vacation or natural and socio - economical conditions suitable for recreation and recovery of population.

- *Historical medico - geographic maps* present geographic position of nosoareals of various diseases during the past times. The maps are very important for determining causes, ways and dynamics of diseases distribution.
- *Teaching medico - geographic maps* are used for education, as well as making independent contributions within medical geography and other disciplines (biology, human ecology).

Level of detailed presentation by mapping, that is - highest permitted reduction of some space survey of the phenomenon is dependent on the scale of topographic base. From this point of view, the following types of medico - geographic maps are distinguished: *large - scale* (1:200,000 and larger), *medium - scale* (less than 1:200,000 to 1:1,000, 000) and *small - scale maps* (less than 1:1,000, 000).

According to *the level of markers generalization and features of the presented phenomenon (process),* analytical, synthetic and complex maps are known.

a) *Analytical medico - geographic maps* present individual elements or features of some phenomenon with non - generalized, concrete markers (for example, map of fluoride content in mg/l in drinking water of springs or other groundwater occurrences, or map of soil temperature for some period).
b) *Synthetic medico - geographic maps* contain total (integral) characteristic of cartographic occurrences (processes), with generalized integral parts of the occurrence.
c) *Complex (special) medico - geographic maps* present several different, but, according to their effects, connected, occurrences or several different characteristics of appropriate phenomenon (each one presented by adequate markers). As an example, map of biogeochemical zoning can serve, presenting level of biologically active macroelements and trace elements at the study area. Complex (universal) medico - geographic maps are also used, integrating series of various, but closely connected occurrences within *environment - population - health level* system, as map of complex medico - geographic zoning of the interested region.

At the end, the following classification of medico - geographic maps *in dependence on level of objectivity (reliability)* is presented:

a) *Documentation medico - geographic maps,* fixing concrete facts, processes, occurrences, acquired during field exploration (measurements);
b) *Medico - forecasting maps,* treating scientific forecast of results related to state, distribution and dynamics of uncertain or not yet expressed

processes and occurrences and their influence on health. They present forecasts in time (for example, decrease in population vulnerability to endemic goiter in dependence on applying iodine to food products) or space (reducing areas with potential risk of some infective disease by performing complex of prevention measures), as well as space - time forecast (for example, weather forecast within the *environment - human health* system, as a result of anthropogenic and other factors);

c) *Recommended medico - geographic maps* are important as a finishing link on a chain of system medico - geographic cartography. Complex (synthetic) map of medico - geographic optimization of the environment is an example.

Medico - geographic atlases present systematic collection of organic inter - connected medico - geographic maps, limited by subject and characteristics of the application. Classification of medico - geographic atlases is very similar to that of medico - geographic maps. From the history of medical geography it can be seen that such kind of presentation was used in all stages of development of the discipline.

Three decades ago, **computing geography**, based on huge amount of statistic data and results of concrete exploration, started to develop. The role of this discipline within medical geography was recognized from the beginning, particularly in domain of analysis of space relationships among occurrence of various infective and other diseases. Maps of malignant tumors distribution are made more and more often. For example, **Atlas of cancer risk in USA** published in 1975, containing 65 maps in color, with outlined territories with relatively high coefficient of mortality. Nowadays, technical capabilities of presenting diseases markers in geographic maps are extremely numerous.

Developed approaches to medico - geographic cartography, mentioned in the previous text, are of high importance for **medical geology**, which is at the initial stage of the development. It has to be understood that it is related to the part presenting natural factors. Task of **medico - geological cartography** is to present in space relationships of certain geological factors as disease carriers and medical consequences for human health. Wide range of different compiled geological maps is available, as: photogeological maps and satellite records, basic geological map and various detailed geological maps, geological maps of the territory of the whole country (geological, geomagnetic, geomorphologic, neotectonic, seismotectonic and seismological, geochemical, metallogenetic, hydrogeological, engineering geological, pedological; map with mineral and thermal waters presented), hydrogeochemical, geobotanical, ecogeological, etc. The following examples show how existing cartographic material is useful.

1. *Basic geological map* can be of invaluable use to solving some research tasks, not only in domain of geology, but also medical geology. It presents geological structure of the terrain, tectonic structures (folds and faults), occurrences and raw mining deposits. According to the presented geological structure, preliminary (orientational) evaluation of living conditions in the natural environment (as suitable or not) can be made. It is clear that, from this point of view, each geological formation presented in the map should be analyzed. That is the way of determining medico - geological categorization of geological units.
2. *Geomorphologic maps* present distribution of different sectors of the study area *according* to shape, genesis and age of the relief elements. Content of geomorphologic map is made with the aim of having multidisciplinary purpose - in geology, civil engineering, agriculture, land - use planning, environmental protection, even in medical geology. It should be underlined that geomorphologic factors are of high importance in determining living conditions and treatment of population. Classic type of the maps can be approached in purpose in the case of outlining medico - geomorphologic (geological) types of the territory on the map, with some positive or negative influence on human health and conditions for organizing health tourism and recreation.
3. *Geochemical, hydrogeochemical and biogeochemical maps* offer important information on chemical content of rocks, soil, ground - and surface waters, plants, animals and basic food products. Significant attention has to be paid to trace elements, because of their role when speaking about health of human beings and animals. The maps, adjusted to the requirements of the discipline, can be very important element of medico - geological cartography and to contribute to solving etiology of endemic and other diseases. In order to achieve the goal, a lot of work referring to increasing content of the map and adapting ways of preparing the map to requirements of computing technology is necessary.
4. *Hydrogeological map* is based on distribution of groundwater and their quality. In that sense, according to their permeability, all rocks are divided into categories of water - bearing, poorly water - bearing and practically arid areas (Fig.8.5.), and data on sources of low - mineralized, mineral and thermal waters, their reserves, physico - chemical features and water quality are predominantly presented in the legend or the proposal. On the basis of hydrogeological relationship, level of natural aquifer vulnerability to contamination from the surface can be evaluated. Because of enormous role of water for life of all human beings, hydrogeological cartography is widely used all over the world (particularly during the past two decades). Hydrogeological map is classifies among very important basis in solving various problems - water supply, soil amelioration, exploitation of mineral deposits, constructing sanitary - resort structures, etc., and the similar development is expected in the field of medical geology. For the needs of medical geology, its content can and has to be significantly changed, in other words - to adjust to the principal task, with a main point on

presentation and interpreting chemical content, mineralization, quality and protection of groundwater sources. Besides, possibilities of restricted hydrogeological and hydrogeochemical zoning of the interesting area, which could be carried out according to space distribution of geological formations interesting from the geochemical viewpoint, that is - according to changes in content of some trace element in water in space, hardness and water mineralization, are formed.

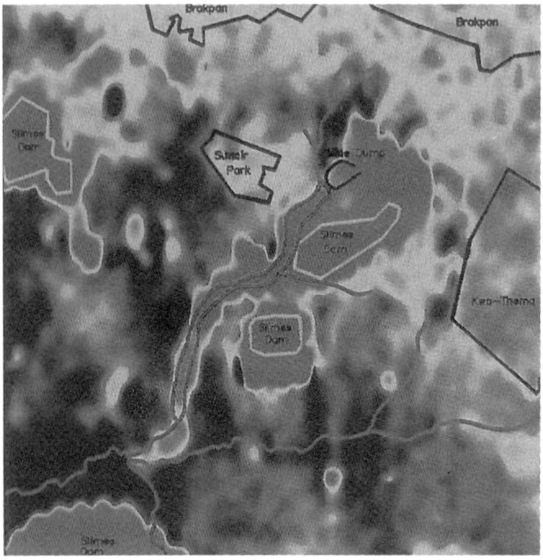

Fig. 8.2. Detailed total - field airborne radiometric color map of an area showing the presence of slimes dams and the movement of mine waste.

In any case, just now, main activity in the field of medico - geological cartography is approaching. It is necessary, in spite of all, to lean on medico - geographic maps and methodology of their compilation, particularly on the following described types of maps: 1a - *medico - geographic maps of natural environment*; 2a - *maps of medico - geographical evaluation of natural environment*; 3 - *nosogeographic maps* (3a - *maps limited principally by natural factors*), but also different special medico - geographical maps, analytical and medico - forecast maps.

Geochemical atlas of Slovak Republic, published in 1997, in scale 1:1000, 000, is an example illustrating importance of highly informative atlas for medical geology.

The atlas consists of maps and legends, referring to groundwater, rocks, soil, distinguished elements from the forest ecosystems and natural radioactivity in rocks and water. Part of the legend related to groundwater is based on results of 16,359 investigated water samples (one sample every 3 km^2), and the maps of: mineralization, Mg - content, fluoride and arsenic, as well as level of groundwater contamination and hydrochemical diagrams for all important geological formations of the state territory are made separately. Part related to rocks (3,839 samples) contains geological - tectonic map and review of chemical content of 64 types of geological formations individually. Soil was investigated along 5,000 lines (9,892 samples) in order to investigate 36 important chemical elements; maps of As, Cd, Cr, Cu and Zn distribution within A soil horizon are attached. Part referred to ecostress maps (3,063 samples) was made on the basis of the following elements content within forest leaves: Al, As, Ba, Be, Ca, Cd, Co, Cr, Cu, F, Fe, Hg, K, Li, Mg, Mn, N, Na, Ni, Pb, Rb, S, Se, Sr, V and Zn, and maps present fluoride, sulfur and mercury concentration. Total natural radioactivity in rocks and water was evaluated by 4,900 samples, gamma - spectrometry was used for determining content of K, U, Th, U_{total} in rocks, but also natural radioactivity in rocks, radon presented by maps of risk, and $U_{natural}$, Ra_{226} and Rn_{222} in water; for rocks, maps of potassium, thorium and uranium were made, as of gamma - radiation; for Rn - map of risk and map of Rn concentration in water.

Fig. 8.3. Rajko cave, Eastern Serbia (Photo: D. Milovanovic).

16. Medico - geographical and medico - geological zoning. Long time ago, experts in **medical geography** realized necessity of distinguishing regions homogenous in natural socio - economical conditions, particularly when health of population over spacey territories has to be studied. Besides, according to practice, distinguishing of such kind of regions - zoning has to be a result of detailed exploration of the territory, and diligent consideration of medico - geographical features. Further, that some deficiencies appear if perform only physico - geographical differentiation of the territory, except in the case of presenting only preliminary information on unknown territory with natural environment lightly degraded by agricultural and other activities.

Nowadays, more and more voluminous collections of maps of medico - geographic zoning are published. Investigated regions are divided into existing dynamical territorial systems - sun ecosystems or medico - geographical comples, from elementary to the most complex - of planetary order. As the main sun ecosystem, *medico - geographic rayon* is distinguished - space unit of the environment with the equal health level of one - type groups of inhabitants and unique local pathology, caused by interior homogeneity of natural and living conditions of population within areas of each particular rayon. By the zoning of one territory, solving the main task of medical geography - making contribution to recovery of the environment, more precisely - regulation of all factors of the environment showing (or which could show) harmful effects on human physical development, health and living ability, is possible.

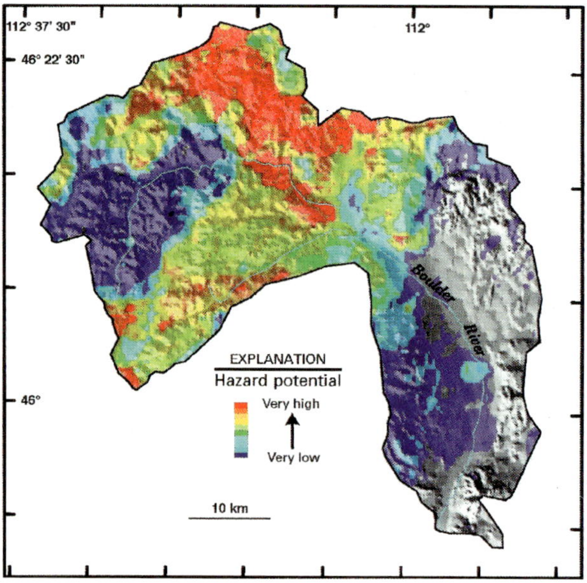

Fig. 8.4. Geoenvironmental hazard potential for acidic, metal - rich drainage in the Boulder River basin, Montana (USGS, 2000).

Numerous approaches to medico - geographical differentiation of the territory are developed, in dependence, first of all, on the research goals and researcher's abilities. So, one - factor, multi - factors and total rayons are recognized, or one - , two - , three - , four - and five - members (full) comples are distinguished. The most complex way of zoning is defining multifactor (complex) systems.

Medical geology needs different approach to zoning of the territory, with minimal consideration of socio - economic factors, and on the basis of one or several geological factors and medical consequences. Zoning has to be carried out not only on the basis of existing geological maps and other documentation (Fig.8.6.), but also of detailed field exploration and laboratory investigations.

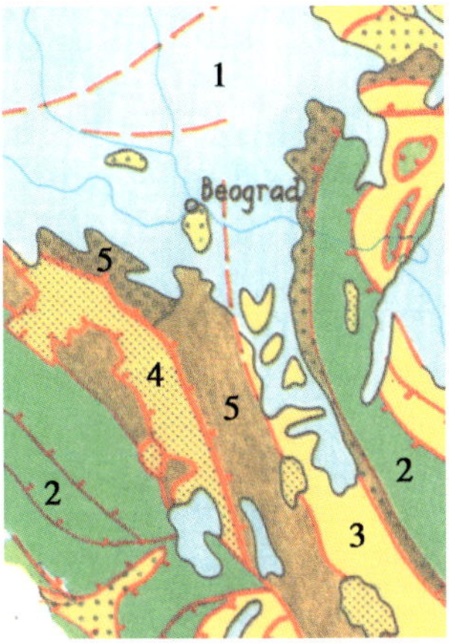

Fig. 8.5. Overview hydrogeological map of Serbia with vicinity. 1 - Neogene and Quaternary sediments of Pannonian basin (*aquifer*); 2 - Mesozoic karstified limestones (*aquifer*); 3 - Tertiary sediments (*complex*); 4, 5 - fractured medium (*low permeable formations*).

Methodology of work and content of the legend are dependent on numerous elements and characteristics within geological structure of the terrain, defined goal (problem), purpose, range of information, etc., and a lot of is based on experience and creativity of the authors in the field of cartography. Generally, as regular and indispensable result of medico - geological exploration, medico - geological zoning may be contained in classification of interesting territory or geological unit into rayons which are separated according to characteristics or intensity of influencing geological conditions, in other words - some geological factors on health of population, genesis and distribution of diseases.

For medico - geologist, one or more factors (in different cases - different) appear in the role of leading ones, as they have the most important effect on human health. Just these factors can significantly determine boundaries of distinguished medico - geological units and their range. However, zoning can become much more complicated because of anthropogenic factors (contaminants), and in such cases, distinguishing subrayons has to be made. In these circumstances, but also in general, it is very important to use as more information obtained in field and laboratory, which can be successfully processed by mathematical methods and computing techniques, as possible.

Possibility of making very different medico - geological zoning, in dependence on the subject, should be underlined. The zoning could be based on:

- *geological structure of the Earth's crust* (rocks, geological formations, tectonic structures, seismic and volcanic activity);
- *characteristics of soil and groundwater;*
- *geomorphologic characteristics of the terrain;*
- *geochemical features of rocks, soil, water, and, indirectly - vegetation and animal world;*
- *natural radioactivity of rocks, soil and water;*
- *Exogenous geodynamical processes* (landslides, rockslides, mud - stony

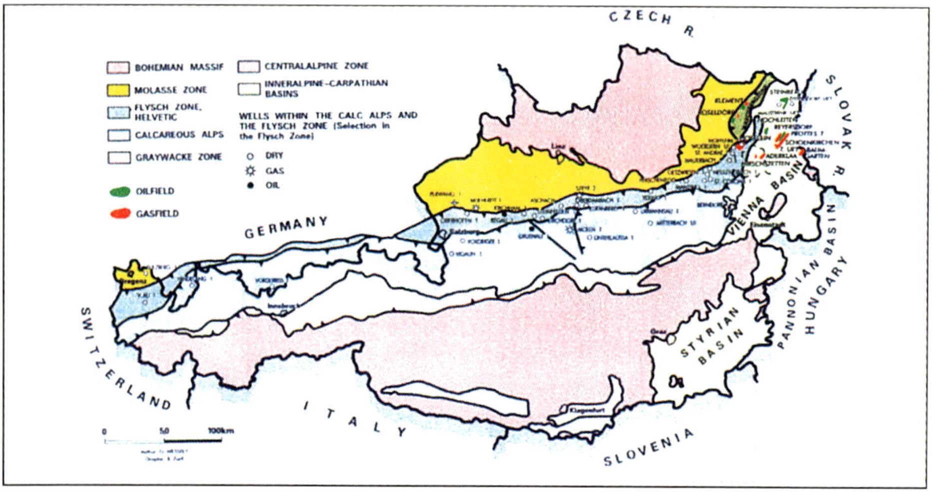

Fig. 8.6. Geological formations in the territory of Austria (G.Wessely, 2000).

Using elements of geological environment, from one side, and leading pathology as one of the basic markers of the factual health level of population, from the other side, a conclusion referring to comfortability of the territory can be made.

In the further text, examples of the two possible types of zoning are discussed.

1. Each *geological formation*, with soil formed by weathering, is characterized by certain mineralogical content and geochemical features, provoking various regional geological and geochemical regularities interesting for medical geology. Their distribution is presented in the Basic geological map and other geological maps. Presented circumstances, beside

data of geochemical exploration, make producing of *general or purpose geochemical zoning of the area* possible. *Biogeochemical zoning* according to presence (deficiency or surplus concentration) by biologically active macro - and microelements is also very useful. For example, schematic map of biogeochemical zoning of the territory of former USSR is characterized by distinguished regions and subregions of biosphere with deficiency, surplus or uneven content of elements), appearing in the form of carriers of development of natural biogeochemical endemic diseases of human beings, animals and plants. Rayons poor in Co, Cu, Ca, Se, Compounds, Zn, F, etc., as well as rayons with uneven Co/P relation, are outlined. Geochemical and biogeochemical zoning of the terrain, in general, could be of outstanding importance in exploration of careers of various diseases, particularly biogeochemical endemic diseases (different microelementoses - selenosis, fluorosis, iodine, iron or zinc deficiency, etc.). As diagnosis of biogeochemical pathology is often very complicated, high - quality medical base, that is - results of detailed medical exploration, should be available for the sake of zoning.

2. *Hydrogeological zoning,* as a method of scientific and practical generalization of data related to groundwater, has been widely distributed and applied during the last three decades. More often, goal is to distinguish rayons characterized by appropriate regularities of groundwater forming and distribution, or regularities important for solving any practical task related to use of the waters, in the areas of spacey territories of the Earth or some continents. For geomedical investigations, it is certainly important that zoning, based on the type and degree of aquifer permeability, has to point out rayons provided by drinking water. Data on drinking water quality (usability) is obtained by *hydrogeochemical zoning* (distribution of some water types according to chemical content, mineralization, hardness, deficiency or surplus of some biogenoactive components in water). All mentioned elements are very important for understanding medical occurrences and consequences. During the 2nd World War, for example, Association of American Geographers started to investigate fluorine content within American water supply systems. Results of the exploration were published in 1996, serving as an impulse for developing the first project of diseases atlas.

17. Medico - geographic and medico - geological cadastre and informatics. Necessity for the work on medico - geographic cadastre was particularly distinguished three decades ago, at the 4th symposium of medico - geographers, which was dedicated to this topic exclusively. It was concluded that cadastre has to contain completely reliable information related to influence of natural, economic - geographic and medico - sanitary conditions of the territory of interest on public health; it is compiled on the basis of complex medico - geographic study of the terrain and data of previous exploration of some parts of the territory.

Nowadays, noting, processing and forming of a cadastre (data bases, information systems) of all data referring to certain natural resources, but also data in domain of medical research and health care, are outlined as very important throughout the world. In Norway, for example, multidisciplinary group for medical geography was formed in 1982, consisting of geographers, physicians, geologists, geochemists and epidemiologists, which made geographic information system of mortality in the country. This data base contained all cases in the country starting from 1966. Authomatized data search and unique processing programms in the system make quick obtaining of different information necessary for medico - geographic analyses possible.

Introducing of unique health information systems in many countries, makes international research cooperation as well as world studies in various domains of medicine in general easier. It is noticeable that the information gets biohuman meaning and significance. It can be concluded that, with more developed information technique is, particularly information processing, biological cognition of human being is stronger (Z. Jovicic, 1997).

When speaking about contemporary data bases, territorial register of malignant diseases certainly is the most known and relatively widespread type of medical records. The main task of the cadastre is offering reliable information on malignant tumors risk. The presented form of cadastre has been suddenly developed since fifties of the last Century. Role of the cadastre in science and health care was extremely expressed in Finland, Sweden, Norway, Denmark and Eastern Germany, countries known as ones with the most regular registration of population, hence - possibilities to perform research by automatized compilation of information on each inhabitant.

Information on natural conditions, beside information on socio - economical and living conditions and material containing data on public health level and diseases dangerous for man and animals, is inevitable in making medico - geographic base of the study area. Reliability of such information can be significantly improved by medical geology with data on geological, and more, natural environment. **Geological information system (cadastre) of the state territory**, as a rule, contains all important data on abiotic components of natural environment, as are: geological and geomorphologic characteristics, surface soil (pedological soil), mineral raw materials, hydrogeological and hydrological conditions, engineering - geological features of the terrain, geochemical characteristics of rocks and soil; hydrogeochemical features of groundwater, etc. Generally, during the last several decades, noting, processing and forming data funds related to the mentioned components of the natural environment are known throughout the world as of high importance[103].

Information on geological environment is of complex significance, not only in the field of rational use of the territory, but also in domain of efficient protection of the environment and human health. In Serbia, for example, acquiring of information started even in 1853 (when Josif Pancic introduced a new subject *Mineralogy with geology* at the Royal Serbian lycee), but suddenly intensifying during the last five decades. Basis of information system related to geological environment contains basic geological maps of the state territory, issued in the scale 1: 50,000 or 1:100,000, but also of high importance can be voluminous documentation on detailed exploration of mineral resources, groundwater, soil, engineering - geological features of soil, etc.

18. Statistic methods. Even at the end of seventeenth Century, English researchers concluded that tables of vital statistics, acquired by J. Graunt, are of a decisive significance for epidemiology. Graunt was not only the first scientist who measured difference between mortality in urban and rural environment, and determined a method of population evaluation according to samples collection, but also the first constructive critic of medico - statistical data. Later, importance of morbidity statistics was distinguished in England and USA.

Nowadays, statistic methods are widely used in medicine and medical geography, particularly in monitoring public health, including statistics of diseases frequency, mortality, physical development, but also health cure. Statistical materials can be very useful within the process of medico - geographic evaluation and diagnostics, zoning, modeling, cartography, etc.

Medical statistics studies and defines *quantitative relationships*: (1) level and dynamics of public health in general (of the country, or distinguished administrative - territorial units, etc.) and appropriate qualitatively homogenous groups (according to gender, age, profession, length of service, etc.) in concrete conditions of the place and time, on the basis of complex qualitative analysis of existing and activities of mass processes of medical and health protection; (2) state and activities of institutions and health officials; (3) evaluation of efficiency of performed measures related to public health preservation in general or among the groups (Keller, Shchepin, Chaklin, 1993). Important place and content of medical statistics belongs to methods of acquisition, processing and medico - statistical analyses of different clinical, laboratory and

---

[103] Within American information system, in the part related to groundwater, even in 1977, approximately 650,000 springs and wells were registered. Soon, many countries were included in hydrogeological cadastre, among of which: France, Hungary, countries of northern and western Africa.

experimental medical data, used during research and applied activities within public health care (treatment).

According to Keller, Shchepin and Chaklin, medico - statistical exploration includes five severely limited and interconnected stages: (1) *planning of exploration;* (2) *statistical acquisition of data;* (3) *statistical grouping and analysis of documentation;* (4) *statistical processing of material;* and (5) *scientific - statistical analysis of documentation, with appropriate literature and graphical design.*

By rich statistical literature, choice and use of approaches to solving concrete tasks is possible. Statistics is more and more based on numerous opportunities of computing technique and specially developed packages of suitable programs, overcoming the field of complex methods of the tasks statistical solving.

Mathematical - statistical analysis has important role in methodology of statistical research. The following typical tasks are solved in this way: (1) compiling empirical distribution and finding appropriate mathematical relationships (models), with checking validity of their agreement; (2) tasks, evaluations and checking hypotheses; (3) testing a link or mutual links; (4) methods of poli - measuring classification and factor analysis; (5) investigation methods of occurrences and processes dynamics; (6) prognosis of occurrences and processes; (7) statistical planning of experiments; (8) iteration methods of statistical model; (9) representative exploration and evaluation of obtained parameters, etc.

Among chapters of mathematical statistics, *theory of correlation* makes actual discovering of links between factors and human health possible. Within statistics, well developed mathematical - statistical apparatus for testing various correlation links (classification methods, correlation and regression analysis, dispersion analysis, etc.) is present. *Correlation coefficient* is most often used method in evaluating statistical relationships.

Graphical analysis is widely applied for processing of acquired empirical documentation, as well as for their analysis. The widest use is of graphical presentation of frequency distribution (histograms or column diagrams, distribution polygon, cumulative curve, linear graphs referring to distribution of two or more occurrences, circular and sector diagrams - available for presentation of complex distributions of the study occurrences).

In medical geography, cartograms and cartodiagrams are widely used, suitable for presenting various statistical data of medico - geographical character (level of diseases frequency and mortality for appropriate types of diseases and their groups; coverage of a territory by physicians and other medical staff, etc.) for some area on geographic map. P.P.Reshetnikov with associates (1965), for example, developed for Celinia area (Russia) relationship between occurrence of dysentery and air temperature during August by making cartograms for administrative rayons and determining *correlation coefficient* r between intensity of the studied disease and temperature in August (r_{yxl}=0.63).

Methods of medical statistics will certainly have significantly role in solving tasks in the field of **medical geology**, providing it by material referring to public health, necessary for obtaining medico - geological evaluation of different territories (geological units, geological formations). Special possibilities are present in the field of founding statistical links between diseases markers and mortality and factors of geological medium, while we can use the numerous tests performed by medico - geographers. It is recommended, in spite of all that, to start with points (geological formations), where health problems are most frequent and with maximal use of data on geochemical and other features of rocks, soil and groundwater[104].

19. Methods of mathematical modeling. Initial stage in application methods of mathematical modeling in order to study links between environment and public health started during sixties of the last century, which was period of collecting significant experience in the field of use methods of mathematical statistics. During the first stage, uniqueness of application of mathematical statistics in medical geography was particularly in the fact that principal information related to the environment and level of public health had a space expression, which was performed by use of special geographic and medical maps.

Later, improving of simple ways of study even links, applied during the initial stage of the mathematical modeling methods development, was made simultaneously in several directions. First of all, ways of application of principal information for mathematical analysis were improved. The way out of an unpleasant situation (because of various natural and other occurrences, unhomogeneity of information, different time, different groups of inhabitants according to dimensions and status), was found in transformation of starting information through change of different concrete territorial units, by homogeneous, but conditional or formal units. The presented approach was called *formally - territorial* by B.B.Kucheruk (1965). In this way, it was attained that information on medico - biological occurrences obtained more objective character, and it's linking with information on the environment by mathematical methods become more correct. Directions of development of mathematical modeling methods according to Keller, Shchepin and Chaklin (1993), are presented in the further text, with examples of modeling shown in Fig. 8.7.

First of all, it has been concluded that determining stochastic relation between medico - biological occurrence and distinguished factors of the environment makes outlining the study area into the sectors (parcels) with equal probability of relationship of the analyzed pair of occurrences possi-

104 For example, several authors are dealing with analysis of correlation link between gastric cancer and magnesium content in soil (W.W. Alpatov, 1960; W.A. Chervyakov, 1964.; P.P. Reschetnikov et al., 1965.).

ble. Compiling such map, with distinguished even links of the studied medico - biological occurrence with environmental factors, enables persistent analysis of the role of distinguished factors in formation of areal and structure of the event.

Character of transition from the even links analysis, including numerous different occurrences and factors, can be clearly illustrated by example of introducing ***information analysis method*** into medico - geographical exploration. The method is characterized by a large number of advantages, enabling wide use of mathematical modeling in forecasting.

During the beginning of the seventh decade of the last century, methods of ***factoral and regression analysis*** were widely applied for solving tasks of determining links and relations between human health and the environment. Advantage of the factor analysis method in comparison to information analysis method was in the fact that it enables more adequate introducing of complex systems exploration, which depends on plenty of variables. The method was intensively used in exploration of hidden links with the environment of originally non - infectious medico - biological occurrences of not so clear etiology. Links of various environmental factors with cardiovascular pathology, but also with malignant neoplasms, are particularly studied, and appropriate risk factors investigated.

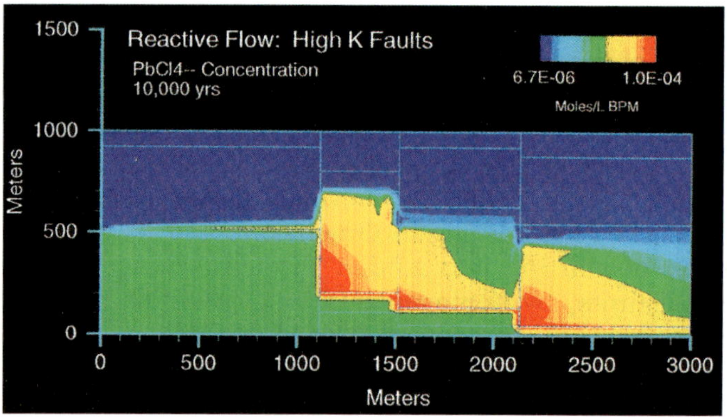

Fig. 8.7. Isotherms for the deposit - scale reactive flow simulation of Lisheen, Ireland. Flow is from South to North, driven by topography.

Regression analysis is principally used for solving the first part of medico - geographic task - to study form of the link among variety of unexpected occurrences, which in any case show effects on level of human health. The second part - informative analysis, which aim is to evaluate how the studied form of the link appears in geographic area, as a rule, has not been solved. There are numerous papers with the basic attention directed to solving only the first part of the task. In cases when informative medico - geographic analysis of obtained data doesn't lead to its logical end, good prospects are that exploration is considerably of epidemiological, hygienic, in one word - more medico - biological, than medico - geographic character.

Basic idea of regression analysis use within medico - geographic exploration consists of presenting characteristics describing conditions of medico - biological occurrences in form of function of total action of numerous environmental factors, and regression coefficient as importance coefficient or their effects. For example, during study of some somatic symptoms of population in the northern part of Chile, it was defined that, among factors showing effect on large number of these symptoms, the first place belongs to the hypocsic factor, expressed through height of the settlement above sea level.

Mathematical modeling is used in medical geography as a method of realization of different forms of expression of medico - geographic differentiation. The following three basic types of zoning of the territory by mathematical methods are distinguished:

I type presents a form of expression of space differentiation of medico - biological occurrences, characterizing level of public health;

II type presents a form of medico - geographic evaluation of physico - geographical, economic geographical and other types of territorial differentiation;

III type presents a form of systematic analysis of factually present ecological links of population and the environment.

Sometimes, zoning according to disease preconditions (or according to potential conditions of disease distribution), corresponds to the *type II*, and zoning according to factors assigning disease presence (or according to potential risk in some territories in relation to distinguished diseases and their effect on human health) - to the *type III*.

For zoning the territory of the *I* and *II type*, factor analysis is most widely used, and of the *type III* - information analysis.

20. Medico - geographic and medico - geological diagnostics. Relatively new approach. Under the term, complexity of the approach and methods directed to defining medico - geographic type of the territory and determining basic present or potential medico - geographic occurrences. Development of medico - geographic events characteristic for some type of the territory is preconditioned by objectively present complex of sanogenic and pathogenic factors. Determining the leading factors in the areas of some territory is also subject of medico - geographical diagnostics. Computing technique can significantly contribute to operative acquiring necessary data related to indicators. Potential possibilities of distinguishing territorial preconditions of diseases applying methods of remote sensing and means of biosphere investigation are also enormous.

The methodical approach contains use of the following two ways:

analytical (information - documentary), based on study, systematization and generalization of available information sources and (2) *expedition* (field exploration). According to Keller and others (1993), practical performing of the diagnostics in the areas of interesting region, the following stages are included:

- *defining diagnostic goals and tasks;*
- *study of present information sources;*
- *determining deadlines, itineraries, rayons, objects and exploration methods in the areas of the territory, as well as plan of exploration;*
- *field exploration and data processing;*
- *making a medico - geographical diagnosis* (basic forms of medico - geographic terms and degree of their presence; medico - geographic types of the territory; leading risk factors).

However, under real conditions, necessity for fast and simplified determining of only main medico - geographic features of the concrete area appears more often.

Expedition method of studying medical geography of certain territory is not so applied, because of other goals of scientific - exploratory expeditions[105]. One of a few of expeditions of this type, expedition in Celinuja area (Russia), was performed in two stages - in 1963 and 1964. The expedition was based on very useful contribution of medico - geographers, biogeochemists and biochemists after the first stage - compiling medico - geographic map, collecting water, soil and vegetation samples for further chemical investigations in laboratory located in Moscow (A.P. Avtsin, V.D. Arutjunov, 1965); significant attention was paid to investigation of fluoride content in drinking water and food products, as well as liquids in organism tissues of population. During the second stage, fluorosis diagnostics was stated precisely (dental fluorosis was registered at 54.3% of inhabitants), and rayons under endemic fluorosis risk (marked by high fluoride content in drinking water, up to 4.5 mg/l), were outlined at the special map. Within this part of job, dedicated to endemic fluorosis as a big problem, pathologists, stomatologists and radiologists were involved.

By the described example, where endemic fluorosis caused by geological factors was treated, it can be concluded significance of medical geology. Namely, participation of hydrogeologists and geochemists in the expedition would be certainly justified, and in making a diagnosis, particularly within the part referring to natural factors, medico - geological base and zoning would be very useful.

Medico - geological diagnostics gets a leading role in case of territories with expressed microelementoses, raised natural radioactivity and radon risk, inconvenient physico - chemical characteristics of drinking

---

[105] A.E. Beliaev with associates (1965), developed methodology of expedition, exploration on example of Hissar administrative rayon in Tajikistan.

water or soil, geochemical, geomorphologic, engineering geological and other geological factors influencing negatively or positively on human health and life (Fig. 8.8.). Distinguished geological medium morphologically formed and rich in mineral and thermal waters, for example, can be suitable for carriers' reproduction of numerous dangerous diseases of human beings and animals.

21. Medico - geographic and medico - geological forecast. Presents one of the most important stages of the research work of medico - geographers and medico - geologists[106]. It is closely connected to economical and land - use planning. This stage can be analyzed as potential characteristic of the general level of public health (during *universal forecast*) or distinguished markers of that level (during *branch* or *component forecasts*) in some region for certain period in the future, with pointing to factors of the environment affecting it. Pointing out to positive and negative factors of the environment on health is indispensable, because study of the factors enables to cross from research forecast to program or organization prognosis.

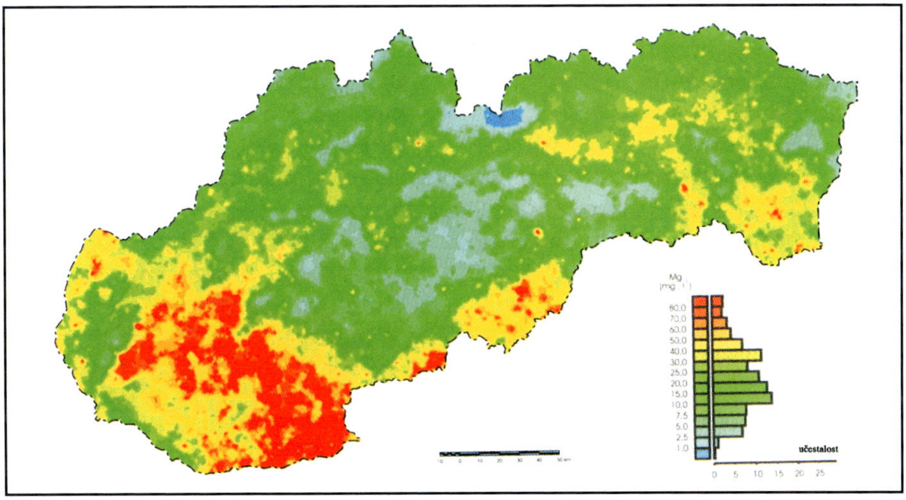

Fig. 8.8. Geochemical atlas of Slovak Republic. Distribution of magnesium in groundwater (K. Vrana, S. Rapant, 1999).

Integral (complex) medico - geographic prognoses are predominantly made for large economical rayons.

---

106 Short review of medico-geographical forecasts is presented by Keller and associates (1993). In dependence on the defined goal, the authors classified all medico-geographic prognoses into *integral* (complex), branch and component.

Branch medico - geographical forecast presents forecast in changing health level, related to occurrences studied by certain parts of medical geography. As an example, wide range of exploration made in order to forecast malignant tumors risk, can serve. According to A.V. Chaklin (1984), the exploration are contained of: studying regional characteristics of distinguished forms of malignant tumors; discovering groups of high risk; analysis of combined effect of endogenous and exogenous factors, followed by development of differentiated mark system; making diseases and mortality forecast of distinguished population groups, taking into consideration occurrence of new or rejecting environmental factors which were active earlier. Test of *branch medico - geographical forecast* is not finished with detailed development of prognosis in domain of oncology. Interesting investigations are, for example, performed in domain of biogeochemical endemics forecasting.

Component medico - geographic forecast present scientific predicting of consequences of influence of various components of geographic medium or their combinations on public health. In dependence on number of simultaneously treated components, the prognoses are classified into: one - , two - or polyfactor.

As a direct auxiliary method for giving the solution, *restricted maps*, as: (1) maps presenting contemporary natural opportunities of the prognosed territory; (2) maps referring to perspective economical development of the interesting territory; (3) maps - reviews of the future demographic structure of population; (4) maps of infrastructure; (5) maps related to up - to - date medico - geographical situation (maps of diseases frequency, maps of natural and anthropogenic preconditions of biogeochemical endemics; maps of components of the natural environment with negative influence on public health; maps of complex medico - geographic evaluation of the territory) are used. Role of the ending cartographic model for the forecasting has a map of medico - geographic differentiation of the territory, with legend. Legend of the map contains all factors of geographic medium, from negative or positive influence on human health, as well as a list of diseases with preconditions existent in each rayon.

Knowing components of the environment influencing on human health, *prognosing evaluation of public health level* can be made. In addition, evaluations given by experts, particularly of competent geographers and geologists, who studied certain environmental components from the standpoint of their influence on public health, are of outstanding significance. *Extrapolation method*, principally treated as searching method of standard situation, is also of great importance. After performing previous exploration, possibility of compiling hypothetical *nosological profile* appears. Making such profiles, work on making medico - geographical prognosis is mainly finished, because nosoprofile presents public health level sufficiently.

The next step after finishing work on making research forecast is *program and organizational prognosis.* On the basis of studied factors of the environment, lowering health level, as well as of character of pathology which can appear within population, *perspective program* can be made on the basis of the following questions: (1) prophylaxis which has to be taken during subduing the rayon; (2) which are the factors of the external environment dangerous for population. Concrete measures defined by organizational forecast are - *the first,* prophylaxis; *the second* - reclaiming of the territory.

Medical geology can be of high importance in forecasting level of public health for the interesting territory. That is particularly valid in domain of *branch and component prognoses,* as well as of *restricted maps,* presenting factors of the natural (geological) environment. For making prognoses, medico - geographic or medico - geological, good knowledge on geological structure, geomorphologic features, hydrogeological and engineering - geological characteristics of the terrain, but also geochemical features of rocks, soil, groundwater and vegetation of the territory, is one of very important preconditions. The factors will be also of outstanding contribution in discovering possibility of forming processes and occurrences related to anthropogenic activity and influence on the nature and society. So, for example, zoning of the territory according to the aquifer vulnerability to contamination can be very helpful in evaluating drinking water quality and proposing measures of the aquifer sanitary protection (S.Komatina, 1997).

22. Monitoring of the (natural) environment. As we could conclude from the previous text, more and more important precondition of environmental management is prompt prognosis, which is prophylactic activity and predicting possible changes in the natural environment, with necessary measures as a result of the changes. Necessary data are obtained by **monitoring of the (natural) environment,** as a system of observing, control and management of the environment. Monitoring can be performed in different scales (also in global), and has to be directed to the planned goal, interconnected and efficient.

One of the important tasks of monitoring is based on the necessity of distinguishing changes in biosphere conditions which are under influence of anthropogenic activity *(anthropogenic monitoring).* For contemporary environmental monitoring, it is necessary to use experience, present material (documentation) and main monitoring offices dealing with various natural processes and occurrences.

Introducing of control and prognosis functions into modern environmental monitoring makes its content much more complicated. In this way, requirements for representativity of the objects and observatory places and their density are made, because quantity and quality of principal data principally determines level of control efficiency and reliability of any prognosis.

According to I.P. Gerasimov and V.M. Kotl'akov (1983), **bioecological monitoring** is the first (principal) degree or block of contemporary environmental monitoring. The main link of bioecological monitoring is observing environmental conditions from the standpoint of its influence, first of all, on level of human health. Nowadays, for mass use, within the system of bioecological monitoring, toxic indes of soil and water medium contamination - so called indes or scales of norms of permitted concentrations of anthropogenic substances within the environment, are the best developed. In the near future, significant innovations in biological monitoring are expected, among all - in exploration of organic solvents exposure time, because progress of measuring technology made detecting even minimal quantities of some organic solvents in biological material possible. Outstanding progress is also expected in domain of monitoring toxic matters metabolites as well as methods for evaluation of body depots.

The second stage of environmental monitoring has to be **geoecological,** or wider - **geosystemic monitoring**. Task of the monitoring is in observing changes in geosystem level, in that sense - natural, in other words, geoecosystems, but also their transformation into naturally - technical agrosystems, mining basins, urban environment and environment of industrial rayons.

Organizing methodology of bio - and geoecological monitoring is completely different. Bioecological monitoring is based on geophysical, biochemical and biological parameters in the network of *control "points - checkpoints"*, in other words - basically, it is of local character. Geosystemic monitoring is much more based on geophysical, geological, biogeochemical and biological methods, but, beside the network of control points, *key surfaces system* is used, that is - basically, it is of regional character. The network of *key surfaces* has not to be extremely voluminous, but must have representative character.

The third degree is known as **biosphere monitoring.** This type of monitoring has to provide observing, control and prognosis of potential changes, this time not in regional, but in global scale, in other words - in comparison to biosphere in general and its changes caused by activities of society. It contains characteristics of Solar radiation, gas content and atmosphere degradation, as well as direct influence of heat of anthropogenic nature on general energetics of atmosphere, monitoring and measuring global biological productivity of land (continents) and water of the World ocean. Within biosphere monitoring, global effect of anthropogenic influence on climate and particularly gas content of the atmosphere, as also anthropogenic changes of water balance and disturbing global circular motion of moisture, with forecast of future period, is of high importance. The global monitoring is based on the system of geoecological zonal and regional polygons.

Geological monitoring, as one of the important chains of geoecological monitoring, includes numerous factors of the geological environment more completely. Data obtained during the monitoring are used for solving different tasks for scientific and practical purposes, as: exploitation of mineral raw materials; providing drinking water, mineral and thermal water; exploration of geological soil for constructing, etc. State, regime and quality of groundwater but also exogenous and endogenous geological processes and occurrences; natural geophysical, geochemical and radiogeochemical anomalies; geochemical (radiogeochemical) anomalies within pedological soil and leading aquifers, formed by anthropogenic activity in agricultural areas, mining basins and urban environment, are observed by regional network. The mentioned activities are observed by the system of remote and aerocosmogeological monitoring of geological environment.

Nowadays, preservation of groundwater resources, on which water supply of population all over the world is based, and which are, from the other side, significantly contaminated, is particularly studied. Beside developed hydrogeological information systems, numerous countries are famous by well organized groundwater monitoring. So, for example, Geological Institute of Finland started in 1969 with groundwater monitoring of the southern half of the state territory, with 50 monitoring rayons included starting from 1985. (B.Backman et al., 1999).

CHAPTER 9

Regional Medical Geology

There are two ways of organizing work on medico - geological study of a given territory. *In the first place,* medico - geological investigation can be initiated by isolating geological units and then clarifying the diseases that characterize those units. The *second way* is to establish the geographic distribution of diseases and pathological states from reliable data and then characterize certain medico - geological regions (comples) after comparing this information with the geological properties of a location.

For the time being, it is difficult to say which of these ways is the more rational. Use of either the first way or the second can doubtless be justified in practice. The most important thing is to answer the question as to whether certain diseases and pathological states are linked with the geological (natural) environment.

The goal of the chapter on regional medical geology is more in keeping with the first of the two indicated approaches. To be specific, geology can make a considerable contribution in the event that geotectonic units of the first or lower orders are singled out and studied. It is certain that such a method can make it much easier to assess a geological environment from the medical standpoint and draw valid conclusions based on that assessment. Many examples from the practice of research conducted to date indicate the great promise of such an approach. This is especially true because the geological framework of a territory in greater or lesser measure dictates the position and characteristics of pedological, climatic, biogeographical, floristic, geomedical, and other types of ranges.

In further text geotectonic (geomedical) units of the first and second order, are discussed.

GEOTECTONIC UNITS OF THE FIRST ORDER INTERESTING FROM THE GEOMEDICAL STANDPOINT

Many geological bodies and various structures - geotectonic units - have been formed in the Earth's crust that gives it the features of a complex edifice of mosaic nature. *By geotectonic unit, we mean a geological body, unit, or part of the Earth's crust that in its composition, structure, origin, and geological age stands apart and represents a geological environment that is distinct in space and time from other neighboring bodies and stuctures.* Some geotectonic units are large, extensive, and all - inclusive (extending over the entire surface of the crust and embracing continents and oceans), while others are spatially limited and even local in extent.

Because of the complexity of tectonic relations prevailing in the Earth's crust, it has always been difficult to isolate its structural units. It is understandable that the geological literature and terminology reflect many views on the origin of geotectonic units and contain numerous divisions and terms that are mutually incompatible and subject to different interpretations. Still, the first glance at a geological map of the world reveals isolated units of interest to us from the geomedical standpoint. We shall rely on the modern views synthesized in the book "**The Earth, Plate Tectonics, and Magmatism**" by D. Milovanovic and B. Boev (2001). Geologists in their research today for the most part create models based on the *theory of plate tectonics* as a foundation of geology. That theory explains many geological forms and processes: mid - oceanic rifts, island and volcanic arcs, collision of continental plates, location of volcanoes and earthquakes on the Earth, creation of mountain systems, etc. The plates themselves were discussed in the chapter on geological factors.

On the basis of geological and geophysical forms, the continental crust can be divided into several typical formations: continental and pre - Cambrian shields, continental platforms, Paleozoic orogenic belts, Mesozoic—Cenozoic orogenic (mountain) belts, and terrains (Fig. 9.1.) (Table 9.1).

1. *Continental* and *pre - Cambrian shields* represent stable parts of continents built of pre - Cambrian (metamorphic and acidic magmatic) rocks with or without a thin sedimentary cover. They are from 500 million to 3.5 billion years old. Regions of shields exhibit slight variations of relief and are tectonically stable over longer periods of time. They occupy about 12% of total volume of the continental crust. The largest among them are the African, Canadian, and Antarctic shields.
2. *Continental platforms* are also stable parts of the continental crust. Like shields, they are built of pre - Cambrian rocks, but differ from shields in being covered by a layer of sedimentary rocks up to 5 km deep.

TABLE 9.1.
Geotectonic units of the Earth's crust.

Categories of the units	Geotectonic units	Main geological features
1. Consolidated areas of the Earth's crust	*1.1. Continental shields* African, Canadian, Antarctic, South-American, Fine Scandinavian, Indian, Australian, Chinese, etc.	Very old stable parts of the continent made of Precambrian metamorphic and acid magmatic rocks. Number of shields is predominantly covered by moraine material or postglacial lacustrine clays. Expressed flatness of relief, with spacious plains or deserts, is characteristics of this type.
	1.2. Continental platforms Eastern European platform (Russian plain, German-polish plain), Western European plate (plain), Arabian platform, Siberian platform (Western Siberian depression, Middle Siberian plateau), Saharo - sudanian platform, Northern American platform, etc.	Stable parts of the continents made of metamorphic rocks, covered by 1-5 km thick (or more) sedimentary rocks. Spacious covers of glacial formations, with expressed flatness of relief and deserts. Huge artesian basins with mineralized waters. Parts of platforms on the northern hemisphere with developed criolite - zone (permanently frozen soil and eternal ice).
2. Relatively consolidated areas	*2.1. Paleozoic orogenic belts* Central massif (France), Rhaina-middlegerman mountains, Czech massif, mountains of British isles, Scandinavian mountains, Altai-Baikal mts., Mongolian - Chinese plateau, Manjuria - Korean area, etc.	Mountainous massifs made predominantly by metamorphic and magmatic rocks. Intensive tectonic sinking and uplifting of block and long-term erosion processes resulted in low and vaulted enormous mountains, with troughs and erosion surfaces.

3. Mobile areas of the Earth's crust	*3.1. Active continental margins* 3.1.1. Alpine high-mountain ranges: Andes, Cordilleras, Alpine - Himalayan system (Pyrenees, Alps, Apennines, and Carpathian - Balkan mts., Dinaridic mts., Atlas, G. Caucasus, Tian-Shan, Tibet - Himalayan mts.), Urals, mountains of NE Asia, Appalachian mts., mountains of Eastern Australia, etc.). 3.1.2. Spacious plateaus: Tibet, Colorado, etc...	Mountains formed in places of two plates bordering. Made of thick deposits of sedimentary, magmatic and metamorphic rocks, very disturbed tectonically; among intrusive rocks, granodirites are predominant, and among volcanites - andesites. Formed chains of high marked mountains. Expressed volcanic and seismic activity. Outcrops of low - mineralized, mineral and thermal waters are numerous.
	3.2. Rifts and rift valleys 3.2.1 Inside of continents: African - Arabian rift (Syria - Lebanon, Dead Sea, Ethiopia - Kenyan or Eastern African), Niassa - Tanganyika, rift of Rhine, Baikal rift, rift of Southeastern Australia, Rio Grande rift, plateau basalts of Paranna (eastern part of Southern America), Yellowstone park (USA).	Basalt flows and plateau basalt; numerous volcanic structures (domes); intensive uplifting and sinking of tectonic blocks; high heat flow.
	3.2.2. Middle - ocean rifts, that is - horsts. Iceland, Galapagos, etc.	Islands marked by volcanic domes made of basaltic lava. Hydrothermal occurrences are numerous.
	3.3 Island arcs: Japan, Fillipines, New Hebrides, New Guinea, Indonesia, New Zealand, Antilles, Curile arc, Isu - Bonin arc, Mariana arc, Tonga arc, etc..	Islands predominantly made of basalt and andesite. Marked domes of active volcanoes. Very strong earthquakes. Outcrops of high - thermal waters overheat gas and steams are numerous.
	3.4. Basins in back of arcs: Mariana islands, Tonga and Isu - Bonnin in Pacific, Scotia in Atlantic.	Islands made of basalt and sedimentary rocks. High heat flow from the Earth's interior.
	3.5. Oceanic islands and plateaus: Volcanic chain of Hawaii (more than 2,000 km long) and Tahiti in Pacific; Canaries and Azorean islands in Atlantic; islands in Indian ocean.	Islands present volcanoes under the water surface, partly over sea level. Ocean plateaus are existent, also of volcanic origin. Expressed recent volcanic activity. High heat flow.

NOTE: Geomedical features of distinguished units presented in the text.

3. *Paleozoic orogenic belts* are several hundreds to several thousands of kilometers long. They are built of various kinds of magmatic, metamorphic, and sedimentary rocks. These belts are deeply eroded mountain ranges with weakly expressed relief and relative tectonic stability.
4. *Mesozoic—Cenozoic orogenic (mountain) belts* are similar in size and pattern to Paleozoic orogenic belts. Being younger, however, they are tectonically unstable and geomorphologically much more diverse. The three greatest Cenozoic orogenic belts are the Cordilleras, the Andes, and the Alpine—Himalayan belt. Blocks and present - day plateaus (Colorado and Tibet) that have not undergone the deformations characteristic of orogenic belts occur in them.
5. *Terranes* are tectonically limited segments of continents that have a geological structure, history of formation, composition, and pattern different from those of the surrounding region or surrounding terrains, with which they are in contact in most cases. Fused terrains indicate great movement from their primary position, which is confirmed by paleomagnetic investigations.

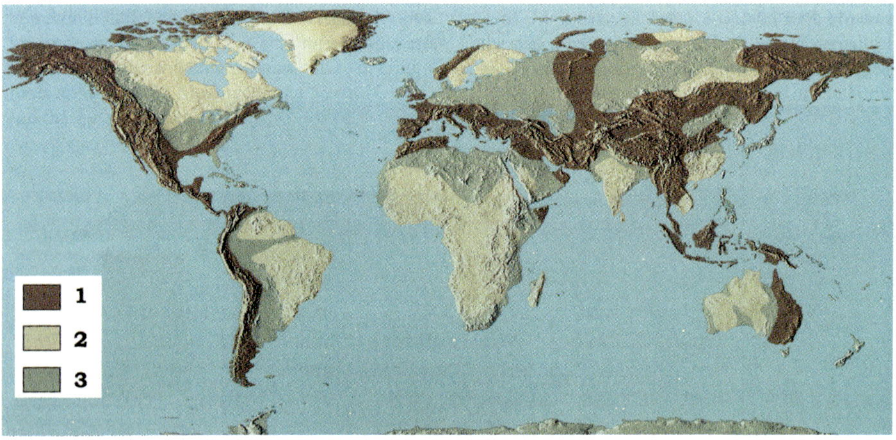

Fig. 9.1. Position of Alpine orogenic belts(1), continental shields(2) and platforms(3) in the Earth (D. Milovanovic, B. Boev, 2001).

The most intensive geological processes take place on the boundaries of plates (Fig. 2.12.). In addition to complex tectonic activity, formation of the majority of magmatic rocks occurs there. This applies to mid - ocean - rift basalts; basic and alkaline rocks within continental rifts; volcanic and intrusive rocks of island and volcanic arcs; and basalts of basins behind the arcs. Plate boundaries are also zones of current volcanic activity and earthquakes (Fig. 2.20.). The next part of the text will therefore be devoted to these interesting boundary regions (Figs. 9.2. and 9.3.).

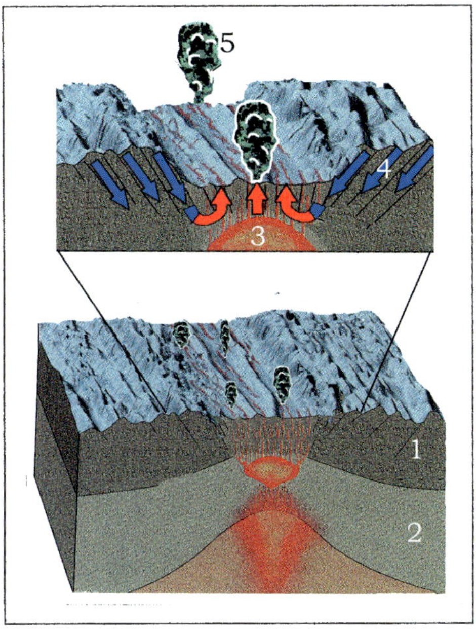

Fig. 9.2. Origin of "black chimneys" in rifts with fast plate opening. 1 - Oceanic crust; 2 - Oceanic lithosphere; 3 - Warm sea water; 4 - Cold sea water; 5 - black chimneys (D. Milovanovic, B. Boev, 2001).

Plate Divergence Boundaries or Rifts

Divergence of two plates in the ocean (in a *mid-ocean rift*) results in the outflow of a large amount of basalt, creating a *mid - ocean* ridge. Representing enormous underwater mountain ranges built of basalt, mid - ocean ridges are more than 20,000 km long (in the Atlantic and Pacific Oceans) and 1,500 to 2,000 km wide. They reach a height of 2 to 3 km above the ocean floor and are marked by numerous volcanic structures, which very rarely rise above the surface of the sea (as in the case of Iceland). Also characteristic are hydrothermal occurrences formed by sinking of seawater along fissures and faults near rifts, heating of that water, and the appearance of hot springs rich in dissolved sulfur and metals separated from basaltic lava with hydrothermal solution temperatures of up to 350°C and more (Fig. 9.2.).

Rift valleys are formed if divergence along rifts occurs within continents. Basaltic magma flows out along faults, resulting in the formation of a basaltic dome or volcanic plate (plateau). Together with intensive volcanic activity, considerable heat is generated due to rising of the ground, and weak earthquakes with foci at shallow depths occur. The rift can be hundreds of kilometers long and several tens of kilometers wide. The greatest structure of this kind on Earth is the African—Arabian Rift, which extends from Mozambique to Turkey over a length of about 6,500 km. Rifts of high volcanism occur in Kenya and Ethiopia (the East - African Rift, which is a constituent part of the African—Arabian Rift) and North America (the Rio Grande Rift).

In contrast to continental rift basalt (which is effused in the rift itself), *continental plateau basalts* are formed due to an overflow from the given structure (rift valley) of material that covers large surfaces of continents and builds volcanic provinces. A volcanic province is composed of subhorizontal to horizontal outflows or plateaus of basaltic lavas with an area of from several tens to several hundreds of square kilometers lying on the Earth's crust.

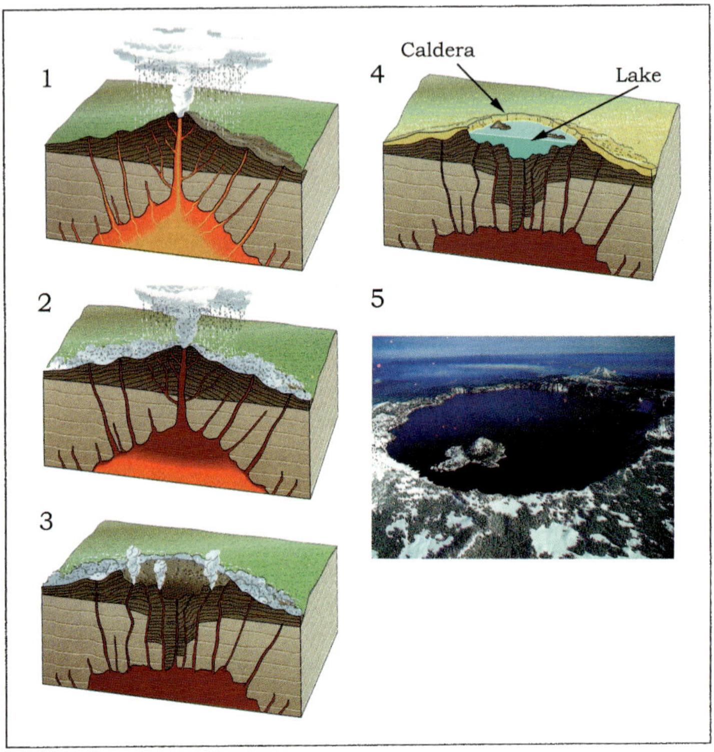

Fig. 9.3. Formation of caldera (Milovanovic, Boev, 2001).

Boundaries of Undertow or Subduction

Zones of subduction (the process of one plate descending beneath another) represent the most complex tectonic regions on Earth and one of the most significant phenomena of global tectonics. About 80% of active volcanoes are found in regions of plate subduction, in island arcs, and on active margins of continents; about 15% of them are found in regions of plate divergence (rift formation); and only about 5% occur within plates. The majority of powerful earthquakes on the Earth are also linked with the edges of subducted plates.

Island arcs are formed by the descending of one oceanic plate beneath another one. Due to conduction of heat from the surrounding layers as well as mutual friction, subducted rocks are progressively metamorphosed and in the end become melted material. Being lighter, this material emerges on the surface to form an island arc. Among volcanic island arcs, it is possible to distinguish ones that are young (Tonga), "middle - aged" (New Hebrides), and old (Japan, the Philippines, etc.). Volcanism linked with subduction zones usually begins suddenly and powerfully, a volcanic front of basaltic and andesite magma forming at a distance of 150 to 300 km away. Volcanoes themselves are built of lava and pyroclastic material, while collapse of their walls leads to the formation of calderas (Fig. 9.3.). Volcanic ash and dust can be deposited as far as 100 km away from the volcano.

Even before appearance of the tectonic plate theory, the Andes were regarded as an example of mountains formed by subduction of an oceanic plate beneath a continental one, i.e., as an *active continental margin.* The structure of a volcanic or magmatic arc formed in this way is very similar to that of an island arc. In exceptional cases, if subduction is subhorizontal, great nappes and a complex tectonic area are created without the formation of a volcanic or magmatic arc.

Because the great majority of active volcanoes are found in zones of plate subduction or divergence, *oceanic (volcanic) islands* are very numerous in these regions, especially in the Pacific. More than 2,000 km long and built exclusively of basalt, the volcanic chain of Hawaii represents the most thoroughly studied of oceanic islands (Fig. 9.4.).

An oceanic phenomenon, *marginal basins* are formed behind island chains. They are built primarily of basalt.

Collision of Two Continental Plates

When two continental plates collide instead of undergoing subduction, a contact zone is created with intensive tectonic, metamorphic, and magmatic activity. The area becomes a zone of crust thickening marked by great mountain ranges such as the Himalayas, Alps, Appalachians, Urals, etc.

Fig. 9.4. Volcanic chain of Hawaii (D. Milovanovic, B. Boev, 2001).

BASIC GEOMEDICAL CHARACTERISTICS OF FIRST - ORDER GEOTECTONIC UNITS AND THEIR EXTREME ZONES

Geographic zonality is generally known. At the boundary of the atmosphere and lithosphere, horizontal and vertical zones are a result of effects and mutual influences among many factors, viz., solar heat, rocks, soil, water, relief, vegetation, animal life, and man. The areas of these zones on the Earth are given according to A. Maksimovic (1955) in Table 9.2.

Generally speaking, climatic characteristics in large measure determine the natural conditions for development of life, and several latitudinal zones are discernible from the pole to the equator:

- The *arctic region*;
- A *zone of temperate humid climate*;
- A *zone of desert climate*; and
- The *equatorial zone.*

Hydrological zones correspond to the geographical zones listed in Table 9.2.: the zones of the tropics, forests, and tundras are humid and excessively humid; those of steppes and semi - deserts are variably humid and semi - arid; and desert zones are arid.

Specific vegetation and animal life develop in the given conditions, which are largely determined by soil and morphology. All of this dictates the appearance of natural foci of disease. For example, it is known that the number of parasitic, transmissible, and other infectious diseases increases as we approach the tropical zone, which is recognizable from a number of natural properties. In addition to this, the degree of epidemiological danger increases, as does the index of linkage of foci. On the other hand, mass incidence of sclerosis increases as we move away from the equator, in both northern and southern latitudes.

TABLE 9.2.
Geographic zones.

	Geographic zones	Surface (10^6 km^2)	%
1	Polar and high mountains ice	16.08	10.8
2	Tundra	5.9	4.0
3	Forest zones	18.08	12.1
4	Stepes and forest - stepe areas	32.39	21.8
5	Deserts and half - deserts	27.21	18.3
6	Mountainous zones	22.62	15.2
7	Alluvial area	4.26	2.9
8	Continental water (excluding Caspian sea)	0.97	0.7
		148.63	100.0

Keller et al. (1993) created a complex epidemiologo - geographic regionalization on the basis of analyzing the results of studying geographic conditions in the world, the level of incidence and structure of diseases, the degree of risk from infection, and the spatial distribution (endemism) of illnesses. Eighteen epidemiologo - geographic regions, described in the book "**Handbook of Medical Geography**", are isolated. Greater or lesser agreement with geotectonic regionalization is evident. For example, the Canadian—Greenlandic, Eastern South American, West African, Central African, Australian, and European regions[107] for the most part coincide with tectonically passive zones of the Earth's crust (shields and platforms); while the Central American, Western South American, Mediterranean, and Southeast Asian regions, together with certain parts of other regions, coincide with mobile zones.

Already in 1936, A. A. Grigor'ev noted the necessity of creating and using a parallel regionalization based on geological facts. Grigor'ev then indicated that the time was ripe for physical geography to divest itself of its "physical" traditions: "*Illumination and investigation of the chemical aspect of processes must now be raised to the level needed.*" Geochemical factors of the environment were thus acknowledged to be of great significance in what can probably also be considered an unconscious assertion of the advantages of geological (geotectonic) regionalization[108].

Justification for geotectonic (geomedical) regionalization can be found in many facts uncovered in the long course of medico - geographical research. For example, the need for them follows from information about various microelementoses and numerous other illnesses and data on the spatial distribution and natural settings of previously described rare and strictly localized diseases. The influence of geological factors (rocks, geological formations, soil, relief, endogenous forces, geophysical fields, geochemical properties of the geological environment and groundwater, etc.) on human health can be discerned from their discussion in Chapter 2 of the book. Because geotectonic units integrate the given factors, it is entirely proper to consider them from the geomedical standpoint. To be specific, geotectonic units represent settings with special geological, geophysical, geomorphologic, geochemical, pedological, climatic, hydrogeological, biogeographic, floristic[109], and other natural characteristics and conditions for human and animal life.

---

[107] As well as the Siberian region, which the authors did not include in their regionalization.

[108] It was completely logical that A.E. Fersman (1922, 1931) indicated the coincidence of *geochemical provinces* - by which he meant geochemically homogeneous regions with a certain association of elements - with geotectonic units (shields, platforms, or mobile belts).

[109] Correlation of floristic and geotectonic regions is evident.

Geomedical characteristics of geotectonic units have not been the subject of special study to date. We rely in our treatment on the geotectonic regionalization presented in Table 9.1. Because the influence of morphological and climatic factors (apart from geological ones) is very pronounced in the case of some expansive units, there are grounds for distinguishing so - called *extreme zones* within the given units. Extreme zones can be particularly interesting from the medical standpoint. For example, the Siberian Platform consists of the West Siberian Lowland (with an area of about 2 million km^2) - the most ideally level plain in the world, which lies predominantly below 100 m above sea level; and the Central Siberian plateau (with an area of 3.5 million km^2), with broadly level watersheds at an average elevation of 500 - 700 m above sea level. Well - known extreme zones, including the tundra, taiga (forests of temperate climate) and steppe zones are developed in both cases, albeit with specific features. Within such expansive zones, moreover, many features significant for human, animal, and plant life are dictated by the type and geochemical characteristics of the lithological substrate and by geochemical characteristics of rocks, soil, and water. It is therefore sometimes expedient to devote special study to such lower - ranking geomedical units.

Geomedical study of the isolated geotectonic units and their extreme zones is especially significant for a number of reasons, primarily because the pathology, physiology, and hygiene of those areas exhibit a fairly great number of specific characteristics, which are largely dictated by geological, geomorphologic, geochemical, hydrogeological, and climatic features. Anyway, it has long been considered necessary to study the influence of extreme environmental conditions on the human organism. For precisely this reason, geographers, doctors, and biologists have organized numerous expeditions into areas of high mountains, polar regions, waterless deserts, and humid tropical regions, thereby enriching the medical and biological sciences with much valuable new information. Not only have new forms and variants of diseases been discovered in this way, but remarkable examples of the adaptive capacity of the human organism have been revealed as well. The arsenal of medical and biological research methods has also been enriched.

Consolidated Regions and Their Extreme Zones

Consolidated regions of the Earth's crust (old shields and platforms) embrace enormous areas with relatively homogeneous geological and geochemical composition, tectonic framework, and hydrological and hydrogeological conditions. The soil is also homogeneous, of predominantly limited fertility, and deficient in many elements important for life

(Ca, Mg, Na, K, P, Co, Cu, etc.)[110]. Many acidic swamp soils are difficult to drain and unfavorable for plants, animals, and man. In contrast to acidic soil, soil formed on limestone (carbonate) terrains is rich in calcium carbonate and thereby favorable not only for plants and animals, but for agricultural production as well[111].

Several types of extreme zones are developed on shields and platforms: a polar region with tundra zone, a forest belt, and a steppe zone on the Canadian Shield (with the Northeast Platform) and the Eastern European and Siberian Platforms; arid zones (deserts and semi - deserts) on the Sahara—Sudan, Arabian, and Australian Plates; and a tropical belt in equatorial regions, i.e., on parts of the South American, Chinese, and Indian Plates. Human life in different climato - geographical or climato - geological regions of the Earth is characterized by various physiological reactions and adaptations to the given environment. Several geomedical characteristics of extreme zones are singled out below.

Polar (Arctic and Antarctic) Regions

These regions are located north or south of 66° in both hemispheres of the globe. They are typical representatives of extreme zones with geographically isolated populations. The vast Antarctic and arctic deserts are ice - covered and characterized by the polar type of climate (shortage of sunlight, especially severe chill factors in regions of high mountains due to low air temperatures and strong winds), infertile soil, and low content of mineral compounds in water. Polar regions are rightly considered to be harsh and unfavorable for plant, animal, and human life. Apart from the short growing season, high water levels caused by potential ice melting make farming difficult.

The adaptive responses of man in these regions are very diverse. In the area of medical protection, a specific approach is required in dealing with the native population and newcomers. A new branch of science has been developed to meet these needs, namely *arctic* or *polar medicine* (cryo - pathology, epidemiology, and hygiene), supported by geographic pathology and bioclimatology of man under conditions of high latitudes.

The danger of contracting diseases appears especially with the arrival of people from other territories, when the otherwise pure bacterio - viral setting is contaminated by various agents of infectious diseases. In addition to the psycho - emotional stress syndrome described earlier, polar regions are characterized by the following disease structure:

- General cooling of the body with numerous consequences, up to and including death;
- Diseases of the nervous system and senses;
- Chronic bronchitis; long - term and chronic nonspecific pneumonias;

---

[110] Calcium deficiency is manifested especially clearly on the organism of animals: animals suffer from skeletal diseases (rickets), birds are poor layers, their eggs have thin shells, etc.

[111] The given regions are characterized by the presence of very extensive artesian basins, mainly with water that is highly mineralized.

- Rheumatism;
- Diseases of the teeth (high frequency of caries) and the oral cavity;
- Diseases of the circulatory system; hypertonic sickness, sometimes in combination with atherosclerosis;
- Anemia due to iron deficiency in the blood; and
- Biogeochemical endemias (microelementoses).

Epidemics of poliomyelitis and influenza have been recorded (in Eskimos of the Canadian Arctic), together with tuberculosis infections (in the Eskimos of Alaska).

The high - mountain terrains of the Arctic and Central Antarctica are characterized by low atmospheric oxygen content in proportion to elevation above sea level, and this causes severe or chronic mountain sickness in the inhabitants of those territories.

The Forest Belt

This belt encompasses vast areas of the Northern Hemisphere with a damp, moderately warm climate. Climatic conditions were exceptionally favorable for development of forests, especially in Siberia. Landscapes of *transitional* (from acidic to calcic) and *calcic classes* were created, depending on the rock substrate. The former are prevalent in areas of the Siberian taiga and deciduous forests of the East European Plain and West Siberian Lowland, the latter in the Central Asian part of the belt.

In contrast to climatic conditions, the properties of soils of the *transitional class* of landscapes are far less favorable for life. Soluble compounds important for plant and animal life are readily leached from soils and transported into river courses by groundwater. For this reason, acidic soils poor in mobile mineral compounds are best represented. Sandy ground often consists of 98% residual quartz. It follows that the transitional type of landscape from the geochemical point of view is poor in elements important for vegetation. Calcium deficiency is manifested especially visibly on animals of these regions, which are shorter in stature, have fragile bones, and suffer from osteoporosis, rickets, and other diseases. Cobalt deficit occurs frequently, with the result that sheep and (more rarely) cattle are subject to the serious disease B - 12 hypovitaminosis. Deficiency of copper also causes a number of diseases in domestic animals. Iodine deficit sometimes leads to breakdown in functioning of the thyroid gland and development of endemic goiter in domestic animals and humans.

Landscapes of the *calcic class* are formed on terrains with limestones and dolomites, as well as in carbonate moraine material. A carbonate weathering crust with its soil cover is formed here. In comparison with the weakly mineralized water of the transitional class of landscapes, surface and groundwater is appreciably richer in minerals and sometimes hard. Generally speaking, calcium is the typomorphic element of these regions: it erts decisive influence on organisms, soil, and local water and causes such landscapes to differ significantly from acidic taigas, for example. The very favorable climatic conditions, fertile soil, and water of good quality have played an important part in making flatland regions of this type suitable for life and agriculture. For this reason, such regions were the first ones in the European Arctic to be opened up for farming, and they have to date been completely reclaimed. The diseases of animals and humans characteristic of landscapes of the transitional class are far rarer under such conditions.

Arid Regions

Formation of arid regions was primarily dictated by unevenness in the distribution of heat and moisture and zonation of the globe's geographic envelope. Such regions are widespread on territories of shields and platforms (in the northern half of the African Shield, on the Australian Shield, and on the Arabian Platform), but also in mountainous regions of North America, South America, and Asia at elevations of the order of 3 - 4 thousand m above sea level (Tibet and elsewhere).

Deserts are characterized by high summer air temperatures, small amounts of precipitation, absence of surface watercourses, prevalence of a sandy substrate, salinization of shallower horizons of groundwater, and migration of soluble salts in the soil. In spite of the abundance of heat, light, and fertile soil, a poverty of vegetation is evident.

About 4% of the world's population lives in the harsh extreme conditions of arid regions[112] (Fig. 9.5.). Because water is the basic factor for life in these regions, irrigation of deserts has played a leading role in the history of mankind, and many civilizations with a high degree of culture sprang up on irrigated lands: ancient Egypt, Babylon, and Mesopotamia[113].

Despite many common features, deserts can differ significantly. These differences determine characteristics of regional medical geography (geology) and geographic pathology. Every desert has its own medical geography (geology), determined by its natural (geological) and socio - economic conditions.

In examining the nosogeography of arid regions, Keller et al. (1993) single out the following three groups of diseases:

1. *Diseases caused by the influence of arid climate;*
2. *Diseases linked with the action of geochemical factors; and*
3. *Pathology determined by the existence of endemic vegetation and animal life.*

The *climatically - determined* pathology of arid zones is formed under conditions of high temperatures and dry air, extended periods of their action, spectrally varied solar radiation, and sharp daily and seasonal oscillations of meteorological factors. All of this places high demands on the adaptive system of the human organism. The following diseases are typical:

- *Heat stroke and other forms of overheating of the organism;*

---

[112] Arid regions today occupy an area of the order of 28 million km^2 or 18.8% of the world's land (59% of the African continent, 83% of Australia, the whole of the Arabian Peninsula, 22.7% of the Asian continent, etc.). One of the largest deserts on the Earth, Sahara extends for a length of 5,000 and width of 2,000 km. It is characterized by the highest air temperatures (up to 58-59°C) and ground heating (up to 70-80°C), the lowest level of precipitation (up to 22 mm annually), and the greatest oscillations of daily air temperature (even more than 30°C).

[113] In addition to surface watercourses, large amounts of the groundwater of well-known artesian basins have also been used and are being used. For example, the Saharan Artesian Basin occupies parts of the territory of Morocco, Algeria, Tunisia, and Libya.

- *Gastro - intestinal diseases (some parasitic diseases are very widespread);*

Fig. 9.5. Desert ambient.

- *Formation of stones in the upper and lower urinary pathways;*
- *Skin cancer and dermal pathology as a whole; and*
- *Cardiovascular diseases (according to some authors, while others hold that they are atypical).*

In arid regions, a number of characteristics of diseases linked with the action of *geochemical factors* are expressed. The unfavorable action of these factors is amplified by high temperature of the surrounding atmosphere, complex resistance of organisms to chemical substances, and increased permeability of sweaty skin. Let us dwell on the influence of soil and water on human health. In natural conditions, this influence is manifested in the form of the following things (among others):

- *High incidence of goiter and fluorosis due to iodine or fluorine deficiency in some arid regions;*
- *Endemism of arid regions for cancer of the esophagus due to the high share of quartz in desert dust, action of nitrates and nitrites of the soil, specific diet, and shortage of good - quality drinking water and low content in it of mineral substances and microelements (the main foci are in Southeast Africa and the republics of Central Asia, viz., Turkmenistan and Kazakhstan); and*

- *Urolithiasis (according to certain authors, arid regions represent endemic foci of urolithiasis, depending on the degree of mineralization of drinking water).*

Diseases caused by specific characteristics of the *arid flora* are primarily manifested in the high presence of allergic ailments. On the other hand, the following groups of diseases are caused by specific characteristics of the *arid fauna*:

- *The group of diseases linked with non - living nature (climate, soil, etc.): non - transmissible diseases whose agents must experience certain transformations under prevailing environmental conditions in order to become invasive; and*
- *The group of endemic infections (invasions), which belong to the category of natural - focus diseases and occupy a significant place in the pathology of inhabitants of hot zones of the Earth.*

Tropical Regions

The damp tropics occupy enormous areas of Equatorial Africa (the southern half of the African Shield), South and Southeast Asia (the Indian and Chinese Shields and island arcs of Southeast Asia), Central America, and South America. More than half of the world's population lives in this space. The greatest number of diseases (including the most serious ones) is disseminated here. We shall dwell on territories of old shields marked by extensive lowlands and plains (the very flat lowlands of Amazonian, for example, with an area of as much as 5 million km^2).

A characteristic of the damp tropics is the constant coincidence of a hot climate and high humidity. The presence of great heat and abundant moisture dictates high annual production of living matter, from which such regions are recognizable. Also characteristic is an abundance of plant species[114].

The soil provides diverse food products for animals and man. Differences are discernible here between the *acidic* and *acidic - swampy classes* of landscapes, which are prevalent.

1. Typical representatives of the *acidic class* of landscapes of the damp tropics are formed on magmatic, metamorphic, and sedimentary silicate rocks, where a thick acidic weathering crust is created under acidic soil, the indicated crust being enriched with inert elements such as iron, aluminum, residual quartz, and certain rare elements of the inert group (Fig. 9.6.). Especially characteristic is the low content of calcium, which is sometimes less than 0.1% (in contrast to Fe_2O_3, with 30% and higher, or Al_2O_3, with even as much as 50%). Richness of the soil and weathering crust in hydroxides of iron gives them a red, orangish, or yellow color.

---

[114] The colossal amount of biological information not only is a consequence of climatic conditions favorable for organisms, but also results from geological circumstances. To be specific, landscapes of the damp tropics belong to regions of old land, where evolution has taken place continuously for many millions of years, starting with the Paleogene and even the Mesozoic.

Characteristics of the soil and weathering crust also determine the geochemistry of soil and groundwater. The water in question is moderately fresh, with mineralization below 100 mg/l. It is understandable that plant and animal life is likewise characterized by a deficit of mineral substances. For this reason, use of nitrogen, phosphorus, potassium, and other fertilizers is necessary to ensure realization of the enormous potential for farming. Use of chemicals in ranching is also effective.

2. Soils of landscapes of the *acidic - swampy class* are formed on plains and in valleys of the tropics. An incomparably smaller number of plant species are developed on them, and biological productivity as a whole is lower. This class of landscapes is characterized by an acute deficit of oxygen in the soil, accumulation of black humic matter in it, and a very acidic reaction of its iron - rich water. Forest mud occupies large areas in the basins of the Amazon, Orinoco, and Congo Rivers.

As stressed above, tropical regions are unique for extreme precipitation and droughts, which lead to strong geochemical responses of elements. Excesses or deficits of certain elements occur, and this has consequences for human beings and their health in both cases. These consequences are magnified by the virtual non - existence of preventive protection of the population, since the countries in question are among the poorest in the world. The great dependence of man on nature permits geomedical studies by provinces.

Fig. 9.6. Intensive colors of tropical birds principally are result of aluminum content in soil.

It is a well - known fact that **tropical medicine** is very highly developed. We cite two groups of diseases by way of illustration: diseases linked with the action of geochemical factors; and parasitic and infectious diseases. It is also important to stress problems connected with high air temperature and very high atmospheric humidity. Chronic diseases undergo worsening in such conditions, the functioning of various systems (cardiovascular, digestive, nervous, and endocrine) deteriorates, skin diseases with high incidence of skin cancer are recorded (in arrivals from lands with a temperate climate), conjunctivitis is common, etc.

Geochemical characteristics of the acidic damp tropics leave a clear mark on the flora and fauna, agricultural production, and human health [115]. These characteristics are primarily manifested through deficits of individual elements:

1. Sodium content in water and food products is so low that it does not satisfy requirements of the human organism. The deficit of sodium is amplified by sweating, since a man in the tropics daily gives off up to 12 l of sweat. Decline of sodium content in the blood causes exhaustion of the nervous system and reduces the capacity for work;
2. Calcium deficiency is probably manifested in smaller stature of animals in regions where it occurs compared with animals from other extreme zones (for example, the okapi in Equatorial Africa has a height of 1.5 to 2.0 m, whereas the related giraffe from Ca - rich savannas is about 6.0 m tall). Thus, animals in the acidic damp tropics adjust to a deficit of some element or elements by being smaller. Diseases caused by Ca deficiency (rickets) occur rarely in the damp tropics because the strong sunlight favors formation of vitamin D, which stores calcium and phosphorus in the organism, with the result that animal skeletons are fairly strong;
3. Despite the fact that vegetation of the damp tropics contains much iron, this element is poorly assimilated by the human organism from plant food. Owing to the fact that the inhabitants of the tropics eat little meat (the main source of iron), anemia is widely disseminated here;
4. Due to deficiency of calcium and iodine in the soil, dental caries and endemic goiter are widespread in West Africa. Consequences of fluorine deficiency are recorded in some areas.

The extremely great richness of tropical pathology is primarily based on very numerous *infections* and *parasitic diseases* caused by viruses, bacteria, and fungi. Especially favorable conditions for propagation of diseases prevail in lowlands due to high density of the human population, abundance of animal species, increased humidity, high temperature stability, and abundance of vegetation. Lowlands are characterized by the epidemic type of propagation of infectious diseases (jungle yellow fever, etc.), while flatland relief is characterized by the epidemic - focal type of propagation. Among tropical diseases, generally the most prevalent are ones whose spread is accomplished by carriers that are lovers of damp or dry environments (malaria, etc.). Also characteristic are diet - related diseases (pellagra, beriberi, etc.).

Soils of tropical regions play an important role in the propagation of various diseases. For example, the acidic reaction of these soils is favorable for the development of lower fungi, and different skin and fungal diseases are very widespread here.

The very numerous aquatic basins (rivers, lakes, reservoirs, and irrigation canals) represent good places for multiplication of mosquitoes and carriers of various intestinal infections and other diseases. Moreover, irrigation canals are the home

[115] Water of good quality is available to only a third of the population.

of snails in whose bodies the larvae of parasitic disease agents develop, and lesions associated with cancer occur in some forms.

The exceptional wealth of the flora is characterized by a multitude of useful plants, including medicinal species. More than 500 species of tropical plants can be used as pharmaceutical raw material. On the other hand, poisonous plants also grow in the tropics.

The animal life of the tropics is diverse and of extremely great significance to human beings. Many representatives of the animal kingdom serve as reservoirs of the agents of infectious diseases, and many are poisonous (the tarantula, black mamba, cobra and other snakes, etc.).

Paleozoic Orogenic Belts

Paleozoic orogenic belts embrace the central zone of Europe and areas of Northern China, Mongolia, and Southeast Russia[116]. These are mountains formed by thrusting during the Hercynian and Caledonian Orogenies and deformed by strong radial movements during the Tertiary, when the great present - day mountains and valleys were formed. Owing to long - term erosive processes, the mountains are for the most part of low to medium height (with maximum elevation of 4,507 m, the Altai Range represents an exception created by very strong uplifting during the Neogene and Quaternary) and marked by extensive areas. Viewed seismically, they are today relatively consolidated regions.

The natural conditions of *low mountains* in the damp part of Paleozoic orogenic belts ert mainly favorable influence on human health. In exceptional cases, measures are needed to prevent the occurrence of plague, malaria, scrub typhus, woodcutter's encephalitis, and other diseases in certain regions.

The influence erted by natural conditions of *medium - high mountain massifs* in regions with a damp climate - in contrast to the continental variant characteristic of the mountains of Central Asia - is of a positive nature and promotes improvement of vital activity of the organism. At the same time, it is often necessary to increase the energy value of food in conjunction with supplementary vitaminization, especially through the use of wild greens. In many regions, it is imperative that preventive measures be taken against endemic goiter. Especially needed are organizational measures for prevention of gastro - intestinal diseases, worm invasions, malaria, woodcutter's encephalitis, scrub typhus, etc.

---

[116]The Altai-Baikal or South Siberian Mountains cover the greatest area (more than 2 million km^2). The neighboring Mongolian-Chinese Plateau for the most part coincides with the extent of the Gobi Desert.

Mobile Regions

In geological structure and many other characteristics of significance to human health, mobile regions of the Earth's crust differ greatly from the described old tectonic units. Above all, the lithological (petrological) diversity of mobile belts is far more pronounced, their structural framework is incomparably more complex, volcanic and seismic activity is constantly present, and geomorphologic forms are represented by striking young mountains. Without question, the geochemical properties of rocks, soil, and groundwater are also different, risks are presented by current volcanic and seismic activity, and the danger of landslides and rock and mud torrents exists in regions with active fault structures. As for groundwater, its distribution in rock masses is complex, and the water as a rule is of better quality in comparison with that of old shields and platforms. Springs of thermo - mineral water, superheated vapor, and gases are numerous. Together with the listed geological factors, the high - mountain relief and harsher climate ert great influence on human health.

We have noted that regional tectonic structures of standard origin are distinguishable within mobile belts (Table 9.1.). These tectonic units must be taken into account in the further development of medical geology. For the present, we note certain characteristics of two groups of such structures: 1) structures of alpine orogenic belts (active continental margins); and 2) groups of other types of boundary structures.

Alpine Mountain Massifs

Mountain ranges of enormous dimensions were created by subduction of one plate beneath another or by collision of two continental plates. Such ranges include the Andes, Cordilleras, Himalayas, and Alps. The Andes Range is 9,000 km long, and a number of its peaks (volcanic cones) exceed an elevation of 6,000 m above sea level. The system of the Cordilleras is about 6,000 km long and climbs to an altitude of 6,193 m above sea level (Mt. McKinley). The Tibetan - Himalayan mountain region is the most extensive and highest in the world: it has an area of about 3 million km^2, Mt. Everest (8,882 m above sea level) being the highest peak in it. The Tien Shan chain is about 2,450 km long with altitudes up to 7,439 m above sea level. The belt of the Carpathian - Balkan Mountains is 2,100 km long and 100 to 35 km wide with elevations up to 2,663 m above sea level... The relief of the given regions is very sharply expressed (with steep slopes), the soil cover is developed in places, the atmosphere is rarified, and the climate of higher zones is harsh. For all of these reasons and for certain positive effects on human health, the regions in question represent a special type of extreme zones (Fig. 9.7.).

The natural environment of high mountains affords predominantly unfavorable conditions for life. Geological, climatic, and biological factors can be distinguished for the influence they ert in this sense.

Fig. 9.7. Mountainous ambient.

As we have said, alpine mountain massifs are characterized by diverse **geological factors**. We note the negative and positive action of the more important among them:

1. Risks presented by the destructive force of *faults*, *earthquakes*, *landslides*, and *rock and mud torrents*, followed by various risks associated with *karst regions*, were the subject of the book's chapter on geological factors (Fig. 9.8). To them must be added the potential danger of high - mountain glaciers, snow avalanches, and slope failures. The indicated risk factors of the natural environment, like the positive influence of some of them (faults as pathways for circulation of low - mineral, mineral, and thermal waters, for instance), are primarily characteristic of alpine mountain massifs;
2. Diversity of the lithological composition of terrains, soils, and geological formations strongly affected the *geochemical characteristics* of mountainous regions and is manifested on the environment in different ways. Generally speaking, a higher concentration of vari-

ous macro - and microelements - with predominantly positive effects on public health - is discernible in the geological environment;

3. Shallow layers of more fertile *soils* of the transitional and Ca classes are distributed on the carbonate rocks widely disseminated in the Alps, the Caucasus, and mountainous regions of Central Asia. These soils have different effects on human health. To be specific, elevated or reduced content of microelements acts indirectly on human health, both negatively (toxic action, endemic goiter due to iodine deficiency) and positively (in the cases of elevated content of calcium, magnesium, and other elements). Because contamination of soils can occur due to leaching from higher regions, conditions are created for occurrence and propagation of gastro - intestinal diseases, while microbic pollution of the soil leads to anaerobic infection of wounds, tetanus, etc.;

Fig. 9.8. Cliff near to Petrovac (Montenegro).

4. *Waters* of mountainous regions are characterized by a number of properties interesting from the standpoint of their influence on man. Brooks and streams have cold water that is slightly mineralized in regions with non - carbonate rocks. In the event of heavy rainfall, the lower courses become muddy and most often possess a limited capacity for self - purification and unfavorable sanitary properties (under these conditions, they become sources of intestinal and other types of infections). The more mineralized groundwater of karst springs and springs of other kinds is conspicuous for its good quality;

5. With respect to their *elevation above sea level* and the nature of their *relief*, mountainous regions of this geotectonic unit belong predominantly to the category of medium - high (from 1,500 to 2,500 m above sea level) and high (with elevations of more than 2,500 m above sea level) mountains. High plateaus such as the Eastern Pamirs, Tibet, and the Tien Shan Range form separate morphological units. Geomorphologically emphatic relief has an indirect effect on man that can be negative or positive. More will be said about the effects of relief in conjunction with climate in the text to follow. Here we mention increased physical strain and energy expenditure (a consequence of cardiovascular diseases among other things), fear, apathy, mental derangement, and mountain sickness;
6. In connection with *climate* - which in mountains erts especially profound and comprehensive influence on human health - all medium - high and high - mountain regions can be divided into wet (oceanic) and dry (continental) variants. The former variant includes mountainous regions of Southern Europe, parts of North and South America, the Far East, and Southeast Asia; while mountainous terrains of Asia Minor, Central Asia, and southern parts of the Cordilleras and Andes belong to the latter climatic variant. Natural conditions of the indicated types of mountainous regions and the nature of their influence on human health are different. Generally speaking, reduction of atmospheric pressure of the air and oxygen partial pressure with increase in altitude represents the main climatic factor in mountainous regions, which are also characterized by a harsh (sometimes dry) climate, cold and short summers, and strong winds. Disturbance of nervous activity occurs under the influence of climate in high - mountain regions, together with altitude sickness accompanied by headaches, apathy, drowsiness, weakness of the organism, and reduction of working capacity. Changes occur in cardiovascular activity, with expansion of the heart and blood vessels; functioning of the digestive organs is altered; the function of the liver breaks down; fibrose anemia occurs; etc. Sunburns are frequent, due to enhanced solar radiation. At the same time, the potentials of mountainous regions are very great from the standpoint of balneoclimatology, health tourism, and sport;
7. The *flora* of mountainous regions is diversified, but unequally disseminated. On carbonate rocks and poor soil rich in calcium at elevations greater than 2,000 m, extensive pastures are developed with grass rich in nitrogen, phosphorus, and other nutritive elements, but deficient in iodine, sodium, and other ingredients.

Beech forests grow in the Carpathian and Caucasus Mountains, while the regions of the Tien Shan, Pamir, and Altai Ranges are characterized by broad - leaved forests of ancient origin. A wealth of endemic plant species is also characteristic of the Caucasus region (where nearly 20% of the entire flora is endemic), mountains of the Balkan Peninsula (where endemics can constitute as much as 27% of the flora), and other regions. Environments with such vegetation represent unique natural areas for creation of national parks, tourist zones, and spa facilities. Moreover, the plant life of mountains is rich in resources of medicinal herbs and vitamin - bearing plants;

8. Diversity of the *animal life* of mountainous regions ensures the existence and propagation in nature of many parasitic diseases, infections with natural foci, and transmissible diseases such as plague, leptospirosis of animals and man, malaria, tick - borne diseases, etc.

Other Boundary Structures

This group of boundary structures includes *rift valleys within continents*, *island arcs*, and *oceanic islands and plateaus* (Table 9.1.). The given structures are largely built of effusions of basaltic lava and are marked by the high cones of active volcanoes (Fig. 9.4.). In addition to common characteristics, each type of structure has its own unique origin and specific petrographic, mineralogical, and chemical composition, as well as its own specific content of microelements and isotopes (especially in regions with strong volcanic activity). Under such conditions, characteristic concentrations of fluorine, boron, and certain other elements play an important part in the evolution of organisms.

Mobile regions of the Earth's crust - above all ones with volcanic boundary structures - are marked by both current and ancient hydrogeothermal occurrences. Such occurrences have been studied most thoroughly in countries where much has been achieved in investigation and utilization of geothermal energy: the USA, Iceland, Japan, New Zealand, Italy, Turkey, and the former Soviet Union. Apart from geyserites (masses of amorphous silica), zeolite, travertine, onyx, quartzite, and other rocks, such regions are characterized by occurrences of mercury and its ores (cinnabarite), hematite, lead and zinc sulfides, and other ore mineral products. For example, ancient and current hydrogeothermal occurrences on the territory of Turkey include a mercury mineralization with the Halikoy - Beydag and Kaymar - ci mercury mines, and mercury is separated from contemporaneous hot springs.

The negative influence of geological factors in the given regions stems for the most part from the danger of *volcanic eruptions* and *strong earth-*

quakes, the most impressive, destructive, and terrifying of natural phenomena (Fig. 9.9.). The effects of these phenomena have already been discussed. We particularly stress the danger to the human population and to plant and animal life presented by poisonous gases, fumes, and volcanic mud currents, which can affect large areas during volcanic eruptions. Volcanic eruptions are often accompanied by contamination of springs and famine. The dangers presented by powerful earthquakes followed by rockslides and the damage they cause are also very great in these seismically extremely active regions. In addition to the danger from movement of masses along faults and rift structures, risks are presented by so - called *gas breathing* of the Earth's crust along such structures. Mercury vapors and elevated concentrations of radon are especially harmful[117].

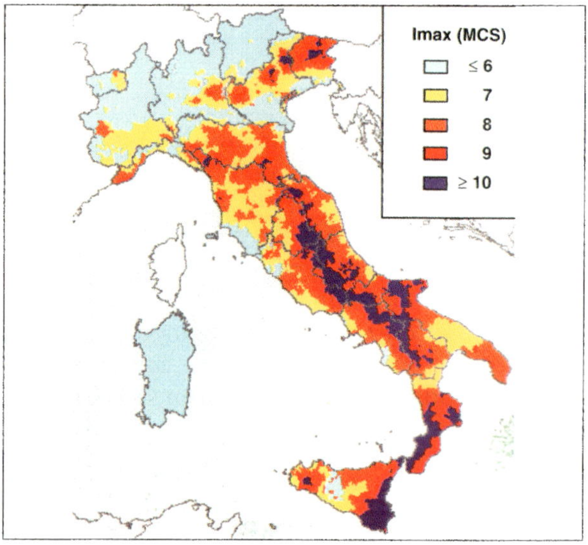

Fig. 9.9. Maximal seismic intensities registered in Italy between 1000 and 2000 (Molin, 2000).

From the other side, because of striking volcanic domes, fumaroles, solfataras, geysers, numerous hot springs and other occurrences, volcanic areas are picturesque and attractive for tourists, convenient for national parks forming. Suitabilities appearing because of volcanic activity are also expressed through emanation of necessary gasses into the atmosphere, forming fertile soil from volcanic ash rich in various trace elements, abundance of groundwater for drinking, etc.

---

[117]*Helium breathing*, *argon breathing*, and formation of high concentrations of methane, carbon dioxide, etc., can also occur. It is interesting to note that there is even an ailment known as *rift fever*, a viral disease fatal for animals and man.

LOWER - RANKING GEOMEDICAL UNITS

In studying geomedical characteristics of any territory, it is desirable to start with a geologically monotonous region, since it is possible in this case to establish basic regularities prevailing in the region and determine its favorability for human life. Such a region is usually composed of a single formation with definite petrological, geochemical, hydrogeological, engineering - geological, and other characteristics. For this reason, we shall examine lower - ranking geomedical units primarily in terms of the distribution of certain geological formations. It goes without saying that the medical aspect of such lower - ranking units has not yet been systematically studied. This will be one of the next tasks of medical geology and medicine.

Fig. 9.10. Process of karst erosion in limestones.

The definition of a geological formation as a lower - ranking geomedical unit need not exclude reliance on some criterion used to isolate geomedical units of the first order. For example, a regionalization in which units of the first order are established on the basis of lower - ranking tectonic structures can be completely justifiable. In any event, geomedical characteristics of every isolated region must be examined in connection with general regional regularities.

Let us dwell on several geological formations, which by the way are very numerous:

Fig. 9.11. Karst (limestone) ambient.

1. Thick plates of *basalt* cover enormous areas of the mobile belts of continents. For example, the cover in the western part of the Indian Peninsula has an area of about 300,000 km^2 and thickness of 1,600 m (Fig. 2.4.). In the chemical properties of rocks, soil, and water, as well as other geological characteristics, this geological (natural) environment differs greatly from others. For this reason, it can be assumed with high probability that definite geomedical conditions prevail on such a monotonous area. This is all the more likely in view of the fact that such regions with basalt are distinctive for the content and abundance of microlements and rare earths present in them, as well as the frequent negative effects of pronounced current volcanic activity;
2. Formations of *granite* and *granodiorite* are very widely disseminated in the world. These are rocks that can have significant concentrations of uranium and other natural radionuclides, with all possible risks to human health. Such formations include the uranium - bearing Phosphoria Formation on the Colorado Plateau (USA) and the Francevilien Formation in Gabon, both of which cover considerable areas. The problem is further complicated by radioactive contamination of soil and groundwater;
3. Attention was called earlier to the large areas covered by limestone (carbonate) formations (Figs.9.10. and 9.11.). Marked by phenomena created by the karst process (resulting in special morphology), superficially waterless over large areas, but possessing significant deep - lying reserves of good drinking water, such geological formations are characterized by specific organization of settlements and conditions favorable for human life, but also by a wide range of risks to human activity (collapsing sinkholes, risks in the sphere of hydrotechnological construction, etc.). With areas measuring several hundreds of thousands of km^2, such territories unquestionably deserve to be the subject of special geomedical study.

Chapter 10

Applied Medical Geology

During the second half of the last century, various medical and geological tasks had to be solved. That is why nowadays, behind applied medicine, from the one side, and applied geology, from the other, voluminous useful materials and attained results are present, while more and more significant part is related to concrete problems concerning *geological environment - natural environment - human ecology* relationship. Parallel development of medical ecology, dealing with complicated interrelationships between human being and factors of the external environment is suitable opportunity. From the other side, methods of applied geological exploration are intensively developed, taking into consideration various geoecological aspects of the environment. Besides, multidisciplinary approach of medical and geological sciences, including technical accomplishments, becomes predominant and more fruitful. In that way, using contemporary methods and practical solutions of biology, chemistry, physics, veterinary medicine, medical geography and other sciences and scientific disciplines, ecological mosaic is closed, with *applied medical geology* as new applied discipline.

GEOMEDICAL RISKS AND TASKS OF APPLIED MEDICAL GEOLOGY

Areas where we grew up as well as land where we live make almost pleasant impression of extraordinary stability and unchangeability on us. But, on the contrary to the stability giving tricky impression of unchangeability, a series of occurrences throughout the world, very often sudden and violent, sometimes disastrous for population, directs to noticeable mobility of the land. Besides, areas destroyed by catastrophic earthquakes, where the coasts are demolished and moved by sea waves, and numerous settlements are under the permanent risk of giant volcanoes eruptions, or where coastal parts of the land are sinking so rapidly, that the coast has to be protected from sea penetration by embankments, are not so rare. "*And in that way, almost everywhere, in all points of the Planet, close to the illusory peace and stability, a series of processes, weaker or more intensive, of local or regional significance, constantly building or demolishing, which started soon or lasting for the hundreds of millions centuries, and which result is today's form of the Earth, distribution of the continents and oceans, relief and all features of its surface, is underway*" (B. Milovanovic, 1949).

During extremely long geological time, nature was, among all, making equilibrium in chemical content of atmosphere, lithosphere and hydrosphere of the Earth. The equilibrium was, however, locally disturbed, not only in natural way, but also as a result of the human activity. Unfortunately, human work was expressed through more and more destructive forms, with serious consequences on the environment and, particularly, human ecology. Note that development of medicine (medical ecology), which succeeded mainly to clear up influence of unhealthy surroundings on development of disease, even these which can be hardly overcame by contemporary medicine, has a very important role in educating related to catastrophic influences of human beings on the environment.

One of the tasks of applied medical geology is to define a role of geological factors within an ecosystem, interconnected system of living organisms and the natural environment, but also to distinguish factors - carriers of some diseases and perform zoning of the interesting area according to degree of negative (or positive) influence of geological environment on human health (Fig. 10.1.). It is understandable that, nowadays, the discipline is also dealing with the role of geological environment expressed to anthrop; orogenic activity, which is also very important task; especially because *contaminant migration predominantly depends on geological properties of rocks, soil and groundwater as media where the process develops.* Geological medium is, namely, under such conditions that all three potentials important for providing human standard and survival are endangered:

1. *soil formation* presents certainly the most important result of rock weathering; global problem is increasing degradation of that geological resource because of intensive chemisation and improving of erosion processes;

2. *groundwater forming,* as the most result of permeability properties of rock masses; by contamination processes and overexploitation, contemporary global problem of providing high - quality drinking water was made;
3. *Forming of important deposits of energetic and other mineral raw materials* - one of the main supports of economic development; global problem is not only overexploitation of the mentioned restorable resources, but also contamination of the environment during exploitation.

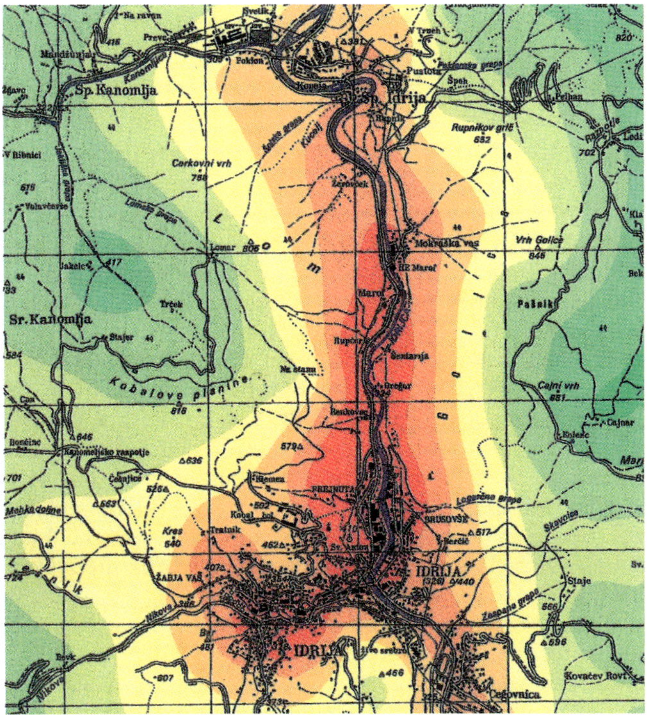

Fig. 10.1. Geochemical map of mercury distribution in soil in the territory of Idrija, Slovenia. Maximal Hg content over 140 mg/kg (M. Gosar, R. Sain, 2001).

Besides, *heavy metals releasing from rocks* presents one of predominant processes in the nature. Geochemical point of view, present in the global scale, with serious medical consequences because of existence of undesirable elements, has to be explored to fulfilling level,serving as a basis for making medico - geological evaluation of the interesting area[118].

It seams to be paradoxical that consciousness on significance of geological medium and necessity to study protection of its values is still stunted, which is in the contrary to real needs of mankind - particularly because under natural conditions, living world is subjected to many geomedical and other risks (Chapter 2). According to the author's opinion, because of expressed intimacy and complexity, relation between geological environment and human health presents very important area of future geomedical research, particularly related to application of medical geology in health cure and economy.

Let us remind of some paragraphs, brief reports and examples, presented in chronological order mainly in the Chapter 2 of the book. Goal is to get a clear picture of importance of the discussed topic.

1. In order to illustrate importance of knowing mineralogical and chemical content of rocks, we noted an example of increased molybdenum content in rock masses of some regions in the world, with increased content of the element in plants and occurrence of molybdenum toxicosis in domestic animals as a consequence.
2. From the health point of view, areas of peridotite massifs are very suitable for living. Because of high magnesium content in rocks, soil and water, for example, mortality caused by cardiovascular diseases is less than in other terrains. However, it doesn't mean that the whole peridotite massif is convenient from that standpoint. For example, indications that increased mortality of cancer is increased along the fault zones, because of influence of certain geochemical factors, potential high concentration of some gasses which migrated from the Earth's interior, etc., are present.
3. Very high selenium content in black shales, hard coal and oil cause visible consequences on animal health.
4. Calcium abundance in carbonate sediments and soil near St. Petersburg is with direct influence on fertility of pedological soil and development of plants and animals.
5. Soil made by weathering of rocks rich in minerals with high copper content is poisonous for plant world.
6. Characteristics of plant and animal world in humid tropics are significantly consequence of lack of calcium and wealth in iron, aluminum and silica in soil (coloring, high Si content in various tropical plants, etc.) (Fig. 9.6.).
7. Positive and negative influences of pedological soil on plants, animals and human beings are extremely different and numerous, with specific

118 Chemical transfer of health risks is performed through with increased or decreased content of elements important for life. Natural wase of elements migration are through soil, water, plants and animals (Fig. 2.40.).

place belonging to the role of microelements (toxic effects; various microelementoses; occurrence of cancer and other diseases caused by deficiency in important elements, etc.).

8. Influence of geomorphologic forms on living conditions and health level of human beings is outstanding, regardless if speaking about either indirect or direct influence (mountain disease and other diseases of humans staying in high mountains; suitability of low mountains for recreation and tourism; treating for tuberculosis, bronchial asthma and other diseases, etc).
9. Dependence of living world on geological (geomorphologic) environment is particularly expressed in spacey karst areas (Figs. 9.10. and 9.11.), with risks in any domain of anthropogenic activity (drinking water shortage; breaks; risks during constructing artificial reservoirs and regulation of groundwater flow, etc.).
10. Extremely different living conditions at the territories of old shields and platforms, from the one side, as well as within surrounding tectonic belts, from the other (Chapter 9). On the contrary to the old terrains, destructive and other risks in zones of active fault structures in the areas of volcanic and seismic activity, expressed at the territories of mobile belts; also positive effects of volcanic eruptions on air, soil and water quality, thermal waters occurrences and large reserves of hidrogeothermal energy.
11. Disturbing normal conditions of life and work by developing exogenous geological processes and occurrences (catastrophic rock masses sliding; intensive erosion of pedological soil; risks and harmful effects of stone - mud flows).
12. Importance of water on living world is extremely high, particularly of high - quality drinking water. In dependence on physico - chemical, microbiological and radiological characteristics and purpose, groundwater influence is visible through irreplaceable physiological, epidemiological and hygienic role, but also some negative influences, predominantly because of surplus or deficit of some elements in water. Effects of skilled use of mineral and thermal waters, gasses and peloids are mainly positive[119].
13. Nowadays, mineral resources, according to their significance in the field of improving living conditions of the whole population on the Planet, rank second place, just after agriculture. Energetic raw materials (oil, natural gas, coal) are the main potential necessary for presence and

---

[119] Giving a lecture related to mineral waters and spas in Vrnjacka banja in 1912, J. Zujovic expressed also risks on carbondioxid (carbon acid), desired gas in mineral water. *"When speaking on the topic, Dog cave near Naples is usually mentioned, on which floor content of carbon acid is so high, that dogs are died out, but man can stand without risk, because that gas doesn't rich his nose. - In village Leskovica near Zepce in Bosnia, there is Otrovnica settlement, where extreme quantity of carbon acid occurs at the ground surface, causing livestock fades away, and that is why the name of the place is Otrovnica. - In Java and in Northern America, "Death valleys" are known, which deserve such name, because even human beings are died out".*

development of the contemporary society, the most valued mineral product. On the contrary to advantages, deposits of mineral raw materials, with their aureoles of scattering and fields of exploitation and refining, are the areas of disturbed environment. Under the certain conditions, human health is endangered by air, water and soil, polluted by ecologically toxic elements, but also through catastrophes in mines caused by methane.

14. Nowadays, uranium, as energetic raw material, is extremely responsible for previous, present and future radioactive contamination of the Planet, contamination which is the most alarming problem of the present. When taking safety and population health into consideration, for this global problem, beside anthropogenic component of radiation level of biosphere, natural radioactivity of the lithosphere is responsible (effect of small doses in the areas of chemical aureoles (anomalies); radon risks, etc.). Note also positive effects of radioactivity phenomena in health services, including also positive radon effects during balneological treatments.
15. Precondition for presence indispensable macro - and microelements in the food chain of plants, animals and human beings are their existence in rocks, soil and water. Significant part of the text (Chapter 2) is treated this topic, and it could be concluded how wide range of influence of elements on the body is, including also numerous effects of irregular nutrition.

For needs of population and health cure, knowing of mentioned and many other elements of geological medium has to be raised on adequate level. After all, we are witnesses that questions of providing, use and protection of geological potentials, but also consequence of various geological forces, processes and occurrences, biological and medical role of macro - and microelements, practically present in daily life all over the world; about this questions, everyone is concerned, laymen are discussing the matter in various newspapers, without mentioning that such questions are from domain of geology. Geosciences and geomedical discipline have to force on correct viewpoints related to the nature and its relationships, in order to support, by its social mission, education of politicians, economists and businessmen, as well as wide range of population. Also, by appropriate scientific activities and results, to contribute to correcting some approaches in human treatment and health cure in general.

Let us conclude that good knowing of the elements of geological environment is very important, necessary if we want to perform prophylactic activities in the field of fight for healthy population. Among all, anomalous rayons unsuitable for living have to be defined, as well as the ways to improve them, in other words - which areas are ecologically suitable. Task of applied medical geology is to solve such questions, which is possible to do if takes advantage of fund materials, information and knowledge, if outlines concrete territories according to interesting geological components and distinguishes rayons with extreme contents of some elements, perform multidisciplinary exploration in order to make close correlation of geological factors with medical symptoms.

MEDICO - GEOLOGICAL EVALUATION OF THE ENVIRONMENT

Analysis of geomedical characteristics of interesting area within general regional regularities looks very reasonable. In other words, that means that the area has to be studied through belonging to some of the first order geomedical units, that is - through geological characteristics and disease preconditions existing in the regional unit. In Chapter 9, there was a discussion about geotectonic (geomedical) units and their extreme zones, when clear difference between consolidated and mobile areas of the Earth's crust was emphasized (Table 9.1.). It was also stressed that, beside geological, geomorphologic, hydrogeological and climatologically features, each of distinguished units is characterized by appropriate pathology, physiology and epidemiology.

When position of the interesting area was defined within the first order unit to which it belongs, we are one step closer to its basic medico - geological characteristics and regularities. The second important step is determining the state of belonging of the area to corresponding geological formation, that is - tectonic structure, from which characteristics of geological column and texture of the terrain, as well as a series of geological regularities influencing on position and features of geochemical, pedological, hydrogeological, climatologically, biogeographic, floristic, geomedical and other areals are originated. For example, if the interesting area belongs to certain geological formation, appropriate petrologic, geochemical, hydrogeological, engineering - geological and other characteristics can be assumed with high certainty. Distinguishing geological features of the terrain is of outstanding support in planning and performing exploration, which goal is to present geomedical evaluation of the interesting area, in other words - of the environment.

Hence, in order to evaluate the environment, reliable restricted geological base has to be made, that is - to define all significant geological factors - petrologic composition, fault structures (Figs. 2.15. and 8.1.), geochemical features of rocks, soil and groundwater, geomorphologic characteristics of the terrain, geophysical fields, pedological soil characteristics, potential effect of endogenous and exogenous forces, existence of landslides, hydrogeological conditions of the terrain, presence of the sources of low - mineralized, mineral and thermal waters and their quality, presence and characteristics of deposits of mineral raw materials (including energetic ones), natural radioactivity of the lithosphere, presence of radon and risks, etc.

The second stage of exploration is presented by geomedical study of the area, where the methods described in the Chapter 8 are applied. On the basis of the material of the previous exploration and results of complementary studies, it is possible to make basic geomedical map and other restricted maps, as map of medicogeological evaluation of the environment and map of population health, as the most reliable integral criterion for influence of environmental characteristics on human body. Further, different types of zoning, first of all - synthetic medico - geological zoning of the concrete area, that is - classification of the area into rayons with different characteristics and intensity of influence of geological factors on population health, origin and distribution of diseases are very useful. By using elements of the geological environment, from the one side, and leading pathology as one of basic indicators of real health level of population, from the other, conclusion related to degree of living comfortability of the area is possible to make.

Conceptual approach to solving the given task and range of graphical presentation methods depend on concrete geomedical conditions and problems which are to be solved. Distribution of necessary macro - and microelements and presence of anomalies of toxic microelements or increased radioactivity has to be placed on a higher level very often. Within such geomedically risky rayons, detailed exploration is obligatory.

Let us discuss an example of high *radon concentrations*, for which it is realized that causes the strongest radiation. It is known that, after inhaling, decay products of the gas are deposited in bronchi and lungs, influencing by radiation the surrounding tissue, with present risks from lung cancer. In order to define space distribution of the rayons, systematic and very detailed exploration is necessary, because level of the dangerous gas extremely depends on local geological conditions (lithological composition, presence of uranium - bearing formations, and position of fractures). For example, areas with high radon risk in Great Britain are predominantly made of rocks with high uranium content, as granites in southeastern England and parts of Scotland, phosphate rocks near Northampton shire, black shales in Wales, or famous Quaternary coal - bearing series. High radon values are present over the limestones in Derbyshire, because of high permeability (N.H. Woodcock, 1994).

Methodological approach to cartography referring to radon is presented according to experiences of experts from the former CSSR. Wide research activity in the country directed to solving population protection from radon, started in 1989, performed through three stages:

- I stage - regional exploration of radon, in scale 1:200,000;
- II stage - exploration of radon for needs of land - use planning, 1:25,000 to 1:10,000;
- III stage - detailed exploration of radon at localities determined for housing construction (V. Kulajtn, 1990).

As a result of exploratory works, all three stages are *maps of radon risk*, where the following three categories of rayons are distinguished: with high, medium and low radon content; within rayons of medium and high risk, contrast anomalies can occur (Table 10.1.). Special category of extremely high risk, which was not stated in the Table 10.1., is related to presence of anomalies along fault zones or parts of the terrain with macroscopic uranium occurrences, etc.

TABLE 10.1.

Risk category	Total Rn_{222} activity in air from soil (kBq m^{-3}), in soil, in foundations permeable for water and gas.		
	Low	Medium	High
Low	30	20	10
Medium	30 - 100	20 - 70	10 - 30
High	100	70	30

During radon exploration in detailed scales, in the former CSSR started with measurements along the 20 x 20 m network, in order to eliminate parts of the area with low content of the gas, and in the rest of the terrain, detailed exploration in scale 1:1,000 was carried out (at the beginning, along 10 x 10 m network, then with 5 x 5 m network for anomalous zones with extremely high concentrations). In that way, necessary basis for planning appropriate building constructing protection from radon, except for types of ground belonging to category of extremely high risk, where constructing is not recommended.

Geochemical maps of content and distribution of micro - and macro elements in rocks, soil and water can be made in similar way, but also distinguish zones of different risk degree for human health. Maps of space distribution of some interesting microelements, for example selenium, or fluoride or iodine, too. All such maps are significant for evaluation of the concrete environment.

ROLE OF MEDICO - GEOLOGICAL EXPLORATION IN IMPROVING THE ENVIRONMENT AND HUMAN HEALTH PROTECTION

The world presents a global ecological system, where all living creatures, including human beings, are submitted to certain biological regularities. When adapted himself to the influences of the natural (geological) medium, man with his activity caused essential changes within the system without necessary forecast, disturbing more and more ecological links and causing almost irreparable damages to the environment, where biological equilibrium was kept during several millennia until then. That is why knowledge of ecology, one of the fundamental disciplines of multidisciplinary character, acquired outstanding significance for applied activities in domain of preserving and maintaining healthy environment. It is understandable that, in order to protect the environment efficiently, beside a need for intensifying and coordinating activities from all countries on the Planet, necessity for wide action in the field of education and specializing of physicians, geologists and other professions occurred, showing as equal partners in solving complex tasks of protection the environment.

With unseen rapid development of civilization, geological environment becomes very important integral part of the environment, as hydrosphere, atmosphere and biosphere, entering more and more into interaction with anthropogenic activity and producing material basis for direct change of matter and energy between man and nature. So, four qualitatively different groups of relationships between human beings and appropriate geological environment came into the first plan: 1) *basic raw materials and energetic resources necessary for socio - economical development are obtained from the lithosphere;* 2) *natural geodynamical processes are endangering human lives and capital values;* 3) *man is more and more forced to ensure harmonious interaction of building and other constructions with geological environment in advance and rationally;* 4) *human beings are more and more aware of necessity for defending geological environment from contamination and disturbing* (M. Matula, 1990)[120].

---

[120] During seventies of the last century, in accordance with forming geological tasks in producing and protection of the environment, in the former CSSR, the following four groups of *geofactors of the environment* were distinguished:

1. *geofactors enabling suitable development of the society* (mining and non-mining raw materials, energy resources, groundwater, cultivable soil, high-quality ground for constructing, building materials, convenient areas for waste disposing, etc.);

Significance of *geofactors* was distinguished during sixties of the 20th Century, when, as a result of higher level of knowledge related to necessity to protect the environment, need to evaluate anthropogenic influence on the environment (Environmental Impact Assessment) occurred. At that time, USA Government concluded that it was necessary to make a mechanism of decision making, ensuring some kind of analysis of all large projects from the standpoint of influence on the environment. In January 1970, the law called **National Environmental Policy Act (NEPA)** was accepted by the administration[121].

Evaluation of potential influences on the environment is made by application of the most convenient methodology. According to regulations in Yugoslavia, it is necessary to evaluate the following influences from the domain of geology:

1. loss and damage of geological, palaeontological and other objects of geological heritage;
2. influence on rocks and soil (changes in local morphology, soil erosion, rock masses and soil sliding; influence of emissions and precipitation on soil; waste disposal);
3. influence on water (changes in groundwater and surface water regime; influence of contamination, waste and waste disposal);
4. Influence of mineral raw materials presence and exploitation (consequences formed by raw materials, water, energy and other geological potentials exploitation).

Before start to analyze role of medico - geological exploration, experiences of medical geography are discussed.

Medico - geographic evaluation of the territorial systems of the surrounding environment is made according to classifications of the

---

2. *geofactors endangering human lives and their products* (volcanic eruptions, destructive earthquakes, catastrophic landslides, floods, toxic and radioactive working, etc.);
3. *geofactors reducing efficiency and safety of technical objects functioning* (foundation ground of low capacity, instable rocks - chemically and in volume, unstable and labile slopes, softening of foundation ground, seismic active terrains, etc.);
4. *geofactors damaging the environment by negative anthropogenic activity* (terrain sinking caused by underground exploitation and pumping, raging because of surface exploitation, material disposing, groundwater and rocks pollution caused by agricultural production, landfills, soil softening or drainaging because of constructing, etc.).

[121] From the middle of 1970, in Netherlands evaluation of influence on the environment is obligatory, and from 1986, there is also a law. Complex regulation was issued in Canada in 1973. European economic union issued **Directives for evaluation of influence of certain public and private projects on the environment** in July, 1985. The similar legislation was issued by many countries over the world.

components (factors) of the environment defined according to various standpoints: according to periodicity of influence on living activity of human beings; according to possibility of rejecting influences of the environmental factors to living activity of human beings; according to possibility of making interactions with other components of the environment and reinforcing or weakening their effect on living activity; according to the degree of direct participation in influencing on work, rest and population health (Keller et al., 1993). Influence intensity of the environmental components on population health can be: 1) *weak*; 2) *medium*; 3) *strong*; 4) *critical or limiting*. Critical or limiting parameters of the environment, which can complicate or even make impossible work, everyday life and rest of the human beings, are particularly interesting. That is why special systems of life insurance are required.

Nowadays, it can be stated that two types of comparative evaluations of regional characteristics referring to work, everyday life, rest, population health are distinguished within medical geography. *The first* - quantity or half - quantity evaluation of factors with enough correct numerical parameters. This type of evaluation was significantly taken on from hygiene and is referring to distinguished sides of population life. *The second type of evaluation* - quantitative, that is - relative evaluation given over point number, applicable for cases of integral sanoecological processes and occurrences. The mentioned two approaches are closely interconnected.

Among sides of population living activity, population nutrition, quantity of heat isolation of apartments and clothes, constructing expenditures, etc. The questions were treated by Keller with assistants (1993).

Among traditional methods, *comparative method*, with comparative evaluation of different geographical processes and occurrences according to population health level as an integral part, is used within medical geography for a long time. According to Keller and associates, medico - geographic evaluation can be treated as scientifically based defining of the role and degree of influence intensity of some environmental factor of the concrete territory or their reflection on population health level, character of distribution of different diseases, human productivity or rest. It is understandable that, within evaluation of a series of factors, important contribution can be obtained by appropriate experts. Making *medico - geographic evaluation*, as a product of exploration activity for the concrete area, is one of very important functions of medical geography, which should have in mind.

Majority of medico - geographical papers contains preliminary, first of all - oral evaluations of some occurrence. So, in evaluation of natural conditions to which population has to be adapted, it is very difficult to say how they would influence on each man. But, tendency on the basis of comparative analysis of different territorial systems according to population adaptation can be discovered and evaluated. As a result of similarity analyses, complex or simpler evaluation scale appears. Note the next, very simple scale: 1) territories with very bad adaptation conditions; 2) territories with bad adaptation conditions; 3) territories with satisfactory adaptation conditions; 4) territories with convenient adaptation conditions; 5) territories with very convenient adaptation conditions. Or, even much simpler, if oral characteristics are changed by score characteristics, and adaptation conditions of new coming population would be: 1, 2, 3, 4, and 5 scores.

For example, in similar way, circulation intensity of naturally - focal infections and invasion carriers can be evaluated. In a series of cases, evaluation is based on 2 - , 3 - score scales.

According V.A.Bakalov (1976), several *natural factors* are evaluated - climate, relief, hydrography, soil and engineering geological conditions, as also living conditions of Baikal area according to degree of their suitability for living. On the basis of the developed methodology and analysis of evaluated parameters, evaluating tables for each observed component of natural environment are presented in scores, in comparison to their suitability for temporary living of population. As a result, significant part of exploratory area was evaluated with low score mark, and that is why it was classified into unsuitable and low - suitable rayons for permanent stay of human beings.

When speaking about *anthropogenically contaminated territory*, highest allowed concentrations of the mostly distributed and chemically active substances involved into the environment are used as standards of environmental contamination. From the other side, evaluation of potential ability of the territorial complex for self - cleaning is made. Comparing potential capabilities of the geosystem for self - protection (self - cleaning) with factual contamination of the natural (living) environment, makes evaluation of ecological conditions for observed important group of factors possible. On the basis of general stability according to anthropogenic interventions, natural comples are evaluated with one of six scores: 1) extremely stable; 2) stable; 3) poorly stable; 4) decreased stability; 5) stable; 6) very stable. It is reasonable that extremely inconvenient demoecological conditions are formed during high factual contamination of natural comples and low ability of self - cleaning.

If start from the role of numerous geofactors, it can be assumed that, in domain of improving the environment and preservation of human health, high contribution is expected from **medical geology**. That goes without saying that that geological environment is previously explored in detail and that influence of appropriate factors is specially studied and spacey determined. Also, in application of *comparative method*, experience and results of medical geography are used. Comparative method will certainly find a place in evaluating medico - geological influences of different geological processes and occurrences on population health level of the interesting area, too.

By application of medico - geological exploration methods, consequences of different risks can be evaluated and their areals cartographically distinguished. Role of fundamental and applied methods in outlining certain risks, that is - improving the environment and preservation of human health, is illustrated by several examples, while example related to compiling maps of radon risk has already been presented.

1. For evaluation of *aquifer vulnerability to contamination*, inevitable raw materials for drinking water supply, several ways are known, and it is possible to perform satisfactory evaluation of terrain zoning when speaking about contamination from the terrain surface very often (S. Komatina, 1995, 1998). However, in order to protect aquifers endangered by contaminant front (plume), in USA, numerous methods of exploration and technological measures of rehabilitation are developed. Principal goal is to destroy focus, in other words - to prevent further penetration of the plume, by application of technologies based on solving the problem without contaminant extraction from the subsurface (various barriers in front of the plume, technologies with gas injecting, bioremediation, etc.). Preserved quality of drinking water is a main precondition for population health protection. Besides, results of hydrogeological exploration, as well as of hydrogeochemical and hydrogeomedical investigations, will give the answer to questions referring to how much groundwater is present and of which quality, where potential aquifers are, in other words - how to evaluate effects of consumed water from the health standpoint, etc.
2. Exploration of *risk of amphibole asbestos, tremolite, wollastonite, zeolite and other fibrous mineral and their dust* are of medico - mineralogical character[122]. They directed to stronger or weaker influence on malignant lungs tumor occurrence, but also lungs fibrosis (asbestosis), so - more and more attention is paid to them. By geomedical methods, not only potentially risky geological formations, as well as specially risky rayons (mine deposits, mines, aureoles of scattering), can be distinguished, but also safe rayons outlined, consequences of living in risky areas explored and measures for reducing risks of that type of contamination suggested.
3. It has been already mentioned that *microelements* act within numerous and different functions in all organisms and that their deficiency or surplus causes diseases. Contribution of geochemistry is which has been suddenly developed starting from the sixties of the last century, to classify these elements in rocks, soil and water into ethiological factors of the primary importance. That is why it is possible to define geochemically risky rayons, as well as rayons suitable from the medical standpoint and to determine measures for improving the environment. Bioinorganic research can be very helpful in this domain, particularly when speaking about correlation of toxic elements with medical symptoms and prevention of contaminating the environment by toxic elements and their compounds.
4. Numerous human victims and financial losses caused by *earthquakes* gave to the scientists a pretext for study of possibilities to construct

---

[122] Among all, it was determined that risk of diseases depends significantly on the type of asbestos, more precisely - amphibole asbestos is much dangerous than serpentine chrysotile asbestos.

safely from the aseismic standpoint, so the science nowadays contains wide range of methods. As geological, hydrogeological and engineering geological feature of rocks (soil) influence to a greatest extent on degree of the objects damaging, necessity of making map of microseismic zoning, that is - outlining rayons characterized by different degree of seismic intensity, has been adopted for the terrains from the mobile belts of the Earth's crust which are interesting from the standpoint of urbanization[123].

By the right choice of medico - geological exploration methods, as well as by successful use of cartography, it is possible to direct to significant part of medical problems endangering the environment and health of the population. Medicine must and can to obtain contemporary (high quality) *medico - geological basis*, with enough precise evaluation of negative and positive geological factors and rayons outlined according to degree of suitability for living. By such base, the right way to prophylactic action and performing other measures necessary within population health cure will be found.

---

[123] Starting from seventies of the last century, in China, former USSR, Japan and USA, organized work on earthquake forecast started; however, with partial success. It was realized that very important information can be obtained by application of hydrogeoseismological methods (hydrogeochemical and hydrogeodynamical methods), while important indicator can be electromagnetic changes characterizing stage of tectonic movements. For example, as hydrogeochemical indicators, gasses (radon, helium, argon, etc.) are used, as also various isotope relations. Trace elements and all other parameters of groundwater chemical content, if there changes before and during the earthquake are interesting (G.A. Mavlianov et al., 1983).

CHAPTER 11

Use of Applied Medical Geology Results in Economy and Health Care

By rapid development of the society, because of numerous values, geological environment became more and more important part of the environment, gaining nowadays the first - class significance (regardless if speaking about positive or negative role of geofactors). It is not acceptable to understand protection of the environment as only out - and - out defense of the human body from diseases, but as conservation of all values in the human beings surrounding, making their life more pleasant and to be handed down to next generations. In that way, tasks which can be solved by applied geology and applied medical geology for the sake of well - being of mankind, are noticeably broadened, more complex, different and very important. Subject of the further text are the two groups of practical tasks: - (1) tasks useful for economy and improving of living standard and (2) tasks important for organizing population health care.

USE OF RESULTS OF MEDICO - GEOLOGICAL EXPLORATION IN ECONOMY

Situation, in which the humanity was found at the early beginning of the 21st Century, is very complicated and alarming. Exponential growth of the population, followed by rapid decrease of natural resources reserves and simultaneous increase in various contaminants, are dramatically reminding the mankind to seriousness of the newly created conditions and directing to the necessity of fundamental analysis of our total behavior, activity and strategy of in regard to the surrounding environment and the whole biosphere. Beside the mentioned, other global ecological problems, as progressive destruction of biodiversity, damaging the ozone layer, global climate disturbances, reducing cultivable soil, global water problem, global energetic problem, waste disposing problem, etc., are also producing worse situation.

In solving greater and greater disharmony between assigning for higher degree of the nature and environment and further progressive development of the modern society, introducing a concept of *coordinated (balanced, sustainable, permanent) development* is not only beneficial, but also only rational solution (I. Savic et al., 2000)[124]. Naturally, if we behave in that spirit. In that domain, geological and medico - geological practice can be very helpful. Some of numerous potential examples of taking measures in order to overcome the cited global problems, regardless tasks of defining and providing sources of raw or energetic material can be support for the mentioned attitude.

Measures for geodiversity conservation. Geological environment is the basic support of the environment. That is why (including other reasons, which were treated later) international behavior related to geodiversity, in other words - to conservation of geological elements (factors) diversity, has to be much more responsible in comparison to nowadays situation. General development of the mankind has to be balanced completely with strategic goal of protecting and improving quality of the living (natural, geological) environment for the future, for next generations. Some trials to attain the goal are performed by geological institutions from a number of countries, becoming engaged in the field of protection of geological heritage (ProGEO).

Measures for erosion prevention and soil recultivation. Soil erosion is one of the main forms of destruction and contamination of the

---

[124]According to A.I.Perelman (1975), one of important tasks of the science related to landscape is to develop optimization theory of the cultural landscape, that is - determine *optimal cultural landscapes* for different natural rayons. For each rayon, plan of natural resources management, based on complex study of the nature, particularly in evaluating link between integral parts of the landscape, has to be made. One of the ways of perception such links is geochemical study of landscape.

Within an optimal cultural landscape, by use positive geochemical characteristics and additional measures, geochemical conditions which can't be found in the nature can be made. So, after draining peat mud, landscape with significant content of dead organic substances in the form of peat coal uniting with very oxide medium, is formed, this is not possible to be found in the biosphere.

Naturally, optimal cultural landscape must be the most convenient in hygienic sense, and, finally - not contaminated. Modern hygiene defines certain requirements regarding quality of water, food, air, and made optimal standards of consumption for some compounds. Within the cultural landscape, using natural resources of not only studied rayon, but also others, very often of distant rayons, forming optimal conditions for life of the whole human population is possible, fulfilling standards of medical science.

environment, particularly within the process of intensive agricultural activity, felling of trees and destroying vegetation cover in general, constructing and mining activity as well as urbanization (Fig. 11.1). Result of erosion process is losing of the most fertile surface layer of soil, water accumulations filling by deposit and their saturation by nutritious elements from soil and fertilizers. For preventing and reducing erosion, various measures, rational and efficient only if planned according to the results of restricted geological exploration and zoning according to degree of rocks and soil cover erodibility, can be performed.

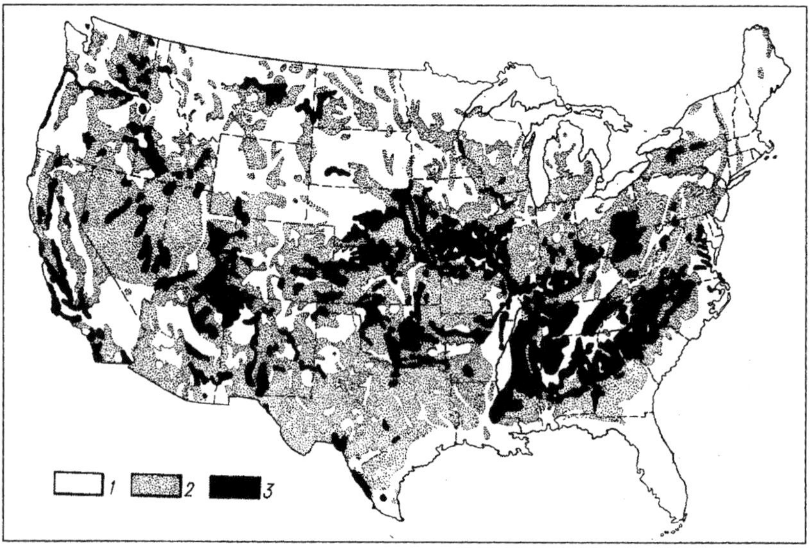

Fig. 11.1. Soil erosion in USA (according to the data of the Ministry of agriculture). 1 - insignificant or absent; 2 - medium (25 - 75 % of surface layer of the ground is destroyed, ravines presence possible); 3 - significant (over 75 % of surface layer of the ground disappeared, numerous deep ravines.

According to historical experiment, increase in biological productivity of soil and geological medium in wide sense was solved by two principally different ways (A.I. Perelman, 1961). *The first way* is based on mobilization of internal resources of the environment, as are: plowing fields (in order to make better conditions for breathing and feeding for vegetation), application of fertilizers (lacustrine sludge, fertilizers, etc.), intensifying oxidation process in swamp soil and their transformation into cultivable soil (with measures of drainaging and desalting, adding dolomite, zeolites, etc.), overcoming problems related to processes of soil erosion and ravining, plant selection. *The second way* means external adding a series of deficitary chemical elements into the soil, which is performing by: wider and wider range of fertilizers (boric, selenic, manganese, copper, zinc, magnesium, etc.), maximal widening of biological circular process of elements within the soil, and, on the contrary, minimal natural removing of useful elements phosphorus, calcium, nitrogen, etc.), removing harmful elements, optimal moistening of pedological soil (irrigation). To make a right choice of efficient measures for soil recultivation, that is - forming cultural soil cover, with optimal geochemical regime and hygienic conditions, it is necessary to make high - quality geological basis, first of all - pedological, geochemical and hydrogeological bases of the interesting area.

Measures for providing high - quality water. In solving water problem of the mankind, providing high - quality drinking water is, certainly, a primary goal. Two principal directions of the activity are known in this domain - performing *measures directed to water protection from contamination and measures for protection of water resources from overexploitation.* Applied hydrogeology can have a very important contribution in solving the problems; *first* - to study vulnerability of so important groundwater resources from contamination from the terrain surface and define zones of sanitary protection. *The second* to contribute to rational use and regulation of groundwater reserves, by measures of artificial recharge of intergranular water - bearing formations, or by regulating groundwater regime, when speaking about karst - fractured formations.

Other ways of providing high - quality drinking water are also known, as well as modern directions for providing water for irrigation, that is - regulation of moisture regime at irrigated territories (M. Komatina, 1990, 1994). Questions referring to providing drinking water of high quality and for irrigation under desert and semi - desert conditions are specific and difficult for solving, but numerous high - yield deep wells direct us to conclude that even such cases have appropriate solutions. In finding solutions for rational management of groundwater resources, it is recommendable to start from *Basic hydrogeological map*, made for territories of the whole countries, in sheets of scale 1:50,000 or 1:100,000, as well as systematic and stage approach to hydrogeological exploration.

Protective measures from landslides, rockfalls, subsidence and soil breaking. The mentioned exogenous geodynamical occurrences are followed by catastrophic consequences for human lives and huge financial losses very often. Beside in USSR, the most complex national program for reducing losses from *landslides*, was issued by Japan, where, starting from the 2nd World War, planned, legally supported, systematic exploratory and rebuilding works are carried out. In France, on the basis of adopted program for zoning rayons with potential rock sliding, whole series of map (1:25,000 or larger scale) is present, outlining unstable zones, but also giving instructions for correct distribution of objects in ground where constructing is planned.

When speaking about *rockfalls*, rehabilitation of the process means only prevention. Among the reclamation measures, artificially activation of rockfalls under controlled conditions and anchoring of mobile monoliths for stable foundation are applied most often. Protective measures at the steep slopes in Rio de Janeiro (Brazil) include concrete and stone walls, supporting concrete columns and metal bars strengthening rock masses bended over the slopes (A. Howard, I. Remson, 1978).

Damages of *subsidence* can attain catastrophic dimensions, particularly in cases of groundwater overexploitation, or oil and gas exploitation, when can spread over large areas and cause sinking low sectors of land and cities, destroying buildings and other structures, deformation of irrigation channels, destruction of underground communications, railroad tracks, bridges, etc. Such areas require special approach to engineering geological and hydrogeological exploration as well as to monitoring, in order to make correlation between water or oil exploitation and subsiding process, and to define exploitation regime.

Rayons with *soil breaking* occurrences, however, predominantly belong to karst terrains. Particularly vulnerable geological formations are very soluble gypsum and anhydrite, and that is reason why several dams in the areas characterized by such type of formations, have been abandoned. High quality geological base enables evaluation of latent danger of the mentioned occurrences. Soil breaking alone is very difficult to be regulated, but can be avoided, if the terrain is previously explored in detail and potential sectors for use outlined.

Protective measures from earthquakes. As human victims and financial losses as a result of this natural event can be catastrophic, voluminous engineering geological and geophysical investigations, as well as numerous preventive activities are performed. Two indispensable bases - regional seismic map and map of microseismic zoning, have been treated in the previous text. In the map of seismic risk for the territory of USA, rayons of potential risk were outlined (Fig. 11.2.), and in California, for precise planning and evaluation of potential destruction, detailed seismic maps and plans of seismic risk are prepared, as integral parts of all general plans in California. All mentioned bases offer possibility to evaluate dimensions of potential destruction, but also to plan precisely and define aseismic constructing (Fig. 11.3.).

Protective measures from volcanic activity. As speaking about geological event which is almost unpredictable and enormous risks to live in the areas of volcanic activity, series of unique exploration methods is performed, as well as monitoring and protective measures of distinguished rayons, in order to mitigate consequences of lava and ash flows. The approach is, first of all, dependent on the type of volcano. Monitoring includes observing potential signals - warning eruptions, earthquakes, changes in slopes of the terrain, occurrences of new hot springs and fumaroles or changes in temperature and chemical content at the existing springs, soil temperature changes, etc., and main monitoring equipment is seismograph (A. Howard, I. Remson, 1978). Among protective measures, let us note lava flow bombing, barriers and trenches for drainaging lava flows, etc. In 1973, experiment was successfully performed on Iceland.

Measures for toxic waste disposal. Factually, almost in all countries of the world, waste dangerous for biosphere, in other words - for its normal survival, is existent. Problem of their disposing becomes more and more important during several last decades; investigations of possibilities are very complex, long - term and diligent, and acquired experiences very different. Water basins are not convenient as disposals, loaded by various unknowns, while in land, each country is looking for solutions within geological structure of appropriate territory - Germany is orientated to salt diapirs, Great Britain to evaporite formations and clay rocks, France - to granites, schists, clays and salt formations, etc. (B.Ciric, 1990).

In any case, good knowledge on geological structure and hydrogeological conditions of interesting areas and potential geological formations is the best basis to choose the most convenient formations and the most efficient measures which has to be undertaken in solving this very responsible ecological task[125]. It is reasonable that approach to finding the most convenient rayon for toxic waste disposing is based on geological and hydrogeological criteria, to which other criteria, required by valid regulations, are merged. The approach has been applied to example of geologically very complex territory, as the territory of Serbia (M. Komatina, 1990, 1993).

USE OF RESULTS OF MEDICO - GEOLOGICAL EXPLORATION WITHIN HEALTH CARE

Among problems with which mankind is confronting during the last decades, caused by contamination and enormous destruction of the environment, *global medical problem* extended into the first plan. All newly created global problems, in fact, are hiding serious dangerous for human civilization, life and health of human beings.

---

[125] The most sensitive problem is, certainly, disposal of radioactive waste, where time dimension is very important. That is why among basic conditions which have to be fulfilled, the following ones are included: 1) impossibility of contaminated groundwater circulation into the other rayons and 2) permanent stability of distinguished rock masses.

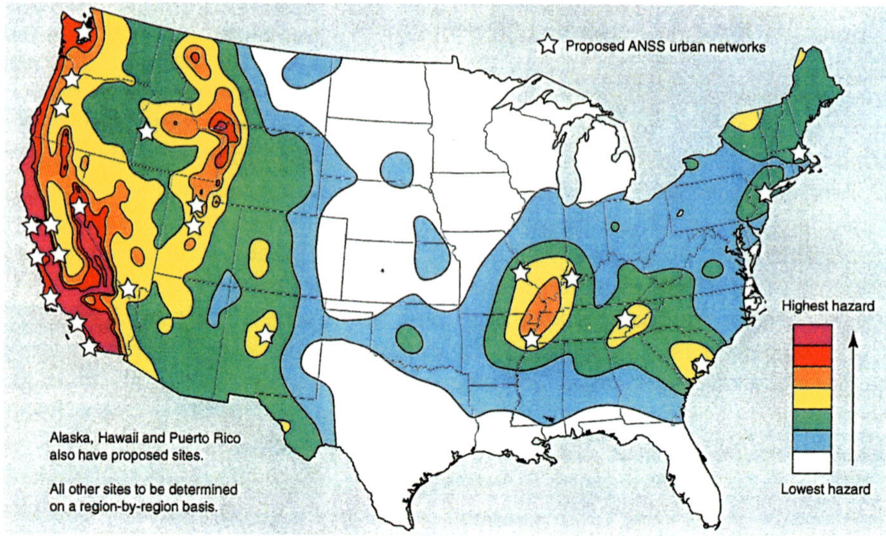

Fig. 11.2. Seismic hazard for the territory of USA (USGS, 2000).

Doubtlessly, ecological problem of protection of the environment formulated very complex and difficult tasks to the contemporary medical science and health care. In that way, within medical science and practice, more and more significance belongs to *global problems of health care*, which is not, according to their character, possible to be solved by individual efforts of the countries, in other words - require united efforts of all countries in the world as well as the international community. After Keller and associates (1993), three following health care problems of global character are known:

1. *programs development and preparing efficient complex systems of health care in countries throughout the world, characterized by different socio - political system and level of economic development, the system capable to provide right of inhabitants to obtain rapid and qualified, urgent and specialist's, prophylactic and medical help;*
2. *organizing fight against such permanent infectious diseases, as plague, cholera, yellow fever, AIDS, or widespread tropic disease, as malaria, for instance;*
3. *problem to fight against cardiovascular diseases, neoplasmes, some chronically and hereditary diseases, widely spread all over the world.*

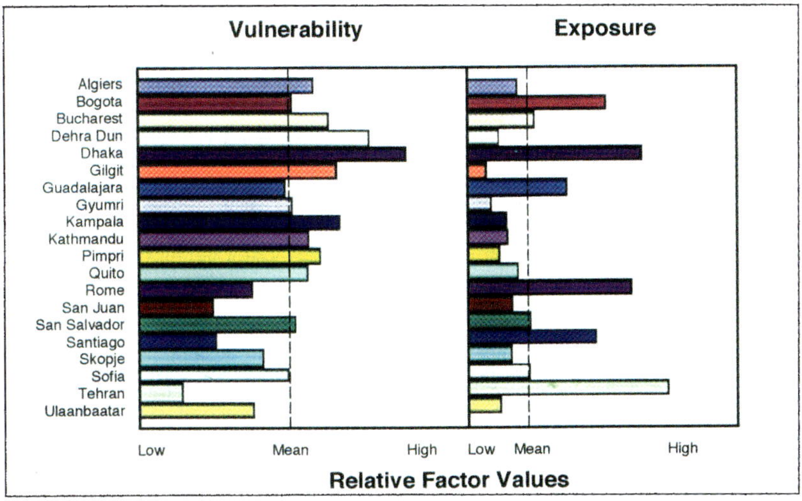

Fig. 11.3. Seismic risks in some urbanized rayons (IDNDR, 2000).

In this domain, significant results have been attained by the World Health Organization, as a leading and coordinating institution in the international activity related to health care. Let us remember successful liquidation of pox during 1958 - 1985, or organized work on prevention, diagnostics and clinical treatment of AIDS patients[126].

Preventive medicine, in other words - performing prompt and rational prophylactic measures, and qualified treatment of sick persons, has very important role within the process of improving level of population health. Conditions of health care in different regions of any state territory can be very heterogeneous, depending on large number of natural and socio - economic factors, among all - urbanization degree and state of transportation network. Unique situation can appear in the territories predisposed to negative elemental natural (geological) occurrences, which can disturb and discontinue all kinds of connections, becoming a source of traumatism and other pathologies (earthquakes, volcanic eruptions, catastrophic mass sliding, rockfalls, stone - sludge torrential flows, floods, fog, etc.).

Further, problem of logical linking of the complex *nature - man (health, health cure) - economy*, as an essence of health care within some country, requires organized multidisciplinary activity directed to development new modern methodology of research as well as ecological approach to health prevention. Medical treatment of the population

---

[126] The first case of AIDS was registered in USA in 1981, suddenly spreading over the whole Planet in the form of infection pandemy.

cannot be carried out according to a stereotype and over administrative units - it has to be based on geographic and geological features of certain regional types of the country's territory, for which unique standards and requirements of health care are defined.

Nowadays, factors of natural (geological) environment are not studied enough yet, as well as influence of numerous anthropogenic factors on human health are difficult to be predicted. That is why new *nosological forms* and appropriate scientific classification of diseases appears to be a very important task of medicine, task of the highest significance for health care organization. Even in 1972, A.P. Avtsin underlined real benefit from correct outlining nosological units, which is of the same significance for the mankind as discovery of new chemical ingredients or new elementary particles in physics. "*Argumentated distinguishing each new disease, and hence, more disease categories, means not only certain victory of human intellect, but also one of the first real preconditions for successful health care*" - concluded Avtsin.

Role of applied medico - geological discipline, as well as importance of use obtained results and cartography in health care organizing is also noticeable in rapid and efficient solving actual complex tasks of medicine and health care. First of all, contribution of medical geology in affirmation of geological preconditions of diseases, in other words - in distinguishing new nosological forms, but also in clearing up natural characteristics, forces, processes and occurrences; degree evaluation of comfortability, conditions and quality of the environment; making optimal conditions necessary for living activities. Some domains of possible use of geological and medico - geological exploration results for the cited purposes are presented in the further text.

1. Link between natural conditions and human health is noticeable. Nutrition, health conditions, capability for work and vitality are dependent on quality of air, climate, soil, plants and animals. In order to be healthy, strong, alert and persistent, human beings tend to interact with preserved nature. From the other side, natural potential of numerous territories of the Planet, particularly mobile tectonic belts, is still enormous. In order to use and preserve that gift of the nature in optimal and coordinated way, for the well - being of this and future generations, high level of geological knowledge on all potential areas when speaking about treatment, vacation, recreation and tourism; is principally required; starting from zoning and evaluation of potlined rayons[127]. Let us note rayons rich in

[127] It is desirable that all aspects of practical contribution of environmental and health care geochemistry are presented in the maps of medico - geochemical zoning; for example, distribution and prognosis of biogeochemical endemies, other geochemical and geological

mineral and thermal waters, which are very significant resources in numerous countries of the world. Use of the water resources makes the spas to be very attractive tourist destinations with a long tradition, not only in domain of tourist, but also of health care services. For rational management of mineral and thermomineral water resources, high degree of information and protection is required, as well as reliable geological - medical evaluation of quality, quantity and usability of the potentials. According to results of various detailed hydrogeological, hydrogeochemical and balneological investigations, desired basis is possible to be obtained. Special type of applied medico - geological and medico - geographic works is analysis of conditions in various rayons, serving for clearing up capabilities of their use for recreation.

2. In order to solve a series of practical tasks of prophylactic medicine and health care, it is necessary to acquire voluminous material referring to geological (geochemical), medical, biological and other investigations. Geochemistry, for example, has a significant contribution in collecting input data for explaining disease carriers, development of prophylactics and finding the best ways of treatment (eradication of biogeochemical endemies, caused by deficiency or surplus of some elements within the system *rock - soil - water,* etc.)[128], but also forming optimal content of chemical elements in food products and water, providing human body in the best way. Leading role in solving such problems belongs to medical sciences, but geochemical study of each interesting area is also of high importance. Finally, for each area, sanitary - hygienic evaluation of appropriate geochemical conditions, with defined ways of forming an optimal geochemical regime (that is - optimization task solved), is necessary. "*Through capability of forming optimal geochemical regime, mankind obtains a new powerful factor of improving health, vitality extending, improving the whole physical nature of human beings, a factor which was not a characteristic of the nature during several billions of years of life presence in the Earth" (Perelman, 1975).* For example, optimal geochemical regime understands optimal level of selenium, as one of trace elements essential for human body. That is why a problem of population appeared because of deficiency this element in rocks,

---

factors of the environment influencing on human health, spa resources of geochemical origin, etc. (A.I. Perelman, 1975).

[128]Numerous diseases of cultivation plants and domestic animals are also caused by deficiency or surplus of chemical elements within the environment.

soil, water and food, determined by preliminary geochemical investigations, can be solved by application of selenium microfertilizers and in other ways, as done (on the basis of program on the national scale) in Finland, China and New Zealand. In that way, important role in prevention of cancer, heart diseases, detoxication during heavy metals contamination is fulfilled; helping also in possible physiological damages caused by diabetes, immunity stimulation, longevity reaching. The situation is similar when speaking about iodine, fluoride and other elements deficiency in food and water as well as their compensation. Problem of fluorine, iodine, boron, and iron or manganese surplus can be also solved by application of some of appropriate technological removing procedures.

3. In the world, areas characterized by extremely high or high content of naturally radioactive elements are known. Outlining such (radioecogeological) areals is of indisputable importance from radiation - epidemiological and radiation - hygienic viewpoint, in order to evaluate radiation risk for human population living in the certain territory or wider[129]. In solving the cited problem, data of geological exploration of nuclear raw materials, as well as radon exploration followed by compiling maps of radon risk, about which we discussed in the previous chapters, and particularly radiogeological investigations of the interesting area intended to needs of medical science and health care, can be of invaluable help.

GEOLOGICAL FACTORS AND HEALTH ASPECT OF STAY IN OFFICE AND AREA OF AN APARTMENT

During their development, human beings expressed tendency of longer and longer stay indoors, office or in apartment. According to available data, nowadays, man spends indoors even over 80% time during a day. By such way of life, he practically deprived himself of all positive influences of the environment in relation to his health and working capability, from one side, and exposed to danger influences present indoors, from the other. Stay indoors, less and less made of natural materials, resulted in consequences on human health, followed by a series of *specific diseases* (allergic alveolitis, fever because of humid air, asthma, allergic diseases of

---

[129] Of course, consequences of soil and water (and, indirectly, food) contamination by radionuclides occurred after damages of nuclear power stations or during nuclear tests, partly because of use phosphorus fertilizers made of apatite and phosphorite should be also kept in mind.

nose and eye, manifestations on skin), but also various bacterial and virus diseases, mild chronic intoxications. Besides, carcinogenesis as a mechanism of diseases origin means possible harmful effect of asbestos from building materials and, particularly, radioactive radon (Rn_{222}) and its descendants - frequent “inhabitants” of indoor space. Beside mentioned *specific diseases*, health disturbances which are not defined medically as diseases, but expressed through non - specific subjective difficulties, are the centre of attention. Group of such disturbances of benignant character has been studied since 1983, as a *sick building syndrome*. The most frequent symptoms of the *sick building syndrome* are: headache, blocked nose, sore throat, eye itch or tearing, lethargy, painful neck, dry eyes, dry skin (Z. Marmut, 1998).

Specific carrier factors causing in some number of persons symptomatology of *sick building syndrome* can be classified into the following categories: 1) *geological,* 2) *biological,* 3) *chemical* and 4) *physical factors of the environment.* We will focus on the part related to *geological factors* and possibilities of reducing or total eliminating their influence on human health. From that point of view, geological composition and characteristics of rocks and soil within the foundation of the building, tectonic disturbance of rocks and particularly presence of fault structures, groundwater existence in rocks and moisture in soil and used building material, various geophysical fields, radioactive radiation from rocks, soil and used geological building materials, are interesting (Fig. 11.4.).

Starting from a whole series of the mentioned geological elements, it is real to presume that, by exploration results, geosciences and applied geomedicine can contribute significantly to analyze negative effects of the cited and other geofactors, hence - to construct on the suitable terrain and, using biologically qualitative (“*healthy*”) materials, to avoid some bad materials. In other words, geology and geomedicine can be of indisputable importance in providing healthy conditions for life and work of human beings, staying (as mentioned) the greatest part of their lives indoors. In the following text, some of illustrations are presented. According to exploration carried out in Great Britain, in percentual relation of *radioactive emission in the environment,* radiation from the terrain is 47%, from soil and objects - approximately 14%, food and beverages - 12%, etc). The other authors direct our attention to crucial role of natural sources of radioactivity within total radiation of population (85%), of which radon is predominant (49.5%), much less from artificial sources (15%) (I. Petrovic, 1998). On the basis of epidemiological studies of American Environmental Protection Agency (EPA), performed in cooperation with various health centers, radon is at the second place as a carrier of lung cancer (after smoking), highlighting very high risk for smoker staying in apartment with high radon concentration. Radon risks can be extremely increased within seismic active areas, because it was determined that during intensive repeated earthquakes, content of the gas in groundwater significantly changes.

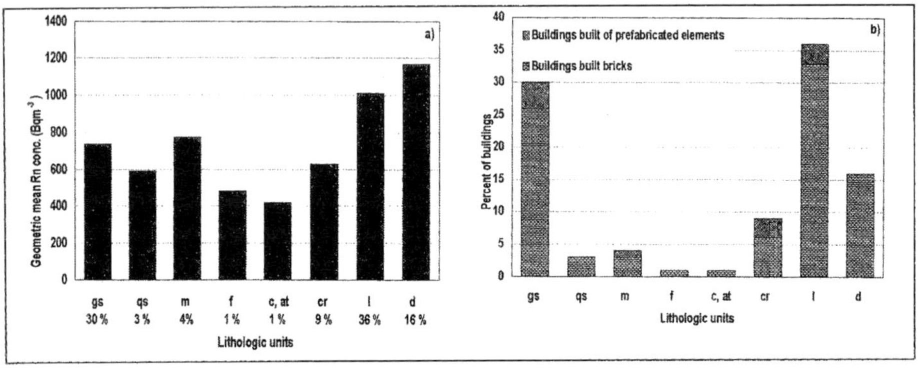

Fig. 11.4. Geometric mean radon concentration on each lithologic unit (**a**); percent of buildings on each lithological unit, constructed of bricks or prefabricated elements (**b**); (gs - glacial sediments; qs - quartz sandstones; m - marl; f - flysch sediments; c, at - claystone with intermediary layers of andesite tuffs; cr - various clastic rocks; l - limestone; d - dolomite (A. Popit and J. Vaupotic, 2002).

Risk of radioactivity is extremely high in the areas made of rocks containing uranium, including their soil cover (granite, basalt, gneiss, tuff, pumice), particularly if the rocks are tectonically damaged and permeable for radon, or contain groundwater as the gas carrier. As a risk factor, geological building materials from the mentioned areas are also known, but also new materials, made of slag from blast furnace, ash from steam power plants, phosphate waste from uranium mines[130, 131, 132], etc. Negative aspect of stay indoors built of such material was presented for the first time in 1972 by the Ministry of science in Germany,

---

130 Main source of radon in the ground floor is gas from geological environment (approximately 60%), while in case of apartments on higher floors, building material has very important contribution (approx. 50%).

131 At the beginning of the fifties of the last century, alum shale with natural radioactivity (according to summed formula) over 20 (allowed value: 1) was used for concrete production in Sweden; during 1952-1966 period, in Colorado and Ontario (USA), as also in Australia, waste from uranium mines were used; in Japan, until 1974, building industry produced 3 million tons of artificial gypsum, made from phosphorite, etc.

132 Large number of used building materials can contain chemically harmful - toxic supplements. Consequences of toxin activity can be malignant excrescences or occurrence of carcinogenic diseases. When speaking about negative effects on health, asbestos, which was intensively used in civil engineering, is mentioned very often, as well as other fibrous non-metallic row materials.

recommending that building materials characterized by high natural radiation, particularly when speaking about space where someone stays for a long time, should be avoided (R. Terzic, 1997).

Geophysical (physical) fields of natural or artificial origin, have a significant influence on human beings and other organisms (S. Komatina, 1998). The topic was discussed in appropriate chapter, and potential risks are analyzed in detail by S. Mares and associates in the book **Applied geophysics in environmental engineering and science** (1997).

In the literature, consequences of various fields of underground flow and geopathogenic zones on the human health, based on the results of measuring by pendulum or divining rods, are also presented. The method itself, known as *radiestesial method*, is considered as non-scientific, and is not treated in this book. For example, it is not real to expect presence of hydrogeopathogenic privileged groundwater flows within isotropic intergranular alluvial medium, which are presented by V. Momcilovic and V. Minic (1998) as areas of high importance in occurrence of endemic nephropathy and malignant neoplasms in southeastern Serbia. In this case, possible effects on human health could be a result of geological events in the alluvion footwall (for example, radon migration or of some other dangerous gas coming from the deeper levels along some fault line).

APPLIED MEDICAL GEOLOGY AND LAND - USE PLANNING

Land - use planning is very important task of far - reaching significance for each country, and has to be realized extremely resreliably and thoroughly. As an example of good organizing, land - use planning referring to health care in Netherlands, where, besides all, public opinion was also taken into consideration is presented. Main approach to the land - use planning has to be ecological, based on widely accepted and established ecological principles and regulars in planning, arranging and protection of the space, in accordance with concept of sustainable (coordinated) development. In essence, instead directing to protection of exclusively natural comples, as a basis for their rational use, nowadays, problem of human health protection, recovery (improving) conditions for life, work and rest of population, is at the first place. In that way, number and significance of questions which has to be solved, beside others, by experts for medical geography and medical geology, is increased.

Characteristics of medico - geographical and medico - geological exploration for needs of planning offices depend significantly on geographical conditions, in other words - on geological structure of the interesting territory, uniqueness of a task, dimensions of the studied event, planning stage, etc. According to Keller and associates (1993), for case of medico - geographical exploration, a series of similar features is the following:

- precise defining of input medico-geographical situation;
- analysis of perspectives of economic and socio - public development of the territory and choice of the most acceptable strategy of the development observed from the position of medical geography;
- development of medico - geographical zoning system of the studied territory, adequately to the task defined in the planning task;
- making medico - geographical forecasts for each distinguished rayon;
- wide use of cartography during all stages of solving the defined task.

The mentioned decisions can be predominantly valid even when speaking about medico - geological exploration for the sake of land - use planning.

It is not so strange if in some of previously made land - use plans, geological environment with appropriate potentials, which are important for health care planning, is not treated enough. For example, Spatial plan of Republic of Serbia, from 1995, predominantly neglected not only mineral raw materials, but also hydrogeological, engineering - geological, pedological, and seismological and geomedical factors. This is the way of underestimating geological wealth or making their rational sustainable use difficult by wrong purpose and space occupying, from one side, and overlooking geomedical factors complicates correct organizing of population health care, from the other. One of the most significant types of collision between geological wealth and space occupation appears during constructing large settlements, cities or traffic arteries in the areas with groundwater sources, cultivable soil, deposits of mineral raw materials, etc. A striking example of collision between town and geological wealth is Tuzla (Bosnia and Herzegovina), with salt deposit, over which a town with over 100,000 inhabitants was formed; the town was endangered by terrain settling over the deposit exploitated by salt water pumping, and value of geological wealth was significantly reduced because of increased expenditures of exploitation (M. Babovic, 1992). Underestimation of geological wealth by urbanization is probably much frequent than it is known, because the ore body under the settlement will be very difficult to be discovered, and much difficult to be exploitated. Just in that, but also on other facts, is significance of some *modern geological base* made for needs of land - use planning, related to management and protection of geological potential.

From the standpoint of planning space use, two basic approaches to classification of geological environment - from the standpoint of danger and from the standpoint of resources - can be distinguished. Dangers has to be taken into consideration, but man has to be diligent to resources, in order to understand natural processes and adapt to living conditions in the areas of limitations restricted by the processes (Howard, Remson, 1978). Knowledge on features of the geological environment and correct defining is one of preconditions for quality of plans, not only spatial, but also plans of development.

For making plans, cartographic methods are most often used. Three types of base: 1) *bases of general type,* with presented geological structure of the terrain (Basic geological map of the territory and others); 2) *analytical bases,* treating certain relevant features of geological environment (for example, metallogenetic map, hydrogeological map, map of seismic activity and terrain stability, geoecological map, radiogeoecological map, etc.); 3) *synthetic bases,* dedicated exclusively for direct use by planning teams, with content dependent on purpose, that is - of the level of planning.

Geological problems of protection the environment through space planning is not equally treated in different countries. According to J. McCall and B. Marker (1989), various contents of bases are more often stated in the following way:

1. *geological maps of the environment* (geoscientific maps), including geological, geomorphologic, hydrogeological, geochemical, geophysical and other maps as a complete set, sometimes with collective restricted interpretation, as well as some necessary non - geological bases;
2. *engineering geological maps,* referring to problems of constructing the objects;
3. *maps of potentials,* presenting resources of geological wealth;
4. *maps of geological resources,* showing mineral raw materials at the surface and in depth, appropriate dimensions and quality;
5. *maps of pedological potential;* presenting possibilities of soil use or potentials of processes which are underway;
6. *thematic maps,* related to special problems;
7. *documentation maps,* containing field and other data referring to the treated subject;
8. *maps of restrictions,* directing to limitations of soil use;
9. *maps of dangers,* presenting types and degree of different dangers at the explored area (Fig. 10.1.);
10. *risk maps,* related to certain type of danger and presenting probability or dimensions of their effects.

It is noticeable that *hydrogeological and hydrogeochemical maps,* which are important because of presenting distribution and chemical content of groundwater (low - mineralized, mineral and thermal water), nowadays so important resources of the Planet, are not treated.

The first bases of multithematic content dedicated to space planning were made in the former CSSR in 1961, and this manner spread to other countries of Europe and USA. In Yugoslavia, *maps of suitability for constructing (urban suitability)* were made in 1972, and Spanish experts included into the set this base five years later[133]. In USA and Canada, for presenting geological problems for the needs of space planning, term *geological map of the environment* is used, with different contents in various parts of the state territories (in dependence on natural conditions, population density and development plans, but also on subjective approach of geologist; for example, in western parts of USA, very important place belongs to seismicity and terrain stability, but in others, soil recultivation after mining activity is important, in the third ones - groundwater contamination, or salt intrusion in coastal parts, etc. In Great Britain, the first *geological map of the environment* was made at the beginning of the eighties of the last century, for one area in Scotland. In France, detailed bases (with presented geomorphologic characteristics, thickness and type of sediments, erosion, foundation conditions, limestone mines, karst terrains) were made for the area of Paris in 1979, as well as zoning according to geotechnical characteristics. Some time later, *regional maps* (with review of conditions of soil and groundwater use, zones of sanitary protection of aquifers, zones of groundwater recharge and accumulation, near - surface deposits of mineral raw materials, interpretation of pedological soil fertility, limitations in sense of constructing because of unsuitable ground characteristics and needs to protect important geological localities)were made in Western Germany. In the former USSR, problems of land - use planning were gradually entered into the content of engineering geological maps.

In the book **Geology and protection of the environment** (1992), as an example of systematic exploration of the whole territory of one country, M. Babovic distinguished *Geological and realized maps of the natural environment and resources, 1:50,000,* based on the project of Geological Survey of the former Czech Republic, started in 1985 (B. Moldan, 1986). This set contains the following nine maps:

1. *Geological map of traditional type,* presents structure of the ground and cover, including Quaternary deposits;

[133] *Maps of groundwater protection* were made for the territory of Illinois (USA) in 1968, for the territory of Czech Republic (1:500,000) in 1967 and 1968; territory of France was covered by map of aquifers contamination (1:1,000,000) in 1970, and similar maps were rapidly made for the countries of Latin America. In the former Yugoslavia, the first map of aquifer vulnerability to contamination (1:500,000) was made in 1976, including territory of Croatia.

2. *Map of mineral resources,* presents all types of economic and subeconomic, proved and assumed reserves;
3. *Applied geophysical map,* contains results of all regional and detailed geophysical investigations;
4. *Hydrogeological map,* predominantly oriented to review of groundwater from the standpoint of water supply;
5. *Engineering geological map,* containing information on various natural conditions from the standpoint of constructing and soil use;
6. *Maps of pedological soil,* presenting kinds of fertile soil, as also composition of the soil, including interpretation of soil fertility and other features important from the viewpoint of agriculture and forestry;
7. *Geochemical map,* showing components vulnerable to react with groundwater, supposed velocities of reactions with groundwater, anomalous content of main biological components, as also microelements;
8. *Hydrochemical map,* made for surface waters;
9. *Settled map* (map of natural potentials), contains six sheets: 1) *elements of lithosphere* deposits and reserves of mineral raw materials, mines, main fault zones; landslides and other geodynamical phenomena, seismic rayons, stone ground); 2) *qualitative soil* (endangered by erosion, endangered by floods); 3) *hydrosphere* (activities in water supply, main surface reservoirs, zones of water resources protection, important occurrences of mineral waters, main sources of contamination and identified contaminants); 4) *biosphere* (natural reservations and similar; the most productive forests, important localities with vegetation out of forests - protection against erosion); 5) *areas of atmosphere contamination;* 6) *the most important objects made by man* (settlements, roads, railroads, devastated areas, etc.).

Lack of information regarding *radioactivity and ionization* (which can be related to conspiracy), as well as *geochemical composition of rocks and soil* - geological elements very important from the health care standpoint, is conspicuous (Fig. 11.5).

In the text referring to cartographic methods, we concluded that very informative atlas important for medical geology can be made. In the form of example, **Geochemical atlas of Slovak Republic** (1997) made of maps in scale 1:1,000,000 and legend referring to groundwater, soil, rocks, distinguished elements from the forest ecosystem and natural radioactivity in rocks and water, was discussed.

From the standpoint of space use, natural environment is consumed the most efficiently, if planner takes into consideration entire forms of relief making landscape, but also natural processes causing its application. Besides, he is obliged to know in which cases element of relief as a resource or a danger has to be treated. Further, resource as soil is should be cultivated in why which provides maximal benefit for society; from the other side, it can present a danger swelling, rackfalls, sliding along a slope - processes followed by destructions of different degree). Let us analyze geological risks and their influence on human life. According to Howard and Remson, there are five types of reactions to geological dangers:

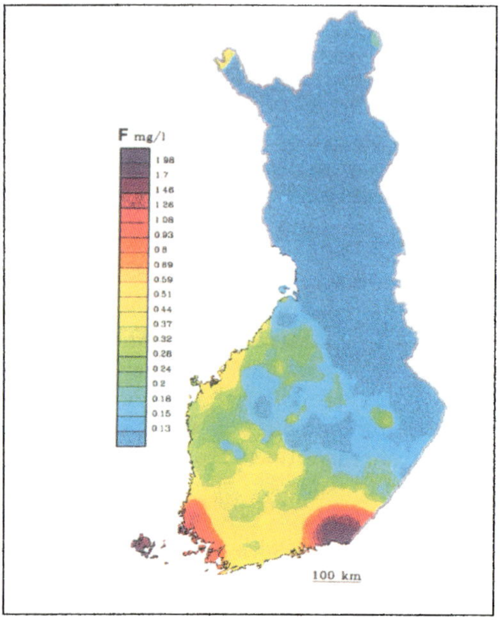

Fig. 11.5. Fluoride concentration in groundwater at the territory of Finland. High values of fluorine correspond to area of Rapakivi granites and other magmatic rocks, while water in crystalline schists are with very low content of that element (P. Lahermo, B. Backman, 2000).

1. *Danger avoiding,* as the most frequent reaction to the present danger (constructing on the territories exposed to floods and landslides, also rayons with active faults);
2. *Stabilization* is related to some geological dangerous which can be stabilized by economical investments;
3. *Providing measures of safety during constructing;* for example, measures of aseismic constructing, with economically justified expenditures;
4. *Introducing limitations in space use and its settling,* that is - possible regulation for agricultural or urbanization needs in accordance with degree of danger;
5. *Making informing system* for some dangers which can be prognosed and where urgent measures can be performed (floods, hurricanes, some parts of volcanic activity).

If there is enough soil for anthropogenic activity, one of the better ways is danger avoiding, that is - reclaiming safe (from that pooint of view) territories. However, many large settlements have already been constructed in the rayons with present geological dangers, and in such cases, the mentioned ways of reactions and measures have to be considered. For example, a lot of houses, hospitals and schools in the state of California (USA) are constructed directly on the active faults, and so dangerous situation was reason why local authorities introduced a law related to geologically dangerous zones in the state (which was previously discussed). Or, use of land vulnerable to sliding, as a very difficult problem, was solved in town Portola - Vella in California by strict use of geological stability map, as a manual for activity and town jobs management (Howard, Remson, 1978).

Fig. 11.6. Plitvice lakes, Croatia.

Regarding plans of environmental protection, approach to making part of general plans of towns and districts in California, concerning reserved areas, is interesting. According to the approach, the following spaces are distinguished: 1) *spaces reserved in order to preserve natural resources* (territories necessary for preservation plant and animal world, for ecological and other investigations, then rivers, brooks, bays and estuaries, beaches; coasts of lakes, rivers, aquifers), 2) *spaces reserved for exploitation of natural resources* (forest regions, pastures, plowed fields and territories of economical significance for food production; areas necessary for groundwater recharge; water objects used for fishing regulation; territories of basic deposits of mineral raw materials); 3) *spaces reserved for making recreation zones in the nature* (picturesque localities or localities with historical and cultural significance (Fig. 11.6.); territories necessary for parks and recreation zones, etc.); 4) *spaces used for needs of health care and population safety* (territories requiring special attention because of possible earthquakes, sectors with unstable land, water tapping, zones necessary for water quality protection, places necessary for protection and improving air medium, etc.)

When speaking about terrains of mobile belts of the Earth's crust, wide range of tasks is put for geology and applied medical geology. Solving the tasks can contribute significantly to land - use planning and protection of agricultural and constructing land from erosion, sliding, rockfalls, settling, cracking; planning water supply and health care in karst terrains; geological heritage protection (Fig. 11.7.); planning protection human lives in the areas with earthquakes and volcanic activity (Fig. 11.8.); outlining potential areas and objects of health recreation; preservation of population in rayons with high radiation risk; health care planning in rayons with geochemical anomalies (deficiency or surplus of certain essential macro - and micro elements in rocks, soil and water, indirectly in food products), etc. Part of the mentioned elements of geological environment is particularly interesting in health care planning, and that is why some of them are discussed in the further text, keeping in mind that they have been already analyzed.

Fig. 11.7. Sjenice exsurgence, Western Serbia (Photo: D. Mijovic).

1. *Providing high - quality drinking water and protection of groundwater resources* for public needs is of varied health significance, and reserving space with practically interesting water reserves and performing measures of sanitary protection is one of principal tasks of the planners. Similarly is valid for relation to pedologically interesting *soil cover*, the second precious gift of the nature, nowadays exposed to erosion, chemization, exhausted in the sense of macro - and micro elements, salinization, degradation and other ways of anthropogenic activity.
2. During exploration of one or other territory, and planning use of natural resources, not only negative characteristics (factors) of

the environment has to be kept in mind, but also "*health stimulators*", and on that basis - to make a choice of zones planned for rest and treatment. Land - use planning in the areas of *health recreation* often had negative effects on the space, and because of insufficient hydrogeological and balneological knowledge, important resources of mineral and thermal waters, gasses and medical non - metals (peloids) were not used enough or were endangered by various contaminants. For example, in the country marked by over 1,200 springs of mineral and thermal waters, as Serbia is, the already discussed Spatial Plan even doesn't mention that natural wealth (with outstanding variety according to chemical composition and usability); only 30 occurrences is used for needs of spas, and only five of spa centers possess Detailed urbanistic plan.

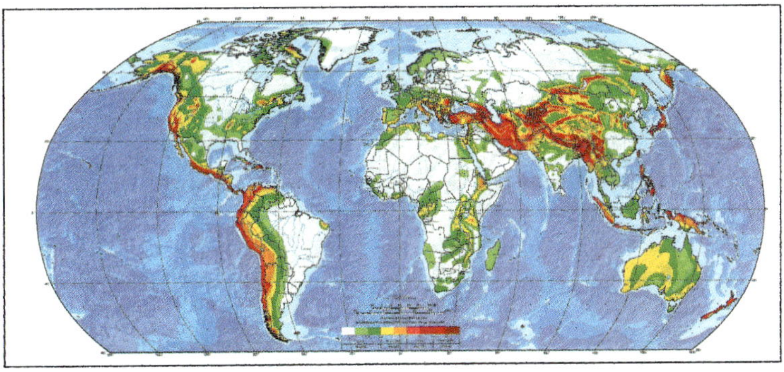

Fig. 11.8. Global seismic hazard map (produced by the Global Seismic Hazard Assessment Program - GSHAP) (USGS, 1999).

3. In planning health care, because of indisputable significance for human health, distinguishing *geochemical anomalies* have to be subject of much greater attention than until now[134]. That is also referred to *rayons characterized by high radiation risk*, first of all - with high risks of radon presence, which should be known in defining locations for new settlements (objects) (Fig. 11.9.). In both cases, main precondition for making optimal plan present detailed restricted geochemical or complex radioecogeological exploration.

[134] Numerous diseases of biogeochemical nature, as we already know, are dangerous for domestic animals and plant cultures.

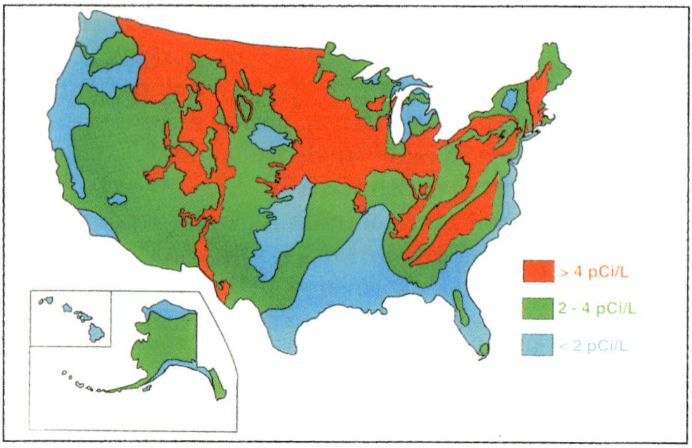

Fig. 11.9. Radon distribution in USA and Hawaii for Sept. 1992 (USGS, 1996).

4. Land - use planning in *areas of volcanic activity* requires unique approach, followed by accepting geological (volcanological) features of the concrete volcanic rayon. Problem of human adaptation on living conditions in such areas is complicated because of outstanding difficulties in domain of forecasting volcanic eruptions in the future (Fig. 11.10.). That is why such terrains are classified into specific zones, used for recreation or as picturesque national parks.

Fig. 11.10. Eruption of St. Helena volcano in the western part of USA at noon, 18th May, 1980 (R. I. Tilling, 1999).

APPENDIX

Base of Medical Geology

Based on numerous natural and social sciences, medical geography has predominantly synthesized general information on the natural conditions of some territory, regardless factors of essential importance – carriers of some disease or positive effect onto the human health. It is interesting to note that, geological factors haven't found much space in the world literature and that geoscience could much contribute to clear up numerous problems. The first and one of rare steps in that field has made by H.Zeiss, introducing a term geomedicine. Under the term, German scientists defined a scientific direction separated from medical geography which, during analysis of disease development in space and time, points out its attention to factors of geosphere (geoenvironment, environmental protection), carriers of one or another form of the development of observed phenomenon. Simultaneously, not only factors of the general geoecology of the human beings (as inhabitants of the Earth) are considered, but also of geoecology of human diseases. Regionally, geofactors important for the health are analyzed at micro - , medium - and macro - level (H.J. Jusatz, 1983). According to the German scientists, leading role in realization of geomedical exploration belongs to interdisciplinar cooperation of experts in the field.

In 1952, under the Heidelberg Academy of Sciences, Center of geomedical exploration was formed, with 40 experts employed. The Center belonged to the Department of mathematical and natural sciences, performing predominantly medical - mapping activities. Beside Atlas of the world distribution of epidemic diseases, made by E. Rodenwaldt and H.J. Jusatz (1983), as a very important event in contemporary medical geography, within the institution, geomedical monographs dedicated to Libya, Afghanistan, Ethiopia, Kuwait, Kenya, Korea, Thailand, Nigeria, Uganda, Tanzania, were also published.

When speaking about geosciences, series of papers of the Netherlands authors dedicated to research of space relation among occurrence and cancer diseases distribution and geophysical and geochemical factors, drinking water quality and peat - mud soils (S.W. Tromp, 1955). Last years, visible contribution was made by H. Woodcock (1994) – book: Geology and environment in Britain and Ireland and K. Vrana et all. (1997) – Geochemical Atlas of Slovak Republic 1: 1000000.

Close causality of living and non - living nature is certain. Whole life and human's activities are performed within geological environment, and rapid development of geosciences (regional geology, tectonics, geophysics, petrography, geochemistry, hydrogeology, engineering geology, exploration of nuclear and other raw materials) during the last decades is inevitable. Question why geology has not been much more represented in study of effects of the environment onto human beings is excused. Or – why specialists, knowledge and voluminous cartography in domain of geology have not been used enough?

The paper is an excerpt from the monograph Medical geology, which is in press (in Serbian), showing only a base of the new discipline of geology, that is – medicine. In the monograph, risk factors in the environment and positive effects of them onto living nature are more universally presented, as well as subject and tasks of geological and medical disciplines in defining quality of the environment and effects onto the human health, methods of medical geology. The following chapters were also prepared: regional pathological geology; medical geology and the environment; aspects of medical geology application in economy and health care.

Goal of Medical Geology

To answer the important question how much is degree of suitability to live at a certain area is very difficult. A complex multi - phase way, and very uncertain, is exploration of carriers of a disease. Only by multidisciplinar approach, the problem can be solved. However, that doesn't mean that it is not necessary to take a maximum from each connecting scientific discipline, particularly in domains where it means a significant contribution. Discovering real (principal) disease carriers, or positive effect onto the human health, would simplify diagnostics of the disease (opening more efficient ways for treating), in other words to perform preventive medicine.

Suitable circumstance is that science about inorganic world – geology, extends different possibilities in that field. Namely, legality of space distribution of some geological units and geological factors, as potential disease carriers or strengthening of a defense capability of the human organism, are well expressed very often. That is reason why conditions for making a correlation between geological factors and data of field and laboratory medical exploration within interesting geotectonic or structural unit are existent. In other words, zoning of the studied area according to the degree of geological factors effects onto the population health (geomedical zoning) might be very useful, not only for further medical research and practice, but also in the field of economy, urbanization and health care.

The main goal of medical geology is to study general relationships of geological environment effects onto the human health, more precisely – to distinguish among geological factors these ones which could be main carriers of some diseases or could have positive effect onto the human health. Medical geology so gives a significant contribution to the environment optimization, to reach the highest level of the population health of the Planet. By presence of the new discipline and appropriate experts, information on voluminous fundamental and applied geological exploration would be used much better, which will be an important step in the development of the world science. By founding medical geology, we mean that accomplishments and experiences of medical geography, medicine, biology and other connecting sciences have to be included.

Geological Factors and Providing Conditions for Healthy Living

Geological structure of an area effects living conditions of the population. In order to live in harmony with continually changing environment, beside geological factors, essence and velocity of changes have to be known and, in accordance with cited, to adapt to them. For example, in California, houses, hospitals and schools have been constructed at active faults directly. So dangerous situation was a reason why governmental authorities in 1972 decided to propose a law on geologically dangerous areas in California, with obligation for geologists to make maps with positions of active faults and present them to the authorities, in order to prevent constructing houses in the fault zones (A.D. Howard, I. Remson, 1978).

Geological environment is characterized by various factors, like: principal rocks, tectonic structures, geomorphological features, loose sediments (including pedological soil), groundwater, engineering geological occurrences, deposits and occurrences of raw materials, macro - and micro - elements in rocks, soil and water, radioactive elements, etc. Processes changing geological environment, as well as following factors – climate and biological, are also included. Degradation processes of rock and minerals, for example, make emission of harmful (medium heavy and heavy) metals possible and helping them to pass from lithosphere to hydrosphere and biosphere. This is one of dominant processes in the nature. Simultaneously with information about negative effects (risks), medical geology defines positive effects of geological environment onto the human health (optimal content of microelements in natural environment, in other words, providing of meat food products by microelements, high - quality drinking water, etc.).

Soil and water are the key links connecting living and non - living nature, because they supply plants with the whole series of nutritious matters, and then, through nutrition chain, to animals and human beings. Petrological, mineralogical and chemical rock content, because of which soil was made through weathering processes, or in which water has moved and accumulated. In 1955, Tromp defined correlation between chemical content of soil and drinking water, on one side, and marker of cancer morbidity. According to the researcher, calcium, magnesium, potassium and sodium salts significantly decelerate cancer development, while silicium oxides are suitable for the development.

Mineral matters, as known, have a very significant role in organism of human being or animal (Ca, Mg, K, Fe, Se, etc.), in various forms of chemical compounds. Variations of the soil chemical content, natural water, vegetation and fauna causes changes in metabolism, but also morphological changes, favor occurrence of endemic diseases, straining natural selection, originating new taxonomic units. Although deviations in metabolism of certain organs have not been known enough, principal characteristics of some diseases or pathological states caused by the deviations are known. Following examples are interesting: 1) endemic goiter is known as a result of iodine shortage (as teeth caries is result of fluorite shortage); 2) "Keshan" disease (endemic cardiomiopatia) of children, in the part of China, caused by selenium shortage; 3) morphologically evident malformations in bones and teeth as a result of reducing or increasing calcium salt content.

Biological importance of numerous microelements is outstanding, because their deficit causes diseases of plants and animals, and shortage in drinking water produces serious pathological processes of animals and human beings (manganese shortage in soil causes a special disease of birds and young pigs; lack of cobalt in soil produces serious anemia of animals; copper deficit in grass of some biogeochemical provinces causes diseases of horned neat cattle; lamb and calf blidness in nickel - rich territory, etc.). It was laso concluded that, by adding matters in short supply into the food, symptoms of numerous diseases (A.P. Avtsin, 1972).

Methodology of Medico - Geological Exploration

During this century, as a great number of other sciences, medicine, biology and geology have very rapid and comprehensive development, and occurrence of numerous new disciplines is of multilateral significance. Besides, rapid technical progress after the 2nd World War, particularly development in the field of electronics and computers, resulted through appearance numerous modern methods of research work, from one side, and new capabilities of quantitative data processing, from the other. That is why medical geology is capable to make solid theoretical base and start with rapid development. First of all, it uses numerous methods of geoscience, developed to such degree that they can answer the major part of questions of this new geological (or medicinal) discipline: remote sensing, different field methods, methods of fundamental geology and geophysics, cartography, geochemistry, pedology, hydrogeology with hydrogeochemistry, engineering geology, nuclear methods, etc. (M. Komatina, 1984). Important part of answers is also expected from methods of biology, bioinorganic chemistry, bioclimatology, pathology, physiology, epidemiology, toxicology, sanitary hygiene, balneology, medical ecology, geomedical cartography and zoning, statistical methods and computing, application of mathematical modeling, medico - geological diagnostics and prognosys.

Today, a lot of spacious geological units in the world is known, classified into biogeochemical provinces. By more precise determining of lithological content and structure, systematical analysis of rocks, soil, water and crops, with making geochemical and geomedical maps and thematic zoning of the area, a good base for discovering important carriers of distinguished diseases, but also positive effects of some terrains can be formed, that is – categorization of the terrain according to living conditions. Method of correlation disease mapping, by which is possible to define not only areas of their existence, but also to give a view of relation between pathological processes and certain factors of the environment, is of special significance.

One of the primary goals of the research work is linking of the natural base of the interesting area and pathological conditions, so it can be understood that, simultaneously with geological, results of medical exploration at the area should be present.

FOURTH USA/CIS JOINT
CONFERENCE ON
ENVIRONMENTAL HYDROLOGY
AND HYDROGEOLOGY

San Francisco (California),
November 7 – 10, 1999.

Miomir Komatina

References

Abelson P.H., 1990: *Uncertainties about health effects of radon.* Science, Vol. 250, N° *4979*, 353 p.

Abernathy C.O., Calderon R. L., Chappell W. R., 1997: *Arsenic exposure and health effects.* Proceedings of the 2nd Internat. Conference »Arsenic exposure and health effects«, San Diego, California, London, Chapman and Hall, 429 p.

Ackerknecth E.H., 1965: *History and geography of the most important diseases.* Hafner, New York, London, 210 p.

Addington W.W., 1967: *The Ecology of Histoplasmosis.* Am. J. med. Sci. 253, 687 p.

Adriano D., 1986: *Trace Elements in the Terrestrial Environment.* Springer – Verlag, 533 p.

Agirović R., Radošević P., Knežević Lj., Petković S., 1995: *Selenium, zinc, copper and iron serum concentrations in patients whit ovarian cancers.* Conference on selenium. Serbian Ac. of Sci. and Arts, No. LXXVIII, Vol. 6, Beograd, 171 - 175.

Alekin O.A., 1970: *Osnovi gidrohimii.* Gidrometeoizdat, Leningrad, 444 p.

Aleksandrov G.V., 1990: *Gipsometričeskij analiz v metallogenii.* Nedra, Leningrad, 1 - 153.

Alexander M., 1977: *Introduction to Soil Microbiology.* John Wiley and Sons, New York.

Allen E.T., Day A.L., 1935: *Hot Springs of the Yellowstone National Park.* Carnegie Inst, riv. 466, Washington, 525 p.

Allison I.S., Palmer D.F., 1980: *Geology. The science of changing Earth.* Seventh edition. Mc Graw - Hill Bock Company. New York - Toronto, 568 p.

Alloway B.J.: *Heavy metals in soils.* Second edition. Blackie Academic & Professional, Glasgow.

Alpatov V.V., 1960: *K tehnike ustanovlenija svjazej meždu priznakami i javlenijami, varirajuščimi v prostranstve.* Počvovedenie No. 17, 108 - 109.

Altura B.M., Altura B.T., 1991: *Cardiovascular risk factors and magnesium: relationships to atherosclerosis, ischemic heart disease and hypertension.* Magnesium Trace Element, 10, 182 - 192.

Amidžić L., 1998: *Zaštita flore i vegetacije u Srbiji.* Zaštita prirode, br. 5, Beograd, 115 - 123.

Anderson J.G., 1979: *Estimating the seismicity from geological structure for seismic risk studies.* Bull. Seismol. Soc. Am, 58 p.

Andjelić M., 1990: *Geomorfologija.* Vojnogeografski institut, Beograd, 276 p.

Andjelković M., 1978: *Stratigrafija Jugoslavije. Paleozoik i mezozoik.* Univerzitet u Beogradu, Minerva, Subotica – Beograd, 1017 p.

Andjelković M., 1988: *Geologija Jugoslavije. Tektonika.* IRO »Građevinska knjiga«, Beograd, 692 p.

Anton J., 1990: *Metodika i tehnika istraživanja radona.* Zbornik radova jugoslovensko - čehoslovačkog simpozijuma ,«Geologija i zaštita životne sredine«, Beograd, 101 - 102.

Antonović A., 1978: *Ležišta urana i torijuma u svetu sa osvrtom na odgovarajuće tipove u Jugoslaviji.* Radovi Geoinstituta, knj. 12, Beograd, 153 - 190.

Antonović A., 1990: *Radioaktivnost u prirodi – značaj za istraživanja u geologiji.* Radovi Geoinstituta, posebno izdanje, knj. 12, Beograd, 191 p.

Appleton J.D. et al., 1998: *An evaluation of the relationships between environmental concentrations of Se and oesophageal cancer in the Lixian Area, Hebei Province, China.* British Geological Survey, Technical Report.

Applications of remote sensing in geology (Guest Eds F. Van der Mer, E.C. Kosters, S.J. Fraikin), 1999. Int. Journ. of the Royal Geol. and Mining Society of the Netherlands, Vol. 77, N° 2, Amsterdam, 201 p.

Armstead H., Christopher H., 1978: *Geothermal energy.* Halsted Press, London.

Armstrong R.M., 1962: *Cancer and soil, review and counsel.* »Profess. Geographer«, 14, 7 - 13.

Arnu M., 1984: *Teoretičeskie osnovi vzaimodejstvija čeloveka i geologičeskoj sredi.* 27 Meždunarodnij geologičeskij kongres, 17, Moskva, 3 - 7.

Arsenic in the environment. Part I: Cycling and characterization. (Ed. J.O. Nriagu), 1994. Adv. Environ Sci., 26, John Wyley & Sons, 430p.

Artemov J.M., Jaroshevskij A.A., 1965: *Geohimia.* N° 9, 1 - 1101.

Arthur J.R., Beckett G.T, 1994: *New metabolic roles for selenium.* Proc. Nutr. Soc, 53, 615 - 624.

Aspects of Geology in Austria (Eds. Neubauer F., Hock V.), 2000. Osterreichische Geologische Gesellschaft, Vienna, 335 p.

Atanacković B., 1982: *Pedološki pokrivač. SR Srbija.* Tom I, Beograd.

Aubert H., Pinta M., 1977: *Trace Elements in Soils.* Elsevier, Amsterdam.

Avtsin A.P., 1959: *O sootnošeni meždu kraevoj i geografičeskoj patologiej.* Arh. pat., 2 - 3.

Avtsin A.P., 1972: *Vvedenie v geografičeskuju patologiju.* Medicina, Moskva, 328 p.

Avtsin A.P., Arutyunov V.D., 1965: *K metodike ekspedicionih issledovanij po geografičeskoj patologii.* V. kn: Metodi mediko - geografičeskih issledovanij, Moskva, 48 - 52.

Avtsin A.P., Zhavoronkov A.A., 1981: Patologija fljuooroza. Nauka, Novosibirsk, 335 p.

Avtsin A.P., Zhavoronkov A.A., Ris M.A., Strochkova L.S., 1991: *Mikroelementozi čeloveka.* Medicina, Moskva, 496 p.

Avtsin A.P., Zhavoronkov A.A., Strochkova L.S., 1983: *Principi klassifikacii zabolevanij biogeohimičeskoj prirodi.* Arh. pat, N° 9, 3 - 11.

Babenko G.A., Reshetkina A.P., 1971. *Primenenie microelementov v medicine.* Kiev.

Babeva I.D., Zenova G.M., 1989: *Biologia počv.* Izdat Moskov univerz.

Babović M., 1991: *Geološki aspekti zaštite životne sredine i njihova uloga u*

planiranju prostora (Doktor.disertacija na Rudarsko - geološkom fakultetu), Beograd.
Babović M., 1992: *Geologija i zaštita životne sredine. Prilog zasnivanju ekogeologije.* Naučna knjiga, Beograd, 161 p.
Backman B. et al., 1999: *Geologian ja ihmisen toiminnan vaikutus pohjave teen.* Seurantatutkimuksen tuloksen vuasilta 1969 - 1996. *Geologian tutkimuskeskus,* Espoo 1999, 261 p.
Bailey R.G., 1998: *Ecoregions. The Ecosystem Geography of the Oceans and Continents.* Springer, 190 p.
Bain R.J., 1979: *Heart disease and geologic setting in Ohio.* Geology 7, 7 - 10.
Bakacs T., 1977: *Kornyezetvedelem.* Medicina, Budapest, 215 p.
Bakacs T., 1980: *Ohrana okružaješćoj sredi* (prevod sa engl.). Medicina, Moskva, 216 p.
Baker B.H., Mohr P.A., Williams L.A.J., 1972: *Geology of the Eaestrn Rift System of Africa.* Geol.Soc.Am.Spec.Pap., 136 p.
Balazs D., 1977: *The geographical distribution of karst areas.* 7th International speleological congress, Sheffield.
Ballo R. et al., 1996: *Mseleni Joint Disease – a molecular genetic approach to defining the aetiology.* S.A.Med.J., 86, 956 - 958.
Balneoklimatologija (Ed.T.Jovanović), 1997. Medicinski fak., Beograd, 216 p.
Barabanov V.F., 1987: *V.I. Vernadskij i razvitie estestvoznanija.* Bjulleten Komisii po razrabotke naučnogo nasledija akademika V. I. Vernadskogo, N° 1, Leningrad, 16 - 26.
Baranov V.I., Titaeva N.A., 1973: *Radiogeologija.* Moskov univerz., Moskva, 242 p.
Baras J., 1999: *Voda, značaj, resursi i vrste.* In: Voda, kvalitet i zdravlje (Eds. I. Gržetić, I. Brčeski). Mol d. d., Beograd, Tuzla, 7 - 36.
Barbier E., 1987: *Geothermal energy in the world energy scenario. Geothermics,* Vol. 15, N° 5/6, Pergamon J., Ltd, Oxford.
Barnet I., et al., 1992: *Radon in Geological Environment.* Czech Geol. Inst, Praha.
Barojan O.V. et al., 1965. *Programma i maket mediko - geografičeskogo atlasa Celinnogo kraja.* V.kn.: Metodi mediko –geografičeskih issledovanij, Moskva, 123 - 133.
Barret E. C., 1974: *Climatology from Sattellites.* Methuen, London, 418 p.
Barth T.E.W., 1950: *Volcanic geology, hot springs and geysers of Iceland.* Carnegie Inst., Washington, Publ, N° 587, 174 p.
Baskov E.A., 1977: *Mineraljnije vodi i paleogidrogeologija Sibirskoj platformi.* Trudi VSEGEI, novaja serija, tom 254, Nedra, Moskva, 148 p.
Baskov E.A., 1983: *Osnovi paleogidrogeologii rudnih mestoroždenij.* Nedra, Leningrad, 263 p.
Baskov E.A., Surikov S.N., 1975: *Gidrotermi Tihookeanskogo segmenta Zemli.* Nedra, Moskva, 172 p.
Baskov E.A., Surikov S.N., 1989: *Gidrotermi Zemli.* Nedra, Leningrad, 244 p.
Beck M.A., et al., 1994: *Benign human enterovirus becomes virulent is selenium – deficient mice.* J.Med.Virol., 43, 166 – 170.

Begon M., Harper J.L., Townsend C.R., 1986: *Ecology (Individuals, populations, communities).* Oxford: Blackwell Publ., 876 p.
Behne D., 1989: *Metabolism and biological function of selenium.* Third Intern. Congr. Trace Elem. Health Disease, Adana Turkey, Abstract Book, p. 13.
Beihler D., Eberhard Falck W., 1999: *Simulation on the effects of geochemical reactions on groundwater quality during planned flooding of the Koningstein uranium mine, Saxony, Germany.* Hydrogeology Journal, Vol.7, N°3, Springer, 284 – 293.
Belovalov A.V., 1971: *Mikroelementi (med, marganec, železo, kremnij) i processi kamneobrazovania vželčnih putjah čeloveka.* AV diss. K.M.N., Doneck.
Bencini A., Duchi V., Martini M., 1981: *Geochemistry of Thermal Springs of Tuscany (Italy).* Chemical Geology, Vol. 19, N° 3, 229 - 252.
Berman E., 1975: *Geothermal energy.* Park Ridge New Jersey, London.
Bernard C., Laverdiere M.R., 1992: *Spatial redistribution of Cs – 137 and soil erosion on Orleans Island, Quebec.* Can. J. Soil Sci. 72, 543 – 554.
Bernardi D. et al., 1995: *Sudden cardiac death rate in area characterized by high incidence of coronary artery disease and low hardness of drinking water.* Anilogy 46, 145 - 149.
Berner E.K., Berner R.A., 1987: *The global water cycle, geochemistry and environment.* Prentice Hall, Englewood Cliffs.
Bethke C.M., 1996: *Geochemical reaction modeling.* Oxford University Press, New York.
Beus A.A., 1981: *Geohimija litosferi.* Nedra, Moskva, 334 p.
Beus A.A., Grabovskaja A.I., Tihonova N.V., 1976: *Geohimija okružajuščej sredi.* Nedra, Moskva, 248 p.
Bilbija N., 1984: *Tehnička petrografija.* Naučna knjiga, Beograd, 239 p.
Billings M.P., 1972: *Structural Geology.* 3rd ed., Prentice - Hall, Inc, Englewood Cliffs, New York.
Binh H.V., Lap N.B., Thang T.T., 1992: *Iodine geochemistry and goitre - cretenism in Southeast Asia.* Proc. Regional Seminar on Environm. Geol, 11 - 13 Nov. Hanoi.
Biogeohimičeskoe rajonirovanie i geohimičeskaja ekologija. 1981. Trudi Biogeohim. laboratorii. Nauka, tom 19, Moskva, 204 p.
Biological and environmental effects of arsenic. Topics in environmental health (Ed. B.A. Fowler), 1983. Elsevier, Amsterdam, 6, 281 p.
Blackwell R.Q., 1961: *Estimated total arsenic ingested by residents in the endemic Blackfoot area.* J. Formosan Med.Assoc. 60, 1143 – 1144.
Boffetta P., 1977: *Cancer mortality among man – made vitreous fiber production workers.* Bolt B.A., et al., 1975: *Geologic Hazards.* Springer – Verlag, New York.
Bogdanov N.A., Gembickii E.V., 1975: *Proizvodstvenij fljuoroz.* Medicina, Leningrad, 96 p.
Bogdanović D., Ubavić M., Hadžić V., 1997: *Teški metali u zemljištu.* In: Teški metali u životnoj sredini (Ed. R. Kastori). Nauč.inst. za ratarstvo i povrtarstvo, Novi Sad.
Bolt B.A., et al., 1975: *Geologic Hazards.* Springer – Verlag, New York.
Bolt B.A., 1978: *Earthquakes: A Primer.* W. H. Freeman and Company, San Francisco.

Bondarenko S.S., Kulikov G.V., 1984: *Podzemnie promišlenie vodi.* Moskva, 358 p.

Borgono J.M., Venturino H., Vincent P., 1980: *Clinical and epidemiological study of arsenism in northern Chile.* Rev. Med. Chile 108, 1039 - 1048.

Bowen H.J.M, 1966: *Trace Elements in Biochemistry.* Academic Press, New York, 241 p.

Bowen H.J.M, 1979: *Environmental Chemistry of the Elements.* Academic Press, 333 p.

Bowen R. 1989: *Geothermal resources.* (Second Edition). Elsevier Applied Science, London, New York, 485 p.

Bowie S.H.U., 1970: *World uranium deposits.* IAEA Panel on uran. explor.geology, Wienna.

Brett R., 1971: *The Earth's core: Speculations of its chemical equilibrium with mantle.* Geochim. Cosmochim. Acta 35, 203 - 221.

British Geological Survey, 1997: *Regional Geochemistry of Parts of North – west England and North – west Wales.* BGS, Keyworth, Nottingham, 128 p.

Brnović R., 1972: *Stroncijum - 90 u životnoj sredini čoveka.* (Magistarska teza), Univerzitet u Beogradu.

Brodine V., Commoner B., 1975: *Radioactive contamination.* Edit. Scientist In. Pu. In. Harcourt Brace Jovanevich, Inc., New York, Chicago

Bugarčić Z., Novokmet S., 1999: *Uloga mikroelementa selena u organizmu.* Ecologica, N°21, god. VI, br.1., Beograd.

Bull A.T., Holt D., Lilly M.D., 1982: *Biotechnology, Trend and Development.*

Bullard F.M., 1962: *Volcanoes: In History, in Theory, in Eruption.* Univ. of Texas Press, Austin.

Buntebarth G., 1980: *Geothermie.* Springer - Verlag, Berlin.

Burić M., 1993: *Termomineralne vode Crne Gore.* Podgorica, 72 p.

Bursać M., 1997: *Korišćenje prostora i regionalizacija Srbije.* In: Geografska struktura i regionalizacija Srbije I, Geograf. institut »Jovan Cvijić«, posebna izdanja, knj. 51, Beograd, 31 - 51.

Butt C.R.M., 1987: *A basis for geochemical exploration models for tropical terrains.* In: Proccedings of an Intern. Seminar on laterites. Chem. Geol, Vol. 60, 5 - 16.

Canter K.J. et al., 1991: *Landscape - ecological mapping of the Netherlands.* Landscape Ecology, 53, 145 - 162.

Carić N., 1989: *Razvoj geografske nauke.* Zbornik radova PMF, Univerz. u Novom Sadu, knj. 19, Novi Sad.

Caroll R.E., 1966: *The relationship of cadmium in the air to cardiovascular disease death rates.* J. Amer. med. Ass., 167, 198.

Castany G., 1967: *Traite pratique des eaux souterraines.* Paris, 661 p.

Castro J. de, 1954: *Geografija goloda* (prevod sa ang.). Moskva.

Cauwet G., 1987: *Influence of sedimentological features on the distribution of trace metals in marine sediments.* Mar. Chem. 22, 221 - 234.

Chaklin A.V., 1977: *Medicinskaja geografia.* Znanie, Moskva, 127 p.

Chaklin A.V., 1984: *O proekte prognoza razvitija problemi „Epidemiologija zlokačestvennih opuholej" do 2000 goda (Prognozirovanie zabolevaemosti zlokačestvennimi opuholjami)*, Moskva, 4 - 10.
Chaklin A.V., 1986: *Geografija zdorovja.* Znanie, Moskva, 151 p.
Chapelle F.H., 1993: *Ground - water microbiology and geochemistry.* Wiley, New York, 405 p.
Chapelle F.H., 2000: *The significance of microbial processes in hydrogeology and geochemistry.* Hydrogeology Journal, Vol. 8, N° 1, Springer, 41 - 46.
Chapellier D., 1996: *Situation and perspectives of environmental and engineering geophysics in Europe.* Europ. Journal of Environmental and Engineering Geophysics, 1, 5 - 14.
Chapman D.S., Pollack H.N., Čermak V., 1982: *Globaljnij teplovoj potok i teplovoe pole Evropi.* V. kn.: Teplovoe pole Evropi, «Mir«, Moskva, 54 - 64.
Chen C.J., Kuo T.L., Wu M.M., 1988: *Arsenic and cancers.* The Lancet 1 (8575/6, 414 - 415).
Chen X., et al., 1980: *Studies on the relation of selenium and Keshan disease.* Biol. Trace Elem. Res. 2, 91 - 107.
Chervyakov V.A., 1964: *Osobenosti opredelenija korreljacionnih svjazej po geografičeskim kartam.* Kand. disertacija. MGU, Geograf. fak., Moskva.
Chu - Ilijskij rudnij pojas. Gidrogeologija i geomorfologija, 1979. Nauka. Alma - Ata, 192 p.
Ciozewski D., 1997: *Source of pollution as a factor controlling distribution of heavy metals in bottom sediments of Chechto River (south Poland).* Environmental Geology. 29/1/2 Springer, 50 - 57.
Clark I.D., Fritz P., 1997: *Environmental isotopes in hydrogeology.* Lewis Publishers, New York.
Collins A.G., 1975: *Geochemistry of oilfield waters.* Elsevier, Amsterdam, 466 p.
Congres XV International d'Hydrologie, de Climatologie et de Geologie Medicales (Belgrade, 29 septembre - 2 octobre 1936). Imprimerie Privredni pregled, Beograd, 1937, 355 p.
Connell D.W., 1997: *Basic Concepts of Environmental Chemistry.* Springer, 560 p.
Coughtrey P.J., Thorone M.C., 1983: *Radionuclide distribution and transport in terrestrial and aquatic ecosystems.* A. A. Balkema, Rotterdam.
Cox A., Hart R.B., 1986: *Plate tectonics.* Palo Alto: Blackwell Scientific, 392 p.
Cox P.A., 1995: *The Elements on Earth.* Oxford University Press, Oxford, 287 p.
Crounce R.G., et al., 1983: *Geochemistry and Man: Health and Disease I. Essential Elements.* In: Applied Environmental Geochemistry. (ed. Thornton I.). Acad. Press, Geol. Ser., London, 267 - 308.
Cvijić J., 1926: *Geomorfologija.* Držav. štampar., Beograd, 506 p.

Čalić - Perišić N. et al., 1981: *Rezultati ispitivanja dece iz porodica sa endemskom nefropatijom.* Zbornik radova I kongr. nefrologa Jugoslavije, Beograd, 129 - 133.
Čebedžić R., 1999: *Mikrobiologija voda.* In: Voda, kvalitet i zdravlje. Mol d.d., Beograd, Tuzla, 51 - 60.

Ćirić B., 1990: *Odlaganje radioaktivnog i drugog opasnog otpada u podzemlju.* Zbornik radova jugosl - čehosl. simpoz. »Geologija i zaštita životne sredine«, Beograd, 239 - 241.
Ćirić B., 1996: *Geologija Srbije – građa i razvoj zemljine kore.* Geokarta, Beograd, 273 p.
Ćirić M., Filipovski Dj., 1972: *Regionalnie počveno - geografičeskie zakonomernosti v Jugoslavii.* Počvovedenie, № 2, Moskva.
Ćurčić B.P.M. et al., 1998: *Biodiverzitet Jugoslavije: Nastanak i istorijski razvoj, sa osvrtom na globalna shvatanja o njegovoj promenljivosti i vrednosti.* Ekološka istina (zbornik radova), 15 - 22, Negotin.

Dalj M.K., 1964.: *Rasprostranennost raka želudka b USSR i vozmožnaja častota ego v svjazi s gidrogeologičeskimi osobennostjami različitih oblastej Ukraini.* In: Problemi geografičeskoj patologii. Moskva, 54p.
Dangić A., 1992: *The environmental geochemistry and health: impact of ore deposits and ore usage.* Exploration of mineral row materials and environmental protection »Igalo 92«, SIT RGMJ, Belgrade, 19 - 23.
Dangić A., Maksimović Z., Mićić I., 1998: *Geohemijske karakteristike granitoidnih stena Kopaonika.* In: Geologija i metalogenija Kopaonika (Zbornik savetovanja). Republ. društv. fond za geol. istr. Srbije, Beograd.
Daniel I., 1932: *Nouvelle observation concernant le traitement du diabete par les eaux minerales sulfureuses.* Revista de hidrologie, No.3.
Danilova N.A., 1971: *Priroda i naše zdorovje.* Mislj, Moskva, 221 p.
Daoxian Y., 1992: *Karst and Karst Water in China.* In: Hydrogeology of Selected Karst Regions (Ed. H. Paloc, W. Back). IAH, Vol. 13, Verlag Heinz Heise, Hannover.
Dark B.P., 1970: *Notes on some aspects of pollution.* Med. J. Austr. 57/11, 595 p.
Darnley A.G., et al., 1995: *A Global Geochemical Database.* Earth Sciences 19, UNESCO Publishing, Paris, 122 p.
Das D., et al., 1994: *Arsenic in ground water in six districts of West Bengal. The biggest arsenic calamity in the world.* The Analyst, Royal Society of Chemistry; UK, Vol.119, No.12.
Das D., et al., 1996: *Arsenic in groundwater in six districts of West Bengal, India.* Environ. Geochem. Health 18, 5 - 16.
Davidovskij I.V., 1969: *Obšćaja patologija čeloveka.* Moskva.
Davies - Brian E., 1983: *Heavy metal contamination from base metal mining and smelting: Implications of man and his environment.* Applied Environm. Geochemistry, Academic Press Geology Series, London, 425 - 459.
Davis S.N., De Wiest R.J.M., 1966: *Hydrogeology.* John Wiley and Sons, Inc., New York, London – Sidney, 463 p.
Dayan A.D., 1993: *Carcinogenicity and Drinking Water.* Pharmacology & Toxicology 72 (1), 108 - 115.
Deak J., 1979: *Environmental isotopes and water chemical studies for groundwater research in Hungary.* In: Isotope Hydrology. Proc. Symp. IAEA, Vol. 1, 221 - 247.

Decker R., Decker B., 1981: *Volcanoes.* Foreeman and Co., San Francisco.

Deming D., 1992: *Catastrophic release of heat and fluid flow in the continental crust.* Geology, 20, 83 - 86.

Deutsch W.J., 1997: *Groundwater Geochemistry. Fundamentals and Applications to Contamination.* Springer, 224 p.

Dhar R.K., et al., 1997: *Groundwater arsenic calamity in Bangladesh.* Current Science, Vol. 73, No1, 48 - 59.

Di Lollo G.C., 1969: *Compendio di idrologia medica.* Cappelli Editore.

Dikun P.P., 1966: *Kancerogennie agenti v srede obitanija čeloveka.* In: Vvedenie v geogigienu. Moskva, Leningrad, 283 p.

Dimitrijević C.N., 1988: *Hidrohemija.* Univ. u Beogradu, Rud - geol. Fak., Beograd, 312 p.

Dimitrijević M.D., 1995: *Geologija Jugoslavije.* Geoinstitut, Beograd, 205 p.

Dimitrijević N., 1973: *Dosadašnja iskustva na proučavanju plemenitih gasova u vodama.* Zapisnici Srpskog geol. društva za 1972. god, Beograd, 57 - 64.

Dinić J., 1997: *Prirodni potencijal Srbije – Ekonomsko - geografska analiza i ocena.* Univerz. u Beogradu, Ekonom. fak. Beograd, 223 p.

Diplock A.T., 1981: *Metabolic and functional defects of selenium deficiency.* Phil. Trans. R. Soc. Lond. B. 294, 105 - 117.

Dissanayake C.B., Chandrajith R., 1999: *Medical geochemistry of tropical environments.* Earth - Science Reviews 47, Elsevier, 219 - 258.

Dissanayake C.B., Weerasooriya S.V.R., 1987: *Medical geochemistry of nitrates and human cancer in Sri Lanka.* Int. J. Environ. Stud. 30, 145 - 156.

Djerković B., 1971: *Geološki i hidrogeološki odnosi područja srednje Bosne.* Inst. za geol. istr. Sarajevo, poseb. izdanja Geol. glasnika, knj. X, Sarajevo, 170 p.

Djerković B., 1986: *Geneza i hemizam visokomineralnih voda svijeta.* In: Zbornik radova posvećen akademiku Zariji Bešiću. Crnog. Akad. nauka i umjetnosti, Titograd, 219 - 230.

Djordjević – Milošević S. i dr., 1998: *Proizvodnja visokovredne hrane kao mogući faktor održanja interesa savremene poljoprivrede za očuvanje biodiverziteta prirodnih travnjaka.* Zaštita prirode, br. 5, Beograd, 83 - 89.

Djordjević M. i dr., 1992: *Periodical and geographical variations of malignant diseases in Serbia.* Instit. of Oncology and Radiology - Beograd, ICN Galenika, Beograd, 176 p.

Djordjević M. i dr., 1998: *Epidemiological and time variations of cardiovascular diseases in Serbia.* KBC »Bežanijska kosa«, Belgrade.

Djordjević V., Djordjević P., Milovanović D., 1991: *Osnovi petrologije.* Nauka, Beograd, 223 p.

Djordjević Z., 1980: *Medicinska geografija malignih tumora mokraćne bešike u Srbiji.* Magistarski rad, Medic. fakultet, Beograd

Djukanović M., 1995: *Epidemiologija životne sredine: mesta rizika i uticaj na zdravlje stanovnika.* Ekosistem i zdravlje. Zbornik, G. 333, Beograd.

Djukanović M., 1998: *Održivi razvoj raspoloživih vodnih potencijala: ekološki pristup u upravljanju i korišćenju za 21. vek.* In: Vodni resursi sliva Vel. Morave i njihovo korišćenje. Zadužbina Andrejević, 376-381.

Dobrovol'skii V.V., 1983: *Geografija mikroelementov. Globaljnoe rassejanie.* Mislj, Moskva, 160 p.

Dokmanović P., 1999: *Hidrogeologija tercijarnih basena Srbije.* Zadužbina Andrejević, Beograd, 131 p.

Dokuchaev V.V., 1899: *Mesto i rolj sovremennogo počvovedenija v nauke i žizni.* SPb, 190 p.

Dokuchaev V.V., 1948: *Učenie o zonah prirodi.* Moskva.

Dolinić N., 1996: *Uticaj vode na zdravlje.* Ecologica. Posebno izdanje, br. 3, Beograd, 230 - 234.

Dolterov N.I., 1939.: *Urovskaja Kašin - Beka bolezn.* Moskva, Leningrad.

Donasy V., Olujić M., Tomašegović Z., 1983: *Daljinska istraživanja u geoznanostima.* Zagreb, 333 p.

Draganić J., 1962: *Radioaktivni izotopi i zračenja.* Knj. 1. Naučna knjiga, Beograd.

Dragišić V., 1992: *Hidrogeologija ležišta bakra istočne Srbije.* Rud - geol. fakultet, Instit. za hidrogeol., Beograd, 211 p.

Dragišić V., 1997: *Opšta hidrogeologija.* Univerz. u Beogradu, Rudarsko - geol. fak., 434 p.

Dragutinović S., 2000: *Konzervacija zidnih slika na arheološkim nalazištima.* Univ. Umetn., Beograd, 142 p.

Driscoll C.T., Schecher W.D., 1990: *The chemistry of alluminium in the environment.* Environmental Chemistry and Health 12, 28 - 50.

Driver I.J., 1982: *The geochemistry of natural waters.* Prentice – Hall, Englewood Cliffs, New York.

Drumya A.V., Shebalin N.V., 1985: *Zemletrjasenie: gde, kogda, počemu?* Štiinca, Kišinev.

Dukić D., 1984: *Hidrologija kopna.* Naučna knjiga, Beograd, 498 p.

Durlach J., Bara M., Dier - Bara A., 1989: *Magnesium level in drinking water: its importance in cardiovascular risk.* In: Magnesium in health and disease (Eds. Y. Itokawa, J. Durlah), London, Paris: John Libbey, 173 - 182.

Durrance E.M., 1986: *Radioactivity in geology (principles and applications).* Ellis Horwood Ltd. Halsted Press (John Wiley & Sons), 441 p.

Dworkin B.M., 1994: Selenium deficiency in HIV infection and the acquired immuno - deficiency syndrome (AIDS). Chemobiological interactions 91 (2 - 3), 181 - 186.

Eckert R., 1986: *Animal physiology.* Third edition. Freeman and comp., USA.

Ecologija i geohimičeskaja dejateljnost mikroorganizmov (Ed. V. V. Ivanov), 1976. Puščino, 179 p.

Ecotoxicology. A. Hieararchical Treatment (Eds. M.C. Newman, C.H. Jagoe). Springer, 448 p.

Eiby G.A., 1980: *Earthquakes.* Heinemann, 260 p.

Eisenbud M., 1987: *Environmental Radioactivity.* Academic Press, New York.

Elkin I.I., Jashkulj V.K., 1964: *Osnovnie problemi epidemiologičeskoj geografii. 2. Ponjatie o nozoarealah.* Žurn. mikrobiol. epidemiol. i immunologii, № 11, 48 - 54.

Emelyanova E.P., 1972: *Osnovnije zakonomernosti opolznevih procesov.* Moskva.

Environment and health: themes in medical geography, 1991. (Ed. by R. Akhtar). New Delhi. Ashish Publishing House, 649 p.

Environmental Geochemistry and Health (Eds. S.H.U. Bowie, I. Thornton), 1985. Reidel Publishing Dodrecht, The Netherlands, 140 p.

Environmental interactions of Clays. Clays and the Environment (Eds. A. Parker, J.E. Rae), 1998. Springer, 210 p.

Environmental medicine, 1976. (Ed. by G.M. Howe and J.J. Loraine – L. Heinemann), 271 p.

Environmental Problems in Medicine (Ed. Mc Kee W.D.). Charles C. Tomas, Springfield, III.

Epstein S.G., 1988: *Alluminium intake and its effects.* In: Thornton I (Ed.), Geochemistry and Health. Science Reviews, 189 - 201.

Ermakov V.V., 1974: *Biologičeskoe značenie selena.* Nauka, Moskva, 296 p.

Feder G.L., Radovanović Z., Finkelman R.B., 1991: *Relationship between weathered coal deposits and the etiology of Balkan endemic nephropathy.* Kidney international, Vol. 40, Suppl. 34, 59 - 511.

Feldman E.S., 1968: *O vozmožnoj zavisimosti neravnomernogo rasprostranenija raka želudka v Moldavii ot urovnja obespečennosti landšaftov makro i mikroelementami i, v časnosti, magniem.* Materiali nauč. konfer. po itogam isledovatelj. rabot za 1967. g, Kišinev.

Feldman E.S., 1977: *Mediko - geografičeskoe issledovanie territorii Moldavii.* Štiinca, Kišinjev, 169 p.

Fergusson J.E., 1990: *The heavy elements: chemistry, environmental impact and health effects.* Pergamon Press, New York, 614 p.

Fersman A.E., 1922: *Geohimia Rosii.* Pg.

Fersman A.E., 1931: *Geohimičeskie problemi Sojuza. Očerk pervij. Osnovnie čerti geohimii Sojuza.* Moskva.

Fersman A.E., 1953 - 1959: *Izbrannie trudi.* T. 1 - 5. Moskva.

Fichier des Sources d'Eaux Minerales Francaises. (Ed. H. Denian), 1983. Annales des Mines, an. 190, N° 8 - 9, 204 p.

Filipović B., Dimitrijević N., 1991: *Mineralne vode.* Inst. za hidrogeol., RGF, Beograd, 274 p.

Filipović B., Vujasinović S., 1982: *Zaštita podzemnih voda.* Grupa za hidrogeol. RGF, Beograd, 358 p.

Filipovski Dj., 1974: *Pedologija.* Skopje.

Flint R.F., 1971: *Glacial and Quaternary Geology.* John Wiley & Sons, Inc, New York

Ford D., Willams P., 1989: *Karst Geomorphology and Hydrology.* Unwin Hyman, London.

Forman D., 1989: *Are nitrates a significant risk factor in human cancer?* Cancer Surv. 8, 445 - 458.

Forstner U., Wittmann G.T.W., 1981: *Metal pollution in the aquatic environment,* Springer – Verlag.

Fraser G.B., Williams D.D., Howard K.W.F., 1996: *Monitoring biotic and abiotic processes across the hyporheic/groundwater interface.* Hydrogeology Journal, Vol. 4, N° 2, Verlag Heinz Heise, Hannover, 36 - 50.

Frausto da Silva J.J.R., Willams R.J.P., 1993: *The Biological Chemistry of the Elements*. Clarendon Press, Oxford, 561 p.

Freeze R.A., Cherry J.A., 1979: *Groundwater*. Prentice Hall Inc., p. 604.

Fricke K., Michel G., 1974: *Mineral – und thermalwasser der Bundesrepublik Deutschland*. Der Mineral brunnen, 3, Bad Godesberg.

Fried J.J., 1975: *Groundwater Pollution*. Amsterdam – Oxford – New York, 304 p.

Frolov N. M., 1976: Gidrogeotermija. Izd. vtoroe, Nedra, Moskva, 280 p.

Fuge R., Johnson C.C., 1986: *The geochemistry of iodine – a review*. Environment Geochemistry and Health 8, 31 - 54.

Gaciri S.J., Davies T.C., 1993: *The occurrence and geochemistry of fluoride in some natural waters of Kenya*. Journal of Hydrology, Vol.143, 395 - 412

Galeev M. A., 1968: *Soderžanie nekotorih mikroelementov v biljnih želčnokamennoj boleznju*. UFA.

Ganrot P. O., 1986: *Metabolism and possible health effects of alluminium*. Environ. Health Perspect. 65, 363 - 441.

Garlick J.P., Keay R.W.J., 1970: *Human Ecology in the Tropics*. Symposia of the Society for the Study of Human Biology, Vol. 9, Pergamon Press.

Garven G., et al., 1999: *Hydrogeologic modeling of the genesis of carbonate - hosted lead - zinc ores*. Hydrogeology Journal, Vol. 7, N° 1 , Springer, 108 - 126.

Gavrilović D., 1978: *Regionalna geomorfologija*. Prirod. - matem. fakultet, Beograd, 258 p.

Geographical aspects of health (Ed. by N.D. Mc Glashan and J.R. Blunden), 1983, Academic Press, London, New York, 392 p.

Geological hazards (B. Bolt., W.Z. Horn, G.A. Macdonald, G.F. Scott). Springer Verlag, Berlin, Heidelberg, New York.

Geologie et Confinement des Dechets Toxiques (Red/Eds. M. Arnold, M. Barres, B.Come), 1993. Geoconfine 93/Montpellier/France, A.A. Balkema/Rotterdam/Brookfield/1993, Vol. 1, 2, 770 p.

Geologija Srbije (uEd. K. Petković), 1975. Rudarsko - geol. fakultet, Univerzitet u Beogradu, knjige I - VIII.

Geomedical aspects in present and future research, 1980 (Ed. by J. Lag). Oslo etc.: Univ. forl, 226 p.

Geomedizin in Forschung und Lehre: Beitrage zur Geookologie des Menschen (Hrsg. von H. J. Jusatz), 1979: Steiner, Wiesbaden, 122 p.

Geomorfološka karta Jugoslavije, 1: 500 000 (Ed. M. Zeremski), 1992. Savez geografskih društava Jugoslavije, Beograd.

Georgievskii B.U., Anneenkov N.B., Samohin I.B., 1979: *Mineralnoe pitanie životnih*. Kolos, Moskva.

Geothermal Energy. Resources, Production, Stimulation (Ed. by P. Kruger, C. Otte), 1973. Stanford University Press, Stanford, California, 360 p.

Gerasimov I.P. i dr., 1980: *Sovremennie trendi razvitija sovetskoj fizičeskoj geografii*. In: Sovremennie tendencii i metodi v geografii (Materiali sovetsko - britanskogo simpoziuma). Progres, Moskva, 14 - 21

Gerasimov I.P., Kotl'akov V.M., 1983: *Monitoring okružajuščej sredi i rolj v nem aerokosmičeskih metodov. V. kn: Sovremenaja problematikadistancionih issledovanij geosistem.* Akad. Nauk. SSSR. Institut geografii An SSSR, Moskva, 3 - 8.

Germanov A.I., Panteleev V.M., Shvec V.M., 1975: *Genetičeskie svjazi organičeskogo veščestva i mikrokomponentov podzemnih vod.* Nedra, Moskva, 136 p.

Gertik S., 1978: *Metalogenetsko - geološke karakteristike ležišta urana kod Kalne.* Zbornik radova Rud - geol. fakulteta u Boru, Vol. XX.

Gertik S., Ilić B., Kovačević J., 1990: *Uranske mineralizacije istočne Srbije.* XII Kongres na geolozi na Jugosl., knj: III, Ohrid, 589 - 598

Ghassemi F.A., Jakeman A.J., Nix H.A., 1995: *Salinization of land and water resources.* University of New South Wales Press.

Gibson S., 1992: *Effects of fluoride on immune system function.* Complimentary Med. Res. 63, 11 - 113.

Gidrogeologičeskie issledovanija za čivezom (Ed. N.A. Marinov), 1982. Nedra, Moskva, 426 p.

Gidrogeologičeskije i inženernogeologičeskie uslovija teritorii gorodov. Metodi izučenija i prognoza izmerenij (redaktori A.A. Konopljancev i G.L. Koff), 1989. A. N. SSSR., Nauka, Moskva, 117 p.

Gidrogeologija Afriki (Ed. N.A. Marinov), 1978. Nedra, Moskva, 372 p.

Gidrogeologija Azii (Ed. N. A. Marinov), 1974. Nedra, Moskva, 575 p.

Gil'arov A.M., 1990: *Populjacionnaja ekologija.* Izdat. Moskov. univers, Moskva, 191 p.

Gillmore G.K. et al., 2000: *Hazards, Geology and Exposure of Cave Users: A Case Study and Some Theoretical Perspectives.* Ecotoxicology and Environmental Safety, 46 (3), 279 - 288.

Gimpeljson E. I., 1956: *Kamni poček i močetočnikov.* Moskva.

Gippokrat, 1936: *O vozduhah, vodah i mestnostjah.* In: Gippokrat. Izbranie knjigi, Moskva, 275 p.

Glasse R., 1967: *Cannibalism in the Kuru Region of New Guinea.* Trans. N. Y. Acad. Sci, 29, 748 p.

Gleeson M. J., 1990: *Effect of unprocessed wheat bran on calciuria and oxaluria in patients with urolithiasis.* Urology, 35:3, 231 - 234.

Gleich P.H., 1995: *Global Water Resources in the 21st Century: Where we go and how should we get there?* Proc. Stocholm Water Symposium, 73 - 78.

Global epidemiology. A geography of disease and sanitation (Ed. by J.S. Simmons et al.), London etc. , Lippincott, 1944. Vol. 1, 504 p; 1951, Vol. 2, 652 p; 1954, Vol. 3, 357 p.

Global Geocancerology, 1986 (Ed. by Howe – Edinburgh etc.) Churchill Livingstone, 350 p. Gruško J.M., Lazarev N.V., 1966: *Čelovečestvo kak geologičeskij faktor.* In: Vvedenie v geogigienu. Moskva – Leningrad. p. 14.

Goguel J., 1975: *La geothermie.* DOIN, editeurs, Paris, 170 p.

Goldman M., Dacre J.C., 1991: *Inorganic arsenic compounds: are they carcinogenic, mutagenic, teratogenic?* Environ. Geochem. Health 13, 179 - 191.

Goldschmidt V. M., 1954: *Geochemistry.* Oxford and the Clarendon Press, Oxford.

Goleva G. A., 1968: *Gidrogeohimičeskie poiski skritogo orudnenija.* Nedra,

Moskva, 292 p.
Goodall B., 1987: *Dictionary of Human Geography*. Penquion Books, London.
Goodhart R., Shils M., 1973: *Modern Nutrition in Health and Disease*. Lea Febiger, Philadelphia.
Gordanić V., Jovanović G., Žunić Z., 1998: *Radionuklidi u građevinskom kamenu sa osvrtom na nivo koncentracije radona u stambenim prostorijama*. 13. kongres geologa Jugoslavije. knj. IV (Mineralne sirovine), Herceg Novi, 328 - 333.
Gordeev P.V., Shemelina V.A., Shulyakova O.K., 1990: *Gidrogeologija*. Vissšaja škola, Moskva, 448 p.
Gosar M., Šajn R., Biester H., 2002: *Zvrsti živega srebra u theh in podstrešnem prahu na Idrijskem*. Geologija, knj. 45/2, Geološki Zavod Slovenije, Ljubljana, 373 - 378.
Gostuški R., 1970: *Lečenje lekovitim biljem*. Sarajevo.
Govedarica M., Jarak M., 1995: *Mikrobiologija*. Univerz. u Novom Sadu, Poljoprivredni fakultet, Novi Sad.
Govedarica M., Jarak M., Milošević N., 1993: *Mikrobiološke karakteristike zemljišta Vojvodine*. In: Teški metali i pesticidi u zemljištu.
Govedarica M., Milošević N., Jarak M., 1997: *Teški metali i mikroorganizmi u zemljištu*. In: Teški metali u životnoj sredini (Ed. R. Kastori). Instit. za ratarstvo i povrtarstvo, Novi Sad.
Grachev A.F., 1977: *Riftovie zoni Zemli*. Nedra, Leningrad, 247 p.
Grezhdev A.F., Belovalova A.V., Soroka V.R., 1970: *Mikroelementi i procesi kamneobrazovanij v želčnih putjah čeloveka*. Klinič. girurgija, N° 10. Moskva.
Grigor'ev A.A., 1936: *O himičeskoj geografii*. In: Akademiku V.K. Vernadskomu k pjatidesjatiletiju naučnoj i pedagogičeskoj dejateljnosti. Moskva, 1231 - 1236.
Groundwater in water resources planning, 1983. In three volumes, UNESCO, IAH, Nat. Comit., of the Fed. Rep. Germany, Koblenz.
Gruntorad J., Mazach O., 1995: *Impact of Subtle Dynamic Geofactors on Environment*. Acta Universitatis Carolinae, Environmentalica 8, 3 - 53.
Gruško J.M., Lazarev N.V., 1966: *Čelovečestvo kak geologičeskij faktor*. In: Vvedenie v geogigienu. Moskva – Leningrad. 14 p.
Gržetić I., 1999: *Deponije otpada i zagađivanje voda*. In: Voda, kvalitet i zdravlje. Mol d. d., Beograd, Tuzla, 71 - 93.
Gržetić I., Jelenković R., 1994: *Uranska mineralizacija Stare planine (Kalna) i procesi kontaminacije*. Ecologica, god. I, br. 3, Beograd, 14 - 23.
Gržetić I., Jelenković R., 1995: *Prirodni radioaktivni elementi, geološko poreklo i oblici pojavljivanja i migracije*. In: Jonizujuće zračenje iz prirode (Ed. M. Kovačević et al.). Jugoslov. društvo za zaštitu od zračenja, Vinča – Beograd, 3 - 41.
Gudzenko V.V., Dubinchuk V.T., 1987: *Izotopi radija i radon v prirodnih vodah*. Nauka, Moskva.

Hails J., 1977: *Applied Geomorphology*. Elsevier.
Harrington C.R., et al., 1994: *Alzheimer's disease – like changes in tan protein*

processing: association with alluminium accumulation in brains of renal dialysis patients. Lancet 343, 993 - 997.
Harvey R., et al., 1996: *A review of the geochemical factors linked to podoconiosis.* In: Appleton J.D., Fuge R., Mc Call G.J.H. (Eds.), Environmental Geochemistry and Health, Geological Society Special Publication, UK, N° 113, 255 - 260.
Heancy R.P., 1992: *Calcium in the prevention and treatment of osteoporosis.* J. Intern. Med. 231, 169 - 180.
Heavy Metals in Soil, 1995. Second edition (ed. by B.J. Aloway). Blackie Academie & Professional, London, Madras, 282 p.
Heavy Metals. Problems and Solutions (Eds. W. Salomons, U. Forstner, P. Mader), 1995. Environmental Science, Springer, 414 p.
Heimann H., 1961: *Effects of air pollution on human health.* In: Air pollution. WHO Monograph Series N° 46, Geneva, 159 - 220.
Heinrichs G., Udluft P., 1999: *Natural arsenic in Triassic rocks: A source of drinking - water contamination in Bavaria, Germany.* Hydrogeology Journal, Vol. 7, N° 5, Springer, 468 - 476.
Heiskanen W.A., Vening Meinesz F.A., 1958: *The Earth and its Gravity Field.* Mc Graw - Hill book, New York, Toronto, London.
Hem J.D., 1985: *Study and interpretation of the chemical characteristics of natural water.* P. S. Geological Survey, Water - Supply Paper 2254, 263 p.
Hirsch A., 1881 - 1886: *Handbuch der historich - geographischen Pathologie.* Stuttgart.
Hlebovich I.A., 1972: *Mediko - geografičeskaja ocenka prirodnih kompleksov.* Nauka, Leningrad, 123 p.
Hobday D.K., Galloway W.E., 1999: *Groundwater processes and sedimentary uranium deposits.* Hydrogeology Journal, Vol. 7, N° 1. Springer, 127 - 138.
Holmgren G.G.S., et al., 1993: *Cadmium, Lead, Zinc, Copper and Nickel in Agricultural Soils of the United States of America.* Journal of Environm. Quality, Vol. 22, Iss. 2, 335 - 348.
Holting B., 1984: *Hydrogeologie,* Stutgart, 370 p.
Hotchin J., 1966: *Kuru as a persisting tollerated infection.* Lancet, 2., 28 p.
Hove K., Ekern A., 1988: *Combating radiocesium contamination in farm animals.* In: Lac, J. ed. Health problems in connection with radiation from radioactive matter in fertilizers and rocks. Oslo, Norwegian University Press, 139 - 153.
Howard A.D., Remson I., 1978: *Geology in environmental planning.* McGraw - Hill Book Company, New York, 583 p.
Howe G.M., 1970: *A national atlas of disease mortality in the United Kingdom.* Revised edition, Nelson, London.
Howe G.M., 1980: *Statistical and Cartographic Methods of Analyzing Spatial Patterns of Mortality.* University of Strathclyde, Glasgow.
Huang W. T., 1967: *Petrology.* Mc Graw - Hill Book Co., New York, 411 p.
Hueper W.C., 1959: *Carcinogenic studies on water - soluble and insoluble macromolecules.* Arch. Path. 67, 589 p.
Huges J., 1992: *Diet and calcium stones.* Can. Med. Associat. J. 146:2, 137 - 143.
Hughes M.N., 1973: *Inorganic Chemistry of biological processes.* John Wiley.
Human Bioclimatology (Ed. A. Auliciems), 1997. Environm. Sciences and

Environmental Engineering. Advances in Bioclimatology, Vol. 5, Springer, 184 p.

Humans as Components of Ecosystems (Eds. M.J. McDonnell, S.T.A. Pickett), 2nd printing 1997. Environm. Sciences and Environmental Engineering. Springer, 364 p.

Hun J.D., Chery R.N., 1974: *Mannual of Remote Sensing.* American Society of Fotogrametry, Falls Church, 526 p.

Hutchinson J., 1887: *Arsenic cancer.* British Medical J. 2, 1280 - 1281.

Hutt M.S. R, Burkitt D.P., 1986: *The geography of non - infections disease.* Oxford etc.: Oxford Univ. Press, 164 p.

Hydrogeology and the Environment (Editor: Yanxin Wang), 2000. China Environmental Science Press, Beijing, 658 p.

Hydrogeology of great sedimentary basins (Editor: A. Ronai), 1978, IAH, Memoires, Vol. XI, Budapest, 829 p.

Hydrogeology of Salt Water Intrusion (Editor - in - Chief W. De Brenck), 1991. IAH, Vol. 11, 422 p.

Hydrogeology of Selected Karst Regions (Editors - in - Chief: W. Back, J.S. Herman, H. Paloc), 1992, IAH, Vol. 13, Hannover, 493 p.

Hydrogeology of the Dinamic Karst (Ed. B. Mijatović). IAH. Publication, Vol. 4, Heisse.

Ignat'ev E.I., 1964: *Principi i metode mediko-geografičeskogo izučenija komponentov geografičeskoj sredi.* V.kn.: *Medicinskaja geografija,* Irkutsk, 20 – 42.

Ilić B., Kovačević J.,1996: *Radiohidrogeoekologija kroz dva primera koja upozoravaju i opominju.* Zbornik referata XI Jugoslovenskog simpozijuma o hidrogeologiji i inženjerskoj geologiji, Budva, 585 – 593.

Il'in B.B., 1977: *Počvoobrazovanije i mikroelementi.* In: Problemi Sibirskogo počvovedenija. Nauka, Novosibirsk,16 – 39.

Ingebritsen S.E., Sanford W.E.,1998: *Groundwater in geologic processes.* Cambridge University Press, Cambridge.

Integrated Risk Information System (data base). IRIS. EPA – Office of research and development , Washington DC.

Introduction to Applied Geophysics (ed. S. Mareš), 1984. Reidel Publ. Co., Dordrecht, 581 p.

Introduction to Environmental Geotechnology (Ed. H.Y. Fang), 1997. Springer, 765 p.

Ivankin P. F., 1991: *Morfostrukturi i petrogenezis glubinnih razlomov.* Nedra, Moskva, 256 p.

Ivanov V.V., Nevreev A.G., 1964: *Klasifikacija podzemnih mineralnih vod.* Nedra, Moskva.

Izrael Ju.A., 1974: *Globalnaja sistema nabljudenij. Prognoz i ocenka izmenenij sostojanija okružajuščoj prirodnoj sredi.* Gidrometeoizdat, Moskva, 560 p.

Izrael Ju.A., 1984: *Ekologija i kontrolj sostojanija prirodnoj sredi.* Gidrometizdat, Moskva.

Jakovljević M. et al., 1997: *Zavisnost između sadržaja različitih oblika teških metala i nekih parametara plodnosti zemljišta.* Zbornik radova Kongresa Jugoslovenskog društva za proučavanje zemljišta: «Uređenje, korišćenje i očuvanje zemljišta», Novi Sad.

Jakovljević M., Pantović M., 1991, *Hemija zemljišta i vode.* Naučna knjiga, Beograd, 252 p.

Jakšić B., Sofrenović Đ., 1989: *Specijalna patološka morfologija.* Naučna knjiga. Beograd, 709 p.

Jakushevskaya I.V., 1973: *Mikroelementi v prirodnih landšaftah.* Moskva.

Janjić M.,1985: *Inženjerska geologija sa osnovama geologije.* Naučna knjiga, Beograd, 403 p.

Janković M.M., 1990: *Fitogeografija.* II izdanje, Naučna knjiga, Beograd, 425 p.

Janković S., 1981: *Ležišta mineralnih sirovina – Geneza rudnih ležišta.* Rudarsko - geološki fakultet, Beograd, 529 p.

Janković S.,1990: *Rudna ležišta Srbije.* Rudarsko – geološki fakultet, Beograd, 760 p.

Jaredić M., Vučetić J.,1982: *Mikroelementi u biološkom materijalu* (prvo izdanje), Privredni pregled, Beograd, 110 p.

Jean J.S., 1999: *Outbreak of enteroviruses and groundwater contamination in Taiwan: Concept of biomedical hydrogeology.* Hydrogeology Journal, Vol. 7, N°4, Springer, 339 – 340.

Jeftić N., Petrović B., 1987: *Mikroelementi u biološkom materijalu.* Privredni pregled, Beograd.

Jelenković R., 1991: *Mineralizacija urana Šumadije: genetski i morfostrukturni tipovi.* Katedra Ekon. geol., pos. izdanja 2, Gf, 40, Beograd, 187 p.

Jerant – Pantić V., 1995: *Medicinska virusologija.* Medicinska knjiga, Beograd.

Johnson C. C., Ge X., Green K, Liu X., 1996: *Studies of selenium distribution in soils, grains, drinking water and human hair samples from the Keshan Disease Belt of the Zhangjiakom, District, Hebei Province, China.* British Geol. Survey Overseas Series. Tech. Report, WC/96/52.

Johnson W. R., Proctor J., 1977: *Metal concentrations in plants and soils from two British serpentine sites.* Plant Soil 46, 275 – 278.

Jones F.O., 1973: *Landslides of Rio de Janeiro and the Serra das Araras escarpment, Brazil.* U. S. Geological Survey Professional Paper 697, 42 p.

Jorga V., 1993: *Selen i druge fizičkohemijske osobine vode za piće i maligna oboljenja.* Medicinska istraživanja, Vol. 26, Beograd.

Jorga V., 1999: *Voda i zdravlje.* In: Voda, kvalitet i zdravlje, Mol d.d., Beograd, Tuzla, 61 – 69.

Jovanović V., Djordjević G., Ivetić M., 1998: *Geološke podloge u prostornom planiranju.* 13. kongres geologa Jugoslavije, knj. II (Regionalna geologija, stratigrafija i paleontologija), Herceg Novi, 91 – 95.

Jovanović M., 1989: *Fiziologija domaćih životinja.* Medicinska knjiga, Beograd – Zagreb, 687 p.

Jovanović R., Ilić B., 1996: *Geoekologija.* In: Geološke karakteristike i potencijalnost geoloških resursa regije Kolubare, Mačve i Podrinja (Ed.I. Filipović), Beograd, 153 – 159.

Jovanović T., Maksimović Z., 1999: *Magnesium in some mineral waters from Serbia and their influence on health.* Serbian Academy of Sciences and Arts, Vol. XCII, book 1, Belgrade, 39 – 47.

Jović V., 1989: *Geohemija bioesencijalnih mikroelemenata u procesima površinskog raspadanja kredno - tercijarnih vulkanita Srbije.* Doktorska disertacija, Rudarsko - geološki fakultet, Beograd, 380 p.

Jović V., 1990: *Geohemija Cr, Co, Ni i V u procesima površinskog raspadanja kredno -tercijarnih vulkanita Srbije.* XII Kongres na geolozi na Jugoslavija, kn. III, Ohrid, 663 – 672.

Jović V., 1990: *Površinsko raspadanje kredno – tercijarnih vulkanita u Srbiji.* XII Kongres na geolozi na Jugoslavija. Ohrid, 278 – 288.

Jović V., 1998: *Geohemija magnezijuma u životnoj sredini.* Prvi simpozijum o magnezijumu (Apstrakt), Srpska Akademija nauka i umetnosti, Beograd, 15 – 17.

Jović V., 1999: *Environmental geochemistry of magnesium.* In: Magnesium in the environment and in organisms. Serbian Acad. of sciences and arts, Vol. XCII, book 1, Beograd, 3 – 12.

Jović V., Milić S., Popadić D., 1995: *Selenium in some metamorphic rocks in Serbia.* Conference on selenium. Serbian Academy of Sciences and Arts.,Vol. LXXVIII, book 6, Belgrade, 21 – 27.

Jovičić Ž., 1996: *Geografizacija geografije – jedan od fundamentalnih zadataka njenog daljeg razvoja.* Zbornik radova, Geografski institut «Jovan Cvijić», knj. 46, Beograd, 277 – 290.

Jovičić Ž., 1997: *Naša planeta Zemlja. Stvarnost i vizija.* SANU, Geografski institut «Jovan Cvijić», posebno izdanje, knj. 50, Beograd, 188 p.

Jovičić Ž., 1998: *Osnove medicinske geografije Srbije. Teorijsko – metodološki koncept.* Srpsko geografsko društvo, Beograd, 132 p.

Kabata – Pendias A., Pendias H., 1985: *Trace Elements in Soil and Plants.* CRC Press, Boca Raton, 315 p.

Kaim W., Schwederski B., 1994: *Bioinorganic chemistry: Inorganic elements in the chemistry of life – An introduction and guide.* John Wiley & Sons Chichester, 269 –285.

Kappelmeyer O., Haenel R., 1974: *Geotermics with Special Reference to Application.* Geoexploration monographs, Series 1, N° 4, Gebrüder Berntraeger, Berlin, Stuttgart, 238 p.

Karamata S., 1967: *Petrogeneza.* Građevinska knjiga, Beograd, 324 p.

Karamata S., 1974: *Geohemijske, petrološke i metalogenetske provincije kredno – tercijarne starosti dela Balkanskog poluostrva i Male Azije.* SANU, pos. izd. 42, Beograd, 55 p.

Kastori R., 1994: *Izvori zagađenja agroekosistema.* Kvalitet, 3 – 4, 24 – 28.

Kastori R., Petrović N., Arsenijević – Maksimović I., 1997: *Teški metali i biljke.* In: Teški metali u životnoj sredini (Ed. R Kastori). Institut za ratarstvo i povrtarstvo, Novi Sad.

Kathren R.L., 1984: *Radioactivity in the environment : sources, distribution and surveillance.* Harwood Academic Publishrs. Chur, Switzerland.

Katzer F., 1919: *K poznavanju mineralnih vrela Bosne.* Glasnik Zemaljskog muzeja u Bosni i Hercegovini,Sarajevo,191 – 264.

Kirjuhin V. K., Shvec V. M., 1980: *Procesi formirovanija jodnih vod.* Nedra, Moskva, 94 p.

Kiryuhin V.A., Korotkov A.I., Pavlov A.N., 1988: *Obšćaja gidrogeologija.* Nedra, Leningradskoe otdelenie, Leningrad, 359 p.

Kiss Z., Varkoyi T., Zajkas G., 1994: *The role of magnesium in the human organism.* In: Magnesium in Biological Systems (Ed. by T. Fazekas, B. Selmeczi). Academiai Kiado, Budapest, 135 – 174.

Klijn F., Witte M. Jan - Philip, 1999: *Eco – hidrology: Groundwater flow and site factors in plant ecology.* Hydrogeology Journal, Vol.7, N⁰ 1, Springer, 65 – 77.

Knežević –Djordjević V., Djordjević P., 1975: *Osnovi petrologije.* IGS, Beograd, 199 p.

Kobars F.J., Nystuen D.J., 1974: *Human Geography.* Mc Graw – Hill Book Company, Hague.

Kobayashi J., 1971: *Relation between the «itai – itai» disease and the pollution of river water by cadmium from mine.* Pergu. Press.

Kolarski D., 1995: Osnovi ishrane domaćih životinja. Naučna knjiga, Beograd, 420 p.

Koljonen T., 1975: *The behavior of selenium in Finnish soils.* Ann. Agric. Fenn. 14, 240 -247.

Kolomeiceva M. G., Gabovich R.D., 1970: *Mikroelementi v medicine.* Ivanovo – Frankovsk, 312 – 313.

Komatina D., 2001: *Analiza tečenja gustih mešavina – teorija i eksperimenti.* Zadužbina Andrejević, Beograd, 144 p.

Komatina M., 1984: *Hidrogeološka istraživanja. Metode istraživanja I.* Geozavod, Beograd, 375 p.

Komatina M., 1990: *Hidrogeološka istraživanja. Primenjena hidrogeologija III.* Geozavod, Beograd, 391 p.

Komatina M., 1993: *Geološka osnova za izbor potencijalnih rejona za deponovanje opasnog otpada.* Zaštita voda '93. Jugoslovensko društvo za zaštitu voda, Aranđelovac, 85 – 92.

Komatina M., 1994: *Vodni problem današnjice i putevi njegovog rešavanja.* Ecologica, god. I, br.1, Beograd, 11 – 15.

Komatina M., 1995: *Gazdovanje podzemnim vodama na teritoriji Srbije.* In: Vodni mineralni resursi litosfere Srbije. Rudarsko – geološki fakultet, Univerzitet u Beogradu, Beograd, 57 – 76.

Komatina M., 1995: *Hidrogeološka istraživanja. Primenjena hidrogeologija, IV.* Geozavod, Beograd, 427 p.

Komatina M., 1997: *Istorijat i perspektiva razvoja metoda hidrogeoloških istraživanja u Jugoslaviji.* Monografija «100 godina hidrogeologije u Jugoslaviji «, Beograd, 273 -287.

Komatina M., 1998: *Hidrogeologija krasa.* Speleološki atlas Srbije. SANU, Geografski institut «Jovan Cvijić», posebno izdanje, knj. 52, Beograd, 23 – 30.

Komatina M., 1999: *Base of medical geology.* Fourth USA/CIS Joint Conference on Environmental Hydrology and Hydrogeology (San Francisco, November

1999). American Institute of Hydrology (AIH).
Komatina M., 1999: *Osnove medicinske geologije.* Vesnik Geozavoda, knj. 49, Beograd, 399 – 406.
Komatina M., 1999: *Podzemne vode u prirodi, resursi, značaj.* In: Voda, kvalitet, zdravlje (Ed. I. Gržetić, I. Brčeski). Beograd, Tuzla, 37 – 50.
Komatina M., 2000: *Influence of geological factors on the human health.* 31st International geological congress. Rio de Janeiro (Brasil), (6 – 17 August 2000).
Komatina M., 2000: *Vodni resursi Planete i značaj za razvoj društva.* Vesnik Geozavoda, knj.50, Beograd.
Komatina M., 2001: *Medicinska geologija.* Tellur, Beograd, 429 p.
Komatina M., Božović B., Navala M., 1997: *Geološka podloga za obezbeđenje zdravih uslova proizvodnje i življenja.* In: Zaštita potrošača, tržišna privreda i ekologija (Zbornik radova I Balkanskog centra potrošača) Beograd, 51 – 56.
Komatina S., 1995: *Uloga geofizike u proceni prirodne zaštićenosti podzemnih voda od zagađenja.* Ecologica, N0 7, god II, br. 3., Beograd, 18 – 20.
Komatina S., 1996: *Dometi ekogeofizike u svetu i pravci daljeg razvoja.* Savremeni trendovi razvoja geofizike (Zbornik). Rudarsko – geološki fakultet, Beograd, 249 – 256.
Komatina S., 1996: *Primena geofizičkih metoda u izučavanju prirodne zaštićenosti podzemnih voda od zagađenja (Doktorska disertacija).* Tempus centar, Univerzitet u Novom Sadu.
Kongshaung G. et al., 1992: *Proc. Chemical Climatology and Geomedical Problems.* Norsk Hydro, Oslo, Norway.
Kononov V. I., 1983: *Geohimija termaljnih vod oblastej sovremenogo vulkanizma.* Nauka, Moskva 216 p.
Korcenshtain V.N., 1984: *Rastvorenie gazi podzemnoj gidrosferi Zemli.* Nedra, Moskva, 230 p.
Korn'eiev N.A., Sirotkin A.N., 1987: *Osnovi radioekologiji seljskohozjajstvenih životnih.* Energoatomizdat, Moskva.
Koronkevich N.I., 1972: *Preobrazovanie vodnogo balansa v SSSR.* Znanie, Moskva.
Korte N., 1991: *Naturally occuring arsenic in ground waters of the mid – western United States.* Environ Geol Water Sci, 18/2 , 137 – 141.
Kostenko N. P., 1972: *Razvitie skladčatih i razrivnih deformacij v orogennom reljefe.* Nedra, Moskva, 320 p.
Kotevski G., 1987: *Hidrogeologijata na mineralnite, termalnite i termomineralnite vodi vo teritorijata na SR Makedonija.* Samoupravnaja praktika, Skopje, 352 p.
Kotlov F.V., 1978: *Izmenenie geologičeskoj sredi pod vlijanijem dejateljnosti čeloveka.* Nedra, Moskva.
Kovalevskij A.L., 1969: *Osnovnie zakonomernosti formirovanija himičeskogo sostava rastenij.* In: «Biogeohimija rastenij», Ulan – Ude.
Koval'skii V. V., 1982: *Geohemičeskaja sreda i* žiznj. Nauka, Moskva, 77 p.
Koval'skii V.V., 1957: *Rolj mikroelementov v žizni životnih v različnih zonah SSSR.* Moskva.
Koval'skii V.V., 1963: *Geohimičeskaja ekologija i jejo evoljucionnije napravlenija.*

Izv.AN SSSR , ser. biologičeskaja, 830 p.
Koval'skii V.V., 1974: *Geohimičeskaja ekologija: Očerki.* Nauka, Moskva, 280 p.
Koval'skii V.V.,1979: *Geochemical ecology and problems of health.* Phil. Trans. R. Soc. London, B 288, 185 - 191.
Kovda V.A., 1973: *Osnovi učenija o počvah.* Nauka, Moskva, kn. 1, 448 s; kn. 2, 468 p.
Kovda V.A., 1977: *Aridizacija suši i borba s zasuhoj.* Nauka, Moskva.
Kowalski J., 1987: *Hydrogeologia z podstawami geologii. Państwowe wydaw - nictwo nawkowe,* Warszawa, 349 p.
Kralik M., 2000: *Environmental Geology in Austria.* In: Aspects of Geology in Austria, Vienna, 313 - 335.
Krauskopf K.V., 1979: *Introduction to geochemistry.* Mc Graw - Hill, New York.
Kravchenko L.F., 1965: *Bolezn Kašina - Beka* . Aftoref diss. dokt., Rostov na Donu.
Krepkogorskii L.N., 1953: *K probleme fljuoroza.* Gig. i San., 11, 7.
Krivoluckij A. E., 1971: *Žiznj zemnoj poverhnosti.* Mislj., Moskva, 407 p.
Kronberg B.I. et al., 1979: *The chemistry of some Brazilian soils: element mobility during intense weathering.* Chem. Geol. 24, 211 - 229.
Kulajta V., 1990: *Istraživanje radona i njegovo korišćenje.* Zbornik radova Jugosloven. - čehoslov. simpoz. «Geologija i zaštita životne sredine», Beograd, 109 - 113.
Kulier I., 1986: *Znate li šta jedete? Kako se dobro hraniti da bismo živjeli duže i zdravije.* Nakladni Zavod , Znanje, Zagreb, 463 p.
Kvalitet vode za piće. Problemi i rešenja. (Ed. B. Dalmacija), 1998, Univerzitet u Novom Sadu, 214 p.

L'vovich M.M., 1974: *Mirovie vodni resursi i ih budušćee. Mislj, Moskva.*
L'vovich M.M., 1986: *Voda i žiznj.* Mislj, Moskva.
La Maja Ž.,1987: *Medicina stanovanja.* Akvarijus, Beograd.
La Moreaux et al., 1997: *Environmental Hydrogeology.* Springer, 370 p.
La Moreaux P. E., Newton J. G., 1986: *Catastrophic subsidence: An environmental hazard, Shelby County, Alabama.* Environ Geol.Water Sci 8 (½).
La Moreaux P. E., Powell W. J., Le Grand H. E., 1997: *Environmental and legal aspects of karst areas.* Environmental Geology 29 (1/2). Springer - Verlag.
Lahermo P., Backman B., 2000: *The occurrence and geochemistry of flourides with special reference to natural waters in Finland.* Geological Survey of Finland, Espoo 2000, 40 p.
Langguth H.R., Voigt R., 1980: *Hydrogeologische Methoden.* Springer Berlin, Heildberg, New Jork, 468 p.
Larson J., 1984: *Groundwater in hard rocks.* UNESCO, 228 p.
Lave L.B., Seskin E. P., 1970: *Air pollution and human health.* Science, 169, 732 p.
Lawter P.J., 1959: *Cancerogene in der städtischen Luft.* Zbl. Bakt. I. Abt. Orig., 176, 187 p.
Lazić M., 1994: *Polazne osnove izučavanja faktora rizika procesa nestabilnosti terena.* Vesnik, ser. A, B, knj. 46, Beograd, 417 - 426.
Lead, mercury, cadmium and arsenic in the environment. (Eds. T.C. Hutchinson,

K.M. Meema), 1987, SCOPE, John Wiley & Sons.
Learmonth A., 1975: *Patterns of disease and hunger.* Newton Abbot etc.: David and Charles, 256 p.
Lebedev A.D., 1967: Geografičeskoe rasprostranenie i regionaljnie osobenosti prirodnih očagov gemorragičeskogo nefrozonefrita. *Aftoref dis. dokt., Moskva.*
Lebedev A.D., Avtsin A.P., 1965: *Zadači medicinskoj geografii i geografičeskoj patologii.* V. kn.: Metodi mediko – geografičeskih isledovanij, Moskva, 9 - 22.
Legget R., 1976: *Goroda i geologija.* Mir, Moskva.
Leko T. M., Shcherbakov A., Joksimovich M. H., 1922: *Lekovite vode i klimatska mesta u kraljevini Srba, Hrvata i Slovenaca.* Ministarstvo Narodnog zdravlja, Beograd, 276 p.
Levander O.A., 1986: *Selenium.* In: Mertz, W., (Ed.) Trace Elements in Human and Animal Nutrition, Academic, Orlando, Fith Edition, Vol.2, 209 - 279.
Lindenmayer E., 1856: Opis mineralni i lekoviti voda i njino opredelenije voobšte a ponaosob lekoviti voda u Kneževstvu Srbije dosad poznati. *Praviteljstvo Knjigopeč. Knjaž. Srbskog, Beograd.*
Lobava E.V., Habarov A. V., 1983: *Počvi.* Mislj, Moskva, 303 p.
Logar M., 1998: *Serpentinski minerali Jugoslavije sa osvrtom na ekološki značaj hrizotil – azbesta.*13. Kongres geologa Jugoslavije, In: III (Petrologija, mineralogija i geohemija). Herceg Novi, 156 – 160.
Lomnitz C., 1974: *Global Tectonics and Earthquake Risk.* Elsevier Scientific Publishing Company, New York.
Lostih E., 1959: *O roli vulkanov i term v vnose tepli iz nedra Zemli.* V. kn.: Problemi geotermii i praktičeskogo ispolzovania tepli Zemli. Tom I. Izv. AN SSSR, Moskva.
Loukovitch M., 1937: *Hydrologie des stations hydrominerales Yougoslaves (Resume du raports).* XV Congres International d'Hydrologie, de Climatologie et de Geologie Medicales, 261 - 352.
Lucas J., 1975: *Our Food – A Survey of the Risks.* Charle Knight Comp. Limit, London – Tonbridge.

Macdonald G.A., Abbott A.T., Peterson F.L., 1983: *Volcanoes in the sea.* The geology of Hawaii, 2nd edn. University of Hawai Press, Honolulu.
Mahdi G.S., 1991: *Role of chromium in barley in modulating the symptom of diabetes.* Annals of Nutrition and Metabolism, 35, 65 – 70.
Maksimovich G. A., 1955: *Himičeskaja geografija vod suši.* Geografgiz, Moskva, 328 p.
Maksimović Z. et al., 1998: *Selenium in soils, grass and human serum in the Zlatibor area, (Serbia): Geomedical Aspects.* Journal of Environmental Pathology and Oncology, 17 (384): 221 - 227.
Maksimović Z., 1975: *Geohemijske karakteristike alpinotipnih ultrabazita Jugoslavije.* Geološki anali Balkanskog Poluostrva, knj. 39, Beograd, 231 – 302.
Maksimović Z., 1985: Geohemijska proučavanja i endemska nefropatija. SANU, knj. XXIII, Odelj. Medic. Nauka, knj.3, Beograd, 123 – 130.
Maksimović Z., 1987: *Trace element deficiency and Balkan Endemic*

Nephropathy (BEN). In.: Etiology of Endemic (Balkan) Nephropathy (Eds: Strahinjić S., Stefanović V.), Niš, 43 – 47.

Maksimović Z., 1990: *Savremeni problemi geohemije životne sredine*. XII Kongres na geolozi na Jugoslavia, kn. III, Ohrid, 673 – 680.

Maksimović Z., Dangić A., Ršumović M., Jović V., 1986: *Selen u kredno – tercijarnim eruptivnim stenama Srbije*. XI Kongres geologa Jugoslavije, knj. 3. (Petrologija, geohemija, tektonika). Tara, 97 – 107.

Maksimović Z. et al., 1989: *Deficit selena u životnoj sredini u istočnoj Srbiji i moguće posledice na zdravlje*. Čovek i životna sredina, 4 – 5, Beograd.

Maksimović Z., Đujić I., Jović V., Ršumović M., 1992: *Selenium deficiency in Yugoslavia*. Biol. Trace Elem. Res. 33, 187 – 196.

Maksimović Z., Jovanović T. Ršumović M., Đorđević M., 1999: *Magnesium and calcium in drinking water in relation to cardiovascular mortality in Serbia*. Serbian Academy of Sciences and Arts, Vol. XCII, book 1, Beograd, 21 – 27.

Maksimović Z., Ršumović M., 1988: *Problemi koncentracije teških metala u sedimentima i vodama reka Dunavskog sliva*. Geološki anali Balkanskog Poluostrva, tom LII. Geološki Zavod Univerziteta u Beogradu, Beograd, 409 – 428.

Maksimović Z., Ršumović M., Jovanović T., 1997: *Vode iz ultramafita Zlatibora i njihov uticaj na zdravlje*. 100 godina hidrogeologije u Jugoslaviji (Monografija), Beograd, 249 - 258.

Maksimović Z. et al., 1998: *Magnezijum i kalcijum u pijaćoj vodi i smrtnost od kardiovaskularnih bolesti u Srbiji*. Prvi simpozijum o magnezijumu (Abstracts), SANU, Beograd, 1 – 3.

Malberg H., 1997: *Meteorologie und Klimatologie. Ein Einführung*. Springer, 340 p.

Manahm S.E., 1991: *Environmental Chemistry*. Lewis Publishers, Chelsey. Michigan, 583 p.

Mandal B.K., et al., 1997: *Chronic arsenic toxity in West Bengal*. Current Science, Vol. 72, N° 2, 114 – 117.

Manson – Bahr Ph., 1966: *Manson's Tropical Diseases*. Bailliare, London.

Maresh S., Kelly E.W., Mazach O., 1997: *Applied geophysics in environmental engineering and science*. Praha, 98 p.

Marhinin E.K., 1985: *Vulkanizm*. Nedra, Moskva, 288 p.

Marić L., 1974: *Minerali, stijene i rudna ležišta u našoj zemlji od prehistorije do danas*. Zagreb, 342 p.

Marier J.R., Rose D., Belanger L.F., 1963: *Hardwater and heart disease*. Br. Med. J. 14, 686 – 693.

Marković D. et al., 1996: *Fizičkohemijski osnovi zaštite životne sredine*. Knj. II: Izvori zagađivanja, posledice i zaštita, Univerzitet u Beogradu, 510 p.

Marković D., Djarmati Š., Gržetić I., Veselinović D., 1996: *Fizičkohemijske osnove zaštite životne sredine*. Knj. II:. Izvori zagađivanja, posledice i zaštita. Univerzitet u Beogradu, Beograd, 552 p.

Marković M., 1983:*Osnovi primenjene geomorfologije*. Geoinstitut, knj. 8, Beograd,174 p.

Marković M., 1993: *Neotektonska karta: Osnovni pojmovi, principi i metodika*. Vesnik, ser.A, B, knj. 45, Beograd, 6 – 18.

Marković M., Menković Lj., 1996: *Geomorfološka karta: koncepcija, izrada i primena.* Zbornik radova Geografskog instituta «Jovan Cvijić», knj. 46, Beograd, 91 - 106.
Marković P., 1993: *Poljoprivredni atlas Srbije, prvi tom. Dugoročna regionalna strategija agroindustrijske proizvodnje Republike Srbije.* Beograd, 236 p.
Marmut Z., 1998: *Zdravstveni aspekti boravka i rada u zatvorenom prostoru – sindrom «nezdravih» zgrada.* Ecologica, posebno izdanje, br.5, Beograd, 9 – 14.
Marot R., 1958: *Pathologie regionale de la France.*T. 1,2, Paris.
Martinović S. et al., 2000: *Geofizički karotaž – obrada i interpretacija.* Naftagas, Novi Sad, 383 p.
Martyn C.N. et al., 1989: *Geographical relation between Alzheimer's disease and Al in drinking water.* Lancet I, 59 – 62.
Masirioni R., 1979: *Geochemistry and cardiovascular diseases.* Environmental Geochemistry and Health. Phil. Trans. R. Soc. London B 288, 193 – 201.
Matić S., 1994: *Procena uticaja životne sredine na zdravlje ljudi.* Ecologica, god.I, br. 2, Beograd, 9 – 13.
Matschullat J., 2000: *Arsenic in the geosphere – a review.* The Science of the Total Environment 249, 297 – 312.
Matschullat J., et al., 2000: *Human and environmental contamination in the Iron Quadrangle, Brazil.* Applied Geochemistry, 15.
Matschullat J., Tobschall H. J.,1997: *Geochemie und Umwelt. Relevante Prozesse in Atmo -Pedo – und Hydrosphäre.* Springer, 443 p.
Matula M., 1990: *Zadaci geologije u racionalnom iskoriščavanju i zaštiti geološke sredine.* Zbornik radova Jugoslovensko – čehoslovačkog simpozijuma «Geologija i zaštita životne sredine», Beograd, 15 – 17.
May J.M., 1950: *Medical geography: its metods and objectives.* Geogr. Rev., Vol. 51, 9 – 41.
May J.M., 1958: *The ecology of human disease.* New York: M D publ., 328 p.
May J.M., 1974: *Medical geography: a theory of medicinae.* J. Biosoc.Sci,6.2, 187 - 192.
Mc Call J., Marker B., 1989: *Earth Science Mapping.* Graham & Trotman, London – Dordrecht – Boston.
Mc Gilvery R.W., 1970: *Biochemistry – a functional approach.* WB Saunders, Philadelphia.
Mc Grath S.P., Smith S., 1990: *Chromium and nickel.* In: B.J. Alloway (Editor), Heavy metals in soils. Wiley, New York, 125 – 150.
Mc Milan G., 1978: *Distribution of trace elements related to certain cancers, cardiovascular diseases and urolithiasis.* In «Geochemistry of the Environment», Nation. Acad. Sci. Washington DC, v.3: 114 - 132.
Mc Neely J.A, et al., 1990: *Conserving World's Biological Diversity.* IUCN, WRI, CI, WWF – US, World Bank, Wschington D. C., 193 p.
McMachon P.B., 2001: *Aquifer/aquitard interfaces: mixing zones that enchance biogeochemical reactions.* Hydrogeology Journal. Springer, Vol. 9, N°1, 34 – 43.
Meade M.S., Florin J. W., Gesler W. M., 1988: *Medical geography.* The Guilford Press. New York, London, 340 p.

Medicinskaja geografija i zdorovie (Ed. A.A. Keller), 1989. Nauka, Leningrad, 218 p.
Meinzer O.E., 1923: *The occurrence of ground water in the United States, with a discussion of principles.* US Geol. Surv. Water – Supply, 489 p.
Melnick J.L., Gerba C.P. Wallis C., 1978: *Viruses in water.* Bull WHO, 56, 499 – 508.
Mengel K., Kirkby E.A., 1978: *Magnesium.* In: Principles of Plant Nutrition. Int. Potash Inst., Worblaufen – Bern, Switzerland.
Menković Lj., 1997: *Geomorfološko kartiranje u funkciji geografske regionalizacije.* V. knj.: Geografska struktura i regionalizacija Srbije, I. Geografski institut «Jovan Cvijić», posebno izdanje, knj. 51, Beograd, 109 – 117.
Merkov A.M., Guseva I.S., 1965: *Ispoljzovanie statističeskih dannih v mediko – geografičeskih issledovanijah.* V. kn.: Metodi mediko – geografičeskih issledovanij. Moskva, 85 – 89.
Meshchenko V.M., 1964: *Izučenie geografičeskogo rasprostranenija biohimičeskih endemij.* V. kn.: Medicinskaja geografija. Irkutsk, 129 p.
Metelev V.V., Kanaev A.I., Dzasohova N.G., 1971: *Vodnaja toksikologija.* Izdat Kolos, Moskva, 247 p.
Methods for Assessment of Soil Degradation (Eds. R. Lal, V.H. Blum, C. Valenti), 1998. Springer, 352 p.
Metodičeskie rekomendacii po giohimičeskim issledovanijam dlja ocenki vozdejstvija na okružajušćuju sredu proektiruemih gornodobivajuščih predprijatij. Institut mineralogii, geohimii i kristallohimii redkih elementov, 1986. Moskva, 98 p.
Metzuler E.D., 1987: *Biochemistry. The chemical reaction of living cells.* Vol. I, II, III. Academic Press, New York.
Mihailović D., 1998: *Praistorijska pećinska nalazišta.* Speleološki atlas Srbije. SANU, Geogr. inst. «Jovan Cvijić», poseb. izd., knj. 52, Beograd, 85 – 88.
Mihajlović M., 1996: *Metabolizam – promet materije i energije.* In: Veterinarska fiziologija. (Ed. V. Stojšić).
Mijatović B., 1990: *Kras. Hidrogeologija kraških vodonosnika.* Geozavod, Beograd, 304 p.
Mijović D., 1997: *Hidrogeologija granodiorita Srbije.* Zadužbina Andrejević, Beograd, 136 p.
Milanović P., 1979: *Hidrogeologija karsta i metode istraživanja.* Trebinje, 302 p.
Milanović P., 1999: *Geološko inženjerstvo u karstu.* Energoprojekt, Beograd, 291 p.
Milanovskii E.E., 1987: *Geologija SSSR. Čast 1. Vvedenie. Drevnie platformi i metaplatformennie oblasti.* Izdat. Moskovskogo Universiteta, 414 p.
Milenković M., Vitorović S., 1994: *Osnovi toksikologije sa elementima ekotoksikologije.* Naučna knjiga, Beograd.
Milivojević M., 1982: *Paleohidrogeotermalne pojave Avale.* Rudarsko – geološki fakultet, Beograd, 203 p.
Mills C.F., 1996: *Geochemical aspects of the aetiology of trace element related diseases.* In: Appleton, J.D., Fuge R., McCall, G.J.H. (Eds.) Environmental Geochemistry and Health, Geological Society, London, Special Publication №113, 1 - 5.

Milosavljević S., 1997: *Lekovite vode Vojvodine.* Stručni časopis DIT, Naftagas, god XIV, br. 25, Novi Sad, 103 – 113.
Milovanović B., 1949: *Opšta geologija.* Izdavačko preduzeće Ministarstva rudarstva FNRJ, Beograd, 451 p.
Milovanović D., Boev B., 2001: *Zemlja, tektonika ploča i magmatizam.* Nauka, Beograd, 174 p.
Mindel E., 1997: *Vitaminska biblija* (prevod sa engleskog). Familet, Novi Sad, 256 p.
Mineral – Water interface geochemistry (eds. M.F. Hochella and A.F. White), 1990. Reviews in mineralogy, Vol. 23., Washington, 603 p.
Mironenko V.A., Mol'skii E.B., Ruminin B.G., 1988: *Izučenie zagrjaznenija podzemnih vod v gorno – dobivajuščih rejonah.* Nedra, Leningrad, 279 p.
Mirvish S., 1983: *The etiology of gastric cancer.* J. Nation, Cancer. Inst., 71, 629 – 647.
Mokranjac M., 1990: *Toksikološka hemija.* Univerz. u Beogradu, Beograd.
Moldan B., 1986: *Acidifation in Central Europe: its causes and effects.* Geology for environmental planing. Norges geol. undersokelse, Trondheim, 82 – 83.
Momčilović V., 1995: *Lokacije stambenih objekata na hidropatogenim trasama aluvijalnih terena odgovorne za pojavu endemske nefropatije.* Ecologica, br.1, Beograd, 10 – 13.
Montgomery C.W., 1986: *Environmental geology.* Wm. C. Brown Publishers, Dubuque, 454 p.
Moore J.W., Ramamoorthy S., 1984: *Heavy metals in natural waters.* Springer – Verlag, New York, 268 p.
Morozov N.P., 1969: *Geohimia.* N° 6, Nauka, Moskva, 729 p.
Morse R.H., 1976: *Radon counters in uranium explorations.* IAEA – SM – 208 / 55, Explorat. of uranium ore deposits, Vienna, 229 – 239.
Moss R.P., 1974: *The ecology of human comunities.* Inaugural Lecture, University of Birmingham.
Mossman B.T. et al., 1990: *Asbestos: Scientific Development and Implication for Public Policy.* Science, V 247, N° 4940, 299 p.
Moynihan D.P., 1970: *La polltion de l'air.* Science et Vie, 117, 632 p.
Mueller R.F., Sana S. K., 1977: *Chemical Petrology.* Springer, New York, 394 p.
Muntanjola – Cvetković M., 1990: *Opšta mikologija.* Naučna knjiga, Beograd.
Murray R.K. et al., 1988: *Harper's Biochemistry.* Twenty – first Edition. International Edition. Apleton and Lange, USA.
Muškatirović J., 1998: *Primena savremenih informacionih tehnologija u domenu upravljanja vodnim resursima.* In: Vodni resursi sliva Velike Morave i njihovo korišćenje. Zadužbina Andrejević, Kruševac, 341 – 348.

Nakamura H., 1969: *Mineral and thermal waters in Japan.* In: Mineral and thermal waters of the World. B. Overseas countries. Prag, 45 – 62.
Nappi G., 1996: *Medicina e clinica termale.* Milano.
Navara J.G., Weisberg J. S., Mele F. M., 1971: *Earth science.* New York.
Nazarenko I.I., Kislova I.V.,1981. *Selen kak element – indikator obšćej*

zagrjazennosti prirodnih vod. In: Novie oblasti primenenija geohimičeskih metodov. Institut minerologii, geohimii i kristalohimii redkih elementov, Moskva, 33 - 37.

Nazaroff W.W., 1992: *Radon transport from soil to air.* Review of Geophysics 30 (2), 137 - 160.

Nejgebauer B., 1952: *Činioci stvaranja zemljišta Vojvodine.* Zbornik Matice srpske, sveska 2, Novi Sad.

Nenadović L., 1936: *Banje, morska i klimatska mesta u Jugoslaviji.* Beograd, 411 p.

Nian - Feng L., 1991: *Medical environmental geochemistry.* Jilin Science and Technology Publishing House, Changchun City.

Nian - Feng L., Jie T, 1995: *Some problems of ecological environmental geology in arid and semiarid areas of China.* Environ. geol., 26, 44 - 67.

Niaragu J.O., 1980: *Cadmium in the environment.* Wiley, New York, 682 p.

Niaragu J.O., 1994: *Arsenic in the environment. Part I. Cycling and characterization.* Wiley, New York.

Nicolas D.J.D., Egan A.R., 1975: *Trace Elements in Soil – Plant – Animal Systems.* Academic Press, New York.

Niggli P., 1954: *Rocks and mineral deposits.* San Francisco.

Nikitin N.B., Sudarikov S. M., 1990: *Režim gidrogeohimičeskih oreolov i potokov rassejanija rudoprojavlenij (jug Daljnego Vostoka).* In: Gidrohimičeskie poiski mestoroždenij poleznih iskopaemih. Nauka, Novosibirsk, 33 - 41.

Nikolaev N.I., 1962: *Neotektonika i ee viraženie v strukture i reljefe teritorii SSSR.* Gosgeoltehizdat, Moskva, 392 p.

Nikolić P., 1984: *Osnovi geologije.* Naučna knjiga, Beograd, 628 p.

Nikolić P., 1989: *Geotektonika.* Naučna knjiga, Beograd, 596 p.

Nikolić S., Dimitrić A., 1998: *Aseizmička gradnja u svetlu novih saznanja.* 13 Kongres geologa Jugoslavije, knj. VI (Geofizika i seizmologija). Herceg Novi, 209 - 218.

Nitzsche O., Merkel B., 1999: *Reactive transport modeling of uranium 238 and radium 226 in groundwater of the Königstein uranium mine, Germany.* Hidrogeology Journal, Vol.1 7, N⁰ 5, Springer, 423 - 430.

Nockolds S. R., 1954: *Average chemical composition of some igneous rocks.* Bull. Geol. Soc. Am., 65, 1007 - 1032.

Ockerse T. 1953: *Chronic endemic dental fluorosis in Kenya, East Africa.* British Dental Journal, Vol. 95, 57 - 60.

Odum Ju., 1986: *Ekologija.* Mir, Moskva, T1. 328 s; T2. 376 p.

Odum P.E., 1971: *Fundamentals of ecology.* Philadelphia - Toronto - London.

Ohsawa G., 1971: *Cancer and the philosophy of the Far East.* Oroville, California, George Ohsava Macrolistic Fondation.

Okamoto S., 1973: *Introduction to earthquake engineering.* University of Tokio Press.

Ollier K., 1984: *Tektonika i reljef.* Nedra, Moskva, 460 p.

Omaljev V., 1987. *Geohemijsko polje.* Geoinstitut, poseb. izdanje, knj.10, Beograd,169 p.

Omaljev V., 1996: *Radioaktivni elementi sa osvrtom na magmatske stene Srbije.*

Radovi Geoinstituta, knj. 32, Beograd, 5 – 24.
Omaljev V., Antonović A., 1989: *Dva veka od otkrića hemijskog elementa urana.* Radovi Geoinstituta, knj. 23, Beograd, 7 – 17.
Onishi H., Sandell E. B., 1955: *Geochemistry of arsenic.* Geochim. Cosmochim. Acta, 7, 1 – 33.
Orlov M.S., Shestakov V.M., 1990: *Gidrogeoekologija – nužna li ona i kakvo jejo soderžanie.*Vestnik Moskovskogo Univerz.; Ser. 4 Geologija, N^{0} 6, Moskva, 62 – 66.
Osnovi gidrogeologii. Gidrogeohimija (Ed. S.L. Shvarcev), 1982. Nauka, Novosibirsk.
Osnovi gidrogeologii. Obšćaja gidrogeologija (Ed. E.V. Pinneker), 1980. Nauka, Novosibirsk, 229 p.
Oster O., Prellwitz, 1990: *Selenium and Cardiovascular Disease.* Biological Trace Element Research, 91 – 101.
Ovchinnikov A.M., 1963: *Mineraljnie vodi.* Izdanie vtoroe, Moskva, 375 p.
Ozerova N., Karamata S., 1998: *Zakonitosti pojavljivanja žive u rudnim ležištima Balkanskog Poluostrva.* 13. Kongres geologa Jugoslavije, knj. IV (Mineralne sirovine), Herceg Novi, 423 – 431.
Ozerova N.A., 1977: *O degazacii rtuti iz glubin Zemli. Geohimija procesov migracii rudnih elementov.* Nauka, Moskva, 15 – 35.

Pantelić M., Drašković R., Aleksić S., 1996: *Mineralne vode i peloidi Gornje Trepče.* Gornja Trepča, Čačak, 127 p.
Pantić N., 1984: *Priroda i čovek.* Čovek i priroda, SANU, Beograd, 9 – 48.
Pantić N., 1990: *Racionalno iskorišćavanje neobnovljivih prirodnih resursa sa aspekta zaštite životne sredine.* Perspektiva, ciljevi, zadaci i mere u zaštiti i unapređenju životne sredine u Srbiji, SANU. Beograd, 42 – 51.
Pantić N., 1996: *O jedinstvu prirodnog i duhovnog.* Nezav. istraž. centar Plavi Zmaj, Novi Sad.
Pantić N., Belij S., Mijović D., 1998: *Geo – nasleđe u sistemu prirodnih vrednosti i njegova zaštita u Srbiji.* Zaštita prirode, br.5, Beograd, 407 – 413.
Pantić N., Komatina M., 1998: *Voda – osnovni resurs življenja.* In: Vodni resursi sliva Velike Morave i njihovo korišćenje. Zadužbina Andrejević, Beograd, 15 – 17.
Papworth D., 1997: *A need to reduce the radon gas hazard in the UK.* Journal Royal Society of Health, 117(2), 75 – 80.
Passwater R.A., 1980: *Selenium as food and medicine.* Pivot Original Edition published, New Canaan, 240 p.
Pavlov I. M., Jakubov N. T., 1991: *Aerofotografija.* Nedra, Moskva, 335 p.
Pavlović R. et al., 1998. *Daljinska detekcija u istraživanju prirodnih vrednosti.* 13. Kongres geologa Jugoslavije, knj. II (Regionalna geologija, stratigrafija i paleontologija), Herceg Novi, 283 – 393.
Pavlović S., Maksimović Z., Nikolić D., 1973: *Prirodni izvori mineralnog porekla koji ugrožavaju životnu sredinu.* Čovek i životna sredina, SANU, Beograd.
Pavlovskij E. N., 1955: *Sostojanije učenija o prirodnoj očagovosti boleznej*

čeloveka. In: Prirodnaja očagovost boleznej čeloveka i kraevaja epidemologija. Leningrad, 17 p.

Peive A., 1945: *Glubinie razlomi v geosinklinalnih oblastjah.* Izv.An. SSSR, ser. geol., br 5, Moskva.

Perelman A.I., 1956: *Istorija atomov i geografija.* Gosud. Izdatelj. Geograf. Literaturi, Moskva, 62 p.

Perelman A.I., 1973: *Geohimija biosferi.* Moskva.

Perelman A.I., 1975: *Geohemija landšafta.* Vissaja škola, Moskva, 341 p.

Perelman A.I., 1979: *Geohimija.* Vissaja škola, Moskva, 423 p.

Perić J., 1976/79 : *O potrebi hitnog i neodložnog rešenja problema vodosnabdevanja seoskih naselja sa pojavom žarišta endemske nefropatije.* Zbornik radova Rudarsko – geološkog fakulteta, sv. 21, Beograd.

Perić J., 1985: *Hidrogeološko istraživanje terena i endemska nefropatija.* III Simpozijum o endemskoj nefropatiji. SANU, knj.XXIII, Odeljenje medicinskih nauka, knj.3, Beograd, 105 – 122.

Perić J., Milivojević M.,1990: *Studija geotermalne potencijalnosti teritorije SR Srbije van teritorija SAP (elaborat).* Rudarsko – geološki fakultet, Grupa za hidrogeologiju, 402 p.

Perić J., Stefanović D., 1980: *Nitriti i bor u vodi u kopanim bunarima kao eventualni uzročnici nastanka endemske nefropatije.* Zbornik radova Rudarsko – geološkog fakulteta, sv.22, Beograd.

Perišić M., 1997: *Eutrofikacija i tretman vode za piće.* Poseb. izdanja Geoinstituta, knj. 22, Beograd, 187 p.

Petrishcheva P.A., 1965: *O nekotorih napravljenijah izučenija geografie bolezni s prirodnoj očagovostju.* V. knj.: Metodi mediko – geografičeskih issledovanij. Moskva, 22 – 31.

Petrović B., 1975: *Transkurentni rased Jordana i sekundarne strukture.* Geološki anali Balkanskog poluostrva, knj. 39, Beograd, 107 – 133.

Petrović B., Mitrović R., 1991: *Radijaciona higijena u biotehnologiji.* Naučna knjiga, Beograd, 333 p.

Petrović I., 1998: *Radon u zatvorenim prostorijama.* Ecologica, posebno izdanje, br 5, Beograd, 45 – 49.

Petrović M., 1996: *Modeli proračuna seizmičkog hazarda.* Savremeni trendovi razvoja geofizike (Zbornik). Rudarsko – geološki fakultet, Beograd, 169 – 175.

Pierard D, Lauwers S., 1998: *Enteric infections due to Escherichia coli.* Clin. Microbiol. Infect., 4, 414 – 415.

Pinneker E.V., 1966: *Rassoli Angaro -Lenskogo artezijanskogo bassejna.* Nauka, Moskva, 332 p.

Pinneker E.V., 1984: *Podzemnaja gidrosfera.* Nauka, Novosibirsk, 159 p.

Pinneker E.V.,1977: *Problemi regionaljnoj gidrogeologiji. Zakonomernosti rasprostranenia i formirovanija podzemnih vod.* Nauka, Moskva, 196 p.

Pit'eva K.E., 1878: *Gidrogeohimija. (Formirovanie himičeskogo sostava podzemnih vod).* Izdat. Moskovskogo Universiteta, Moskva, 328 p.

Plant J., Thornton I., 1986: *Geochemistry and health in the United Kingdom.* In: Thornton I. (Ed.), Proc. First International Symposium on Geochemistry and Health. Science Reviews, Northwood, 5 – 15.

Plant J.A. et al., 2000: *Environmental geochemistry at the global scale.* Journal of the Geological Society, London, Vol.157, 837 - 849.

Plotnikov N.I., 1989: *Tehnologičeskie izmenenija gidrogeologičeskih uslovij.* Nedra, Moskva, 268 p.

Plotnikov N.T., Kraevskij S., 1983: *Gidrogeologičeskie aspekti ohrani okružajuščoj sredi.* Nedra, Moskva, 207 p.

Paul R.D., 1972: *Klinička epidemologija.* Beograd. Izdav. Grafič. zavod, Beograd.

Polanski A., 1988: *Podstawy geochemii.* Warszawa: Wyd Geolog., 633 p.

Polić P., Blagojević S., 1997: *Teški metali u vodama.* In: Teški metali u životnoj sredini (Ed. R. Kastori). Naučni inst. za ratarstvo i povrtarstvo, Novi Sad.

Polk C., Postow E., 1986: *Hanbook of biological effects of electromagnetic fields.* CRC Press, 503 p.

Pool R., 1990a: *Is there En F – Cancer Connections?* Science 249, 1096 - 1098.

Pool R., 1990b: *Electromagnetic Fields: The biological evidence.* Science 249, 1378 – 1380.

Popit A., Vaupotič J., 2002: *The influence of geology on elevated radon concentrations in Slovenian schools and kindergartens.* Geologija, knj. 45/2, Geološki Zavod Slovenije, Ljubljana, 499 - 504.

Popov I.V., 1959: *Inženernaja geologija.* Vtoroe izdanie, Moskov. Univerzitet, 510 p.

Potoki rasejanija uranovih mestoroždenij, 1979: Atomizdat, Moskva, 136 p.

Price E.W., 1988: *Non – filarial elephantiasis – confirmed as a geochemical disease and renamed podoconiosis.* Ethiopian Medical Journal 26, 15 – 153.

Prodanović D., 1998: *Ekologija i zaštita vode od zagađenja.* Beograd, Aranđelovac,187 p.

Prostorni plan Srbije. (Grupa autora), 1996. Službeni glasnik RS, br 13/96, Beograd.

Protić D., 1990: *Hidrohemijske anomalije radioaktivnosti radonovog tipa.* XII Kongres na geolozi na Jugoslavija, kn.IV, Inženerska geologija, hidrogeologija i geotermija, Ohrid, 202 – 209.

Protić D., 1995: *Mineralne i termalne vode Srbije.* Geoinstitut, posebno izdanje, knj.17, Beograd, 269 p.

Putnik S., Marić S., 1982: *Nekonvencionalni izvori urana sa posebnim osvrtom na ležišta urانonosnih fosforita.* X jubil. kongres geologa Jugoslavije, knj. II, Budva, 283 – 297.

Pyle G.F., 1979: *Applied medical geography.* Washington etc.: Wiley, 282 p.

Radić A., 1969: *Osnovi medicinske geografije.* Medicinska knjiga, Beograd – Zagreb.

Radioaktivni izotopi i zračenja (Ed. I. Draganić), knj I (Opšti pojmovi) 1981., Univerzitet u Beogradu, Institut za nuklearne nauke «Boris Kidrič», Vinča. Beograd, 272 p.

Radovanović S., 1987: *Podzemne vode – izdani, izvori, bunari, terme i mineralne vode.* Srpska književna zadruga, 42, Beograd, 152 p.

Radovanović Z. Et al., 1995: *Epidemiologija II za IV razred Medicinske škole.* Zavod za udžbenike i nastavna sredstva, Beograd, 121 p.

Radovanović Z., 1985: *Epidemiologija endemske nefropatije.* (III simpozijum o endemskoj nefropatiji). SANU, knj. XXIII. Odelj. medicinskih nauka, knj. 3, Beograd, 3 - 27.

Radović I., Mandić R., 1998: *Zaštićena prirodna dobra i zaštita i unapređivanje biodiverziteta - osnova prirodne baštine Srbije.* Zaštita prirode, br 5, Beograd, 13 - 32.

Radulović V., 1994: *Geološki aspekti zaštite životne sredine.* Ecologica, god. I, br. 4, Beograd, 15 - 19.

Raghunath H.M., 1987: *Ground Water.* John Wiley & Sons, New York, Chichester, Brisbane, Toronto, Singapore, 563 p.

Rajh E. L., 1984: *Modelirovanie v medicinskoj geografii.* Nauka, Moskva, 157 p.

Raković - Savčić Lj., Čigoja M., 1998: *Makrobiotika ili tradicionalna ishrana u savremenom životu.* Beograd, 193 p.

Ramesam V., Rajagopalan K., 1985: *Fluoride ingestion into the natural waters of hard - rock areas, Peninsular India.* J. Geol. Soc. India 26, 125 - 132.

Rathery F., 1937: *Le diabete et les etats diabetiques dand leurs rapports avec les cures thermales.* XV Congres International d'Hydrologie, de Climatologie et de Geologie Medicales, 53 - 56 p.

Reed C.B., 1975: *Fuels, Minerals and Human Survival.* Ann Arbor Science Publishers, Inc., Ann Arbor.

Reimann C., deCaritat P., 1998: *Chemical elements in the environment.* Springer, Berlin, 398 p.

Reshetnikov P.P. et al., 1965: *Vijavlenie korreljacionnih svjazej v mediko - geografičeskih issledovanijah.* In: Metodi mediko - geologičeskih issledovanij, Moskva, 106 - 113.

Rethati L., 1983: *Groundwater in Civil Engineering.* Akademiai Kiado, Budapest, 432 p.

Reuter F., Klengel K., Pashek J., 1980: *Ingenieur - geologie.* VEB Deutscher Verlag für Grundstoffindustrie, Leipzig.

Riedl H., 1994: *The human impact on soil erosion in Greece.* Bulletin of the Geological Society of Greece, Vol. XXX/ 4, 15 - 27.

Risk Assessment Guidance for Superfund - Human health evaluation manual. US EPA Environmental Protection Agency, Washington DC.

Ristić M., 1965: *Genetsko - strukturni tipovi uranskih mineralizacija Jugoslavije.* Doktor. disert., Rud.- geol. fakultet, Beograd.

Rodenwaldt E., 1956: *Welt - seuchen -Atlas.* Hamburg.

Rogers E.S., 1960: *Human Ecology and Health.* An Introduction for Administrators, NewYork.

Ronov A., 1980: *Osadočnaja oboločka Zemli.* Količestvennie zakonomernosti stroenija, sostava i evoljucii). Nauka, Moskva, 80 p.

Rozenbaum E. et al., 1994: *Breast Cancer in man exposed to fields.* American Journal of Epidemiology, Vol. 33 - 36, 139 p.

Rubenowitz E., Alsson G., Rylander R., 1996: *Magnesium in drinking water and death from acute myocardial infarction.* Am. J. Epidemiol., 143, 456 - 462.

Rukovodstvo po medicinskoj geografii, (Pod. Ed. A.A. Kellera, O.P. Shchepina, A.V. Chaklina), 1993. Izdat Gippokrat, S. Petersburg, 352 p.

Sabins F.E.Jr., 1978: *Remote Sensing, Principles and Interpretation.* W. H. Freeman and Co., San Francisco, 426 p.

Safronova K.P., Petrova N. G., Fedorova V. S., 1990: *Osobenosti formirovanija osnovnih tipov boronosnih mineraljnih vod Sibirskoj platformi.* V. kn.: Gidrogeohimičeskie poiski mestoroždenij poleznih iskopaemih. Nauka, Novosibirsk, 183 - 189.

Savić I. , Radović I., Mandić R., 1998: *Koncept i strategija zaštite faune na području Srbije.* Zaštita prirode, br 5, Beograd, 177 - 187.

Savić I. et al.: 2000: *Uvođenje metoda teledetekcije u oblasti populacione i predeone ekologije kao i oblast zaštite prirode i biodiverziteta.* Ecologica, N⁰ 25, god VII, br 1, Beograd, 1 -11.

Savićević M., 1997: *Higijena.* Elif. Medica, Beograd.

Schlesinger W. H., 1991: *Biogeochemistry.* An Analysis of Global Change. Academic Press, 443 p.

Schoeller H., 1962: *Les eaux souterraines.* Paris.

Schoen J. 1983: *Petrophysik.* Academia Verlag, Berlin, 405 p.

Schraurer G.N., 1979: *Trace elements of carcinogenesis.* Adv. Nutritional Res. 2 (Ed. H. H. Draper). Plenum Press, 219 – 244.

Schraurer G.N., White D.A., Scheider C.J., 1977: *Cancer mortality correlation studies: III Statistical association with dietary selenium intas.* Bioinorg, Shem. 7, 23 -34.

Sekulić P. et al., 1997: *Osobine pseudogleja lozničko – šabačkog regiona.* Zbornik radova Kongresa Jugosl. društva za proučavanje zemljišta: «Uređenje, korišćenje i očuvanje zemljišta», Novi Sad.

Sekulić P., Vasin J., Satarić M., 1998: *Sadržaj teških metala u zemljištima na području Bečeja.* Ecologica, N⁰ 18, god V, br. 2, Beograd, 12 – 16.

Selen v biologii (Ed. Abdullaev G.V.), 1981. Izdat ELM, Baku, 310 p.

Selenium in biology and medicine (Ed. Wendel A.), 1989. Springer Verlag, Berlin - Heidelberg, 330 p.

Semenenko N., 1975: *Kontinentalna kora.* Naukova dumka, Kiev.

Sergeev E.M., 1997: *Inženernaja geologija i ohrana okružajušćoj sredi.* Vestn. NGU, Ser. geol., N⁰9, 75 – 85.

Shamberger R.J., 1983: *Biochemistry of selenium.* Plenum Press, New York, 334 p.

Shapoval A.N., 1959: *Viljuskij encefalomielit.* Jakutsk.

Shebalin N.B., 1969: *Macroseizmičeskoe pole i očag silnogo zemljetrjasenija.* Disertacija IFZ, AN SSSR, Moskva.

Sheldon R.P., 1959: *Geochemistry of uranium in phosphorites and block shales of the Phosphoria formation.* USA Geol. Surv. Bull. N⁰ 1084 – D.

Shen X. et al., 1980: *Studies on the relation of selenium and Keshan disease.* Biol. Trace Elem. Res. 2, 91 - 107.

Sheriff E.R., 1989: *Geophysical methods.* Prentice Hall, 605 p.

Shoshin A.A. 1962: *Osnovi medicinskoj geografii.* Izdav AN SSSR, Moskva, 146 p.

Shoshin A.A., 1961: *Medicinskaja geografija.*V. kn.: Kratkaja geografičeskaja

enciklopedija. T. 2. Moskva, 557 p.
Shpak A.A., Meljkanovickij I. M., Serezhenikov A. I., 1992: *Metodi izučenija i ocenki geotermaljnih resursov.* Nedra, Moskva, 315 p.
Shtraus B., 1988: *Medicinska biohemija.* Jugosl. medicinska naklada, Zagreb.
Shunping P. et al., 2000: *Geological hazards related to groundwater exploitation. Land subsidence in Suzhou city,China.* In : Hydrogeology and the Environment, Beijing, 546 - 552.
Sichev K.I., 1990: *Geohemičeskoe izučenie teritorii SSSR.* Razvedki i ohrana, Z. Moskva, 3 -10.
Sikošek B., 1982: *Seizmotektonske karakteristike Srbije.* Zbornik radova X jubilarnog kongresa geologije Jugoslavije, Budva.
Sikošek B., 1986: *Seizmotektonski sklop i seizmogena područja Jugoslavije.* In: Zbornik radova posvećen akademiku Zariji M. Bešiću. Crnogorska akad. nauka i umetnosti, Titograd, 165 - 185.
Sikošek B., 1989: *Seizmologija.* Beograd, 127 p.
Sikošek B., Petrović M., 1992: *Seizmička ugroženost Srbije.* Zbornik radova X jubilarnog kongresa geologije Jugoslavije, Budva. Srpsko geološko društvo, Jubilarna knjiga, Beograd, 151 - 158. Kongres geologa Jugoslavije, knj. I, Budva, 687 - 700.
Simić B., 1977: *Mineralne materije.* In: Medicinska dijetetika. Medicinska knjiga, Beograd - Zagreb, 104 - 108.
Simić D., 1988: *Mikrobiologija.* Naučna knjiga, Beograd, 191 p.
Siria T.I., 1979: *Nekotorie mikroelementi želčno – kamenoj bolezni v požilom i starčeskom vozraste.* Aktualjnie problemi biologiji i medicini. Izdat Mecniereba, Tbilisi, 173 -175.
Skinner B.J., 1986: *Earth resources.* Prentice–Hall Inc., Englewood Cliffs, New Jersey, 262p.
Skurlatov Ju. I., Duka G.G., 1989: *Himia i žizn vodi.* Kartja Moldavenjaske, Kišinjev, 125 p.
Sladachek V., 1973: *System of Water Quality from the Biological Point of View.* Arsh. Hidrobiol., 7, 218 p.
Smirnov S.I., 1971: *Proishoždenie solenosti podzemnih vod sedimentacionnih bassejnov.* Nedra, Moskva, 216 p.
Smislov A.A.,1974: *Uran i torij v zemnoj kore.* Nedra, Leningrad, 231 p.
Smith B. et al., 1996: *The environmental geochemistry of aluminium in tropical terrains and its imlications to health.* In : Appleton et al. Eds, Environmental, Geochemistry and Health, Geol. Society Special Publications. UK, N[o] 113, 141 - 153.
Sofrenović Đ., 1993: *Opšta patologija.* Naučna knjiga, Beograd, 256 p.
Soils of Tropical Forest Ecosystems. Characteristics, Ecology and Management (Eds. A. Schulte, D. Ruhiyat), 1997, Springer, 224 p.
Sorre M., 1929: *L'organisme humain et le climat.* Bull.Soc.Geogr. Lille, N[o] 1, 37 - 56.
Sorre M., 1933: *Comples pathogenes et geographie medicale.* Ann. Geogr., 42. N 235, 1 - 18.
Sorre M., 1971: *Les fondaments biological de la geographie humaine. Essai d'une ecologie de la homme.* Paris: Collin, 447 p.

Spallhalz J.E., 1981: *Selenium in Biology and Medicine.* AVI Publishing, 103 p.
Speleološki atlas Srbije (Ed. P. Đurović), 1998. SANU, Geograf. inst. «Jovan Cvijić», posebno izdanje, knj. 52, Beograd.
Stamp L.D., 1964a: *Some aspects of medical geography.* New York,Toronto, London.
Stamp L.D., 1964b: *The geography of life and death,* London.
Stanković S., Stanković A., Krainčanić M., 1993: *Komparativno ispitivanje nivoa kontaminacije radiocezijumom 137 i 134 nekih bioindikatora.* Zbornik XVII Jugosl. simpoz. za zaštitu od zračenja, Beograd, 65 – 68.
Starčević M., 1991: *Gravimetrijske metode istraživanja.* Nauka, Beograd, 226 p.
Stefanović S., 1999: *Pojam zdrave hrane i teški metali.* Ecologica, N°21, god.VI, br. 1., Beograd, 23 – 28.
Strahinjić S. et al., 1985: *Kliničke i patogenetske karakteristike stadijuma endemske nefropatije.* III Simpozijum o endemskoj nefropatiji. SANU, knj. XXIII, Odelj. medic. nauka, knj. 3., Beograd, 209 –216.
Strahov N.M., 1963: *Tipi litogeneza i ih evoljucija v istorii Zemli.* Gosgeoltehizdat, Moskva, 536 p.
Stryer L., 1981: *Biochemistry.* Vol. I, II, III, W. H. Freeman and Company. San Francisco.
Suša S., 1979: *Endemska nefropatija.* Savremena administracija, Beograd.

Škorić M., 1996: *Podzemne vode u Vojvodini.* Zadužbina Andrejević, Beograd, 137 p.

Tadolini T., Spizzico M., 1998: *Relation between "terra rossa" from the Apulia aquifer of Italy and the radon content of groundwater: Experimental results and their applicability to radon occurrence in the aquifer.* J. Hydrogeology Journal, Vol. 6, N° 3, Springer, 450 – 454.
Tassitch V., Militch T., 1937: *Traitement des diabetiques a Vrnjci les - Bains.* XV Congres International d'Hzdrologie, de Climatologie et de Geologie Medicales, 107 - 114.
Tatu C.A., Orem W.H., Finkelman R.B., Feder G.L., 1998: *The Etiology of Balkan Endemic Nephropathy: Still More Questions than Answers.* Environmental Health Perspectives, Vol. 106, N° 11, 689 – 700.
Taylor E.W. et al., 1994: *Basis for new approaches to the chemotherapy of AIDS: Novel genes in HIV – 1 potentially encode selenoproteins expressed by ribosomal frameshifting and termination suppression.* J. Med. Chem. 37, 2637 – 2654.
Taylor E.W. et al., 1995: *Do some viruses encode selenoproteins ? - Assessment of the theory in the light of current theoretical, experimental and clinical data.* Antiviral Research 26.
Tchijevsky A.L., 1937: *L'ionisation artificielle de l'air comme facteur therapeutique.* XV Congres International d'Hzdrologie, de Climatologie et de Geologie Medicales, 217 - 226.
Teitge J.E., 1990: *Incidence in myocardial infarct and mineral content of the drinking water.* Z. Gesamte. Inn. Med. 45, 478 –485.

Tektonosfera Zemli. 1978. Nauka, Moskva, 531 p.

Telford W.M.. Geldart L.P., Cheriff R.E. 1990: *Applied Geophysics.* Cambridge, University Press, 770 p.

Teofilović M., 1994: *Geoekologija – jedan od zaštitnika boljem životu i dužem opstanku čoveka na planeti Zemlji.* Ecologica, god. I, br.4, Beograd, 25 – 28.

Teofilović M., 1998: *Banjske i mineralne stone vode – njihov značaj za život i ekologiju čoveka.* Ecologica. N⁰ 20, Beograd, 40 – 45.

Teofilović M., Obrenović A., Tešić D., 2000: *Mineralne vode sa povećanim sadržajem fluora – njihov raspored na teritoriji Srbije.* Ecologica, N⁰ 25, god. VII, br.1, Beograd, 23 – 25.

Terrestrial heat flow in Europe (ed. by V. Čermak and L. Rybach), 1979. Springer – Verlag, Berlin, Heidelberg, New York.

Terzić R., 1997: *Bioizgradnja: biološko – ekološka izgradnja i sanacija građevinskih objekata.* Ekostan, Beograd, 293 p.

Terzić R., 1998: *Biološki zdravi građevinski materijali.* Ecologica, posebno izdanje, br. 5, Beograd, 24 – 31.

Teški metali u životnoj sredini (Ed. R. Kastori), 1997. Nauč. inst. za ratarstvo i povrtarstvo, Novi Sad, 301 p.

The atlas of endemic diseases and their environments in the People's Republic of China, 1989. Beijing, Science Press, 194 p.

The atlas of Plague and Its Environment in the People's Republic of China (Chief Compilers: Liu Yun – peng, Tan Jian – an, Chen Er – li), 2000. Science Press, Beijing.

Third international conference on arsenic exposure and health effects (Ed. Lin et al. in SEGH). Bock of abstracts. San Diego, California, 173 p.

Thornes J. B., 1976: *Semi – arid Erosional Systems.* L.S.E. Department of Geography Papers, N⁰ 9, 98 p.

Todd D.K., 1980: *Groundwater Hydrogeology.* John Wiley & Sons. New York, Chichester, Brisbane, Toronto, Singapore, 535 p.

Todorović M., 1983: *Moguća rešenja u sistemu čovek – društvo – životna sredina.* Biblioteka «Priroda se znanjem brani», Beograd, 196 p.

Tokin B.P., 1967: *Celebnie jadi rastenij.* Moskva.

Tomanec R., Milovanović D., Vučinić D., 2001: *Nemetalični mineralni resursi Srbije kao faktor zaštite sredine.* III međunarodni simpozijum Rudarstvo i zaštita životne sredine (Zbornik radova). Rudarsko – geološki fakultet u Beogradu, 45 – 52.

Tomanec R., Milovanović D., Vučinić D., 2001: *Nemetalični mineralni resursi Srbije kao faktor zaštite sredine.* III međunarodni simpozijum Rudarstvo i zaštita životne sredine (Zbornik radova). Rudarsko – geološki fakultet u Beogradu, 45 – 52.

Toropov N.I., 1864: *Opit medicinskoj geografii Kavkaza otnositeljno peremežajušćihsja lihoradek.* SPb., 416 p.

Trescases, J.J., 1992: *Chemical weathering.* In. Butt, C.R.M., Zeegers H. (Eds.) Regolith Exploration Geochemistry in Tropical and Subtropical Terrains. Handbook of Geochemistry, Elsevier, Amsterdam, 25 - 40.

Trifunović N., Komatina S., 2002: *Role of enormous changeable Earth magnetic fields intensities in etiopathogenesis of mental disturbances and illness.* Balkan

Geophysical Congress, Sofia
Trimmer E., 1987: *The Magic of Magnesium.* Thorsons Publ. Group. Rochester, Vermont, U.K, 7 – 94.
Tromp S.W., 1955: *Possible effects of geophysical and geochemical factors on development and geographic distribution of cancer.* Schweiz. Z. allg. Path., 18, 929 p.
Tseng W.P. et al., 1968: *Prevalence of skin cancer in an endemic area of chronic arsenism in Taiwan.* J. Natl Cancer Inst. 40, 453 – 463.
Turekian K. K., Wedepohl K. H., 1961: *Distribution of the elements in some major units of the Earth's crust.* Bull. Geol. Soc. Amer., 72, 175 – 192.
Turner F. J., Verhoogen J., 1960: *Igneous and Metamorphic Petrology.* 2nd ed., Mc Graw – Hill Book Company, New York.

Ubavić M. et al., 1993: *Teški metali u zemljištima Vojvodine.* «Teški metali i pesticidi u zemljištu Vojvodine», Poljoprivredni Fakultet, Institut za ratarstvo i povrtarstvo, Novi Sad.
Uiljams D., 1975: *Metali žizni.* Mir, Moskva.
Undervood E.J., 1970: *Mercury in the environment.* Geol. Surv. Proff. Paper 713, Washington. D C, Govern. Print. off.
Undervood E.J., 1977: *Trace Elements in Human and Animal Nutrition.* Academic Press, New York, 302 – 346.
Undervood E.J., 1981: *The incidence of trace element deficiency diseases.* Phil. Trans. R. Soc. Lond., 294, 3 – 8.
Undervood E.J., Wright F. C., 1977: *Trace Elements in human and animal nutrition. 4 th.* Academic Press, New York, 533 p.
Uzumasa Y., 1965: *Chemical Investigations of Hot Springs in Japan.* Tsukiji Shokan Co Ltd., Tokyo, 189 p.

Valberg P.A. et al., 1998: *Likelihood ratio analysis of skin cancer prevalence associated with arsenic in drinking water in the USA.* Environment Geochem. Health 20, 61 – 66.
Valić F., 1994: *Zdravstvena ekologija.* Medicinski fakultet Sveučilišta u Zagrebu.
Vapa M., Vapa LJ., 1997: *Teški metali i životinjski svet.* In: Teški metali u životnoj sredini (Ed. R. Kastori). Institut za ratarstvo i povrtarstvo, Novi Sad.
Vartanyan G.S., 1984: *Poiski i razvedka mineraljnih vod dlja lečebnih celej.* In: Osnovi gidrogeologii. Metodi gidrogeologičeskih issledovanij. Nauka, Novosibirsk, 112 – 124.
Vasković N., Jović V., 1993: *Geohemijske karakteristike jugoistočnog dela Surduličkog granitoidnog masiva.* Vesnik, ser. A, B, knj. 45, Beograd, 97 – 112.
Vasović M., 1997: *Regija i regionalizacija u geografiji.* V. knj.: Geografska struktura i regionalizacija Srbije, I, Geografski institut «Jovan Cvijić», posebna izdanja, knj. 51, Beograd, 5 –15.
Veizer J., 1983: *Trace elements and isotopes in sedimentary carbonates.* In ed. Carbonate mineralogy and chemistry. Mineralog. Society of America Reviews in Mineralogy, Vol.11, 265 – 300.

Vernadskii V.I. , 1954: *Izbrannie sočinenija,* T. 1., AN SSSR, Moskva, 696 p.
Vernadskii V.I., 1965: *Himičeskoe stroenie biosferi Zemli i jejo okruženija.* Nauka, Moskva, 374 p.
Vernadskii V.I., 1926: *Biosfera.* Nauč. him. tehn. izd. - vo, T ½, 146 p.
Vernadskii V.I., 1940: *Biogeohimičeskie očerki.* Moskva, Leningrad.
Veselinović D. et al., 1995: *Fizičkohemijski osnovi zaštite životne sredine. Knj.I: Stanja i procesi u životnoj sredini.* Fakultet za fiz. hem., Univ. u Beogradu, 176 p.
Veselinović D., Gržetić I., Djarmati S., Marković D., 1995: *Fizičkohemijski osnovi zaštite životne sredine.* Fakultet za fizičku hemiju, Beograd, 177 p.
Vinogradov A.P., 1938: *Biogeohemičeskie provincii i endemii.* Dokl. AN SSSR, t 18, N° 4/5, 283 - 284.
Vinogradov A.P., 1949: *Biogeohimičeskie provinciji.* Trudi jubil. sessii, posvjašćenoj 100 - letiju so dnja roždenija V.V. Dokučaeva, Moskva, 42 p.
Vinogradov A.P., 1959: *Himičeskaja evoljucija Zemli.* Izd. vo AN SSSR, Moskva, 44 p.
Vinogradov A.P., 1962: *Average contens of chemical elements in the principal types of igneous rocks of the Earth's crust.* Geokhymia 7, 551 - 571.
Vizzo De Santo A., 1992: *Denitrification in forest soils.* In: Responses of forest ecosystems to environmental changes. Elsevier, New York, 208 - 222.
Vogelsang D., 1995: *Environmental Geophysics.* Springer - Verlag, Berlin, Heidelberg.
Voiner A.O.,1960: *Biologičeskaja rolj mikroelementov v organizme životnih i čeloveka.* Moskva.
Voitkevich et al., 1990: *Spravočnik po geohimii.* Nedra, Moskva, 475 p.
Volcanism and archaeology in Mediterranean area (Eds. M. Cortini, B. De Vivo), 1997. Res. Signpost Publ. Co., Trivandrum, India, 191 p.
Volkov G., Lipin V., Chereskov D, 1964: *Radiobiologija.* Moskva.
Volpe et al., 2000: *Volcanology and chemistry of the Earth's interior. Italian research activity (1995 - 1998) report to IAVCEI.* Bolletino di Geofisica - Teor.Appl. 40, N° 2, supplement. Osservatorio Geofisico Sperimentale, Trieste (Italy), 163 - 298.
Voronov A.G., 1965: *Opit klasifikacii boleznej čeloveka po stepeni i harakteru ih zavisimosti ot osobennostej prirodnoj sredi.* In: Metodi mediko - geografičeskih issledovanij, Moskva, 32 - 47.
Voronov A.G.,1981: *Medicinskaja geografija.* Izd - vo MGU im. M. V. Lomonosova, 161 p.
Voskresenskii S.S., 1957: *Geomorfologia Sibiri.* Kurs lekcij, Izd. Mosk. univ., 312 p.
Vouk V., 1937: *Une classification biologique des eaux thermales a l'egard special des eaux thermales en Yougoslavie.* XV Congres International d'Hzdrologie, de Climatologie et de Geologie Medicales, 265 - 276.
Vrana K., 1997: *Geochemical Atlas of Slovak Republic at a scale 1:1,000,000. An Introduction.* Slovak geological magazine, vol. 3, N° 9, Geol. Survey of Slovak Republic, Bratislava, 77 p.
Vrvić M. M., Vujin S., Matić V., Spasić B., 1999: *Role of magnesium in biology - chemical basis.* Serbian Academy of Sciences and Arts, Vol. XCII, Book 1,

Beograd, 13 – 20.
Vučetić J., et al., 1987: *Mikroelementi i njihov značaj na aktivnost enzima, vitamina i hormona.* Hem. pregled Srpskog hemijskog društva, br. 1 i 2, Beograd.
Vujanović V., Teofilović M., 1983: *Banjske i mineralne vode Srbije.* Kulturni centar RJ «Privredna knjiga», Gornji Milanovac, 295 p.
Vujasinović S., Matić I., Ložajić A., 1999: *Hidrogeološka problematika zagađivanja i zemljište beogradskog izvorišta.* Univerzitet u Beogradu, Rudarsko – geološki fakultet, 146 p.
Vukašinović S., 1976: *Anomalno magnetsko polje i geološka građa unutrašnjih Dinarida (na bazi aeromagnetskih podataka).* Institut za geološko – rudarska istraživanja i ispitivanja nuklearnih i drugih mineralnih sirovina, posebno izdanje, knj. 5, Beograd, 92 p.
Vvedenije v geogigenu (Ed. N.V. Lazarev), 1966. Moskva, Leningrad.

Wadsworth R., Treveek J., 1999: *Geographical Information Systems for Ecology.* Longman, Edinburgh, 184 p.
Waring G. A., 1965: *Thermal Springs of the USA and Other Countries of the World.* A. Summary, Geol. Surv. Proff., Paper 492, Washington.
Warnecke G., 1997: *Meteorologie und Umvelt. Eine Einführung.* Springer, 354 p.
Webber J., 1981: *Trace metals in agriculture.* In: Effects of heavy metal pollution on plants, Vol. 2, Pollut. Monitoring Ser., ed. N. W. Lepp, Applied Sci. Pubs., 159 – 184.
Wedepohl K.H., 1978: *Handbook of geochemistry.* Springer, Berlin, Heidelberg, New York.
Wedepohl K.H., 1995: *The composition of the continental crust.* Geochim Cosmochim Acta 59 (7), 1217 – 1232.
Weigel R. L., 1970: *Earthquake engineering.* Prentice - Hall, Englewood Cliffs, New Yersey.
White D.E., 1969: *Thermal and mineral waters of the United States, Brief rewiev of possible origins.* In: Mineral and thermal waters of the world. B – Oversea countries. Prag, 269 – 286 (23rd Intern. Geol. Congres, Vol 19).
White D.E., Hem J. D., Warring G. A., 1963: *Chemical Composition of Subsurface Waters.* Geol Surv. Professional Paper 440 – F. Data of Geochemistry US Government Printing Office, Washington, 68 p.
White F., 1983: *The vegetation of Africa.* UNESCO, Paris, 468 p.
White L.P., 1977: *Aerial photography and remote sensing for soil survey.* Oxford, 104 p.
Willett W.C., Stampfer M.J., 1988: *Selenium and Cancer. Whether selenium protects against cancer is still unknown.* B. M. J. 297, 573 – 574.
Winkler H.G.F., 1974: *Petrogenesis of metamorphic rocks* (3rd edn.). Springer Verlag, Berlin, Heidelberg, New York, 320 p.
Woodcock H.N., 1994: *Geology and environment in Britain and Ireland.* UCL Press, London, 155 p.
Wright, Peters, Mark, 1982: *Leukemia* in *workers exposed to electrical and*

magnetic fields. Lancer, 1160 p.

Yanchin A.L., 1988: *On the time of the life apperance on the Earth.* Bulletin, Ser. A, tome 44, Geozavod, Belgrade, 5 –15.

Yang C.Y. et al., 1997: *Magnesium and calcium in drinking water and cerebrovascular mortality in Taiwan.* Magnesium Res. 10, 1, 51 – 57.

Yang T.H., Tsai J., 1961: *Preliminary report on selenium levels in water, soil, and food from the endemic Bleckfoot area.* J. Formasan. Med. Assoc., 60, 80 –81.

Yanxin W. et al., 2000: *Fluoride in shallow groundwaters from southern Siberia and northwestern China.* In: Hydrogeology and the Environment, Beijing, 247 – 257.

Yaron B., Calvet R., Prost R., 1996: *Soil Pollution. Processes and Dynamics.* Springer, 313 p.

Yevjevich V., 1976: *Karst Hydrogeology and Water Resources.* W.R.P., Colorado, USA.

Zabl'udovskij E.P. et al., 1981: *Istorija medicini.* Medicina, Moskva.

Zaruba K., Mencl V., 1976: *Engineering geology.* Academia, Prague.

Zavtur L.V., et al., 1968: *Izučenie vzaimosvjazi meždu rasprostraneniem zlokačestvennih novoobrazovanij i geohimičeskimi osobennostjami vnešnoj sredi landšaftno - kartografičeskim metodom.* V. knj.: Metodi izučenija epidemologii zlokačestvennih opuholej. Moskva, 135 p.

Zeremski M., 1990: *Geomorfološka (morfostrukturna) karta Srbije 1 : 500,000,* Beograd.

Zhang X.M., Cassells C.J.S., van Genderen J.L., 1999: *Multisensor data fusion for the detection of underground coal fires.* In: Geologie en mijnbouw (Guest eds. F. Van der Meer, E.C. Kosters, S.J. Fraikin), Vol. 77, N[o] 2, KNGMG, Kluwer Acad. Publishers. Printed in the Netherlands, 117 – 127.

Zhavoronkov A.A., 1968: *Geografija endemičeskogo fljuoroza na zemnom šare.* Vestn. AMN SSSR, 1, Moskva, 230 p.

Ziegler J.L., 1993: *Endemic Kaposi's sarcoma in Africa and local volcanic soils.* Lancet 342, 1348 – 1351.

Zonneveld I.S., 1995: *Land ecology.* SPB Academic Publishing, Amsterdam.

Zotl J.G., 1974: *Karsthydrogeologie.* Wien.

Žujović J.M. 1922: *Pouke iz geologije.* Knjige Srpskog geološkog društva, knj. I, 87 p.

Author Index

Index of Geographical Terms

Subject Index